Digital Systems

Principles and Applications

NINTH EDITION

Digital Systems
Principles and Applications

Ronald J. Tocci
Monroe Community College

Neal S. Widmer
Purdue University

Gregory L. Moss
Purdue University

PEARSON
Prentice
Hall

Upper Saddle River, New Jersey
Columbus, Ohio

Library of Congress Cataloging-in-Publication Data
Tocci, Ronald J.
    Digital systems: principles and applications/Ronald J. Tocci, Neal S. Widmer, Gregory
L. Moss.--9th ed.
      p. cm.
    Includes bibliographical references and index.
    ISBN 0-13-111120-5
    1. Digital electronics.   I. Widmer, Neal S. II. Moss, Gregory L. III. Title.

TK7868.D5T62 2004
621.381--dc21   2003051763

Editor in Chief: Stephen Helba
Acquisitions Editor: Dennis Williams
Development Editor: Kate Linsner
Production Editor: Alexandrina Benedicto Wolf
Design Coordinator: Karrie Converse-Jones
Cover Designer: Ali Mohrman
Cover Art: Digital Vision
Production Manager: Pat Tonneman
Marketing Manager: Ben Leonard

This book was set in Times Roman by Carlisle Communications, Ltd. It was printed and bound by
R.R. Donnelley & Sons Company. The cover was printed by Phoenix Color Corp.

Multisim 2001® is a trademark of Electronics Workbench™.

MAX+PLUS® II is a registered trademark of Altera Corp. Altera™ is a trademark and servicemark of
Altera Corporation in the United States and other countries. Altera products are the intellectual property
of Altera Corporation and are protected by copyright laws and one or more U.S. and foreign patents and
patent applications.

Pearson Education Ltd.
Pearson Education Singapore Pte. Ltd.
Pearson Education Canada, Ltd.
Pearson Education—Japan

Pearson Education Australia Pty. Limited
Pearson Education North Asia Ltd.
Pearson Educación de Mexico, S.A. de C.V.
Pearson Education Malaysia Pte. Ltd.

10 9 8 7 6 5 4 3 2 1
ISBN 0-13-111120-5

To you, Cap, for loving me for so long; and for the million and one ways you brighten the lives of everyone you touch.

—RJT

To my wife, Kris, and our children, John, Brad, Blake, Matt, and Katie: the lenders of their rights to my time and attention that this revision might be accomplished.

—NSW

To my family, Marita, David, and Ryan.

—GLM

PREFACE

This book is a comprehensive study of the principles and techniques of modern digital systems. It teaches the fundamental principles of digital systems and covers thoroughly both the traditional and the most modern methods of applying digital design and development techniques, including how to manage a systems-level project. The book is intended for use in two- and four-year programs in technology, engineering, and computer science. Although a background in basic electronics is helpful, most of the material requires no electronics training. Those portions of the text that utilize electronic concepts can be skipped without adversely affecting the comprehension of the logic principles.

General Improvements

The ninth edition of *Digital Systems* recognizes that the practical methods of developing digital systems and the tools used to implement those methods have changed, but it also recognizes that the underlying principles have *not* changed. Furthermore, the students' need to understand the principles so they can make the application has not diminished at all. Primarily, the new developments have brought new ways of describing the operation of digital logic circuits and digital systems. This situation begs the question, "At what point should we teach the new methods exclusively, rather than using the old methods to introduce the new?" We believe the answer to that question clearly is when the old methods are no longer practiced, will not be encountered in the field, and serve no pedagogical purpose other than offering historical insight.

As for the hardware implementation of digital systems, it is true that the exclusive use of standard SSI and MSI logic circuits is no longer practiced. These devices are still encountered in one form or another, however, by professionals in the digital field and they continue to be a viable means of studying the operation of the fundamental building blocks of digital systems. The wiring and testing of these standard parts is still a vital part of the pedagogy in many introductory, lab-based courses. These topics solidify concepts such as binary inputs and outputs, physical device operation, and practical limitations using a very simple platform. Consequently, we have chosen to continue to introduce the conceptual descriptions of digital circuits and to offer examples using conventional standard logic parts. For those who continue to teach the fundamentals using SSI and MSI circuits, this edition retains those qualities that have made the text so widely accepted in the past.

We believe that graduates will actually apply the concepts presented in this book using higher-level description methods and more complex programmable devices, but they will also be exposed to the simpler technologies. The major shift in the field is toward a greater need to understand the description methods rather than focusing on the architecture of an actual device. Software tools have evolved to the point where there is little need to be concerned with the inner workings of the hardware and much more need to focus on what goes in, what comes out, and how the designer can describe what the device is supposed to do. We also believe that graduates will be involved with projects using state-of-the-art design tools and hardware solutions.

The main improvements in the ninth edition involve the introduction of two hardware description languages (HDL), the concepts of project development using top-down system design techniques, and a broader survey of the technology used to implement digital systems today. The HDL techniques are introduced earlier (starting in Chapter 3) and are covered more thoroughly throughout the remainder of the text compared to the eighth edition. This edition offers over 40 AHDL examples, over 40 VHDL examples, and 10 examples of simulation testing. All of these design files are available on the enclosed CD. Problems involving HDL have been added to the end of each chapter to reinforce the concepts. A new chapter (Chapter 10) on project development and top-down system design teaches practical methods that will be useful in other areas, such as analog circuits and microprocessor-based systems. Finally, the last chapter has been rewritten to give more insight into current PLD architecture and, perhaps more important, to help delineate and choose among many categories of digital system solutions. Making room for the new material has required the following changes from the eighth edition: the CUPL language has been replaced, and Appendix A (which covered microprocessors) and Appendix B (which provided sample data sheets) have been deleted.

The new material does not attempt to teach a particular hardware platform or the details of the use of a software development system. We have used the Altera MAX+PLUS® II software environment to develop the examples, but we have tried to show what this tool can do rather than train the reader how to use it. These details are best learned through hands-on experience and a lab manual, such as the design approach manual by Greg Moss that accompanies this text. Many laboratory hardware options are available to those who use this book. Many educational development boards are also available, such as the UP2 board from Altera's university program (see Figure P-1). It offers 7-segment displays that are already wired to the CPLD IC, and individual switches and LEDs that can be connected point-to-point. Some schools have developed their own laboratory vehicle, such as the DeVry eSOC board shown in Figure P-2. It offers many standard input and output devices with a flexible means of connecting them to the Altera CPLD IC. The PLDT-2 from RSR Electronics (http://www.elexp.com) is a similar experimental board (see Figure P-3). A smaller, more economical board is available from HVW Technologies (http://www.HVWTech.com). It is designed to straddle two standard breadboard strips, as shown in Figure P-4. These boards and a host of others

FIGURE P-1   Altera's UP-2
board.

FIGURE P-2   The DeVry
University eSOC board.

FIGURE P-3   The RSR
Electronics PLDT-2 board.

FIGURE 4 The HVW
Technologies board.

offer an excellent way to teach and demonstrate the practical implementation of the concepts presented in this text.

SEQUENCING Our approach to HDL and PLDs gives instructors several options: (1) the HDL material can be skipped in its entirety without affecting the continuity of the text; (2) HDL can be taught as a separate topic by skipping the material initially and then going back to the last sections of Chapters 3, 4, 5, 6, 7, and 9 and then covering Chapter 10; (3) HDL and the use of PLDs can be covered as the course unfolds—chapter by chapter—and woven into the fabric of the lecture/lab experience. As far as specific hardware description languages are concerned, VHDL is clearly the standard language of the industry and is most likely to be used by graduates in their careers. We have always felt that it is a bold proposition, however, to try to teach VHDL in an introductory course. The nature of the syntax, the subtle distinctions in object types, and the higher levels of abstraction can pose obstacles for a beginner. For this reason, we have included Altera's AHDL as the recommended introductory language for freshman courses. We have also included VHDL as the recommended language for more advanced classes or introductory courses offered to more mature students. We do not recommend trying to cover both languages in the same course. Sections of the text that cover the specifics of a language are clearly designated with a color bar in the margin. The HDL code figures are set in a color to match the color-coded text explanation. The reader only has to focus on the language of his or her choice and skip the other. Obviously, we have attempted to appeal to the diverse interests of our market, but we believe we have created a book that can be used in multiple courses and serves as an excellent reference after graduation.

It is a rare instructor who uses the chapters of a textbook in the sequence in which they are presented. This book was written so that, for the most part, each chapter builds on previous material, but it is possible to alter the chapter sequence somewhat. The first part of Chapter 6 (arithmetic operations) can be covered right after Chapter 2 (number systems), although this would produce a long interval before the arithmetic circuits of Chapter 6 are encountered. Much of the material in Chapter 8 (IC characteristics) can be covered earlier (e.g., after Chapter 4 or 5) without causing any serious problems.

This book can be used either in a one-term course or in a two-term sequence. When used in one term, it may be necessary, depending on available class hours, to omit some topics. Obviously, the choice of deletions will depend on factors such as program or course objectives and student background. Here is a list of sections and chapters that can be deleted with minimal disruption:

Chapter 1: All
Chapter 2: Section 6
Chapter 3: Sections 15–20
Chapter 4: Sections 7, 10–13
Chapter 5: Sections 3, 24–28
Chapter 6: Sections 5–7, 11, 13, 16–23
Chapter 7: Sections 10, 14–18, 23–24, 26–29
Chapter 8: Sections 11, 14–21
Chapter 9: Sections 5; 9, 15–20
Chapter 10: All
Chapter 11: Sections 7, 14–18
Chapter 12: Sections 17–21
Chapter 13: All

PROBLEM SETS The eighth edition contained five categories of problems: basic (**B**), challenging (**C**), troubleshooting (**T**), new (**N**), and design (**D**). The ninth edition adds the category of HDL (**H**) to designate problems that involve hardware description language. Undesignated problems are considered to be of intermediate difficulty, between basic and challenging.

PROJECT MANAGEMENT AND SYSTEM-LEVEL DESIGN Several of the real-world examples that were in previous editions have been incorporated in a new chapter (Chapter 10) that describes the techniques used to manage projects. These applications are generally familiar to most students studying electronics, and the primary example of a digital clock is familiar to everyone. Many texts talk about top-down design, but this edition demonstrates the key features of this approach and how to use the modern tools to accomplish it.

DATA SHEETS The accompanying CD-ROM is now the primary source of manufacturers' data sheets. The information on this single CD is equivalent to an entire shelf full of data books covering all TTL, CMOS, and high-speed bus interface logic ICs.

SIMULATION FILES This edition also includes simulation files that can be loaded into Electronics Workbench Multisim®. The circuit schematics of many of the figures throughout the text have been captured as input files for this popular simulation tool. Each file has some way of demonstrating the operation of the circuit or reinforcing a concept. In many cases, instruments are attached to the circuit and input sequences are applied to demonstrate the concept presented in one of the figures of the text. These circuits can then be modified as desired to expand on topics or create assignments and tutorials for students. All figures in the text that have a corresponding simulation file on the CD-ROM are identified by this icon:

IC TECHNOLOGY This new edition continues the practice begun with the last three editions of giving more prominence to CMOS as the principal IC technology in the small- and medium-scale integration applications. This depth of coverage has been accomplished while retaining the substantial coverage of TTL logic.

Specific Changes

The major changes in the topical coverage are listed here:

- **Chapter 1.** Many explanations covering digital/analog issues have been updated and improved.

- **Chapter 2.** The terms *nibble* and *word* are defined. Several problems are added to help in hex/binary conversion.

- **Chapter 3.** The title of this chapter has been changed to reflect its new focus on the methods of describing logic circuits. In previous editions, Chapter 3 introduced the truth table, logic symbols, and Boolean algebra. Now, it adds the next logical step of introducing languages that describe the circuits' operation. The initial and brief introduction to PLD technology has also been moved to Chapter 3.

- **Chapter 4.** The process of PLD development is expanded, and the hardware and software tools are introduced. Students are presented with the vision that complex systems like those they use every day are made of the simple circuits they are currently learning. The initial introduction to hierarchical design paves the way for the examples given in Chapter 10. The critical topic of representing data in HDL is introduced. This chapter shows how the tremendous capabilities of HDL allow the designer simply to present a truth table describing a circuit's operation, thereby avoiding all the simplification and algebraic steps required by conventional techniques. Finally, the power of describing the operation of a digital circuit in terms that people use every day to make logical decisions takes the mystery out of the book's objectives, and it motivates the reader to learn more about what HDL can do.

- **Chapter 5.** Methods of describing logic circuits using Boolean equations are extended to the representation of sequential, level-triggered latch circuits in HDL. The edge-triggered devices introduced earlier in the chapter are implemented using the structural methods of AHDL and also the more abstract VHDL methods. The concept of developing code in HDL and then using it as a building block or component in other designs is introduced. This approach opens the way for the students to envision larger systems by taking advantage of their previous work.

- **Chapter 6.** The library of standard components is introduced in the Altera development system. The MSI chips described earlier in the chapter to perform arithmetic and logic functions are demonstrated using the HDLs and library functions. Adder circuits are also designed using HDL *without* library functions, but with a focus on the powerful use of data objects in HDL. The code examples lead to several easy methods available in HDL to reuse basic pieces of code that simply need to be fit to the current application by changing the number of bits in the design.

- **Chapter 7.** This chapter has been restructured. HDL techniques for implementing the basic counter circuits are presented in Part I of the chapter; other applications of HDL clocked logic, including shift registers, shift counters, and digital one-shots, are presented in Part II.

- **Chapter 8.** This chapter remains a very technical description of the technology available in standard logic families and digital components. The troubleshooting tools and techniques as well as information about logic families and technologies have been updated. The latest Texas Instruments life cycle curve shows their relative position between introduction and obsolescence.

- **Chapter 9.** The many different building blocks of digital systems are still covered in this chapter and then—one by one—they are demonstrated using HDL. Many other HDL techniques, such as tristate outputs and various HDL control structures, are also introduced.

- **Chapter 10.** This new chapter showcases several systems-level examples. The key feature of this chapter is the presentation of methods for actually planning and building these circuits as projects. These examples are used to demonstrate techniques that allow a complex project to be broken down into simple pieces. The modular pieces are combined using the graphics computer-aided design (CAD) techniques available in MAX+PLUS II, and then the same task is accomplished using only HDL text descriptions in a modular hierarchical approach.

- **Chapter 11.** This chapter covers the same material as Chapter 10 of the eighth edition. Several concepts regarding analog/digital conversions are improved, particularly the section on sigma/delta modulation. This method is widely used in consumer audio equipment and telecommunications and often has been misunderstood.

- **Chapter 12.** This chapter covers the same material as Chapter 11 of the eighth edition. Only minor updates were necessary in this chapter.

- **Chapter 13.** This chapter replaces Chapter 12 of the eighth edition and was rewritten by Greg Moss, who has played an active role in introducing PLDs and HDLs into classrooms and laboratories for the past 15 years. This chapter gives an excellent synopsis of the hardware choices for implementing digital systems. It then covers the fundamentals of PLD circuitry and describes the various architectures that are available. The GAL 16V8 is covered in enough detail for students to understand the inner workings of a popular, simple PLD. The Altera EPM7128 CPLD, used most frequently in laboratory development boards, is described in detail. Finally, the Altera FLEX10K RAM-based HCPLD architecture is covered.

Retained Features

This edition retains all of the features that made the previous editions so widely accepted. It utilizes a block diagram approach to teach the basic logic operations without confusing the reader with the details of internal operation. All but the most basic electrical characteristics of the logic ICs are withheld until the reader has a firm understanding of logic principles. In Chapter 8, the reader is introduced to the internal IC circuitry. At that point, the reader can interpret a logic block's input and output characteristics and "fit" it properly into a complete system.

The treatment of each new topic or device typically follows these steps: the principle of operation is introduced; thoroughly explained examples and applications are presented, often using actual ICs; short review questions are posed at the end of the section; and finally, in-depth problems are available at the end of the chapter. Ranging from simple to complex, these problems provide instructors with a wide choice of student assignments. These problems are often intended to reinforce the material without simple repetition of the principles. They require the student to demonstrate comprehension of the principles by applying them to different situations. This approach also helps the student develop confidence and expand his or her knowledge of the material.

The material on programmable logic devices (PLDs) and hardware description languages (HDLs) is distributed throughout the text, with examples that emphasize key features in each application. These topics are located at the end of each chapter, making it easy to relate each topic to the general discussion earlier in the chapter or to address the general discussion separate from the PLD/HDL coverage.

The IEEE/ANSI standard for logic symbols is introduced and discussed with minimum disruption of the topic flow, and, if desired, can be omitted completely or in part. The extensive troubleshooting coverage is spread over Chapters 4 through 12 and includes presentation of troubleshooting principles and techniques, case studies, 25 troubleshooting examples, and 60 *real* troubleshooting problems. When supplemented with hands-on lab exercises, this material can help foster the development of good troubleshooting skills.

The ninth edition offers over 200 worked-out examples, more than 400 review questions, and over 450 chapter problems/exercises. Some of these problems are applications that show how the logic devices presented in the chapter are used in a typical microcomputer system. Answers to a majority of the problems are available immediately following the glossary. The glossary provides concise definitions of all terms in the text that have been highlighted in boldface type.

An IC index is provided at the back of the book to help the reader easily locate material on any IC cited or used in the text. The back endsheets contain tables of the most often used Boolean algebra theorems, logic gate summaries, and flip-flop truth tables for quick reference when doing problems or working in the lab.

Supplements

An extensive complement of teaching and learning tools has been developed to accompany this textbook. Each component provides a unique function, and each can be used independently or in conjunction with the others.

- Two CD-ROMs are packaged with each copy of the text.

  The first CD is:

  - **The Texas Instruments' logic data sheets CD**, which includes all TTL series, CMOS, and bus interface parts.

  The second CD contains:

  - **MAX+PLUS® II Educational Version software from Altera.** This is a fully functional, professional quality, integrated development environment for digital systems. Students can use it to write, compile, and simulate their designs at home before going to the lab. They can use the same software to program and test an Altera CPLD.

  - **Design files from the textbook figures.** Over 40 design files in each language are presented in figures throughout the text. Students can load these into the MAX+PLUS II software and test them.

  - **Solutions to selected problem HDL design files.** A few of the end-of-chapter problem solutions are available to the student. (All of the HDL solutions are available to instructors in the *Instructor's Resource Manual*.)

- **Circuits from the text rendered in Electronics Workbench Multisim®.** Students can open and work interactively with approximately 100 circuits to increase their understanding of concepts and prepare for laboratory activities.

- **Multisim 2001® Enhanced Textbook Edition software from Electronics Workbench.** The Enhanced Textbook Edition of Multisim is all that is required to open and work with the Multisim circuit files on the CD. However, anyone who wishes to order the full version of Multisim may do so through Prentice Hall by visiting http://www.prenhall.com, phoning 1-800-282-0693, or sending a fax request to 1-800-835-5327.

STUDENT RESOURCES

- **Lab Manual: A Design Approach.** This lab manual, written by Gregory Moss, contains topical units with lab projects that emphasize simulation and design. It utilizes the Altera MAX+PLUS II software in its programmable logic exercises. The new edition contains new projects and examples, revised PLD coverage to match textbook revisions, and some new screen captures. (ISBN 0-13-111128-0)

- **Lab Manual: A Troubleshooting Approach.** This manual, written by James DeLoach and Frank Ambrosio, is presented with an analysis and troubleshooting approach and is fully updated for this edition of the text. (ISBN 0-13-111127-2)

- **Student Study Guide.** The study guide, written by Frank Ambrosio, provides reinforcement of all topics presented in the text. The new edition includes updated coverage to match the new edition of the textbook and features entirely updated diagrams. (ISBN 0-13-111805-6)

- **Companion Website (www.prenhall.com/tocci).** This site offers students a free, online study guide with which they can review the material learned in the text and check their understanding of key topics.

- **Electronics Supersite (www.prenhall.com/electronics).** Students will find additional troubleshooting exercises and links to the Prentice Hall Electronics Technology Companion Websites.

INSTRUCTOR RESOURCES

- **Companion Website (www.prenhall.com/tocci).** For the instructor, this website offers the ability to post a syllabus online with Syllabus Manager. This is a great solution for classes taught online, self-paced, or in any computer-assisted manner.

- **Electronics Supersite (www.prenhall.com/electronics).** The *Prentice Hall Electronics Technology Journal*, extra classroom resources, and all of the supplements for this text are available to instructors online for easy access. Instructors can register at this website to receive a user name and passcode.

- **Online Course Support.** If your program is offering your digital electronics course in a distance-learning format, please contact your local Prentice Hall sales representative for a list of product solutions.

- **Lab Solutions Manual.** This manual contains worked-out lab results for both lab manuals. (ISBN 0-13-114555-X)

- **PowerPoint Transparencies CD-ROM.** Figures from the text are featured, in addition to Lecture Notes for each chapter. (ISBN 0-13-111124-8)

- **TestGen,** a computerized test bank. (ISBN 0-13-111126-4)

ACKNOWLEDGMENTS

We are grateful to all those who evaluated the eighth edition and provided answers to an extensive questionnaire: Mohamad H. Ahmadian, Eastern New Mexico University; John Blankenship, DeVry Institute of Technology; Rex N. Fisher, Brigham Young University–Idaho; Richard L. Henderson, DeVry University; Vincent A. Scotto, New England Institute of Technology; and Theresa Stanton, University of Nebraska–Omaha. Their comments, critiques, and suggestions were given serious consideration and were invaluable in determining the final form of the ninth edition.

We also are greatly indebted to several of our colleagues: Professor Frank Ambrosio, Monroe Community College, for his usual high-quality work on the indexes, the *Instructor's Resource Manual,* and the *Student Study Guide*; Professor Thomas L. Robertson, Purdue University, for providing his magnetic levitation system as an example; and Professor Anthony Oxtoby, Purdue University, for his technical review of topics relating to digital signal processing. We also appreciate suggestions and corrections sent to us by Don Holden of Austin Community College and George Musisca of Stark State College. We appreciate the cooperation of Mike Phipps and the Altera Corporation for their support in granting permission to use their software package and their figures from technical publications.

A writing project of this magnitude requires conscientious and professional editorial support, and Prentice Hall came through again in typical fashion. We thank Dennis Williams, acquisitions editor; Kate Linsner, development editor; Alex Wolf, production editor; and Marianne L'Abbate, copy editor, for all their help to make this publication a success.

And finally, we want to let our wives and our children know how much we appreciate their support and their understanding. We hope that we can eventually make up for all the hours we spent away from them while we worked on this revision.

Ronald J. Tocci
Neal S. Widmer
Gregory L. Moss

BRIEF CONTENTS

CONTENTS

CHAPTER 10   Digital System Projects Using HDL   682

CHAPTER 11   Interfacing with the Analog World   724

CHAPTER 12   Memory Devices   794

Digital Systems

Principles and Applications

CHAPTER 1

INTRODUCTORY CONCEPTS

■ OBJECTIVES

Upon completion of this chapter, you will be able to:

- Distinguish between analog and digital representations.
- Cite the advantages and drawbacks of digital techniques compared with analog.
- Understand the need for analog-to-digital converters (ADCs) and digital-to-analog converters (DACs).
- Recognize the basic characteristics of the binary number system.
- Convert a binary number to its decimal equivalent.
- Count in the binary number system.
- Identify typical digital signals.
- Identify a timing diagram.
- State the differences between parallel and serial transmission.
- Describe the property of memory.
- Describe the major parts of a digital computer and understand their functions.
- Distinguish among microcomputers, microprocessors, and microcontrollers.

■ INTRODUCTION

In today's world, the term *digital* has become part of our everyday vocabulary because of the dramatic way that digital circuits and digital techniques have become so widely used in almost all areas of life: computers, automation, robots, medical science and technology, transportation, telecommunications, entertainment, space exploration, and on and on. You are about to begin an exciting educational journey in which you will discover the fundamental principles, concepts, and operations that are common to all digital systems, from the simplest on/off switch to the most complex computer. If this book is successful, you should gain a deep understanding of how all digital systems work, and you should be able to apply this understanding to the analysis and troubleshooting of any digital system.

We start by introducing some underlying concepts that are a vital part of digital technology; these concepts will be expanded on as they are needed later in the book. We also introduce some of the terminology that is necessary when embarking on a new field of study, and add to this list of important terms in every chapter.

1-1   NUMERICAL REPRESENTATIONS

In science, technology, business, and, in fact, most other fields of endeavor, we are constantly dealing with *quantities.* Quantities are measured, monitored, recorded, manipulated arithmetically, observed, or in some other way utilized in most physical systems. It is important when dealing with various quantities that we be able to represent their values efficiently and accurately. There are basically two ways of representing the numerical value of quantities: **analog** and **digital**.

Analog Representations

In **analog representation** a quantity is represented by a voltage, current, or meter movement that is proportional to the value of that quantity. An example is an automobile speedometer, in which the deflection of the needle is proportional to the speed of the auto. The angular position of the needle represents the value of the auto's speed, and the needle follows any changes that occur as the auto speeds up or slows down.

Another example is the common mercury thermometer, in which the height of the column of mercury is proportional to the room temperature. As the temperature goes up or down, the mercury goes up or down proportionally, so that the level of the mercury represents the value of the temperature.

Still another example of an analog quantity is found in the familiar audio microphone. In this device an output voltage is generated in proportion to the amplitude of the sound waves that impinge on the microphone. The variations in the output voltage follow the same variations as the input sound.

Analog quantities such as those cited above have an important characteristic: *they can vary over a continuous range of values.* The automobile speed can have *any* value between zero and, say, 100 mph. Similarly, the microphone output might have any value within a range of zero to 10 mV (e.g., 1 mV, 2.3724 mV, 9.9999 mV).

Digital Representations

In **digital representation** the quantities are represented not by continuously variable indicators but by symbols called *digits.* As an example, consider the digital watch, which provides the time of day in the form of decimal digits that represent hours and minutes (and sometimes seconds). As we know, the time of day changes continuously, but the digital watch reading does not change continuously; rather, it changes in steps of one per minute (or per second). In other words, this digital representation of the time of day changes in *discrete* steps, as compared with the representation of time provided by an analog ac line-powered wall clock, where the dial reading changes continuously.

The major difference between analog and digital quantities, then, can be simply stated as follows:

$$\text{analog} \equiv \text{continuous}$$
$$\text{digital} \equiv \text{discrete (step by step)}$$

Because of the discrete nature of digital representations, there is no ambiguity when reading the value of a digital quantity, whereas the value of an analog quantity is often open to interpretation. In practice, when we take a measurement of an analog quantity, we always "round" to a convenient level of precision. In other words, we digitize the quantity. The digital representation is the result of assigning a number of limited precision to a continuously variable quantity. For example, when you take your temperature with a mercury (analog) thermometer, the mercury column is usually between two graduation lines, but you would pick the nearest line and assign it a number of, say, 98.6°F.

EXAMPLE 1-1

Which of the following involve analog quantities and which involve digital quantities?

(a) Ten-position switch
(b) Current flowing out of an electrical outlet
(c) Temperature of a room
(d) Sand grains on the beach
(e) Automobile speedometer

Solution
(a) Digital
(b) Analog
(c) Analog
(d) Digital, since the number of grains can be only certain discrete (integer) values and not every possible value over a continuous range
(e) Analog, if needle type; digital, if numerical readout type

---

REVIEW QUESTION *

1. Concisely describe the major difference between analog and digital quantities.

1-2   DIGITAL AND ANALOG SYSTEMS

A **digital system** is a combination of devices designed to manipulate logical information or physical quantities that are represented in digital form; that is, the quantities can take on only discrete values. These devices are most often electronic, but they can also be mechanical, magnetic, or pneumatic. Some of the more familiar digital systems include digital computers and calculators, digital audio and video equipment, and the telephone system—the world's largest digital system.

   An **analog system** contains devices that manipulate physical quantities that are represented in analog form. In an analog system, the quantities can vary over a continuous range of values. For example, the amplitude of the output signal to the speaker in a radio receiver can have any value between zero and its maximum limit. Other common analog systems are audio amplifiers, magnetic tape recording and playback equipment, and a simple light dimmer switch.

---

*Answers to review questions are found at the end of the chapter in which they occur.

Advantages of Digital Techniques

An increasing majority of applications in electronics, as well as in most other technologies, use digital techniques to perform operations that were once performed using analog methods. The chief reasons for the shift to digital technology are:

1. *Digital systems are generally easier to design.* The circuits used in digital systems are *switching circuits,* where *exact* values of voltage or current are not important, only the range (HIGH or LOW) in which they fall.

2. *Information storage is easy.* This is accomplished by special devices and circuits that can latch onto digital information and hold it for as long as necessary, and mass storage techniques that can store billions of bits of information in a relatively small physical space. Analog storage capabilities are, by contrast, extremely limited.

3. *Accuracy and precision are easier to maintain throughout the system.* Once a signal is digitized, the information it contains does not deteriorate as it is processed. In analog systems, the voltage and current signals tend to be distorted by the effects of temperature, humidity, and component tolerance variations in the circuits that process the signal.

4. *Operation can be programmed.* It is fairly easy to design digital systems whose operation is controlled by a set of stored instructions called a *program.* Analog systems can also be *programmed,* but the variety and the complexity of the available operations are severely limited.

5. *Digital circuits are less affected by noise.* Spurious fluctuations in voltage (noise) are not as critical in digital systems because the exact value of a voltage is not important, as long as the noise is not large enough to prevent us from distinguishing a HIGH from a LOW.

6. *More digital circuitry can be fabricated on IC chips.* It is true that analog circuitry has also benefited from the tremendous development of IC technology, but its relative complexity and its use of devices that cannot be economically integrated (high-value capacitors, precision resistors, inductors, transformers) have prevented analog systems from achieving the same high degree of integration.

Limitations of Digital Techniques

There are really very few drawbacks when using digital techniques. The two biggest problems are:

The real world is analog.
Processing digitized signals takes time.

Most physical quantities are analog in nature, and these quantities are often the inputs and outputs that are being monitored, operated on, and controlled by a system. Some examples are temperature, pressure, position, velocity, liquid level, flow rate, and so on. We are in the habit of expressing these quantities *digitally,* such as when we say that the temperature is 64° (63.8° when we want to be more precise), but we are really making a digital approximation to an inherently analog quantity.

To take advantage of digital techniques when dealing with analog inputs and outputs, three steps must be followed:

1. Convert the real-world analog inputs to digital form.
2. Process *(operate on)* the digital information.
3. Convert the digital outputs back to real-world analog form.

Figure 1-1 shows a block diagram of this for a typical temperature control system. As the diagram shows, the analog temperature is measured and the measured value is then converted to a digital quantity by an **analog-to-digital converter (ADC)**. The digital quantity is then processed by the digital circuitry, which may or may not include a digital computer. Its digital output is converted back to an analog quantity by a **digital-to-analog converter (DAC)**. This analog output is fed to a controller, which takes some kind of action to adjust the temperature.

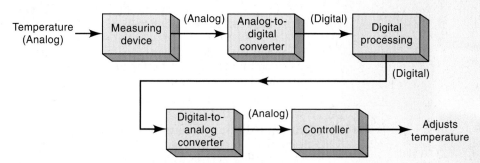

FIGURE 1-1 Block diagram of a temperature control system that requires analog/digital conversions to allow the use of digital processing techniques.

Another good example where conversion between analog and digital takes place is in the recording of audio. Compact disks (CDs) have taken the recording industry by storm because they provide a much better means for recording and playing back music. The process works something like this: (1) sounds from instruments and human voices produce an analog voltage signal in a microphone; (2) this analog signal is converted to a digital format using an analog-to-digital conversion process; (3) the digital information is stored on the CD's surface; (4) during playback, the CD player takes the digital information from the CD surface and converts it into an analog signal that is then amplified and fed to a speaker, where it can be picked up by the human ear.

The second drawback to digital systems is that processing these digitized signals (lists of numbers) takes time. And we also need to convert between the analog and digital forms of information, which can add complexity and expense to a system. The more precise the numbers need to be, the longer it takes to process them. In many applications, these factors are outweighed by the numerous advantages of using digital techniques, and so the conversion between analog and digital quantities has become quite commonplace in the current technology.

There are situations, however, where use of analog techniques is simpler or more economical. For example, several years ago, a colleague decided to create a control system demonstration for tour groups. He planned to suspend a metallic object in a magnetic field, as show in Figure 1-2. An electromagnet was made by winding a coil of wire and controlling the amount of current through the coil. The position of the metal object was measured by passing an infrared light beam across the magnetic field. As the object drew closer to the magnetic coil, it began to block the light beam. By measuring small changes in the light level, the magnetic field could be controlled to

(a)   (b)

FIGURE 1-2 A magnetic levitation system suspending: (a) a globe with a steel plate inserted and (b) a hammer.

keep the metal object hovering and stationary, with no strings attached. All attempts at using a microcomputer to measure these very small changes, run the control calculations, and drive the magnet proved to be too slow, even when using the fastest, most powerful PC available at the time. His final solution used just a couple of op-amps and a few dollars' worth of other components: a totally analog approach. Perhaps today, with a much faster processor, a digital system could handle the problem.

It is common to see both digital and analog techniques employed within the same system to be able to profit from the advantages of each. In these *hybrid* systems, one of the most important parts of the design phase involves determining what parts of the system are to be analog and what parts are to be digital. The trend in most systems is to digitize the signal as early as possible and convert it back to analog as late as possible as the signals flow through the system.

The Future Is Digital

The advances in digital technology over the past three decades have been nothing short of phenomenal, and there is every reason to believe that more is coming. Think of the everyday items that have changed from analog format to digital in your lifetime. Digital audio has moved us to the compact disk. Digital video brought the DVD. Digital video and still cameras; digital cellular phones; and digital imaging in x-ray, magnetic resonance imaging (MRI), and ultrasound systems in hospitals are just a few of the applications that have been taken over by the digital revolution. As soon as the infrastructure is in place, telephone and television systems will go digital. The growth rate in the digital realm continues to be staggering. Maybe your automobile is equipped with an Auto PC that turns your dashboard into a hub for wireless communication, information, and navigation. You may already be using voice commands to send or retrieve e-mail, call for a traffic report, check on the car's maintenance needs, send a fax, order takeout, or just switch radio stations or CDs—all without taking your hands off the wheel or your eyes off the road. In the coming years, your right and left cuff links or earrings may communicate with each other via low-orbiting satellites and have more computer power than your present home or office computer. Telephones will be able to receive, sort, and maybe respond to incoming calls like a well-trained secretary. When you watch television for an hour, what you see will have been delivered to your home in less than a second and stored in your TV's (computer's) memory for viewing at your convenience. Reading about a place 5,000 miles away may include the sensory experience of being there. And that's only the tip of the iceberg.

In other words, digital technology will continue its high-speed incursion into current areas of our lives, as well as break new ground in ways we may not even have thought about. These are exciting times and the current technology and tools for implementing these very complex digital systems are readily available to you and me today. This book will provide an introduction to these tools and concepts that will allow you to grow with this fascinating field. All we can do is try to learn as much as we can about this technology and hang on and enjoy the ride.

1. What are the advantages of digital techniques over analog?
2. What is the chief limitation to the use of digital techniques?

1-3 DIGITAL NUMBER SYSTEMS

Many number systems are in use in digital technology. The most common are the decimal, binary, octal, and hexadecimal systems. The decimal system is clearly the most familiar to us because it is a tool that we use every day. Examining some of its characteristics will help us to understand the other systems better.

Decimal System

The **decimal system** is composed of *10* numerals or symbols. These 10 symbols are 0, 1, 2, 3, 4, 5, 6, 7, 8, 9; using these symbols as *digits* of a number, we can express any quantity. The decimal system, also called the *base-10* system because it has 10 digits, has evolved naturally as a result of the fact that people have 10 fingers. In fact, the word *digit* is derived from the Latin word for "finger."

The decimal system is a *positional-value system* in which the value of a digit depends on its position. For example, consider the decimal number 453. We know that the digit 4 actually represents 4 *hundreds,* the 5 represents 5 *tens,* and the 3 represents 3 *units.* In essence, the 4 carries the most weight of the three digits; it is referred to as the *most significant digit (MSD).* The 3 carries the least weight and is called the *least significant digit (LSD).*

Consider another example, 27.35. This number is actually equal to 2 tens plus 7 units plus 3 tenths plus 5 hundredths, or $2 \times 10 + 7 \times 1 + 3 \times 0.1 + 5 \times 0.01$. The decimal point is used to separate the integer and fractional parts of the number.

More rigorously, the various positions relative to the decimal point carry weights that can be expressed as powers of 10. This is illustrated in Figure 1-3, where the number 2745.214 is represented. The decimal point separates the positive powers of 10 from the negative powers. The number 2745.214 is thus equal to

$$(2 \times 10^{+3}) + (7 \times 10^{+2}) + (4 \times 10^{1}) + (5 \times 10^{0})$$
$$+ (2 \times 10^{-1}) + (1 \times 10^{-2}) + (4 \times 10^{-3})$$

FIGURE 1-3 Decimal
position values as powers
of 10.

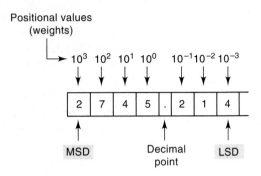

In general, any number is simply the sum of the products of each digit value
and its positional value.

Decimal Counting

When counting in the decimal system, we start with 0 in the units position
and take each symbol (digit) in progression until we reach 9. Then we add a
1 to the next higher position and start over with 0 in the first position (see
Figure 1-4). This process continues until the count of 99 is reached. Then we
add a 1 to the third position and start over with 0s in the first two positions.
The same pattern is followed continuously as high as we wish to count.

FIGURE 1-4 Decimal
counting.

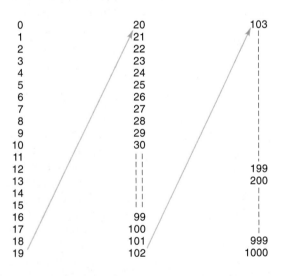

It is important to note that in decimal counting, the units position (LSD)
changes upward with each step in the count, the tens position changes upward every 10 steps in the count, the hundreds position changes upward
every 100 steps in the count, and so on.

Another characteristic of the decimal system is that using only two decimal places, we can count through $10^2 = 100$ different numbers (0 to 99).*
With three places we can count through 1000 numbers (0 to 999), and so on.
In general, with N places or digits, we can count through 10^N different numbers, starting with and including zero. The largest number will always be
$10^N - 1$.

*Zero is counted as a number.

Binary System

Unfortunately, the decimal number system does not lend itself to convenient implementation in digital systems. For example, it is very difficult to design electronic equipment so that it can work with 10 different voltage levels (each one representing one decimal character, 0 through 9). On the other hand, it is very easy to design simple, accurate electronic circuits that operate with only two voltage levels. For this reason, almost every digital system uses the binary (base-2) number system as the basic number system of its operations. Other number systems are often used to interpret or represent binary quantities for the convenience of the people who work with and use these digital systems.

In the **binary system** there are only two symbols or possible digit values, 0 and 1. Even so, this base-2 system can be used to represent any quantity that can be represented in decimal or other number systems. In general though, it will take a greater number of binary digits to express a given quantity.

All of the statements made earlier concerning the decimal system are equally applicable to the binary system. The binary system is also a positional-value system, wherein each binary digit has its own value or weight expressed as a power of 2. This is illustrated in Figure 1-5. Here, places to the left of the *binary point* (counterpart of the decimal point) are positive powers of 2, and places to the right are negative powers of 2. The number 1011.101 is shown represented in the figure. To find its equivalent in the decimal system, we simply take the sum of the products of each digit value (0 or 1) and its positional value:

$$1011.101_2 = (1 \times 2^3) + (0\ 2^2) + (1 \times 2^1) + (1 \times 2^0)$$
$$+ (1 \times 2^{-1}) + (0 \times 2^{-2}) + (1 \times 2^{-3})$$
$$= 8 + 0 + 2 + 1 + 0.5 + 0 + 0.125$$
$$= 11.625_{10}$$

Notice in the preceding operation that subscripts (2 and 10) were used to indicate the base in which the particular number is expressed. This convention is used to avoid confusion whenever more than one number system is being employed.

FIGURE 1-5 Binary position values as powers of 2.

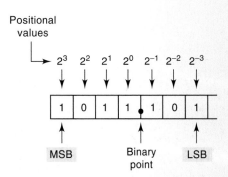

In the binary system, the term *binary digit* is often abbreviated to the term **bit**, which we will use from now on. Thus, in the number expressed in Figure 1-5 there are four bits to the left of the binary point, representing the integer part of the number, and three bits to the right of the binary point, representing the fractional part. The most significant bit (MSB) is the leftmost bit (largest weight). The least significant bit (LSB) is the rightmost bit (smallest weight). These are indicated in Figure 1-5. Here, the MSB has a weight of 2^3; the LSB has a weight of 2^{-3}.

Binary Counting

When we deal with binary numbers, we will usually be restricted to a specific number of bits. This restriction is based on the circuitry used to represent these binary numbers. Let's use four-bit binary numbers to illustrate the method for counting in binary.

The sequence (shown in Figure 1-6) begins with all bits at 0; this is called the *zero count*. For each successive count, the units (2^0) position *toggles*; that is, it changes from one binary value to the other. Each time the units bit changes from a 1 to a 0, the twos (2^1) position will toggle (change states). Each time the twos position changes from 1 to 0, the fours (2^2) position will toggle (change states). Likewise, each time the fours position goes from 1 to 0, the eights (2^3) position toggles. This same process would be continued for the higher-order bit positions if the binary number had more than four bits.

The binary counting sequence has an important characteristic, as shown in Figure 1-6. The units bit (LSB) changes either from 0 to 1 or 1 to 0 with *each* count. The second bit (twos position) stays at 0 for two counts, then at 1 for two counts, then at 0 for two counts, and so on. The third bit (fours position) stays at 0 for four counts, then at 1 for four counts, and so on. The fourth bit (eights position) stays at 0 for eight counts, then at 1 for eight counts. If we wanted to count further, we would add more places, and this pattern would continue with 0s and 1s alternating in groups of 2^{N-1}. For example, using a fifth binary place, the fifth bit would alternate sixteen 0s, then sixteen 1s, and so on.

FIGURE 1-6 Binary counting sequence.

Weights → $2^3 = 8$	$2^2 = 4$	$2^1 = 2$	$2^0 = 1$	Decimal equivalent
0	0	0	0	0
0	0	0	1	1
0	0	1	0	2
0	0	1	1	3
0	1	0	0	4
0	1	0	1	5
0	1	1	0	6
0	1	1	1	7
1	0	0	0	8
1	0	0	1	9
1	0	1	0	10
1	0	1	1	11
1	1	0	0	12
1	1	0	1	13
1	1	1	0	14
1	1	1	1	15

LSB

As we saw for the decimal system, it is also true for the binary system that by using N bits or places, we can go through 2^N counts. For example, with two bits we can go through $2^2 = 4$ counts (00_2 through 11_2); with four bits we can go through $2^4 = 16$ counts (0000_2 through 1111_2); and so on. The last count will always be all 1s and is equal to $2^N - 1$ in the decimal system. For example, using four bits, the last count is $1111_2 = 2^4 - 1 = 15_{10}$.

EXAMPLE 1-2

What is the largest number that can be represented using eight bits?

Solution

$2^N - 1 = 2^8 - 1 = 255_{10} = 11111111_2$.

This has been a brief introduction of the binary number system and its relation to the decimal system. We will spend much more time on these two systems and several others in the next chapter.

REVIEW QUESTIONS

1. What is the decimal equivalent of 1101011_2?
2. What is the next binary number following 10111_2 in the counting sequence?
3. What is the largest decimal value that can be represented using 12 bits?

1-4 REPRESENTING BINARY QUANTITIES

In digital systems, the information being processed is usually present in binary form. Binary quantities can be represented by any device that has only two operating states or possible conditions. For example, a switch has only two states: open or closed. We can arbitrarily let an open switch represent binary 0 and a closed switch represent binary 1. With this assignment we can now represent any binary number as illustrated in Figure 1-7(a), where the states of the various switches represent 10010_2.

Another example is shown in Figure 1-7(b), where holes punched in paper are used to represent binary numbers. A punched hole is a binary 1, and the absence of a hole is a binary 0. Figure 1-7(c) shows a current example of the need for binary representation. The wires are in rows and columns that form a matrix. The intersection of each row and column can be in one of two

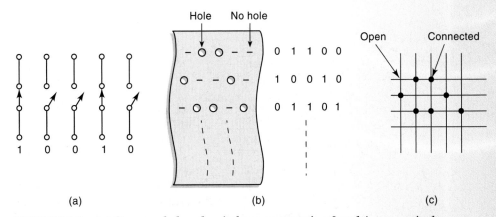

FIGURE 1-7 (a) Open and closed switches representing 0 and 1, respectively; (b) absence or presence of holes in paper tape representing 0 and 1, respectively; (c) connected or not connected junctions in a wiring matrix.

states: connected or open. We will examine this structure in great detail in later chapters when we study modern programmable logic devices.

There are numerous other devices that have only two operating states or can be operated in two extreme conditions. Among these are: light bulb (bright or dark), diode (conducting or nonconducting), relay (energized or deenergized), transistor (cut off or saturated), photocell (illuminated or dark), thermostat (open or closed), mechanical clutch (engaged or disengaged), and spot on a magnetic disk (magnetized or demagnetized).

In electronic digital systems, binary information is represented by voltages (or currents) that are present at the inputs and outputs of the various circuits. Typically, the binary 0 and 1 are represented by two nominal voltage levels. For example, zero volts (0 V) might represent binary 0, and + 5 V might represent binary 1. In actuality, because of circuit variations, the 0 and 1 would be represented by voltage ranges. This is illustrated in Figure 1-8(a), where any voltage between 0 and 0.8 V represents a 0 and any voltage between 2 and 5 V represents a 1. All input and output signals will normally fall within one of these ranges, except during transitions from one level to another.

We can now see another significant difference between digital and analog systems. In digital systems, the exact value of a voltage *is not* important; for example, for the voltage assignments of Figure 1-8(a), a voltage of 3.6 V means the same as a voltage of 4.3 V. In analog systems, the exact value of a voltage *is* important. For instance, if the analog voltage is proportional to the temperature measured by a transducer, the 3.6 V would represent a different temperature than would 4.3 V. In other words, the voltage value carries significant information. This characteristic means that the design of accurate analog circuitry is generally more difficult than that of digital circuitry because of the way in which exact voltage values are affected by variations in component values, temperature, and noise (random voltage fluctuations).

Digital Signals and Timing Diagrams

Figure 1-8(b) shows a typical digital signal and how it varies over time. It is actually a graph of voltage versus time *(t)* and is called a **timing diagram**. The horizontal time scale is marked off at regular intervals beginning at t_0 and proceeding to t_1, t_2, and so on. For the example timing diagram shown

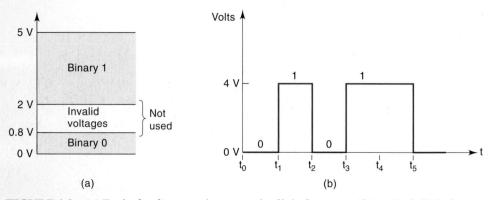

(a)   (b)

FIGURE 1-8   (a) Typical voltage assignments in digital system; (b) typical digital signal timing diagram.

here, the signal starts at 0 V (a binary 0) at time t_0 and remains there until time t_1. At t_1, the signal makes a rapid transition (jump) up to 4 V (a binary 1). At t_2, it jumps back down to 0 V. Similar transitions occur at t_3 and t_5. Note that the signal does not change at t_4 but stays at 4 V from t_3 to t_5.

The transitions on this timing diagram are drawn as vertical lines, and so they appear to be instantaneous, when in reality they are not. In many situations, however, the transition times are so short compared to the times between transitions that we can show them on the diagram as vertical lines. We will encounter situations later where it will be necessary to show the transitions more accurately on an expanded time scale.

Timing diagrams are used extensively to show how digital signals change with time, and especially to show the relationship between two or more digital signals in the same circuit or system. By displaying one or more digital signals on an *oscilloscope* or *logic analyzer*, we can compare the signals to their expected timing diagrams. This is a very important part of the testing and troubleshooting procedures used in digital systems.

1-5 DIGITAL CIRCUITS/LOGIC CIRCUITS

Digital circuits are designed to produce output voltages that fall within the prescribed 0 and 1 voltage ranges such as those defined in Figure 1-8. Likewise, digital circuits are designed to respond predictably to input voltages that are within the defined 0 and 1 ranges. What this means is that a digital circuit will respond in the same way to all input voltages that fall within the allowed 0 range; similarly, it will not distinguish between input voltages that lie within the allowed 1 range.

To illustrate, Figure 1-9 represents a typical digital circuit with input v_i and output v_o. The output is shown for two different input signal waveforms. Note that v_o is the same for both cases because the two input waveforms, while differing in their exact voltage levels, are at the same binary levels.

FIGURE 1-9   A digital circuit responds to an input's binary level (0 or 1) and not to its actual voltage.

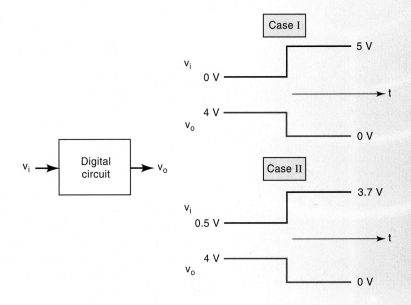

Logic Circuits

The manner in which a digital circuit responds to an input is referred to as the circuit's *logic.* Each type of digital circuit obeys a certain set of logic rules. For this reason, digital circuits are also called **logic circuits**. We will use both terms interchangeably throughout the text. In Chapter 3, we will see more clearly what is meant by a circuit's "logic."

We will be studying all the types of logic circuits that are currently used in digital systems. Initially, our attention will be focused only on the logical operation that these circuits perform—that is, the relationship between the circuit inputs and outputs. We will defer any discussion of the internal circuit operation of these logic circuits until after we have developed an understanding of their logical operation.

Digital Integrated Circuits

Almost all of the digital circuits used in modern digital systems are integrated circuits (ICs). The wide variety of available logic ICs has made it possible to construct complex digital systems that are smaller and more reliable than their discrete-component counterparts.

Several integrated-circuit fabrication technologies are used to produce digital ICs, the most common being CMOS, TTL, NMOS, and ECL. Each differs in the type of circuitry used to provide the desired logic operation. For example, TTL (transistor-transistor logic) uses the bipolar transistor as its main circuit element, while CMOS (complementary metal-oxide-semiconductor) uses the enhancement-mode MOSFET as its principal circuit element. We will learn about the various IC technologies, their characteristics, and their relative advantages and disadvantages after we master the basic logic circuit types.

REVIEW QUESTIONS

1. *True or false:* The exact value of an input voltage is critical for a digital circuit.
2. Can a digital circuit produce the same output voltage for different input voltage values?
3. A digital circuit is also referred to as a _____ circuit.
4. A graph that shows how one or more digital signals change with time is called a _____.

1-6   PARALLEL AND SERIAL TRANSMISSION

One of the most common operations that occur in any digital system is the transmission of information from one place to another. The information can be transmitted over a distance as small as a fraction of an inch on the same circuit board, or over a distance of many miles when an operator at a computer terminal is communicating with a computer in another city. The information that is transmitted is in binary form and is generally represented as voltages at the outputs of a sending circuit that are connected to the inputs of a receiving circuit. Figure 1-10 illustrates the two basic methods for digital information transmission: **parallel** and **serial**.

FIGURE 1-10 (a) Parallel transmission uses one connecting line per bit, and all bits are transmitted simultaneously; (b) serial transmission uses only one signal line, and the individual bits are transmitted serially (one at a time).

Figure 1-10(a) shows how the binary number 10100110 is transmitted from the computer to a printer using *parallel transmission*. Each bit of the binary number is represented by one of the computer outputs and is connected to a corresponding input of the printer, so that all eight bits are transmitted simultaneously (in parallel).

Figure 1-10(b) shows that there is only one connection between the computer and printer when *serial transmission* is used. Here the computer output will produce a digital signal whose voltage level will change at regular intervals in accordance with the binary number being transmitted; that is, one bit is transmitted per time interval (serially) to the printer input. The accompanying timing diagram shows how the signal level varies with time. Note that the LSB is transmitted first; this is typical for serial transmission.

The principal trade-off between parallel and serial representations is one of speed versus circuit simplicity. The transmission of binary data from one part of a digital system to another can be done more quickly using parallel representation because all the bits are transmitted simultaneously, while serial representation transmits one bit at a time. On the other hand, parallel requires more signal lines connected between the sender and the receiver of the binary data than does serial. In other words, parallel is faster, and serial requires fewer signal lines. This comparison between parallel and serial methods for representing binary information will be encountered many times in discussions throughout the text.

REVIEW QUESTION

1. Describe the relative advantages of parallel and serial transmission of binary data.

1-7 MEMORY

When an input signal is applied to most devices or circuits, the output somehow changes in response to the input, and when the input signal is removed, the output returns to its original state. These circuits do not exhibit the property of *memory* because their outputs revert back to normal. In digital circuitry certain types of devices and circuits do have memory. When an input is applied to such a circuit, the output will change its state, but it will remain in the new state even after the input is removed. This property of retaining its response to a momentary input is called **memory**. Figure 1-11 illustrates nonmemory and memory operations.

FIGURE 1-11 Comparison of nonmemory and memory operation.

Memory devices and circuits play an important role in digital systems because they provide a means for storing binary numbers either temporarily or permanently, with the ability to change the stored information at any time. As we shall see, the various memory elements include magnetic and optical types and those that utilize electronic latching circuits (called *latches* and *flip-flops*).

1-8 DIGITAL COMPUTERS

Digital techniques have found their way into innumerable areas of technology, but the area of automatic **digital computers** is by far the most notable and most extensive. Although digital computers affect some part of all of our lives, it is doubtful that many of us know exactly what a computer does. In simplest terms, *a computer is a system of hardware that performs arithmetic operations, manipulates data (usually in binary form), and makes decisions.*

For the most part, human beings can do whatever computers can do, but computers can do it with much greater speed and accuracy, in spite of the fact that computers perform all their calculations and operations one step at a time. For example, a human being can take a list of 10 numbers and find their sum all in one operation by listing the numbers one over the other and adding them column by column. A computer, on the other hand, can add numbers only two at a time, so that adding this same list of numbers will take nine actual addition steps. Of course, the fact that the computer requires only a few nanoseconds per step makes up for this apparent inefficiency.

A computer is faster and more accurate than people are, but unlike most of us, it must be given a complete set of instructions that tell it *exactly* what to do at each step of its operation. This set of instructions, called a **program,** is prepared by one or more persons for each job the computer is to do. Programs are placed in the computer's memory unit in binary-coded form, with each instruction having a unique code. The computer takes these instruction codes from memory *one at a time* and performs the operation called for by the code.

Major Parts of a Computer

There are several types of computer systems, but each can be broken down into the same functional units. Each unit performs specific functions, and all units function together to carry out the instructions given in the program. Figure 1-12 shows the five major functional parts of a digital computer and their interaction. The solid lines with arrows represent the flow of data and information. The dashed lines with arrows represent the flow of timing and control signals.

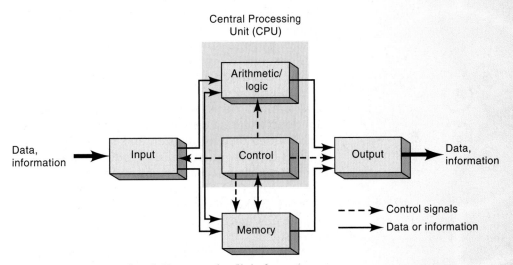

FIGURE 1-12   Functional diagram of a digital computer.

The major functions of each unit are:

1. **Input unit.** Through this unit, a complete set of instructions and data is fed into the computer system and into the memory unit, to be stored until needed. The information typically enters the input unit from a keyboard or a disk.

2. **Memory unit.** The memory stores the instructions and data received from the input unit. It stores the results of arithmetic operations received from the arithmetic unit. It also supplies information to the output unit.

3. **Control unit.** This unit takes instructions from the memory unit one at a time and interprets them. It then sends appropriate signals to all the other units to cause the specific instruction to be executed.

4. **Arithmetic/logic unit**. All arithmetic calculations and logical decisions are performed in this unit, which can then send results to the memory unit to be stored.

5. **Output unit**. This unit takes data from the memory unit and prints out, displays, or otherwise presents the information to the operator (or process, in the case of a process control computer).

Central Processing Unit (CPU)

As the diagram in Figure 1-12 shows, the control and arithmetic/logic units are often considered as one unit, called the **central processing unit (CPU)**. The CPU contains all of the circuitry for fetching and interpreting instructions and for controlling and performing the various operations called for by the instructions.

TYPES OF COMPUTERS   All computers are made up of the basic units described above, but they can differ as to physical size, operating speed, memory capacity, and computational power, as well as other characteristics. Computers are often classified according to physical size, which often, although not always, is an indication of their relative capabilities. The three basic classifications, from smallest to largest, are: *microcomputer, minicomputer (workstation),* and *mainframe.* As microcomputers have become more and more powerful, the distinction between microcomputers and minicomputers has become rather blurred, and we have begun to distinguish only between small computers—those that can fit in an office or on a desktop or a lap—and large computers—those that are too big for any of those places. In this book we will be concerned mainly with microcomputers.

A **microcomputer** is the smallest type of computer. It generally consists of several IC chips, including a **microprocessor** chip, memory chips, and input/output interface chips, along with input/output devices such as a keyboard, video display, printer, and disk drives. Microcomputers were developed as a result of tremendous advances in IC fabrication technology that made it possible to pack more and more digital circuits onto a small chip. For example, the microprocessor chip contains—at a minimum—all the circuits that make up the CPU portion of the computer, that is, the control unit and the arithmetic/logic unit. The microprocessor, in other words, is a "CPU on a chip."

Most of us are familiar with general-purpose personal computers (PCs), which are used in more than half of our homes and in almost all of our businesses. These microcomputers can perform a wide variety of tasks in a wide range of applications depending on the software (programs) they are running. There is a more specialized type of microcomputer called a **microcontroller**, which is not a general-purpose computer. Rather, it is designed to be used as a *dedicated* or *embedded controller* that helps monitor and control the operation of a machine, a piece of equipment, or a process. Microcontrollers are microcomputers because they use a microprocessor chip as the CPU, but they are much smaller than general-purpose microcomputers because the input/output devices they normally use are much smaller. In fact, some of the input/output devices—as well as memory—are usually right on the same chip as the microprocessor. These single-chip microcontrollers are employed in a wide variety of control applications such as appliance control, metalworking machines, VCRs, automated teller machines, photocopiers, automobile ignition systems, antilock brakes, medical instrumentation, and much more.

So you see, even people who don't own a PC or use one at work or school are using microcomputers every day because so many modern consumer electronic devices, appliances, office equipment, and much more are built around embedded microcontrollers. If you work, play, or go to school in this digital age, there's no escaping it: you'll use a microcomputer somewhere.

REVIEW QUESTIONS

1. Explain how a digital circuit that has memory differs from one that does not.
2. Name the five major functional units of a computer.
3. Which two units make up the CPU?
4. An IC chip that contains a CPU is called a _____.

SUMMARY

1. The two basic ways of representing the numerical value of physical quantities are analog (continuous) and digital (discrete).
2. Most quantities in the real world are analog, but digital techniques are generally superior to analog techniques, and most of the predicted advances will be in the digital realm.
3. The binary number system (0 and 1) is the basic system used in digital technology.
4. Digital or logic circuits operate on voltages that fall in prescribed ranges that represent either a binary 0 or a binary 1.
5. The two basic ways to transfer digital information are parallel—all bits simultaneously—and serial—one bit at a time.
6. The main parts of all computers are the input, control, memory, arithmetic/logic, and output units.
7. The combination of the arithmetic/logic unit and the control unit makes up the CPU (central processing unit).
8. A microcomputer usually has a CPU that is on a single chip called a *microprocessor.*
9. A microcontroller is a microcomputer especially designed for dedicated (not general-purpose) control applications.

IMPORTANT TERMS*

analog representation	bit	memory unit
digital representation	timing diagram	control unit
digital system	digital circuits/logic	arithmetic/logic unit
analog system	circuits	output unit
analog-to-digital	parallel transmission	central processing
converter (ADC)	serial transmission	unit (CPU)
digital-to-analog	memory	microcomputer
converter (DAC)	digital computer	microprocessor
decimal system	program	microcontroller
binary system	input unit	

*These terms can be found in **boldface** type in the chapter and are defined in the Glossary at the end of the book. This applies to all chapters.

PROBLEMS

SECTION 1-2

1-1. Which of the following are analog quantities, and which are digital?
(a) Number of atoms in a sample of material
(b) Altitude of an aircraft
(c) Pressure in a bicycle tire
(d) Current through a speaker
(e) Timer setting on a microwave oven

SECTION 1-3

1-2. Convert the following binary numbers to their equivalent decimal values.
(a) 11001_2
(b) 1001.1001_2
(c) 10011011001.10110_2

1-3. Using six bits, show the binary counting sequence from 000000 to 111111.

1-4. What is the maximum number that we can count up to using 10 bits?

1-5. How many bits are needed to count up to a maximum of 511?

SECTION 1-4

1-6. Draw the timing diagram for a digital signal that continuously alternates between 0.2 V (binary 0) for 2 ms and 4.4 V (binary 1) for 4 ms.

SECTION 1-6

1-7. Suppose that the decimal integer values from 0 to 15 are to be transmitted in binary.
(a) How many lines will be needed if parallel representation is used?
(b) How many will be needed if serial representation is used?

SECTIONS 1-7 AND 1-8

1-8. How is a microprocessor different from a microcomputer?

1-9. How is a microcontroller different from a microcomputer?

ANSWERS TO SECTION REVIEW QUESTIONS

SECTION 1-1

1. Analog quantities can take on *any* value over a continuous range; digital quantities can take on only *discrete* values.

SECTION 1-2

1. Easier to design; easier to store information; greater accuracy and precision; programmability; less affected by noise; higher degree of integration   2. Real-world physical quantities are analog. Digital processing takes time.

SECTION 1-3

1. 107_{10} 2. 11000_2 3. 4095_{10}

SECTION 1-5

1. False   2. Yes, provided that the two input voltages are within the same logic level range   3. Logic   4. Timing diagram

SECTION 1-6

1. Parallel is faster; serial requires only one signal line.

SECTION 1-8

1. One that has memory will have its output changed and *remain* changed in response to a momentary change in the input signal.   2. Input, output, memory, arithmetic/logic, control   3. Control and arithmetic/logic   4. Microprocessor

CHAPTER 2

NUMBER SYSTEMS AND CODES

■ OUTLINE

■ OBJECTIVES

Upon completion of this chapter, you will be able to:

- ■ Convert a number from one number system (decimal, binary, octal, hexadecimal) to its equivalent in one of the other number systems.
- ■ Cite the advantages of the octal and hexadecimal number systems.
- ■ Count in octal and hexadecimal.
- ■ Represent decimal numbers using the BCD code; cite the pros and cons of using BCD.
- ■ Understand the difference between BCD and straight binary.
- ■ Understand the purpose of alphanumeric codes such as the ASCII code.
- ■ Explain the parity method for error detection.
- ■ Determine the parity bit to be attached to a digital data string.

■ INTRODUCTION

The binary number system is the most important one in digital systems, but several others are also important. The decimal system is important because it is universally used to represent quantities outside a digital system. This means that there will be situations where decimal values must be converted to binary values before they are entered into the digital system. For example, when you punch a decimal number into your hand calculator (or computer), the circuitry inside the machine converts the decimal number to a binary value.

Likewise, there will be situations where the binary values at the outputs of a digital system must be converted to decimal values for presentation to the outside world. For example, your calculator (or computer) uses binary numbers to calculate answers to a problem and then converts the answers to a decimal value before displaying them.

In addition to binary and decimal, two other number systems find widespread applications in digital systems. The *octal* (base-8) and *hexadecimal* (base-16) number systems are both used for the same purpose—to provide an efficient means for representing large binary numbers. As we shall see, both of these number systems have the advantage that they can be converted easily to and from binary.

In a digital system, three or four of these number systems may be in use at the same time, so that an understanding of the system operation requires the ability to convert from one number system to another. This chapter will show you how to perform these conversions. Although some of them will not be of immediate use in our study of digital systems, you will need them when you begin to study microprocessors.

This chapter will also introduce some of the *binary codes* that are used to represent various kinds of information. These binary codes will use 1s and 0s, but in a way that differs somewhat from that of the binary number system.

2-1 BINARY-TO-DECIMAL CONVERSIONS

As explained in Chapter 1, the binary number system is a positional system where each binary digit (bit) carries a certain weight based on its position relative to the LSB. Any binary number can be converted to its decimal equivalent simply by summing together the weights of the various positions in the binary number that contain a 1. To illustrate, let's change 11011_2 to its decimal equivalent.

$$1 \quad 1 \quad 0 \quad 1 \quad 1_2$$
$$2^4 + 2^3 + 0 + 2^1 + 2^0 = 16 + 8 + 2 + 1$$
$$= 27_{10}$$

Let's try another example with a greater number of bits:

$$1 \quad 0 \quad 1 \quad 1 \quad 0 \quad 1 \quad 0 \quad 1_2 =$$
$$2^7 + 0 + 2^5 + 2^4 + 0 + 2^2 + 0 + 2^0 = 181_{10}$$

Note that the procedure is to find the weights (i.e., powers of 2) for each bit position that contains a 1, and then to add them up. Also note that the MSB has a weight of 2^7 even though it is the eighth bit; this is because the LSB is the first bit and has a weight of 2^0.

REVIEW QUESTIONS

1. Convert 100011011011_2 to its decimal equivalent.
2. What is the weight of the MSB of a 16-bit number?

2-2 DECIMAL-TO-BINARY CONVERSIONS

There are two ways to convert a decimal *whole* number to its equivalent binary-system representation. The first method is the reverse of the process described in Section 2-1. The decimal number is simply expressed as a sum of powers of 2, and then 1s and 0s are written in the appropriate bit positions. To illustrate:

$$45_{10} = 32 + 8 + 4 + 1 = 2^5 + 0 + 2^3 + 2^2 + 0 + 2^0$$
$$= 1 \quad 0 \quad 1 \quad 1 \quad 0 \quad 1_2$$

Note that a 0 is placed in the 2^1 and 2^4 positions, since all positions must be accounted for. Another example is the following:

$$76_{10} = 64 + 8 + 4 = 2^6 + 0 + 0 + 2^3 + 2^2 + 0 + 0$$
$$= 1 \quad 0 \quad 0 \quad 1 \quad 1 \quad 0 \quad 0_2$$

Repeated Division

Another method for converting decimal integers uses repeated division by 2. The conversion, illustrated below for 25_{10}, requires repeatedly dividing the decimal number by 2 and writing down the remainder after each division until a quotient of 0 is obtained. Note that the binary result is obtained by writing the first remainder as the LSB and the last remainder as the MSB. This process, diagrammed in the flowchart of Figure 2-1, can also be used to convert from decimal to any other number system, as we shall see.

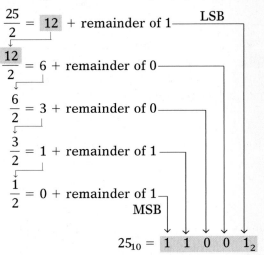

$$\frac{25}{2} = \boxed{12} + \text{remainder of } 1 \text{\qquad LSB}$$

$$\frac{\boxed{12}}{2} = 6 + \text{remainder of } 0$$

$$\frac{6}{2} = 3 + \text{remainder of } 0$$

$$\frac{3}{2} = 1 + \text{remainder of } 1$$

$$\frac{1}{2} = 0 + \text{remainder of } 1$$

$$\text{MSB}$$

$$25_{10} = 1\ 1\ 0\ 0\ 1_2$$

FIGURE 2-1 Flowchart for repeated-division method of decimal-to-binary conversion of integers. The same process can be used to convert a decimal integer to any other number system.

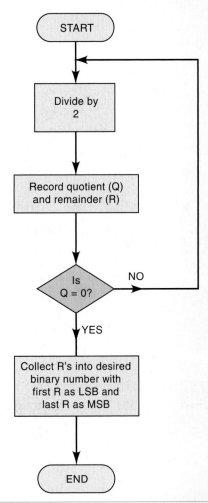

START

Divide by 2

Record quotient (Q) and remainder (R)

Is Q = 0?   NO

YES

Collect R's into desired binary number with first R as LSB and last R as MSB

END

CALCULATOR HINT:

If you use a calculator to perform the divisions by 2, you can tell whether the remainder is 0 or 1 by whether or not the result has a fractional part. For instance, 25/2 would produce 12.5. Since there is a fractional part (the .5), the remainder is a 1. If there were no fractional part, such as 12/2 = 6, then the remainder would be 0. The following example illustrates this.

EXAMPLE 2-1

Convert 37_{10} to binary. Try to do it on your own before you look at the solution.

Solution

$$\frac{37}{2} = 18.5 \longrightarrow \text{remainder of 1 (LSB)}$$

$$\frac{18}{2} = 9.0 \longrightarrow \qquad\qquad 0$$

$$\frac{9}{2} = 4.5 \longrightarrow \qquad\qquad 1$$

$$\frac{4}{2} = 2.0 \longrightarrow \qquad\qquad 0$$

$$\frac{2}{2} = 1.0 \longrightarrow \qquad\qquad 0$$

$$\frac{1}{2} = 0.5 \longrightarrow \qquad\qquad 1 \text{ (MSB)}$$

Thus, $37^{10} = 100101_2$.

Counting Range

Recall that using N bits, we can count through 2^N different decimal numbers ranging from 0 to $2^N - 1$. For example, for $N = 4$, we can count from 0000_2 to 1111_2, which is 0_{10} to 15_{10}, for a total of 16 different numbers. Here, the largest decimal value is $2^4 - 1 = 15$, and there are 2^4 different numbers.
   In general, then, we can state:

Using N bits, we can represent decimal numbers ranging from 0 to $2^N - 1$, a total of 2^N different numbers.

EXAMPLE 2-2

(a) What is the total range of decimal values that can be represented in eight bits?

(b) How many bits are needed to represent decimal values ranging from 0 to 12,500?

Solution

(a) Here we have $N = 8$. Thus, we can represent decimal numbers from 0 to $2^8 - 1 = \textbf{255}$. We can verify this by checking to see that 11111111_2 converts to 255_{10}.

(b) With 13 bits, we can count from decimal 0 to $2^{13} - 1 = 8191$. With 14 bits, we can count from 0 to $2^{14} - 1 = 16,383$. Clearly, 13 bits aren't enough, but 14 bits will get us up beyond 12,500. Thus, the required number of bits is **14**.

REVIEW QUESTIONS

1. Convert 83_{10} to binary using both methods.
2. Convert 729_{10} to binary using both methods. Check your answer by converting back to decimal.
3. How many bits are required to count up to decimal 1 million?

2-3   OCTAL NUMBER SYSTEM

The octal number system is often used in digital computer work. The **octal number system** has a base of *eight*, meaning that it has eight possible digits: 0, 1, 2, 3, 4, 5, 6, and 7. Thus, each digit of an octal number can have any value from 0 to 7. The digit positions in an octal number have weights as follows:

8^4	8^3	8^2	8^1	8^0	8^{-1}	8^{-2}	8^{-3}	8^{-4}	8^{-5}

octal point

Octal-to-Decimal Conversion

An octal number, then, can be converted easily to its decimal equivalent by multiplying each octal digit by its positional weight. For example:

$$372_8 = 3 \times (8^2) + 7 \times (8^1) + 2 \times (8^0)$$
$$= 3 \times 64 + 7 \times 8 + 2 \times 1$$
$$= 250_{10}$$

$$24.6_8 = 2 \times (8^1) + 4 \times (8^0) + 6 \times (8^{-1})$$
$$= 20.75_{10}$$

CALCULATOR HINT:

Use the y^x function to evaluate powers of 8.

Decimal-to-Octal Conversion

A decimal integer can be converted to octal by using the same repeated-division method that we used in the decimal-to-binary conversion (Figure 2-1), but with a division factor of 8 instead of 2. An example follows.

$$\frac{266}{8} = \boxed{33} + \text{remainder of } 2 \rule{0pt}{1em}\text{LSD}$$

$$\frac{33}{8} = 4 + \text{remainder of } 1$$

$$\frac{4}{8} = 0 + \text{remainder of } 4 \rule{0pt}{1em}\text{MSD}$$

$$266_{10} = \boxed{412}_8$$

Note that the first remainder becomes the least significant digit (LSD) of the octal number, and the last remainder becomes the most significant digit (MSD).

CALCULATOR HINT:

If a calculator is used to perform the divisions in the process above, the result will include a decimal fraction instead of a remainder. The remainder can be obtained, however, by multiplying the decimal fraction by 8. For example, 266/8 produces 33.25. The remainder becomes $0.25 \times 8 = 2$. Similarly, 33/8 will be 4.125, and the remainder becomes $0.125 \times 8 = 1$.

Octal-to-Binary Conversion

The primary advantage of the octal number system is the ease with which conversion can be made between binary and octal numbers. The conversion from octal to binary is performed by converting *each* octal digit to its three-bit binary equivalent. The eight possible digits are converted as indicated in Table 2-1.

TABLE 2-1

Octal Digit	0	1	2	3	4	5	6	7
Binary Equivalent	000	001	010	011	100	101	110	111

Using these conversions, we can convert any octal number to binary by individually converting each digit. For example, we can convert 472_8 to binary as follows:

$$
\begin{array}{ccc}
4 & 7 & 2 \\
\downarrow & \downarrow & \downarrow \\
100 & 111 & 010
\end{array}
$$

Thus, octal 472 is equivalent to binary 100111010. As another example, consider converting 5431_8 to binary:

$$
\begin{array}{cccc}
5 & 4 & 3 & 1 \\
\downarrow & \downarrow & \downarrow & \downarrow \\
101 & 100 & 011 & 001
\end{array}
$$

Thus, $5431_8 = 101100011001_2$.

Binary-to-Octal Conversion

Converting from binary integers to octal integers is simply the reverse of the foregoing process. The bits of the binary number are grouped into groups of *three* bits, starting at the LSB. Then each group is converted to its octal equivalent (Table 2-1). To illustrate, consider the conversion of 100111010_2 to octal.

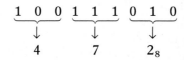

Sometimes the binary number will not have even groups of three bits. For those cases, we can add one or two 0s to the left of the MSB of the binary number to fill out the last group. This is illustrated below for the binary number 11010110.

$$\underbrace{0 \quad 1 \quad 1}_{\downarrow \atop 3} \quad \underbrace{0 \quad 1 \quad 0}_{\downarrow \atop 2} \quad \underbrace{1 \quad 1 \quad 0}_{\downarrow \atop 6_8}$$

Note that a 0 was placed to the left of the MSB to produce even groups of three bits.

Counting in Octal

The largest octal digit is 7, so that in counting in octal, a digit position is incremented upward from 0 to 7. Once it reaches 7, it recycles to 0 on the next count and causes the next higher digit position to be incremented. This is illustrated in the following sequences of octal counting: (1) 65, 66, 67, 70, 71 and (2) 275, 276, 277, 300.

With N octal digit positions, we can count from 0 up to $8^N - 1$, for a total of 8^N different counts. For example, with three octal digit positions, we can count from 000_8 to 777_8, which is 0_{10} to 511_{10} for a total of $8^3 = 512_{10}$ different octal numbers.

EXAMPLE 2-3

Convert 177_{10} to its eight-bit binary equivalent by first converting to octal.

Solution

$$\frac{177}{8} = 22 + \text{remainder of 1 (LSD)}$$

$$\frac{22}{8} = 2 + \text{remainder of 6}$$

$$\frac{2}{8} = 0 + \text{remainder of 2}$$

Thus, $177_{10} = 261_8$. Now we can quickly convert this octal number to its binary equivalent 010110001_2, so that we finally have

$$177_{10} = \mathbf{10110001_2}$$

Note that we chop off the leading 0 to express the result as eight bits.

This method of decimal-to-octal-to-binary conversion is often quicker than going directly from decimal to binary, especially for large numbers. Similarly, it is often quicker to convert binary to decimal by first converting to octal.

1. Convert 614_8 to decimal.
2. Convert 146_{10} to octal, then from octal to binary.
3. Convert 10011101_2 to octal.
4. Write the next three numbers in this octal counting sequence: 624, 625, 626, _____, _____, _____.
5. Convert 975_{10} to binary by first converting to octal.
6. Convert binary 1010111011 to decimal by first converting to octal.
7. What range of decimal values can be represented by a four-digit octal number?

2-4   HEXADECIMAL NUMBER SYSTEM

The most important of the octal and hexadecimal number systems in the digital and computer fields is without question hexadecimal. The octal system was used more in the early days of computer systems, but hexadecimal lives on primarily because digital signals are carried throughout digital systems in groups of bits that are even powers of 2. Consequently, the octal system, which divides binary numbers into groups of 3 bits, is a bit awkward. On the other hand, hexadecimal uses groups of 4 bits, which works out evenly for systems today that commonly have 8-, 16-, 32-, and 64-bit signal paths. The ability to convert between binary and hex *in your head* without even wanting to use a calculator is what makes hexadecimal useful.

The **hexadecimal number system** uses base 16. Thus, it has 16 possible digit symbols. It uses the digits 0 through 9 plus the letters A, B, C, D, E, and F as the 16 digit symbols. Table 2-2 shows the relationships among hexadecimal, decimal, and binary. Note that each hexadecimal digit represents a group of four binary digits. It is important to remember that hex (abbreviation for "hexadecimal") digits A through F are equivalent to the decimal values 10 through 15.

Hex-to-Decimal Conversion

A hex number can be converted to its decimal equivalent by using the fact that each hex digit position has a weight that is a power of 16. The LSD has a weight of $16^0 = 1$; the next higher digit position has a weight of $16^1 = 16$; the next has a weight of $16^2 = 256$; and so on. The conversion process is demonstrated in the examples below:

TABLE 2-2

Hexadecimal	Decimal	Binary
0	0	0000
1	1	0001
2	2	0010
3	3	0011
4	4	0100
5	5	0101
6	6	0110
7	7	0111
8	8	1000
9	9	1001
A	10	1010
B	11	1011
C	12	1100
D	13	1101
E	14	1110
F	15	1111

CALCULATOR HINT:

Again, as with the octal-to-decimal conversions, you can use the y^x calculator function to evaluate the powers of 16.

$$356_{16} = 3 \times 16^2 + 5 \times 16^1 + 6 \times 16^0$$
$$= 768 + 80 + 6$$
$$= 854_{10}$$

$$2AF_{16} = 2 \times 16^2 + 10 \times 16^1 + 15 \times 16^0$$
$$= 512 + 160 + 15$$
$$= 687_{10}$$

Note that in the second example, the value 10 was substituted for A and the value 15 for F in the conversion to decimal.

For practice, verify that $1BC2_{16}$ is equal to 7106_{10}.

Decimal-to-Hex Conversion

Recall that we did decimal-to-binary conversion using repeated division by 2, and decimal-to-octal using repeated division by 8. Likewise, decimal-to-hex conversion can be done using repeated division by 16 (Figure 2-1). The following example contains two illustrations of this conversion.

EXAMPLE 2-4

(a) Convert 423_{10} to hex.

Solution

$$\frac{423}{16} = \boxed{26} + \text{remainder of } 7$$

$$\frac{26}{16} = 1 + \text{remainder of } 10$$

$$\frac{1}{16} = 0 + \text{remainder of } 1$$

$$423_{10} = \boxed{1A7_{16}}$$

(b) Convert 214_{10} to hex.

Solution

$$\frac{214}{16} = 13 + \text{remainder of } 6$$

$$\frac{13}{16} = 0 + \text{remainder of } 13$$

$$214_{10} = \boxed{D6_{16}}$$

Again note that the remainders of the division processes form the digits of the hex number. Also note that any remainders that are greater than 9 are represented by the letters A through F.

CALCULATOR HINT:

If a calculator is used to perform the divisions in the conversion process, the results will include a decimal fraction instead of a remainder. The remainder can be obtained by multiplying the fraction by 16. To illustrate, in Example 2-4(b), the calculator would have produced

$$\frac{214}{16} = 13.375$$

The remainder becomes $(0.375) \times 16 = 6$.

Hex-to-Binary Conversion

Like the octal number system, the hexadecimal number system is used primarily as a "shorthand" method for representing binary numbers. It is a relatively simple matter to convert a hex number to binary. *Each* hex digit is converted to its four-bit binary equivalent (Table 2-2). This is illustrated on the next page for $9F2_{16}$.

$$9F2_{16} = \qquad 9 \qquad\qquad F \qquad\qquad 2$$
$$\phantom{9F2_{16} = } \downarrow \qquad\qquad \downarrow \qquad\qquad \downarrow$$
$$= 1 \ \ 0 \ \ 0 \ \ 1 \quad 1 \ \ 1 \ \ 1 \ \ 1 \quad 0 \ \ 0 \ \ 1 \ \ 0$$
$$= 100111110010_2$$

For practice, verify that $BA6_{16} = 101110100110_2$.

Binary-to-Hex Conversion

Conversion from binary to hex is just the reverse of the process above. The binary number is grouped into groups of *four* bits, and each group is converted to its equivalent hex digit. Zeros (shown shaded) are added, as needed, to complete a four-bit group.

$$1 \ 1 \ 1 \ 0 \ 1 \ 0 \ 0 \ 1 \ 1 \ 0_2 = \underbrace{0 \ 0 \ 1 \ 1}_{3} \ \underbrace{1 \ 0 \ 1 \ 0}_{A} \ \underbrace{0 \ 1 \ 1 \ 0}_{6}$$
$$= 3A6_{16}$$

To perform these conversions between hex and binary, it is necessary to know the four-bit binary numbers (0000 through 1111) and their equivalent hex digits. Once these are mastered, the conversions can be performed quickly without the need for any calculations. This is why hex (and octal) are so useful in representing large binary numbers.

For practice, verify that $101011111_2 = 15F_{16}$.

Counting in Hexadecimal

When counting in hex, each digit position can be incremented (increased by 1) from 0 to F. Once a digit position reaches the value F, it is reset to 0, and the next digit position is incremented. This is illustrated in the following hex counting sequences:

(a) 38, 39, 3A, 3B, 3C, 3D, 3E, 3F, 40, 41, 42

(b) 6F8, 6F9, 6FA, 6FB, 6FC, 6FD, 6FE, 6FF, 700

Note that when there is a 9 in a digit position, it becomes an A when it is incremented.

With N hex digit positions, we can count from decimal 0 to $16^N - 1$, for a total of 16^N different values. For example, with three hex digits, we can count from 000_{16} to FFF_{16}, which is 0_{10} to 4095_{10}, for a total of $4096 = 16^3$ different values.

Usefulness of Hex and Octal

Hex and octal are often used in a digital system as sort of a "shorthand" way to represent strings of bits. In computer work, strings as long as 64 bits are not uncommon. These binary strings do not always represent a numerical value, but—as you will find out—can be some type of code that conveys nonnumerical information. When dealing with a large number of bits, it is more convenient and less error-prone to write the binary numbers in hex or octal and, as we have seen, it is relatively easy to convert back and forth between binary and either hex or octal. To illustrate the advantage of hex or octal representation of a binary string, suppose you had in front of you a printout of the contents of 50 memory locations, each of which was a 16-bit number, and you were checking it against a list. Would you rather check 50 numbers like this one: 0110111001100111, or 50 numbers like this one: 6E67? And which one would you be more apt to read incorrectly? It is important to keep in mind, though, that digital circuits all work in binary. Hex and octal are simply used as a convenience for the humans involved. You should memorize the 4-bit binary pattern for each hexadecimal digit. Only then will you realize the usefulness of this tool in digital systems.

EXAMPLE 2-5

Convert decimal 378 to a 16-bit binary number by first converting to hexadecimal.

Solution

$$\frac{378}{16} = 23 + \text{remainder of } 10_{10} = A_{16}$$

$$\frac{23}{16} = 1 + \text{remainder of } 7$$

$$\frac{1}{16} = 0 + \text{remainder of } 1$$

Thus, $378_{10} = 17A_{16}$. This hex value can be converted easily to binary 000101111010. Finally, we can express 378_{10} as a 16-bit number by adding four leading 0s:

$$378_{10} = 0000 \quad 0001 \quad 0111 \quad 1010_2$$

EXAMPLE 2-6

Convert $B2F_{16}$ to octal.

Solution

It's easiest first to convert hex to binary, then to octal.

$$
\begin{array}{rllll}
B2F_{16} = & 1011 & 0010 & 1111 & \quad \text{\{convert to binary\}} \\
= & 101 & 100 & 101 \quad 111 & \quad \text{\{group into three-bit groupings\}} \\
= & 5 & 4 & 5 \quad 7_8 & \quad \text{\{convert to octal\}}
\end{array}
$$

Summary of Conversions

Right now, your head is probably spinning as you try to keep straight all of these different conversions from one number system to another. You probably realize that many of these conversions can be done *automatically* on your calculator just by pressing a key, but it is important for you to master these conversions so that you understand the process. Besides, what happens if your calculator battery dies at a crucial time and you have no handy replacement? The following summary should help you, but nothing beats practice, practice, practice!

1. When converting from binary [or octal or hex] to decimal, use the method of taking the weighted sum of each digit position.

2. When converting from decimal to binary [or octal or hex], use the method of repeatedly dividing by 2 [or 8 or 16] and collecting remainders (Figure 2-1).

3. When converting from binary to octal [or hex], group the bits in groups of three [or four], and convert each group into the correct octal [or hex] digit.

4. When converting from octal [or hex] to binary, convert each digit into its three-bit [or four-bit] equivalent.

5. When converting from octal to hex [or vice versa], first convert to binary; then convert the binary into the desired number system.

REVIEW QUESTIONS

1. Convert $24CE_{16}$ to decimal.
2. Convert 3117_{10} to hex, then from hex to binary.
3. Convert 1001011110110101_2 to hex.
4. Write the next four numbers in this hex counting sequence: E9A, E9B, E9C, E9D, _____, _____, _____, _____.
5. Convert 3527_8 to hex.
6. What range of decimal values can be represented by a four-digit hex number?

2-5 BCD CODE

When numbers, letters, or words are represented by a special group of symbols, we say that they are being encoded, and the group of symbols is called a *code.* Probably one of the most familiar codes is the Morse code, where a series of dots and dashes represents letters of the alphabet.

We have seen that any decimal number can be represented by an equivalent binary number. The group of 0s and 1s in the binary number can be thought of as a code representing the decimal number. When a decimal number is represented by its equivalent binary number, we call it **straight binary coding**.

Digital systems all use some form of binary numbers for their internal operation, but the external world is decimal in nature. This means that conversions between the decimal and binary systems are being performed often. We have seen that the conversions between decimal and binary can become long and complicated for large numbers. For this reason, a means of encoding decimal numbers that combines some features of both the decimal and the binary systems is used in certain situations.

Binary-Coded-Decimal Code

If *each* digit of a decimal number is represented by its binary equivalent, the result is a code called **binary-coded-decimal** (hereafter abbreviated BCD). Since a decimal digit can be as large as 9, four bits are required to code each digit (the binary code for 9 is 1001).

To illustrate the BCD code, take a decimal number such as 874. Each *digit* is changed to its binary equivalent as follows:

$$
\begin{array}{ccc}
8 & 7 & 4 \quad \text{(decimal)} \\
\downarrow & \downarrow & \downarrow \\
1000 & 0111 & 0100 \quad \text{(BCD)}
\end{array}
$$

As another example, let us change 943 to its BCD-code representation:

$$
\begin{array}{ccc}
9 & 4 & 3 \quad \text{(decimal)} \\
\downarrow & \downarrow & \downarrow \\
1001 & 0100 & 0011 \quad \text{(BCD)}
\end{array}
$$

Once again, each decimal digit is changed to its straight binary equivalent. Note that four bits are *always* used for each digit.

The BCD code, then, represents each digit of the decimal number by a four-bit binary number. Clearly only the four-bit binary numbers from 0000 through 1001 are used. The BCD code does not use the numbers 1010, 1011, 1100, 1101, 1110, and 1111. In other words, only 10 of the 16 possible four-bit binary code groups are used. If any of the "forbidden" four-bit numbers ever occurs in a machine using the BCD code, it is usually an indication that an error has occurred.

EXAMPLE 2-7

Convert 0110100000111001 (BCD) to its decimal equivalent.

Solution

Divide the BCD number into four-bit groups and convert each to decimal.

$$
\underbrace{0110}\ \underbrace{1000}\ \underbrace{0011}\ \underbrace{1001} \\
\ \ \ 6 \qquad 8 \qquad 3 \qquad 9
$$

EXAMPLE 2-8

Convert the BCD number 011111000001 to its decimal equivalent.

Solution

$$
\underbrace{0111}\ \underbrace{1100}\ \underbrace{0001} \\
\ \ \ 7 \qquad \downarrow \qquad 1
$$

The forbidden code group indicates an error in the BCD number

Comparison of BCD and Binary

It is important to realize that BCD is not another number system like binary, octal, decimal, and hexadecimal. In fact, it is the decimal system with each digit encoded in its binary equivalent. It is also important to understand that a BCD number is *not* the same as a straight binary number. A straight binary number takes the *complete* decimal number and represents it in binary; the BCD code converts *each* decimal *digit* to binary individually. To illustrate, take the number 137 and compare its straight binary and BCD codes:

$$137_{10} = 10001001_2 \qquad \text{(binary)}$$
$$137_{10} = 0001\ 0011\ 0111 \quad \text{(BCD)}$$

The BCD code requires 12 bits, while the straight binary code requires only eight bits to represent 137. BCD requires more bits than straight binary to represent decimal numbers of more than one digit because BCD does not use all possible four-bit groups, as pointed out earlier, and is therefore somewhat inefficient.

The main advantage of the BCD code is the relative ease of converting to and from decimal. Only the four-bit code groups for the decimal digits 0 through 9 need to be remembered. This ease of conversion is especially important from a hardware standpoint because in a digital system, it is the logic circuits that perform the conversions to and from decimal.

REVIEW QUESTIONS

1. Represent the decimal value 178 by its straight binary equivalent. Then encode the same decimal number using BCD.
2. How many bits are required to represent an eight-digit decimal number in BCD?
3. What is an advantage of encoding a decimal number in BCD rather than in straight binary? What is a disadvantage?

2-6 PUTTING IT ALL TOGETHER

Table 2-3 gives the representation of the decimal numbers 1 through 15 in the binary, octal, and hex number systems, and in BCD code. Examine it carefully and make sure you understand how it was obtained. Especially note how the BCD representation always uses four bits for each decimal digit.

TABLE 2-3

Decimal	Binary	Octal	Hexadecimal	BCD
0	0	0	0	0000
1	1	1	1	0001
2	10	2	2	0010
3	11	3	3	0011
4	100	4	4	0100
5	101	5	5	0101
6	110	6	6	0110
7	111	7	7	0111
8	1000	10	8	1000
9	1001	11	9	1001
10	1010	12	A	0001 0000
11	1011	13	B	0001 0001
12	1100	14	C	0001 0010
13	1101	15	D	0001 0011
14	1110	16	E	0001 0100
15	1111	17	F	0001 0101

2-7 THE BYTE, NIBBLE, AND WORD

Most microcomputers handle and store binary data and information in groups of eight bits, so a special name is given to a string of eight bits: it is called a **byte**. A byte always consists of eight bits, and it can represent any of numerous types of data or information. The following examples will illustrate.

EXAMPLE 2-9

How many bytes are in a 32-bit string (a string of 32 bits)?

Solution

32/8 = 4, so there are **four** bytes in a 32-bit string.

EXAMPLE 2-10

What is the largest decimal value that can be represented in binary using two bytes?

Solution

Two bytes is 16 bits, so the largest binary value will be equivalent to decimal $2^{16} - 1 = \textbf{65,535}$.

EXAMPLE 2-11

How many bytes are needed to represent the decimal value 846,569 in BCD?

Solution

Each decimal digit converts to a four-bit BCD code. Thus, a six-digit decimal number requires 24 bits. These 24 bits are equal to **three** bytes. This is diagrammed below.

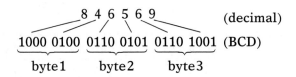

Nibbles

Binary numbers are often broken down into groups of four bits, as we have seen with BCD codes and hexadecimal number conversions. In the early days of digital systems, a term caught on to describe a group of four bits. Because it is half as big as a byte, it was named a **nibble**. The following examples illustrate the use of this term.

EXAMPLE 2-12

How many nibbles are in a byte?

Solution

2

EXAMPLE 2-13

What is the hex value of the least significant nibble of the binary number 1001 0101?

Solution

$$1001\ \boxed{0101}$$

The least significant nibble is 0101 = 5.

Words

Bits, nibbles, and bytes are terms that represent a fixed number of binary digits. As systems have grown over the years, their capacity (appetite?) for handling binary data has also grown. A **word** is a group of bits that represents a certain unit of information. The size of the word depends on the size of the data pathway in the system that uses the information. The **word size** can be defined as the number of bits in the binary word that a digital system operates on. For example, the computer in your microwave oven can probably handle only one byte at a time. It has a word size of eight bits. On the other hand, the personal computer on your desk can handle eight bytes at a time, so it has a word size of 64 bits.

REVIEW QUESTIONS

1. How many bytes are needed to represent 235_{10} in binary?
2. What is the largest decimal value that can be represented in BCD using two bytes?
3. How many hex digits can a nibble represent?
4. How many nibbles are in one BCD digit?

2-8 ALPHANUMERIC CODES

In addition to numerical data, a computer must be able to handle nonnumerical information. In other words, a computer should recognize codes that represent letters of the alphabet, punctuation marks, and other special characters as well as numbers. These codes are called **alphanumeric codes**. A complete alphanumeric code would include the 26 lowercase letters, 26 uppercase letters, 10 numeric digits, 7 punctuation marks, and anywhere from 20 to 40 other characters, such as +, /, #, %, *, and so on. We can say that an alphanumeric code represents all of the various characters and functions that are found on a computer keyboard.

ASCII Code

The most widely used alphanumeric code is the **American Standard Code for Information Interchange (ASCII)**. The ASCII (pronounced "askee") code is a seven-bit code, and so it has $2^7 = 128$ possible code groups. This is more than enough to represent all of the standard keyboard characters as well as the control functions such as the ⟨RETURN⟩ and ⟨LINEFEED⟩ functions.

Table 2-4 shows a partial listing of the ASCII code. In addition to the binary code group for each character, the table gives the octal and hexadecimal equivalents.

TABLE 2-4   Partial listing of the ASCII code.

Character	Seven-Bit ASCII	Octal	Hex	Character	Seven-Bit ASCII	Octal	Hex
A	100 0001	101	41	Y	101 1001	131	59
B	100 0010	102	42	Z	101 1010	132	5A
C	100 0011	103	43	0	011 0000	060	30
D	100 0100	104	44	1	011 0001	061	31
E	100 0101	105	45	2	011 0010	062	32
F	100 0110	106	46	3	011 0011	063	33
G	100 0111	107	47	4	011 0100	064	34
H	100 1000	110	48	5	011 0101	065	35
I	100 1001	111	49	6	011 0110	066	36
J	100 1010	112	4A	7	011 0111	067	37
K	100 1011	113	4B	8	011 1000	070	38
L	100 1100	114	4C	9	011 1001	071	39
M	100 1101	115	4D	blank	010 0000	040	20
N	100 1110	116	4E	.	010 1110	056	2E
O	100 1111	117	4F	(	010 1000	050	28
P	101 0000	120	50	+	010 1011	053	2B
Q	101 0001	121	51	$	010 0100	044	24
R	101 0010	122	52	*	010 1010	052	2A
S	101 0011	123	53	)	010 1001	051	29
T	101 0100	124	54	—	010 1101	055	2D
U	101 0101	125	55	/	010 1111	057	2F
V	101 0110	126	56	,	010 1100	054	2C
W	101 0111	127	57	=	011 1101	075	3D
X	101 1000	130	58	⟨RETURN⟩	000 1101	015	0D
				⟨LINEFEED⟩	000 1010	012	0A

EXAMPLE 2-14

The following is a message encoded in ASCII code. What is the message?

$$1001000 \quad 1000101 \quad 1001100 \quad 1010000$$

Solution

Convert each seven-bit code to its hex equivalent. The results are

$$48 \quad 45 \quad 4C \quad 50$$

Now locate these hex values in Table 2-4 and determine the character represented by each. The results are

$$H \quad E \quad L \quad P$$

The ASCII code is used for the transfer of alphanumeric information between a computer and the external devices such as a printer or another computer. A computer also uses ASCII internally to store the information that an operator types in at the computer's keyboard. The following example illustrates this.

EXAMPLE 2-15

An operator is typing in a BASIC program at the keyboard of a certain microcomputer. The computer converts each keystroke into its ASCII code and stores the code as a byte in memory. Determine the binary strings that will be entered into memory when the operator types in the following BASIC statement:

GOTO 25

Solution

Locate each character (including the space) in Table 2-4 and record its ASCII code.

G	01000111
O	01001111
T	01010100
O	01001111
(space)	00100000
2	00110010
5	00110101

Note that a 0 was added to the leftmost bit of each ASCII code because the codes must be stored as bytes (eight bits). This adding of an extra bit is called *padding with 0s*.

REVIEW QUESTIONS

1. Encode the following message in ASCII code using the hex representation: "COST = $72."
2. The following padded ASCII-coded message is stored in successive memory locations in a computer:

   01010011   01010100   01001111   01010000

   What is the message?

2-9   PARITY METHOD FOR ERROR DETECTION

The movement of binary data and codes from one location to another is the most frequent operation performed in digital systems. Here are just a few examples:

- The transmission of digitized voice over a microwave link
- The storage of data in and retrieval of data from external memory devices such as magnetic and optical disk
- The transmission of digital data from a computer to a remote computer over telephone lines (i.e., using a modem). This is one of the major ways of sending and receiving information on the Internet.

Whenever information is transmitted from one device (the transmitter) to another device (the receiver), there is a possibility that errors can occur such that the receiver does not receive the identical information that was sent by the transmitter. The major cause of any transmission errors is *electrical noise,* which consists of spurious fluctuations in voltage or current that are present in all electronic systems to varying degrees. Figure 2-2 is a simple illustration of a type of transmission error.

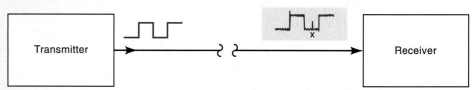

FIGURE 2-2   Example of noise causing an error in the transmission of digital data.

The transmitter sends a relatively noise-free serial digital signal over a signal line to a receiver. However, by the time the signal reaches the receiver, it contains a certain degree of noise superimposed on the original signal. Occasionally, the noise is large enough in amplitude that it will alter the logic level of the signal, as it does at point *x*. When this occurs, the receiver may incorrectly interpret that bit as a logic 1, which is not what the transmitter has sent.

Most modern digital equipment is designed to be relatively error-free, and the probability of errors such as the one shown in Figure 2-2 is very low. However, we must realize that digital systems often transmit thousands, even millions, of bits per second, so that even a very low rate of occurrence of errors can produce an occasional error that might prove to be bothersome, if not disastrous. For this reason, many digital systems employ some method for detection (and sometimes correction) of errors. One of the simplest and most widely used schemes for error detection is the **parity method**.

Parity Bit

A **parity bit** is an extra bit that is attached to a code group that is being transferred from one location to another. The parity bit is made either 0 or 1, depending on the number of 1s that are contained in the code group. Two different methods are used.

In the *even-parity* method, the value of the parity bit is chosen so that the total number of 1s in the code group (including the parity bit) is an *even* number. For example, suppose that the group is 1000011. This is the ASCII character "C." The code group has *three* 1s. Therefore, we will add a parity bit of 1 to make the total number of 1s an even number. The *new* code group, *including the parity bit,* thus becomes

$$1 \quad 1\ 0\ 0\ 0\ 0\ 1\ 1$$

↑————————— added parity bit*

If the code group contains an even number of 1s to begin with, the parity bit is given a value of 0. For example, if the code group were 1000001 (the ASCII code for "A"), the assigned parity bit would be 0, so that the new code, *including the parity bit,* would be 01000001.

The *odd-parity* method is used in exactly the same way except that the parity bit is chosen so the total number of 1s (including the parity bit) is an *odd* number. For example, for the code group 1000001, the assigned parity bit would be a 1. For the code group 1000011, the parity bit would be a 0.

Regardless of whether even parity or odd parity is used, the parity bit becomes an actual part of the code word. For example, adding a parity bit to the seven-bit ASCII code produces an eight-bit code. Thus, the parity bit is treated just like any other bit in the code.

The parity bit is issued to detect any *single-bit* errors that occur during the transmission of a code from one location to another. For example, suppose that the character "A" is being transmitted and *odd* parity is being used. The transmitted code would be

$$1 \quad 1\ 0\ 0\ 0\ 0\ 0\ 1$$

When the receiver circuit receives this code, it will check that the code contains an odd number of 1s (including the parity bit). If so, the receiver will assume that the code has been correctly received. Now, suppose that because of some noise or malfunction the receiver actually receives the following code:

$$1 \quad 1\ 0\ 0\ 0\ 0\ 0\ 0$$

The receiver will find that this code has an *even* number of 1s. This tells the receiver that there must be an error in the code because presumably the transmitter and receiver have agreed to use odd parity. There is no way, however, that the receiver can tell which bit is in error because it does not know what the code is supposed to be.

It should be apparent that this parity method would not work if *two* bits were in error, because two errors would not change the "oddness" or "evenness" of the number of 1s in the code. In practice, the parity method is used only in situations where the probability of a single error is very low and the probability of double errors is essentially zero.

When the parity method is being used, the transmitter and the receiver must have agreement, in advance, as to whether odd or even parity is being used. There is no advantage of one over the other, although even parity seems to be used more often. The transmitter must attach an appropriate parity bit to each unit of information that it transmits. For example, if the transmitter is sending ASCII-coded data, it will attach a parity bit to each seven-bit ASCII code group. When the receiver examines the data that it has received from

---

*The parity bit can be placed at either end of the code group, but it is usually placed to the left of the MSB.

the transmitter, it checks each code group to see that the total number of 1s (including the parity bit) is consistent with the agreed-upon type of parity. This is often called *checking the parity* of the data. In the event that it detects an error, the receiver may send a message back to the transmitter asking it to retransmit the last set of data. The exact procedure that is followed when an error is detected depends on the particular system.

EXAMPLE 2-16

Computers often communicate with other remote computers over telephone lines. For example, this is how dial-up communication over the Internet takes place. When one computer is transmitting a message to another, the information is usually encoded in ASCII. What actual bit strings would a computer transmit to send the message HELLO, using ASCII with even parity?

Solution

First, look up the ASCII codes for each character in the message. Then for each code, count the number of 1s. If it is an even number, attach a 0 as the MSB. If it is an odd number, attach a 1. Thus, the resulting eight-bit codes (bytes) will all have an even number of 1s (including parity).

attached even-parity bits
↓

```
H =  0  1 0 0 1 0 0 0
E =  1  1 0 0 0 1 0 1
L =  1  1 0 0 1 1 0 0
L =  1  1 0 0 1 1 0 0
O =  1  1 0 0 1 1 1 1
```

REVIEW QUESTIONS

1. Attach an odd-parity bit to the ASCII code for the $ symbol, and express the result in hexadecimal.
2. Attach an even-parity bit to the BCD code for decimal 69.
3. Why can't the parity method detect a double error in transmitted data?

2-10   APPLICATIONS

Here are several applications that will serve as a review of some of the concepts covered in this chapter. These applications should give a sense of how the various number systems and codes are used in the digital world. More applications are presented in the end-of-chapter problems.

APPLICATION 2-1

A typical CD-ROM can store 650 megabytes of digital data. Since mega = 2^{20}, how many bits of data can a CD-ROM hold?

Solution

Remember that a byte is 8 bits. Therefore, 650 megabytes is $650 \times 2^{20} \times 8 =$ **5,452,595,200 bits.**

APPLICATION 2-2

An automotive parts shop uses a computer to store all of its parts numbers in 7-bit ASCII code with an odd parity bit. The codes for each part are stored in successive memory locations. List the binary contents of memory that stores the part number JR2-5.

Solution

Take each character of the part number, look up its ASCII code, and attach an odd parity bit as the leftmost bit. Here are the results:

$$J = 01001010$$
$$R = 01010010$$
$$2 = 00110010$$
$$- = 10101101$$
$$5 = 10110101$$

APPLICATION 2-3

A small process-control computer uses octal codes to represent its 12-bit memory addresses.

(a) How many octal digits are required?
(b) What is the range of addresses in octal?
(c) How many memory locations are there?

Solution

(a) Since 3 bits convert to a single octal digit, 12/3 = 4 octal digits are needed.
(b) The binary range is 000000000000_2 to 111111111111_2. In octal, this becomes 0000_8 to 7777_8.
(c) With 4 octal digits, the total number of addresses is $8^4 =$ **4096.**

APPLICATION 2-4

A typical PC uses a 20-bit address code for its memory locations.

(a) How many hex digits are needed to represent a memory address?
(b) What is the range of addresses?
(c) What is the total number of memory locations?

Solution

(a) Since 4 bits convert to a single hex digit, $20/4 = $ **5** hex digits are needed.

(b) 00000_{16} to $FFFFF_{16}$

(c) With 5 hex digits, the total number of addresses is $16^5 = $ **1,048,576.**

APPLICATION 2-5

Most calculators use BCD to store the decimal values as they are entered into the keyboard and to drive the digit displays.

(a) If a calculator is designed to handle 8-digit decimal numbers, how many bits does this require?

(b) What bits are stored when the number 375 is entered into the calculator?

Solution

(a) Each decimal digit converts to a 4-bit code in BCD. Therefore, $8 \times 4 = $ **32** bits are needed.

(b) 375_{10} converts to 0011 0111 0101 (BCD).

SUMMARY

1. The octal and hexadecimal number systems are used in digital systems and computers as efficient ways of representing binary quantities.

2. In conversions between octal and binary, one octal digit corresponds to three bits. In conversions between hex and binary, each hex digit corresponds to four bits.

3. The repeated-division method is used to convert decimal numbers to binary, octal, or hexadecimal.

4. Using an N-bit binary number, we can represent decimal values from 0 to $2^N - 1$.

5. The BCD code for a decimal number is formed by converting each digit of the decimal number to its four-bit binary equivalent.

6. A byte is a string of eight bits. A nibble is four bits. The word size depends on the system.

7. An alphanumeric code is one that uses groups of bits to represent all of the various characters and functions that are part of a typical computer's keyboard. The ASCII code is the most widely used alphanumeric code.

8. The parity method for error detection attaches a special parity bit to each transmitted group of bits.

IMPORTANT TERMS

octal number system
hexadecimal number
  system
straight binary
  coding
binary-coded-decimal
  (BCD) code

byte
nibble
word
word size
alphanumeric code

American Standard
  Code for
  Information
  Interchange
  (ASCII)
parity method
parity bit

PROBLEMS

SECTIONS 2-1 AND 2-2

2-1. Convert these binary numbers to decimal.
- (a) 10110
- (b) 10001101
- (c) 100100001001
- (d) 01011011
- (e) 11111111
- (f) 01110111
- (g) 1111010111
- (h) 10111111

2-2. Convert the following decimal values to binary.
- (a) 37
- (b) 14
- (c) 189
- (d) 1024
- (e) 77
- (f) 405
- (g) 205
- (h) 2313
- (i) 511

2-3. What is the largest decimal value that can be represented by an eight-bit binary number? A 16-bit number?

SECTION 2-3

2-4. Convert each octal number to its decimal equivalent.
- (a) 743
- (b) 36
- (c) 3777
- (d) 2000
- (e) 165
- (f) 5
- (g) 257
- (h) 1204

2-5. Convert each of the following decimal numbers to octal.
- (a) 59
- (b) 372
- (c) 919
- (d) 1024
- (e) 771
- (f) 2313
- (g) 65,536
- (h) 255

2-6. Convert each of the octal values from Problem 2-4 to binary.

2-7. Convert the binary numbers in Problem 2-1 to octal.

2-8. List the octal numbers in sequence from 165_8 to 200_8.

2-9. When a large decimal number is to be converted to binary, it is sometimes easier to convert it first to octal, and then from octal to binary. Try this procedure for 2313_{10} and compare it with the procedure used in Problem 2-2(e).

2-10. How many octal digits are required to represent decimal numbers up to 20,000?

SECTION 2-4

2-11. Convert these hex values to decimal.
- (a) 92
- (b) 1A6
- (c) 37FD
- (d) ABCD
- (e) 000F
- (f) 55
- (g) 2C0
- (h) 7FF

2-12. Convert these decimal values to hex.

(a) 75   (d) 14   (g) 25,619
(b) 314   (e) 7245   (h) 4095
(c) 2048   (f) 389

2-13. Take each four-bit binary number in the order they are written and write the equivalent hex digit without performing a calculation by hand or by calculator.

(a) 1001   (e) 1111   (i) 1011   (m) 0001
(b) 1101   (f) 0010   (j) 1100   (n) 0101
(c) 1000   (g) 1010   (k) 0011   (o) 0111
(d) 0000   (h) 1001   (l) 0100   (p) 0110

2-14. Take each hex digit and write its four-bit binary value without performing any calculations by hand or by calculator.

(a) 6   (e) 4   (i) 9   (m) 0
(b) 7   (f) 3   (j) A   (n) 8
(c) 5   (g) C   (k) 2   (o) D
(d) 1   (h) B   (l) F   (p) 9

2-15. Convert the binary numbers in Problem 2-1 to hexadecimal.

2-16. Convert the hex values in Problem 2-11 to binary.

2-17. List the hex numbers in sequence from 280 to 2A0.

2-18. How many hex digits are required to represent decimal numbers up to 1 million?

SECTION 2-5

2-19. Encode these decimal numbers in BCD.

(a) 47   (d) 6727   (g) 42,689,627
(b) 962   (e) 13   (h) 1204
(c) 187   (f) 888

2-20. How many bits are required to represent the decimal numbers in the range from 0 to 999 using straight binary code? Using BCD code?

2-21. The following numbers are in BCD. Convert them to decimal.

(a) 1001011101010010   (d) 0111011101110101
(b) 000110000100   (e) 010010010010
(c) 011010010101   (f) 010101010101

SECTION 2-7

2-22. (a) How many bits are contained in eight bytes?

(b) What is the largest hex number that can be represented in four bytes?

(c) What is the largest BCD-encoded decimal value that can be represented in three bytes?

2-23. (a) Refer to Table 2-4. What is the most significant nibble of the ASCII code for the letter X?

(b) How many nibbles can be stored in a 16-bit word?

(c) How many bytes does it take to make up a 24-bit word?

SECTIONS 2-8 AND 2-9

2-24. Represent the statement "X = 25/Y" in ASCII code (excluding the quotes). Attach an odd-parity bit.

2-25. Attach an *even*-parity bit to each of the ASCII codes for Problem 2-24, and give the results in hex.

2-26. The following bytes (shown in hex) represent a person's name as it would be stored in a computer's memory. Each byte is a padded ASCII code. Determine the name of the person.

$$42 \quad 45 \quad 4E \quad 20 \quad 53 \quad 4D \quad 49 \quad 54 \quad 48$$

2-27. Convert the following decimal numbers to BCD code and then attach an *odd*-parity bit.

(a) 74   (c) 8884   (e) 165

(b) 38   (d) 275   (f) 9201

2-28. In a certain digital system, the decimal numbers from 000 through 999 are represented in BCD code. An *odd*-parity bit is also included at the end of each code group. Examine each of the code groups below, and assume that each one has just been transferred from one location to another. Some of the groups contain errors. Assume that *no more than* two errors have occurred for each group. Determine which of the code groups have a single error and which of them *definitely* have a double error. (*Hint:* Remember that this is a BCD code.)

(a) 1001010110000

            ↑   parity bit

(b) 0100011101100

(c) 0111110000011

(d) 1000011000101

2-29. Suppose that the receiver received the following data from the transmitter of Example 2-16:

$$
\begin{array}{cccccccc}
0 & 1 & 0 & 0 & 1 & 0 & 0 & 0 \\
1 & 1 & 0 & 0 & 0 & 1 & 0 & 1 \\
1 & 1 & 0 & 0 & 1 & 1 & 0 & 0 \\
1 & 1 & 0 & 0 & 1 & 0 & 0 & 0 \\
1 & 1 & 0 & 0 & 1 & 1 & 0 & 0 \\
\end{array}
$$

What errors can the receiver determine in these received data?

DRILL QUESTIONS

2-30. Perform each of the following conversions. For some of them, you may want to try several methods to see which one works best for you. For example, a binary-to-decimal conversion may be done directly, or it may be done as a binary-to-octal conversion followed by an octal-to-decimal conversion.

(a) $1417_{10} = \underline{\hspace{1cm}}_2$

(b) $255_{10} = \underline{\hspace{1cm}}_2$

(c) $11010001_2 = \underline{\hspace{1cm}}_{10}$

(d) $1110101000100111_2 = \underline{\hspace{1cm}}_{10}$

(e) $2497_{10} = \underline{\hspace{1.5cm}}_8$

(f) $511_{10} = \underline{\hspace{1.5cm}}_8$

(g) $235_8 = \underline{\hspace{1.5cm}}_{10}$

(h) $4316_8 = \underline{\hspace{1.5cm}}_{10}$

(i) $7A9_{16} = \underline{\hspace{1.5cm}}_{10}$

(j) $3E1C_{16} = \underline{\hspace{1.5cm}}_{10}$

(k) $1600_{10} = \underline{\hspace{1.5cm}}_{16}$

(l) $38{,}187_{10} = \underline{\hspace{1.5cm}}_{16}$

(m) $865_{10} = \underline{\hspace{1.5cm}}$ (BCD)

(n) 100101000111 (BCD) $= \underline{\hspace{1.5cm}}_{10}$

(o) $465_8 = \underline{\hspace{1.5cm}}_{16}$

(p) $B34_{16} = \underline{\hspace{1.5cm}}_8$

(q) 01110100 (BCD) $= \underline{\hspace{1.5cm}}_2$

(r) $111010_2 = \underline{\hspace{1.5cm}}$ (BCD)

2-31. Represent the decimal value 37 in each of the following ways.

    (a) straight binary

    (b) BCD

    (c) hex

    (d) ASCII (i.e., treat each digit as a character)

    (e) octal

2-32. Fill in the blanks with the correct word or words.

    (a) Conversion from decimal to $\underline{\hspace{1.5cm}}$ requires repeated division by 8.

    (b) Conversion from decimal to hex requires repeated division by $\underline{\hspace{1.5cm}}$.

    (c) In the BCD code, each $\underline{\hspace{1.5cm}}$ is converted to its four-bit binary equivalent.

    (d) The $\underline{\hspace{1.5cm}}$ code has the characteristic that only one bit changes in going from one step to the next.

    (e) A transmitter attaches a $\underline{\hspace{1.5cm}}$ to a code group to allow the receiver to detect $\underline{\hspace{1.5cm}}$.

    (f) The $\underline{\hspace{1.5cm}}$ code is the most common alphanumeric code used in computer systems.

    (g) $\underline{\hspace{1.5cm}}$ and $\underline{\hspace{1.5cm}}$ are often used as a convenient way to represent large binary numbers.

    (h) A string of eight bits is called a $\underline{\hspace{1.5cm}}$.

2-33. Write the binary number that results when each of the following numbers is incremented by one.

    (a) 0111   (b) 010000   (c) 1110

2-34. Repeat Problem 2-33 for the decrement operation.

2-35. Write the number that results when each of the following is incremented.

    (a) 7777_8 (c) 2000_8 (e) $9FF_{16}$

    (b) 7777_{16} (d) 2000_{16} (f) 1000_{16}

2-36. Repeat Problem 2-35 for the decrement operation.

CHALLENGING EXERCISES

2-37. In a microcomputer, the *addresses* of memory locations are binary numbers that identify each memory circuit where a byte is stored. The number of bits that make up an address depends on how many memory locations there are. Since the number of bits can be very large, the addresses are often specified in hex instead of binary.

    (a) If a microcomputer uses a 20-bit address, how many different memory locations are there?

    (b) How many hex digits are needed to represent the address of a memory location?

    (c) What is the hex address of the 256th memory location? (*Note:* The first address is always 0.)

2-38. In an audio CD, the audio voltage signal is typically sampled about 44,000 times per second, and the value of each sample is recorded on the CD surface as a binary number. In other words, each recorded binary number represents a single voltage point on the audio signal waveform.

    (a) If the binary numbers are six bits in length, how many different voltage values can be represented by a single binary number? Repeat for eight bits and ten bits.

    (b) If ten-bit numbers are used, how many bits will be recorded on the CD in 1 second?

    (c) If a CD can typically store 5 billion bits, how many seconds of audio can be recorded when ten-bit numbers are used?

2-39. A black-and-white digital camera lays a fine grid over an image and then measures and records a binary number representing the level of gray it sees in each cell of the grid. For example, if four-bit numbers are used, the value of black is set to 0000 and the value of white to 1111, and any level of gray is somewhere between 0000 and 1111. If six-bit numbers are used, black is 000000, white is 111111, and all grays are between the two.

    Suppose we wanted to distinguish among 254 different levels of gray within each cell of the grid. How many bits would we need to use to represent these levels?

2-40. Construct a table showing the binary, octal, hex, and BCD representations of all decimal numbers from 0 to 15. Compare your table with Table 2-3.

ANSWERS TO SECTION REVIEW QUESTIONS

SECTION 2-1

1. 2267   2. 32768

SECTION 2-2

1. 1010011   2. 1011011001   3. 20 bits

SECTION 2-3

1. 396   2. 222; 010010010   3. 235   4. 627, 630, 631   5. 1111001111
6. 699   7. 0 to 4095

SECTION 2-4

1. 9422   2. C2D; 110000101101   3. 97B5   4. E9E, E9F, EA0, EA1   5. 757
6. 0 to 65,535

SECTION 2-5

1. 10110010_2; 000101111000 (BCD)   2. 32   3. Advantage: ease of conversion. Disadvantage: BCD requires more bits.

SECTION 2-7

1. One   2. 9999   3. One   4. One

SECTION 2-8

1. 43, 4F, 53, 54, 20, 3D, 20, 24, 37, 32   2. STOP

SECTION 2-9

1. A4   2. 001101001   3. Two errors in the data would not change the oddness or evenness of the number of 1s in the data.

CHAPTER 3

DESCRIBING LOGIC CIRCUITS

■ OUTLINE

■ OBJECTIVES

Upon completion of this chapter, you will be able to:

- ■ Perform the three basic logic operations.
- ■ Describe the operation of and construct the truth tables for the AND, NAND, OR, and NOR gates, and the NOT (INVERTER) circuit.
- ■ Draw timing diagrams for the various logic-circuit gates.
- ■ Write the Boolean expression for the logic gates and combinations of logic gates.
- ■ Implement logic circuits using basic AND, OR, and NOT gates.
- ■ Appreciate the potential of Boolean algebra to simplify complex logic circuits.
- ■ Use DeMorgan's theorems to simplify logic expressions.
- ■ Use either of the universal gates (NAND or NOR) to implement a circuit represented by a Boolean expression.
- ■ Explain the advantages of constructing a logic-circuit diagram using the alternate gate symbols versus the standard logic-gate symbols.
- ■ Describe the concept of active-LOW and active-HIGH logic signals.
- ■ Draw and interpret the IEEE/ANSI standard logic-gate symbols.
- ■ Use several methods to describe the operation of logic circuits.
- ■ Interpret simple circuits defined by a hardware description language (HDL).
- ■ Explain the difference between an HDL and a computer programming language.
- ■ Create an HDL file for a simple logic gate.
- ■ Create an HDL file for combinational circuits with intermediate variables.

■ INTRODUCTION

Chapters 1 and 2 introduced the concepts of logic levels and logic circuits. In logic, only two possible conditions exist for any input or output: true and false. The binary number system uses only two digits, 1 and 0, so it is perfect for representing logical relationships. Digital logic circuits use predefined voltage ranges to represent these binary states. Using these concepts, we can create circuits made of little more than processed beach sand and wire that make consistent, intelligent, logical decisions. It is vitally important that we have a method to describe the logical decisions made by these circuits. In other words, we must describe how they operate. In this chapter, we will discover many ways to describe their operation. Each description method is important because all these methods commonly appear in technical literature and system documentation and are used in conjunction with modern design and development tools.

We will begin by studying Boolean algebra, which has been used for hundreds of years as a tool for the analysis and design of logic systems. Boolean algebra is a relatively simple mathematical tool that allows us to describe the relationship between a logic circuit's output(s) and its inputs as an algebraic equation (a Boolean expression). In this chapter, we will study the most basic logic circuits—*logic gates*—which are the fundamental building blocks from which all other logic circuits and digital systems are constructed. We will see how the operation of the different logic gates and the more complex circuits formed from combinations of logic gates can be described and analyzed using Boolean algebra. We will also get a glimpse of how Boolean algebra can be used to simplify a circuit's Boolean expression so that the circuit can be rebuilt using fewer logic gates and/or fewer connections. Much more will be done with circuit simplification in Chapter 4.

Boolean algebra is not only used as a tool for analysis and simplification of logic systems. It can also be used as a tool to create a logic circuit that will produce the desired input/output relationship. This process is often called synthesis of logic circuits as opposed to analysis. Other techniques have been used in the analysis, synthesis, and documentation of logic systems and circuits including truth tables, schematic symbols, timing diagrams, and—last but by no means least—language. To categorize these methods, we could say that Boolean algebra is a mathematic tool, truth tables are data organizational tools, schematic symbols are drawing tools, timing diagrams are graphing tools, and language is the universal description tool.

Today, any of these tools can be used to provide input to computers. The computers can be used to simplify and translate between these various forms of description and ultimately provide an output in the form necessary to implement a digital system. To take advantage of the powerful benefits of computer software, we must first fully understand the acceptable ways for describing these systems in terms the computer can understand. This chapter will lay the groundwork for further study of these vital tools for synthesis and analysis of digital systems.

Clearly the tools described here are invaluable tools in describing, analyzing, designing, and implementing digital circuits. The student who expects to work in the digital field must work hard at understanding and becoming comfortable with Boolean algebra (believe us, it's much, much easier than conventional algebra) and all the other tools. Do *all* of the examples, exercises, and problems, even the ones your instructor doesn't assign. When those run out, make up your own. The time you spend will be well worth it because you will see your skills improve and your confidence grow.

3-1   BOOLEAN CONSTANTS AND VARIABLES

Boolean algebra differs in a major way from ordinary algebra because Boolean constants and variables are allowed to have only two possible values, 0 or 1. A Boolean variable is a quantity that may, at different times, be equal to either 0 or 1. Boolean variables are often used to represent the voltage level present on a wire or at the input/output terminals of a circuit. For example, in a certain digital system, the Boolean value of 0 might be assigned to any voltage in the range from 0 to 0.8 V, while the Boolean value of 1 might be assigned to any voltage in the range 2 to 5 V.[*]

---

[*]Voltages between 0.8 and 2 V are undefined (neither 0 nor 1) and should not occur under normal circumstances.

Thus, Boolean 0 and 1 do not represent actual numbers but instead represent the state of a voltage variable, or what is called its **logic level**. A voltage in a digital circuit is said to be at the logic 0 level or the logic 1 level, depending on its actual numerical value. In digital logic, several other terms are used synonymously with 0 and 1. Some of the more common ones are shown in Table 3-1. We will use the 0/1 and LOW/HIGH designations most of the time.

TABLE 3-1

Logic 0	Logic 1
False	True
Off	On
Low	High
No	Yes
Open switch	Closed switch

As we said in the introduction, **Boolean algebra** is a means for expressing the relationship between a logic circuit's inputs and outputs. The inputs are considered logic variables whose logic levels at any time determine the output levels. In all our work to follow, we shall use letter symbols to represent logic variables. For example, the letter A might represent a certain digital circuit input or output, and at any time we must have either $A = 0$ or $A = 1$: if not one, then the other.

Because only two values are possible, Boolean algebra is relatively easy to work with compared with ordinary algebra. In Boolean algebra, there are no fractions, decimals, negative numbers, square roots, cube roots, logarithms, imaginary numbers, and so on. In fact, in Boolean algebra there are only *three* basic operations: *OR, AND,* and *NOT.*

These basic operations are called *logic operations.* Digital circuits called *logic gates* can be constructed from diodes, transistors, and resistors connected so that the circuit output is the result of a basic logic operation *(OR, AND, NOT)* performed on the inputs. We will be using Boolean algebra first to describe and analyze these basic logic gates, then later to analyze and design combinations of logic gates connected as logic circuits.

3-2 TRUTH TABLES

A **truth table** is a means for describing how a logic circuit's output depends on the logic levels present at the circuit's inputs. Figure 3-1(a) illustrates a truth table for one type of two-input logic circuit. The table lists all possible combinations of logic levels present at inputs A and B, along with the corresponding output level x. The first entry in the table shows that when A and B are both at the 0 level, the output x is at the 1 level or, equivalently, in the 1 state. The second entry shows that when input B is changed to the 1 state, so that $A = 0$ and $B = 1$, the output x becomes a 0. In a similar way, the table shows what happens to the output state for any set of input conditions.

Figures 3-1(b) and (c) show samples of truth tables for three- and four-input logic circuits. Again, each table lists all possible combinations of input logic levels on the left, with the resultant logic level for output x on the right. Of course, the actual values for x will depend on the type of logic circuit.

Note that there are 4 table entries for the two-input truth table, 8 entries for a three-input truth table, and 16 entries for the four-input truth table. The number of input combinations will equal 2^N for an N-input truth table.

FIGURE 3-1 Example truth tables for (a) two-input, (b) three-input, and (c) four-input circuits.

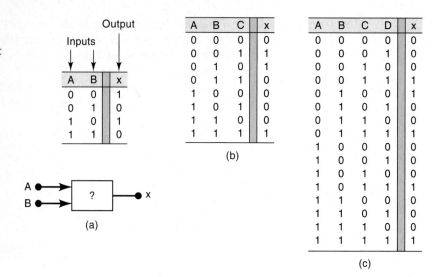

A	B	x
0	0	1
0	1	0
1	0	1
1	1	0

(a)

A	B	C	x
0	0	0	0
0	0	1	1
0	1	0	1
0	1	1	0
1	0	0	0
1	0	1	0
1	1	0	0
1	1	1	1

(b)

A	B	C	D	x
0	0	0	0	0
0	0	0	1	0
0	0	1	0	0
0	0	1	1	1
0	1	0	0	1
0	1	0	1	0
0	1	1	0	0
0	1	1	1	1
1	0	0	0	0
1	0	0	1	0
1	0	1	0	0
1	0	1	1	1
1	1	0	0	0
1	1	0	1	0
1	1	1	0	0
1	1	1	1	1

(c)

Also note that the list of all possible input combinations follows the binary counting sequence, and so it is an easy matter to write down all of the combinations without missing any.

REVIEW QUESTIONS

1. What is the output state of the four-input circuit represented in Figure 3-1(c) when all inputs except *B* are 1?
2. Repeat question 1 for the following input conditions: $A = 1$, $B = 0$, $C = 1$, $D = 0$.
3. How many table entries are needed for a five-input circuit?

3-3   OR OPERATION WITH OR GATES

The **OR operation** is the first of the three basic Boolean operations to be learned. The truth table in Figure 3-2(a) shows what happens when two logic inputs, *A* and *B*, are combined using the OR operation to produce the output *x*. The table shows that *x* is a logic 1 for every combination of input levels where one *or* more inputs are 1. The only case where *x* is a 0 is when both inputs are 0.

The Boolean expression for the OR operation is

$$x = A + B$$

In this expression, the + sign does not stand for ordinary addition; it stands for the OR operation. The OR operation is similar to ordinary addition except for the case where *A* and *B* are both 1; the OR operation produces $1 + 1 = 1$,

FIGURE 3-2 (a) Truth table defining the OR operation; (b) circuit symbol for a two-input OR gate.

A	B	x = A + B
0	0	0
0	1	1
1	0	1
1	1	1

(a)

(b)

not $1 + 1 = 2$. In Boolean algebra, 1 is as high as we go, so we can never have a result greater than 1. The same holds true for combining three inputs using the OR operation. Here we have $x = A + B + C$. If we consider the case where all three inputs are 1, we have

$$x = 1 + 1 + 1 = 1$$

The expression $x = A + B$ is read as "x equals A OR B," which means that x will be 1 when A or B or both are 1. Likewise, the expression $x = A + B + C$ is read as "x equals A OR B OR C," which means that x will be 1 when A or B or C or any combination of them are 1. To describe this circuit in the English language we could say that *x is true (1)* **WHEN** *A is true (1)* **OR** *B is true (1)* **OR** *C is true (1)*.

OR Gate

In digital circuitry, an **OR gate**[*] is a circuit that has two or more inputs and whose output is equal to the OR combination of the inputs. Figure 3-2(b) is the logic symbol for a two-input OR gate. The inputs A and B are logic voltage levels, and the output x is a logic voltage level whose value is the result of the OR operation on A and B; that is, $x = A + B$. In other words, the OR gate operates so that its output is HIGH (logic 1) if either input A or B or both are at a logic 1 level. The OR gate output will be LOW (logic 0) only if all its inputs are at logic 0.

This same idea can be extended to more than two inputs. Figure 3-3 shows a three-input OR gate and its truth table. Examination of this truth table shows again that the output will be 1 for every case where one or more inputs are 1. This general principle is the same for OR gates with any number of inputs.

Using the language of Boolean algebra, the output x can be expressed as $x = A + B + C$, where again it must be emphasized that the + represents the OR operation. The output of any OR gate, then, can be expressed as the OR combination of its various inputs. We will put this to use when we subsequently analyze logic circuits.

FIGURE 3-3 Symbol and truth table for a three-input OR gate.

A	B	C	x = A + B + C
0	0	0	0
0	0	1	1
0	1	0	1
0	1	1	1
1	0	0	1
1	0	1	1
1	1	0	1
1	1	1	1

Summary of the OR Operation

The important points to remember concerning the OR operation and OR gates are:

1. The OR operation produces a result (output) of 1 whenever *any* input is a 1. Otherwise the output is 0.

---

[*]The term *gate* comes from the inhibit/enable operation discussed in Chapter 4.

2. An OR gate is a logic circuit that performs an OR operation on the circuit's inputs.

3. The expression $x = A + B$ is read as "x equals A OR B."

In many industrial control systems, it is required to activate an output function whenever any one of several inputs is activated. For example, in a chemical process it may be desired that an alarm be activated whenever the process temperature exceeds a maximum value *or* whenever the pressure goes above a certain limit. Figure 3-4 is a block diagram of this situation. The temperature transducer circuit produces an output voltage proportional to the process temperature. This voltage, V_T, is compared with a temperature reference voltage, V_{TR}, in a voltage comparator circuit. The comparator output, T_H, is normally a low voltage (logic 0), but it switches to a high voltage (logic 1) when V_T exceeds V_{TR}, indicating that the process temperature is too high. A similar arrangement is used for the pressure measurement, so that its associated comparator output, P_H, goes from LOW to HIGH when the pressure is too high.

FIGURE 3-4 Example of the use of an OR gate in an alarm system.

Since we want the alarm to be activated when either temperature *or* pressure is too high, it should be apparent that the two comparator outputs can be fed to a two-input OR gate. The OR gate output thus goes HIGH (1) for either alarm condition and will activate the alarm. This same idea can obviously be extended to situations with more than two process variables.

Determine the OR gate output in Figure 3-5. The OR gate inputs A and B are varying according to the timing diagrams shown. For example, A starts out LOW at time t_0, goes HIGH at t_1, back LOW at t_3, and so on.

FIGURE 3-5   Example 3-2.

Solution

The OR gate output will be HIGH whenever *any* input is HIGH. Between time t_0 and t_1, both inputs are LOW, so OUTPUT = LOW. At t_1, input *A* goes HIGH while *B* remains LOW. This causes OUTPUT to go HIGH at t_1 and stay HIGH until t_4 because, during this interval, one or both inputs are HIGH. At t_4, input *B* goes from 1 to 0 so that now both inputs are LOW, and this drives OUTPUT back to LOW. At t_5, *A* goes HIGH, sending OUTPUT back HIGH, where it stays for the rest of the shown time span.

EXAMPLE 3-3A

For the situation depicted in Figure 3-6, determine the waveform at the OR gate output.

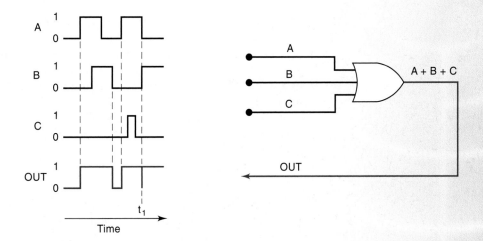

FIGURE 3-6   Examples 3-3A and B.

Solution

The three OR gate inputs *A, B,* and *C* are varying, as shown by their waveform diagrams. The OR gate output is determined by realizing that it will be HIGH whenever *any* of the three inputs is at a HIGH level. Using this reasoning, the OR output waveform is as shown in the figure. Particular attention should be paid to what occurs at time t_1. The diagram shows that, at that instant of time, input A is going from HIGH to LOW while input B is going from LOW to HIGH. Since these inputs are making their transitions at approximately the same time, and since these transitions take a certain amount of time, there is a short interval when these OR gate inputs are both in the undefined range between 0 and 1. When this occurs, the OR gate output also becomes a value in this range, as evidenced by the glitch or spike on the output waveform at t_1. The occurrence of this glitch and its size (amplitude and width) depend on the speed with which the input transitions occur.

EXAMPLE 3-3B

What would happen to the glitch in the output in Figure 3-6 if input C sat in the HIGH state while A and B were changing at time t_1?

Solution

With the C input HIGH at t_1, the OR gate output will remain HIGH, regardless of what is occurring at the other inputs, because any HIGH input will keep an OR gate output HIGH. Therefore, the glitch will not appear in the output.

REVIEW QUESTIONS

1. What is the only set of input conditions that will produce a LOW output for any OR gate?
2. Write the Boolean expression for a six-input OR gate.
3. If the A input in Figure 3-6 is permanently kept at the 1 level, what will the resultant output waveform be?

3-4   AND OPERATION WITH AND GATES

The **AND operation** is the second basic Boolean operation. The truth table in Figure 3-7(a) shows what happens when two logic inputs, A and B, are combined using the AND operation to produce output x. The table shows that x is a logic 1 only when both A and B are at the logic 1 level. For any case where one of the inputs is 0, the output is 0.

The Boolean expression for the AND operation is

$$x = A \cdot B$$

In this expression, the · sign stands for the Boolean AND operation and not the multiplication operation. However, the AND operation on Boolean variables operates the same as ordinary multiplication, as examination of the truth table shows, so we can think of them as being the same. This characteristic can be helpful when evaluating logic expressions that contain AND operations.

The expression $x = A \cdot B$ is read as "x equals A AND B," which means that x will be 1 only when A and B are both 1. The · sign is usually omitted so that the expression simply becomes $x = AB$. For the case when three in-

FIGURE 3-7 (a) Truth table for the AND operation; (b) AND gate symbol.

A	B	x = A · B
0	0	0
0	1	0
1	0	0
1	1	1

(a)   (b)

puts are ANDed, we have $x = A \cdot B \cdot C = ABC$. This is read as "$x$ equals A AND B AND C," which means that x will be 1 only when A and B and C are all 1.

AND Gate

The logic symbol for a two-input **AND gate** is shown in Figure 3-7(b). The AND gate output is equal to the AND product of the logic inputs; that is, $x = AB$. In other words, the AND gate is a circuit that operates so that its output is HIGH only when all its inputs are HIGH. For all other cases, the AND gate output is LOW.

This same operation is characteristic of AND gates with more than two inputs. For example, a three-input AND gate and its accompanying truth table are shown in Figure 3-8. Once again, note that the gate output is 1 only for the case where $A = B = C = 1$. The expression for the output is $x = ABC$. For a four-input AND gate, the output is $x = ABCD$, and so on.

FIGURE 3-8 Truth table and symbol for a three-input AND gate.

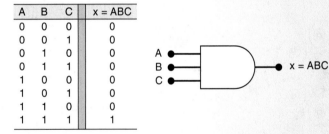

A	B	C	x = ABC
0	0	0	0
0	0	1	0
0	1	0	0
0	1	1	0
1	0	0	0
1	0	1	0
1	1	0	0
1	1	1	1

Note the difference between the symbols for the AND gate and the OR gate. Whenever you see the AND symbol on a logic-circuit diagram, it tells you that the output will go HIGH *only* when *all* inputs are HIGH. Whenever you see the OR symbol, it means that the output will go HIGH when *any* input is HIGH.

Summary of the AND Operation

1. The AND operation is performed the same as ordinary multiplication of 1s and 0s.
2. An AND gate is a logic circuit that performs the AND operation on the circuit's inputs.
3. An AND gate output will be 1 *only* for the case when *all* inputs are 1; for all other cases, the output will be 0.
4. The expression $x = AB$ is read as "x equals A AND B."

EXAMPLE 3-4

Determine the output x from the AND gate in Figure 3-9 for the given input waveforms.

FIGURE 3-9   Example 3-4.

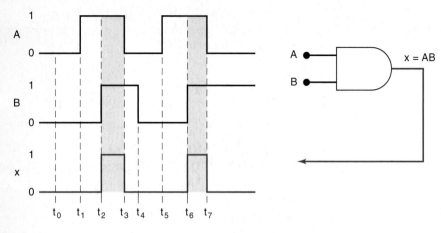

Solution

The output of an AND gate is determined by realizing that it will be HIGH only when all inputs are HIGH at the same time. For the input waveforms given, this condition is met only during intervals t_2–t_3 and t_6–t_7. At all other times, one or more of the inputs are 0, thereby producing a LOW output. Note that input level changes that occur while the other input is LOW have no effect on the output.

EXAMPLE 3-5A

Determine the output waveform for the AND gate shown in Figure 3-10.

FIGURE 3-10   Examples 3-5A and B.

Solution

The output x will be at 1 only when A and B are both HIGH at the same time. Using this fact, we can determine the x waveform as shown in the figure.

Notice that the x waveform is 0 whenever B is 0, regardless of the signal at A. Also notice that whenever B is 1, the x waveform is the same as A. Thus, we can think of the B input as a *control* input whose logic level determines whether or not the A waveform gets through to the x output. In this situation, the AND gate is used as an *inhibit circuit*. We can say that $B = 0$ is the inhibit condition producing a 0 output. Conversely, $B = 1$ is the *enable* condition, which enables A to reach the output. This inhibit operation is an important application of AND gates, which will be encountered later.

EXAMPLE 3-5B

What will happen to the x output waveform in Figure 3-10 if the B input is kept at the 0 level?

Solution

With B kept LOW, the x output will also stay LOW. This can be reasoned in two different ways. First, with $B = 0$ we have $x = A \cdot B = A \cdot 0 = 0$ because anything multiplied (ANDed) by 0 will be 0. Another way to look at it is that an AND gate requires that all inputs be HIGH for the output to be HIGH, and this cannot happen if B is kept LOW.

1. What is the only input combination that will produce a HIGH at the output of a five-input AND gate?

2. What logic level should be applied to the second input of a two-input AND gate if the logic signal at the first input is to be inhibited (prevented) from reaching the output?

3. *True or false:* An AND gate output will always differ from an OR gate output for the same input conditions.

3-5   NOT OPERATION

The **NOT operation** is unlike the OR and AND operations because it can be performed on a single input variable. For example, if the variable A is subjected to the NOT operation, the result x can be expressed as

$$x = \overline{A}$$

where the overbar represents the NOT operation. This expression is read as "x equals NOT A" or "x equals the *inverse* of A" or "x equals the *complement* of A." Each of these is in common usage, and all indicate that the logic value of $x = \overline{A}$ *is opposite to* the logic value of A. The truth table in Figure 3-11(a) clarifies this for the two cases $A = 0$ and $A = 1$. That is,

$$0 = \overline{1} \qquad \text{because 0 is not 1}$$

and

$$1 = \overline{0} \qquad \text{because 1 is not 0}$$

The NOT operation is also referred to as **inversion** or **complementation**, and these terms will be used interchangeably throughout the book. Although we will always use the overbar indicator to represent inversion, it is important to mention that another indicator for inversion is the prime symbol ('). That is,

$$A' = \overline{A}$$

Both should be recognized as indicating the inversion operation.

FIGURE 3-11 (a) Truth table; (b) symbol for the INVERTER (NOT circuit); (c) sample waveforms.

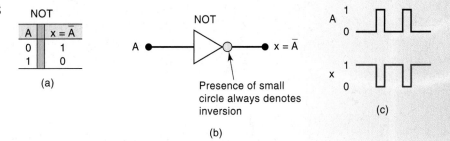

NOT

A	x = $\overline{A}$
0	1
1	0

(a)

NOT

A •——▷○—— • $x = \overline{A}$

Presence of small circle always denotes inversion

(b)

(c)

NOT Circuit (INVERTER)

Figure 3-11(b) shows the symbol for a **NOT circuit**, which is more commonly called an **INVERTER**. This circuit *always* has only a single input, and its output logic level is always opposite to the logic level of this input. Figure 3-11(c) shows how the INVERTER affects an input signal. It inverts (complements) the input signal at all points on the waveform so that whenever the input = 0, output = 1, and vice versa.

Figure 3-12 shows a typical application of the NOT gate. The push button is wired to produce a logic 1 (true) when it is pressed. Sometimes we want to know if the push button is not being pressed, and so this circuit provides an expression that is true when the button is not pressed.

FIGURE 3-12 A NOT gate indicating a button is *not* pressed when its output is true.

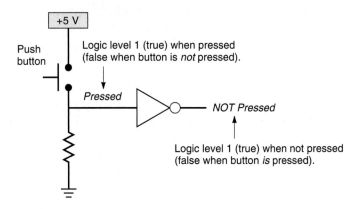

Summary of Boolean Operations

The rules for the OR, AND, and NOT operations may be summarized as follows:

OR	*AND*	*NOT*
$0 + 0 = 0$	$0 \cdot 0 = 0$	$\bar{0} = 1$
$0 + 1 = 1$	$0 \cdot 1 = 0$	$\bar{1} = 0$
$1 + 0 = 1$	$1 \cdot 0 = 0$	
$1 + 1 = 1$	$1 \cdot 1 = 1$	

1. The output of the INVERTER of Figure 3-11 is connected to the input of a second INVERTER. Determine the output level of the second INVERTER for each level of input A.
2. The output of the AND gate in Figure 3-7 is connected to the input of an INVERTER. Write the truth table showing the INVERTER output, y, for each combination of inputs A and B.

3-6 DESCRIBING LOGIC CIRCUITS ALGEBRAICALLY

Any logic circuit, no matter how complex, can be described completely using the three basic Boolean operations because the OR gate, AND gate, and NOT circuit are the basic building blocks of digital systems. For example, consider the circuit in Figure 3-13(a). This circuit has three inputs, A, B, and C, and a single output, x. Utilizing the Boolean expression for each gate, we can easily determine the expression for the output.

FIGURE 3-13 (a) Logic circuit with its Boolean expression; (b) logic circuit whose expression requires parentheses.

(a)

(b)

The expression for the AND gate output is written $A \cdot B$. This AND output is connected as an input to the OR gate along with C, another input. The OR gate operates on its inputs so that its output is the OR sum of the inputs. Thus, we can express the OR output as $x = A \cdot B + C$. (This final expression could also be written as $x = C + A \cdot B$ because it does not matter which term of the OR sum is written first.)

Occasionally, there may be confusion about which operation in an expression is performed first. The expression $A \cdot B + C$ can be interpreted in two different ways: (1) $A \cdot B$ is ORed with C, or (2) A is ANDed with the term $B + C$. To avoid this confusion, it will be understood that if an expression contains both AND and OR operations, the AND operations are performed first, unless there are *parentheses* in the expression, in which case the operation inside the parentheses is to be performed first. This is the same rule that is used in ordinary algebra to determine the order of operations.

To illustrate further, consider the circuit in Figure 3-13(b). The expression for the OR gate output is simply $A + B$. This output serves as an input to the AND gate along with another input, C. Thus, we express the output of the AND gate as $x = (A + B) \cdot C$. Note the use of parentheses here to indicate that A and B are ORed *first*, before their OR sum is ANDed with C. Without the parentheses it would be interpreted *incorrectly*, because $A + B \cdot C$ means that A is ORed with the product $B \cdot C$.

Circuits Containing INVERTERs

Whenever an INVERTER is present in a logic-circuit diagram, its output expression is simply equal to the input expression with a bar over it. Figure 3-14 shows two examples using INVERTERs. In Figure 3-14(a), input A is fed through an INVERTER, whose output is therefore $\overline{A}$. The INVERTER output is fed to an OR gate together with B, so that the OR output is equal to $\overline{A} + B$. Note that the bar is over the A alone, indicating that A is first inverted and then ORed with B.

In Figure 3-14(b), the output of the OR gate is equal to $A + B$ and is fed through an INVERTER. The INVERTER output is therefore equal to $\overline{(A + B)}$

FIGURE 3-14 Circuits using INVERTERs.

$\overline{A}$

$x = \overline{A} + B$

$A + B$

$x = \overline{A + B}$

(a)

(b)

because it inverts the *complete* input expression. Note that the bar covers the entire expression $(A + B)$. This is important because, as will be shown later, the expressions $(\overline{A + B})$ and $(\overline{A} + \overline{B})$ are *not* equivalent. The expression $(\overline{A + B})$ means that A is ORed with B and then their OR sum is inverted, whereas the expression $(\overline{A} + \overline{B})$ indicates that A is inverted and B is inverted and the results are then ORed together.

Figure 3-15 shows two more examples, which should be studied carefully. Note especially the use of *two* separate sets of parentheses in Figure 3-15(b). Also notice in Figure 3-15(a) that the input variable A is connected as an input to two different gates.

(a)

(b)

FIGURE 3-15   More examples.

1. In Figure 3-15(a), change each AND gate to an OR gate, and change the OR gate to an AND gate. Then write the expression for output x.

2. In Figure 3-15(b), change each AND gate to an OR gate, and each OR gate to an AND gate. Then write the expression for x.

3-7   EVALUATING LOGIC-CIRCUIT OUTPUTS

Once we have the Boolean expression for a circuit output, we can obtain the output logic level for any set of input levels. For example, suppose that we want to know the logic level of the output x for the circuit in Figure 3-15(a) for the case where $A = 0$, $B = 1$, $C = 1$, and $D = 1$. As in ordinary algebra, the value of x can be found by "plugging" the values of the variables into the expression and performing the indicated operations as follows:

$$x = \overline{A}BC(\overline{A + D})$$
$$= \overline{0} \cdot 1 \cdot 1 \cdot (\overline{0 + 1})$$
$$= 1 \cdot 1 \cdot 1 \cdot (\overline{0 + 1})$$
$$= 1 \cdot 1 \cdot 1 \cdot (\overline{1})$$
$$= 1 \cdot 1 \cdot 1 \cdot 0$$
$$= 0$$

As another illustration, let us evaluate the output of the circuit in Figure 3-15(b) for $A = 0$, $B = 0$, $C = 1$, $D = 1$, and $E = 1$.

$$x = [D + \overline{(A + B)C}] \cdot E$$
$$= [1 + \overline{(0 + 0) \cdot 1}] \cdot 1$$
$$= [1 + \overline{0 \cdot 1}] \cdot 1$$
$$= [1 + \overline{0}] \cdot 1$$
$$= [1 + 1] \cdot 1$$
$$= 1 \cdot 1$$
$$= 1$$

In general, the following rules must always be followed when evaluating a Boolean expression:

1. First, perform all inversions of single terms; that is, $\overline{0} = 1$ or $\overline{1} = 0$.
2. Then perform all operations within parentheses.
3. Perform an AND operation before an OR operation unless parentheses indicate otherwise.
4. If an expression has a bar over it, perform the operations inside the expression first and then invert the result.

For practice, determine the outputs of both circuits in Figure 3-15 for the case where all inputs are 1. The answers are $x = 0$ and $x = 1$, respectively.

Determining Output Level from a Diagram

The output logic level for given input levels can also be determined directly from the circuit diagram *without* using the Boolean expression. This technique is often used by technicians during the troubleshooting or testing of a logic system because it also tells them what each gate output is supposed to be as well as the final output. To illustrate, the circuit of Figure 3-15(a) is redrawn in Figure 3-16 with the input levels $A = 0$, $B = 1$, $C = 1$, and $D = 1$. The procedure is to start from the inputs and to proceed through each IN-VERTER and gate, writing down each of their outputs in the process until the final output is reached.

In Figure 3-16, AND gate 1 has *all* three inputs at the 1 level because the INVERTER changes the $A = 0$ to a 1. This condition produces a 1 at the AND

FIGURE 3-16 Determining the output level from a circuit diagram.

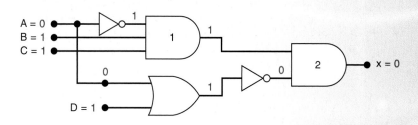

output because $1 \cdot 1 \cdot 1 = 1$. The OR gate has inputs of 1 and 0, which produces a 1 output because $1 + 0 = 1$. This 1 is inverted to 0 and applied to AND gate 2, along with the 1 from the first AND output. The 0 and 1 inputs to AND gate 2 produce an x output of 0 because $0 \cdot 1 = 0$.

EXAMPLE 3-6

Determine the output in Figure 3-16 for the condition where all inputs are LOW.

Solution

With $A = B = C = D = 0$, the output of AND gate 1 will be LOW. This LOW places a LOW at the input of AND gate 2, which automatically produces $x = 0$, regardless of the levels at other points in the circuit. This example shows that it is not always necessary to determine the logic level at every point in order to determine the output level.

REVIEW QUESTIONS

1. Use the expression for x to determine the output of the circuit in Figure 3-15(a) for the conditions $A = 0$, $B = 1$, $C = 1$, and $D = 0$.

2. Use the expression for x to determine the output of the circuit in Figure 3-15(b) for the conditions $A = B = E = 1$, $C = D = 0$.

3. Determine the answers to questions 1 and 2 by finding the logic levels present at each gate input and output as was done in Figure 3-16.

3-8 IMPLEMENTING CIRCUITS FROM BOOLEAN EXPRESSIONS

When the operation of a circuit is defined by a Boolean expression, we can draw a logic-circuit diagram directly from that expression. For example, if we needed a circuit that was defined by $x = A \cdot B \cdot C$, we would immediately know that all that was needed was a three-input AND gate. If we needed a circuit that was defined by $x = A + \overline{B}$, we would use a two-input OR gate with an INVERTER on one of the inputs. The same reasoning used for these simple cases can be extended to more complex circuits.

Suppose that we wanted to construct a circuit whose output is $y = AC + B\overline{C} + \overline{A}BC$. This Boolean expression contains three terms (AC, $B\overline{C}$, $\overline{A}BC$), which are ORed together. This tells us that a three-input OR gate is required with inputs that are equal to AC, $B\overline{C}$, and $\overline{A}BC$. This is illustrated in Figure 3-17(a), where a three-input OR gate is drawn with inputs labeled as AC, $B\overline{C}$, and $\overline{A}BC$.

FIGURE 3-17   Constructing a logic circuit from a Boolean expression.

Each OR gate input is an AND product term, which means that an AND gate with appropriate inputs can be used to generate each of these terms. This is shown in Figure 3-17(b), which is the final circuit diagram. Note the use of INVERTERs to produce the $\bar{A}$ and $\bar{C}$ terms required in the expression.

This same general approach can always be followed, although we shall find that there are some clever, more efficient techniques that can be employed. For now, however, this straightforward method will be used to minimize the number of new items that are to be learned.

EXAMPLE 3-7

Draw the circuit diagram to implement the expression $x = (A + B)(\bar{B} + C)$.

Solution

This expression shows that the terms $A + B$ and $\bar{B} + C$ are inputs to an AND gate, and each of these two terms is generated from a separate OR gate. The result is drawn in Figure 3-18.

FIGURE 3-18   Example 3-7.

NAND Gate

The symbol for a two-input **NAND gate** is shown in Figure 3-22(a). It is the same as the AND gate symbol except for the small circle on its output. Once again, this small circle denotes the inversion operation. Thus, the NAND operates like an AND gate followed by an INVERTER, so that the circuits of Figures 3-22(a) and (b) are equivalent, and the output expression for the NAND gate is $x = \overline{AB}$.

FIGURE 3-22 (a) NAND symbol; (b) equivalent circuit; (c) truth table.

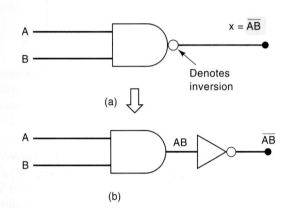

		AND	NAND
A	B	AB	$\overline{AB}$
0	0	0	1
0	1	0	1
1	0	0	1
1	1	1	0

(c)

The truth table in Figure 3-22(c) shows that the NAND gate output is the exact inverse of the AND gate for all possible input conditions. The AND output goes HIGH only when all inputs are HIGH, while the NAND output goes LOW only when all inputs are HIGH. This same characteristic is true of NAND gates having more than two inputs.

EXAMPLE 3-10

Determine the output waveform of a NAND gate having the inputs shown in Figure 3-23.

FIGURE 3-23 Example 3-10.

Solution

One way is to draw first the output waveform for an AND gate and then invert it. Another way utilizes the fact that a NAND output will be LOW only when all inputs are HIGH. Thus, you can find those time intervals during which the inputs are all HIGH, and make the NAND output LOW for those intervals. The output will be HIGH at all other times.

EXAMPLE 3-11

Implement the logic circuit that has the expression $x = \overline{AB \cdot (\overline{C + D})}$ using only NOR and NAND gates.

Solution

The $(\overline{C + D})$ term is the expression for the output of a NOR gate. This term is ANDed with A and B, and the result is inverted; this, of course, is the NAND operation. Thus, the circuit is implemented as shown in Figure 3-24. Note that the NAND gate first ANDs the A, B, and $(\overline{C + D})$ terms, and then it inverts the *complete* result.

FIGURE 3-24   Examples 3-11 and 3-12.

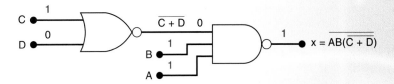

EXAMPLE 3-12

Determine the output level in Figure 3-24 for $A = B = C = 1$ and $D = 0$.

Solution

In the first method we use the expression for x.

$$
\begin{aligned}
x &= \overline{AB(\overline{C + D})} \\
  &= \overline{1 \cdot 1 \cdot (\overline{1 + 0})} \\
  &= \overline{1 \cdot 1 \cdot (\overline{1})} \\
  &= \overline{1 \cdot 1 \cdot 0} \\
  &= \overline{0} = 1
\end{aligned}
$$

In the second method, we write down the input logic levels on the circuit diagram (shown in color in Figure 3-24) and follow these levels through each gate to the final output. The NOR gate has inputs of 1 and 0 to produce an output of 0 (an OR would have produced an output of 1). The NAND gate thus has input levels of 0, 1, and 1 to produce an output of 1 (an AND would have produced an output of 0).

REVIEW QUESTIONS

1. What is the only set of input conditions that will produce a HIGH output from a three-input NOR gate?
2. Determine the output level in Figure 3-24 for $A = B = 1$, $C = D = 0$.
3. Change the NOR gate of Figure 3-24 to a NAND gate, and change the NAND to a NOR. What is the new expression for x?

3-10   BOOLEAN THEOREMS

We have seen how Boolean algebra can be used to help analyze a logic circuit and express its operation mathematically. We will continue our study of Boolean algebra by investigating the various **Boolean theorems** (rules) that can help us to simplify logic expressions and logic circuits. The first group of theorems is given in Figure 3-25. In each theorem, x is a logic variable that can be either a 0 or a 1. Each theorem is accompanied by a logic-circuit diagram that demonstrates its validity.

Case 2. For $x = 0$, $y = 1$,

$$x + xy = x$$
$$0 + 0 \cdot 1 = 0$$
$$0 + 0 = 0$$
$$0 = 0$$

Case 3. For $x = 1$, $y = 0$,

$$x + xy = x$$
$$1 + 1 \cdot 0 = 1$$
$$1 + 0 = 1$$
$$1 = 1$$

Case 4. For $x = 1$, $y = 1$,

$$x + xy = x$$
$$1 + 1 \cdot 1 = 1$$
$$1 + 1 = 1$$
$$1 = 1$$

Theorem (14) can also be proved by factoring and using theorems (6) and (2) as follows:

$$x + xy = x(1 + y)$$
$$= x \cdot 1 \quad \text{[using theorem (6)]}$$
$$= x \quad \text{[using theorem (2)]}$$

All of these Boolean theorems can be useful in simplifying a logic expression—that is, in reducing the number of terms in the expression. When this is done, the reduced expression will produce a circuit that is less complex than the one that the original expression would have produced. A good portion of the next chapter will be devoted to the process of circuit simplification. For now, the following examples will serve to illustrate how the Boolean theorems can be applied. **Note:** You can find all the Boolean theorems on the inside back cover.

EXAMPLE 3-13

Simplify the expression $y = A\overline{B}D + A\overline{B}\overline{D}$.

Solution

Factor out the common variables $A\overline{B}$ using theorem (13):

$$y = A\overline{B}(D + \overline{D})$$

Using theorem (8), the term in parentheses is equivalent to 1. Thus,

$$y = A\overline{B} \cdot 1$$
$$= A\overline{B} \quad \text{[using theorem (2)]}$$

EXAMPLE 3-14

Simplify $z = (\overline{A} + B)(A + B)$.

Solution

The expression can be expanded by multiplying out the terms [theorem (13)]:

$$z = \overline{A} \cdot A + \overline{A} \cdot B + B \cdot A + B \cdot B$$

Invoking theorem (4), the term $\overline{A} \cdot A = 0$. Also, $B \cdot B = B$ [theorem (3)]:

$$z = 0 + \overline{A} \cdot B + B \cdot A + B = \overline{A}B + AB + B$$

Factoring out the variable B [theorem (13)], we have

$$z = B(\overline{A} + A + 1)$$

Finally, using theorems (2) and (6),

$$z = B$$

EXAMPLE 3-15

Simplify $x = ACD + \overline{A}BCD$.

Solution

Factoring out the common variables CD, we have

$$x = CD(A + \overline{A}B)$$

Utilizing theorem (15a), we can replace $A + \overline{A}B$ by $A + B$, so

$$x = CD(A + B)$$
$$= ACD + BCD$$

REVIEW QUESTIONS

1. Use theorems (13) and (14) to simplify $y = A\overline{C} + AB\overline{C}$.
2. Use theorems (13) and (8) to simplify $y = \overline{A}\,\overline{B}C\overline{D} + \overline{A}\,\overline{B}C\overline{D}$.
3. Use theorems (13) and (15b) to simplify $y = \overline{A}D + ABD$.

3-11 DEMORGAN'S THEOREMS

Two of the most important theorems of Boolean algebra were contributed by a great mathematician named DeMorgan. **DeMorgan's theorems** are extremely useful in simplifying expressions in which a product or sum of variables is inverted. The two theorems are:

$$(16) \quad \overline{(x + y)} = \overline{x} \cdot \overline{y}$$
$$(17) \quad \overline{(x \cdot y)} = \overline{x} + \overline{y}$$

Now consider theorem (17):

$$\overline{x \cdot y} = \overline{x} + \overline{y}$$

The left side of the equation can be implemented by a NAND gate with inputs x and y. The right side can be implemented by first inverting inputs x and y and then putting them through an OR gate. These two equivalent representations are shown in Figure 3-27(a). The OR gate with INVERTERs on each of its inputs is equivalent to the NAND gate. In fact, both representations are used to represent the NAND function. When the OR gate with inverted inputs is used to represent the NAND function, it is usually drawn as shown in Figure 3-27(b), where the circles again represent inversion.

FIGURE 3-27 (a) Equivalent circuits implied by theorem (17); (b) alternative symbol for the NAND function.

(a)

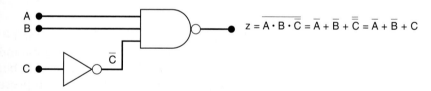

(b)

EXAMPLE 3-17

Determine the output expression for the circuit of Figure 3-28 and simplify it using DeMorgan's theorems.

FIGURE 3-28 Example 3-17.

$z = \overline{A \cdot B \cdot \overline{C}} = \overline{A} + \overline{B} + \overline{\overline{C}} = \overline{A} + \overline{B} + C$

Solution

The expression for z is $z = \overline{AB\overline{C}}$. Use DeMorgan's theorem to break the large inversion sign:

$$z = \overline{A} + \overline{B} + \overline{\overline{C}}$$

Cancel the double inversions over C to obtain

$$z = \overline{A} + \overline{B} + C$$

1. Use DeMorgan's theorems to convert the expression $z = \overline{(A + B) \cdot \overline{C}}$ to one that has only single-variable inversions.
2. Repeat question 1 for the expression $y = \overline{\overline{RST} + \overline{Q}}$.
3. Implement a circuit having output expression $z = \overline{A}\,\overline{B}C$ using only a NOR gate and an INVERTER.
4. Use DeMorgan's theorems to convert $y = \overline{\overline{A} + \overline{B} + \overline{CD}}$ to an expression containing only single-variable inversions.

3-12   UNIVERSALITY OF NAND GATES AND NOR GATES

All Boolean expressions consist of various combinations of the basic operations of OR, AND, and INVERT. Therefore, any expression can be implemented using combinations of OR gates, AND gates, and INVERTERs. It is possible, however, to implement any logic expression using *only* NAND gates and no other type of gate. This is because NAND gates, in the proper combination, can be used to perform each of the Boolean operations OR, AND, and INVERT. This is demonstrated in Figure 3-29.

First, in Figure 3-29(a), we have a two-input NAND gate whose inputs are purposely connected together so that the variable A is applied to both. In this configuration, the NAND simply acts as INVERTER because its output is $x = \overline{A \cdot A} = \overline{A}$.

In Figure 3-29(b), we have two NAND gates connected so that the AND operation is performed. NAND gate 2 is used as an INVERTER to change $\overline{AB}$ to $\overline{\overline{AB}} = AB$, which is the desired AND function.

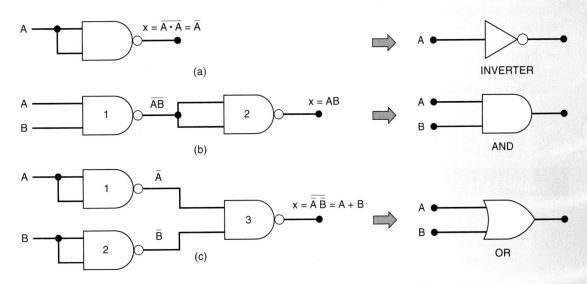

FIGURE 3-29   NAND gates can be used to implement any Boolean function.

the output, and add a bubble to each input. Then change the AND symbol to an OR symbol. The result is an OR symbol with bubbles on its inputs.

We can easily prove that this alternate symbol is equivalent to the standard symbol by using DeMorgan's theorems and recalling that the bubble represents an inversion operation. The output expression from the standard NAND symbol is $\overline{AB} = \overline{A} + \overline{B}$, which is the same as the output expression for the alternate symbol. This same procedure can be followed for each pair of symbols in Figure 3-33.

Several points should be stressed regarding the logic symbol equivalences:

1. The equivalences can be extended to gates with *any* number of inputs.
2. None of the standard symbols have bubbles on their inputs, and all the alternate symbols do.
3. The standard and alternate symbols for each gate represent the same physical circuit; *there is no difference in the circuits represented by the two symbols.*
4. NAND and NOR gates are inverting gates, and so both the standard and the alternate symbols for each will have a bubble on *either* the input or the output. AND and OR gates are *noninverting* gates, and so the alternate symbols for each will have bubbles on *both* inputs and output.

Logic-Symbol Interpretation

Each of the logic-gate symbols of Figure 3-33 provides a unique interpretation of how the gate operates. Before we can demonstrate these interpretations, we must first establish the concept of **active logic levels**.

When an input or output line on a logic circuit symbol has *no bubble* on it, that line is said to be **active-HIGH**. When an input or output line *does* have a *bubble* on it, that line is said to be **active-LOW**. The presence or absence of a bubble, then, determines the active-HIGH/active-LOW status of a circuit's inputs and output, and is used to interpret the circuit operation.

To illustrate, Figure 3-34(a) shows the standard symbol for a NAND gate. The standard symbol has a bubble on its output and no bubbles on its inputs. Thus, it has an active-LOW output and active-HIGH inputs. The logic operation represented by this symbol can therefore be interpreted as follows:

The output goes LOW only when *all* of the inputs are HIGH.

Note that this says that the output will go to its active state only when *all* of the inputs are in their active states. The word *all* is used because of the AND symbol.

FIGURE 3-34 Interpretation of the two NAND gate symbols.

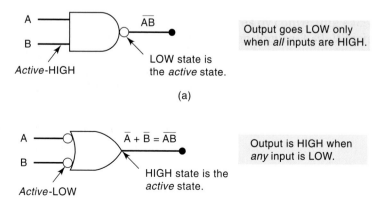

The alternate symbol for a NAND gate shown in Figure 3-34(b) has an active-HIGH output and active-LOW inputs, and so its operation can be stated as follows:

The output goes HIGH when *any input* is LOW.

This says that the output will be in its active state whenever *any* of the inputs is in its active state. The word *any* is used because of the OR symbol.

With a little thought, you can see that the two interpretations for the NAND symbols in Figure 3-34 are different ways of saying the same thing.

Summary

At this point you are probably wondering why there is a need to have two different symbols and interpretations for each logic gate. We hope the reasons will become clear after reading the next section. For now, let us summarize the important points concerning the logic-gate representations.

1. To obtain the alternate symbol for a logic gate, take the standard symbol and change its operation symbol (OR to AND, or AND to OR), and change the bubbles on both inputs and output (i.e., delete bubbles that are present, and add bubbles where there are none).

2. To interpret the logic-gate operation, first note which logic state, 0 or 1, is the active state for the inputs and which is the active state for the output. Then realize that the output's active state is produced by having *all* of the inputs in their active state (if an AND symbol is used) or by having *any* of the inputs in its active state (if an OR symbol is used).

EXAMPLE 3-19

Give the interpretation of the two OR gate symbols.

Solution

The results are shown in Figure 3-35. Note that the word *any* is used when the operation symbol is an OR symbol and the word *all* is used when it includes an AND symbol.

FIGURE 3-35 Interpretation of the two OR gate symbols.

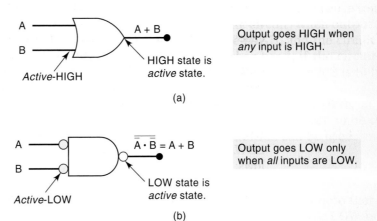

(a)

(b)

lows you to develop building blocks using schematic entry, AHDL, VHDL, and other methods and then interconnect them to form a complete system.

Other HDLs are available that are more suitable for programming simple programmable logic devices. The three most common are CUPL, ABEL, and PALASM. You will find any of these languages easy to use after learning the basics of AHDL or VHDL as covered in this text.

Computer Programming Languages

It is important to distinguish between hardware description languages intended to describe the hardware configuration of a circuit and programming languages that represent a sequence of instructions intended to be carried out by a computer to accomplish some task. In both cases, we use a *language* to *program* a device. However, computers operate much differently than logic circuits. Computers operate by following a laundry list of tasks (i.e., instructions, or "the program"), each of which must be done in sequential order. The speed of operation is determined by how fast the computer can execute each instruction. For example, if a computer were to respond to four different inputs, it would require at least four separate instructions (sequential tasks) to detect and identify if any input changed state. A digital logic circuit, on the other hand, is limited in its speed only by how quickly the circuitry can change the outputs in response to changes in the inputs. It is monitoring all inputs **concurrently** (at the same time) and responding to any changes.

The following analogy will help you understand the difference between computer operation and digital logic circuit operation and the role of language elements used to describe what the systems do. Consider the challenge of describing what is done to an Indy 500 car during a pit stop. If a single person performed all the necessary tasks one at a time, he or she would need to be very fast at each task. This is the way a computer operates: one task at a time but very quickly. Of course, at Indy, there is an entire pit crew that swarms the car, and each member of the crew does his or her task while the others do theirs. All crew members operate concurrently, like the elements of a digital circuit. Now consider how you would describe to someone else what is being done to the Indy car during the pit stop using (1) the individual-mechanic approach or (2) the pit-crew approach. Wouldn't the two English language descriptions of what is being done sound very similar? As we will see, the languages used to describe digital hardware (HDL) are very similar to languages that describe computer programs (e.g., BASIC, C, JAVA), even though the resulting implementation operates quite differently. Knowledge of any of these computer programming languages is not necessary to understand HDL. The important thing is that when you have learned both an HDL and a computer language, you must understand their different roles in digital systems.

EXAMPLE 3-25

Compare the operation of a computer and a logic circuit in performing the simple logical operation of $y = AB$.

Solution

The logic circuit is a simple AND gate. The output y will be HIGH within approximately 10 nanoseconds of the point when A and B are HIGH simultaneously. Within approximately 10 nanoseconds after either input goes LOW, the output y will be LOW.

The computer must run a program of instructions that makes decisions. Suppose each instruction takes 20 ns (that's pretty fast!). Each shape in the flowchart shown in Figure 3-43 represents one instruction. Clearly, it will take a minimum of two or three instructions (40–60 ns) to respond to changes in the inputs.

FIGURE 3-43 Decision process of a computer program.

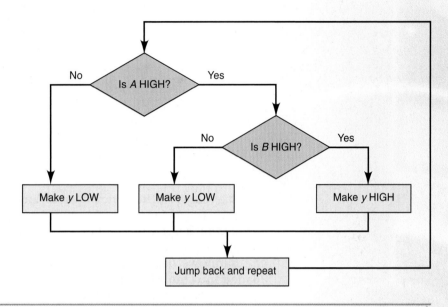

REVIEW QUESTIONS

1. What does HDL stand for?
2. What is the purpose of an HDL?
3. What is the purpose of a computer programming language?
4. What is the key difference between HDL and computer programming languages?

3-18 IMPLEMENTING LOGIC CIRCUITS WITH PLDs

Many digital circuits today are implemented using programmable logic devices (PLDs). These devices are not like microcomputers or microcontrollers that "run" the program of instructions. Instead, they are configured electronically, and their internal circuits are "wired" together electronically to form a logic circuit. This programmable wiring can be thought of as thousands of connections that are either connected (1) or not connected (0). Figure 3-44 shows a small area of programmable connections. Each intersection between a row (horizontal wire) and a column (vertical wire) is a programmable connection. You can imagine how tedious it would be to try to configure these devices by placing 1s and 0s in a grid manually (which is how they did it back in the 1970s).

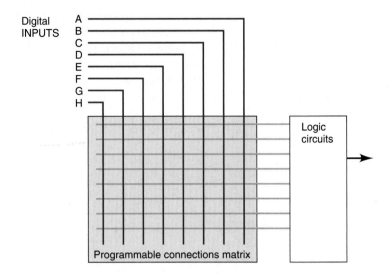

FIGURE 3-44 Configuring hardware connections with programmable logic devices.

The role of the hardware description language is to provide a concise and convenient way for the designer to describe the operation of the circuit in a format that a personal computer can handle and store conveniently. The computer runs a special software application called a **compiler** to translate from the hardware description language into the grid of 1s and 0s that can be loaded into the PLD. If a person can master the higher-level hardware description language, it actually makes programming the PLDs much easier than trying to use Boolean algebra, schematic drawings, or truth tables. In much the same way that you learned the English language, we will start by expressing simple things and gradually learn the more complicated aspects of these languages. Our objective is to learn enough of HDL to be able to communicate with others and perform simple tasks. A full understanding of the intricacies of these languages is beyond the scope of this text and can really be mastered only by regular use.

In the sections throughout this book that cover the HDLs, we will present both AHDL and VHDL in a format that allows you to skip over one language and concentrate on the other without missing important information. Of course, this setup means there will be some redundant information presented if you choose to read about both languages. We feel this redundancy is worth the extra effort to provide you with the flexibility of focusing on either of the two languages or learning both by comparing and contrasting similar examples. The recommended way to use the text is to focus on one language. It is true that the easiest way to become bilingual, and fluent in both languages, is to be raised in an environment where both languages are spoken routinely. It is also very easy, however, to confuse details, so we will keep the specific examples separate and independent. We hope this format provides you with the opportunity to learn one language now and then use this book as a reference later in your career should you need to pick up the second language.

REVIEW QUESTIONS

1. What does PLD stand for?
2. How are the circuits reconfigured electronically in a PLD?
3. What does a compiler do?

3-19 HDL FORMAT AND SYNTAX

Any language has its unique properties, similarities to other languages, and its proper syntax. When we study grammar in our primary and secondary education, we learn conventions such as the order of words as elements in a sentence and proper punctuation. This is referred to as the **syntax** of language. A language designed to be interpreted by a computer must follow strict rules of syntax. A computer is just an assortment of processed beach sand and wire that has no idea what you "meant" to say, so you must present the instructions using the exact syntax that the computer language expects and understands. The basic format of any hardware circuit description (in any language) involves two vital elements:

1. The definition of what goes into it and what comes out of it (i.e., input/output specs)
2. The definition of how the outputs respond to the inputs (i.e., its operation)

A circuit schematic diagram such as Figure 3-45 can be read and understood by a competent engineer or technician because both would understand the meaning of each symbol in the drawing. If you understand how each element works and how the elements are connected to each other, you can understand how the circuit operates. On the left side of the diagram is the set of inputs, and on the right is the set of outputs. The symbols in the middle define its operation. The text-based language must convey the same information. All HDLs use the format shown in Figure 3-46.

FIGURE 3-45 A schematic diagram description.

FIGURE 3-46 Format of HDL files.

In a text-based language, the circuit being described must be given a name. The inputs and outputs (sometimes called ports) must be assigned names and defined according to the nature of the port. Is it a single bit from a toggle switch? Or is it a four-bit number coming from a keypad? The text-based language must somehow convey the nature of these inputs and outputs. The **mode** of a port defines whether it is input, output, or both. The **type** refers to the number of bits and how those bits are grouped and interpreted. If the *type* of input is a single bit, then it can have only two possible values: 0 and 1. If the type of input is a 4-bit binary number from a keypad, it can have any one of 16 different values (0000_2–1111_2). The type determines the

AHDL

range of possible values. The definition of the circuit's operation in a text-based language is contained in a set of statements that follow the circuit input/output (I/O) definition. The following two sections describe the very simple circuit of Figure 3-45 and illustrate the critical elements of AHDL and VHDL.

BOOLEAN DESCRIPTION USING AHDL

Refer to Figure 3-47. The keyword **SUBDESIGN** gives a name to the circuit block, which in this case is *and_gate*. The name of the file must also be and_gate.tdf. Notice that the keyword SUBDESIGN is capitalized. This is not mandatory, but use of a consistent style in capitalization makes the code much easier to read. The style guide that is provided with the Altera compiler for AHDL suggests the use of capital letters for the keywords in the language. Variables that are named by the designer should be lowercase.

FIGURE 3-47   Essential elements in AHDL.

```
SUBDESIGN and_gate
(
    a, b        :INPUT;
    y           :OUTPUT;
)
BEGIN
    y = a & b;
END;
```

The SUBDESIGN section defines the inputs and outputs of the logic circuit block. Something must enclose the circuit that we are trying to describe, much the same way that a block diagram encloses everything that makes up that part of the design. In AHDL, this input/output definition is enclosed in parentheses. The list of variables used for inputs to this block are separated by commas and followed by :INPUT;. In AHDL, the single-bit type is assumed unless the variable is designated as multiple bits. The single-output bit is declared with the mode :OUTPUT;. We will learn the proper way to describe other types of inputs, outputs, and variables as we need to use them.

The set of statements that describe the operation of the AHDL circuit are contained in the logic section between the keywords BEGIN and END. In this example, the operation of the hardware is described by a very simple Boolean algebra equation that states that the output (y) is assigned ($=$) the logic level produced by a AND b. This Boolean algebra equation is referred to as a **concurrent assignment statement**. Any statements (there is only one in this example) between BEGIN and END are evaluated constantly and concurrently. The order in which they are listed makes no difference. The basic Boolean operators are:

    &   AND
    #   OR
    !   NOT
    $ XOR

REVIEW QUESTIONS

1. What appears inside the parentheses ( ) after SUBDESIGN?
2. What appears between BEGIN and END?

BOOLEAN DESCRIPTION USING VHDL

Refer to Figure 3-48. The keyword **ENTITY** gives a name to the circuit block, which in this case is and_gate. Notice that the keyword ENTITY is capitalized but and_gate is not. This is not mandatory, but use of a consistent style in capitalization makes the code much easier to read. The style guide provided with the Altera compiler for VHDL suggests using capital letters for the keywords in the language. Variables that are named by the designer should be lowercase.

FIGURE 3-48 Essential elements in VHDL.

```
ENTITY and_gate IS
PORT (    a, b    :IN BIT;
          y       :OUT BIT);
END and_gate;
ARCHITECTURE ckt OF and_gate IS
BEGIN
        y <= a AND b;
END ckt;
```

The ENTITY declaration can be thought of as a block description. Something must enclose the circuit we are trying to describe, much the same way a block diagram encloses everything that makes up that part of the design. In VHDL, the keyword PORT tells the compiler that we are defining inputs and outputs to this circuit block. The names used for inputs (separated by commas) are listed, ending with a colon and a description of the mode and type of input (:IN BIT;). In VHDL, the **BIT** description tells the compiler that each variable in the list is a single bit. We will learn the proper way to describe other types of inputs, outputs, and variables as we need to use them. The line containing END and_gate; terminates the ENTITY declaration.

The **ARCHITECTURE** declaration is used to describe the operation of everything inside the block. The designer makes up a name for this architectural description of the inner workings of the ENTITY block (*ckt* in this example). Every ENTITY must have at least one ARCHITECTURE associated with it. The words OF and IS are keywords in this declaration. The body of the architecture description is enclosed between the BEGIN and END keywords. END is followed by the name that has been assigned to this architecture. Within the body (between BEGIN and END) is the description of the block's operation. In this example, the operation of the hardware is described by a very simple Boolean algebra equation that states that the output (y) is assigned ($<=$) the logic level produced by a AND b. This is referred to as a **concurrent assignment statement**, which means that all the statements (there is only one in this example) between BEGIN and END are evaluated constantly and concurrently. The order in which they are listed makes no difference.

REVIEW QUESTIONS

1. What is the role of the ENTITY declaration?
2. Which key section defines the operation of the circuit?
3. What is the assignment operator used to give a value to a logic signal?

3-20   INTERMEDIATE SIGNALS

In many designs, there is a need to define signal points "inside" the circuit block. They are points in the circuit that are neither inputs nor outputs for the block but may be useful as a reference point. It may be a signal that needs to be connected to many other places within the block. In an analog or digital schematic diagram, they would be called test points or *nodes*. In an HDL, they are referred to as **buried nodes** or **local signals**. Figure 3-49 shows a very simple circuit that uses an intermediate signal named *m.* In the HDL, these nodes (signals) are not defined with the inputs and outputs but rather in the section that describes the operation of the block. The inputs and outputs are available to other circuit blocks in the system, but these local signals are recognized only within this block.

FIGURE 3-49   A logic circuit diagram with an intermediate variable.

In the example code that follows, notice the information at the top. The purpose of this information is strictly for documentation purposes. It is absolutely vital that the design is documented thoroughly. At a minimum, it should describe the project it is being used in, who wrote it, and the date. This information is often referred to as a header. We are keeping our headers brief to make this book a little lighter to carry to class, but remember: memory space is cheap and information is valuable. So don't be afraid to *document thoroughly!* There are also comments next to many of the statements in the code. These comments help the designer remember what she or he was trying to do and to help any other person to understand what was intended.

AHDL BURIED NODES

The AHDL code that describes the circuit in Figure 3-49 is shown in Figure 3-50. The **comments** in AHDL can be enclosed between % characters, as you can see in the figure between lines 1 and 4. This section of the code allows the designer to write many lines of information that will be ignored by computer programs using this file but can be read by any person trying to decipher the code. Notice that the comments at the end of lines 9, 10, 13, 15, and 16 are preceded by two dashes (--). The text following the dashes to the end of the line is for documentation only. Either type of comment symbol may be used, but percent signs must be used in pairs to open and close a comment. Double dashes indicate a comment that extends to the end of the line.

FIGURE 3-50
Intermediate variables in
AHDL described in
Figure 3-49.

```
1     %  Intermediate variables in AHDL (Figure 3-49)
2        Digital Systems 9th ed
3        NS Widmer
4        MAY 22, 2002              %
5     SUBDESIGN fig3_50
6     (
7        a,b,c       :INPUT;    -- define inputs to block
8        y           :OUTPUT;   -- define block output
9     )
10    VARIABLE
11       m          :NODE;      -- name an intermediate signal
12    BEGIN
13       m = a & b;             -- generate buried product term
14       y = m # c;             -- generate sum on output
15    END;
```

In AHDL, local signals are declared in the VARIABLE section, which is placed between the SUBDESIGN section and the logic section. The intermediate signal m is defined on line 11, following the keyword **VARIABLE**. The keyword **NODE** designates the nature of the variable. Notice that a colon separates the variable name from its node designation. In the hardware description on line 13, the intermediate variable is assigned (connected to) a value ($m = a \& b;$) and then m is used in the second statement on line 14 to assign (connect) a value to y ($y = m \# c;$). Remember that the assignment statements are concurrent and, thus, the order in which they are given does not matter. For human readability, it may seem more logical to assign values to intermediate variables before they are utilized in other assignment statements, as shown here.

REVIEW QUESTIONS

1. What is the designation used for intermediate variables?
2. Where are these variables declared?
3. Does it matter whether the m or y equation comes first?
4. What character is used to limit a block of comments?
5. What characters are used to comment a single line?

VHDL LOCAL SIGNALS

The VHDL code that describes the circuit in Figure 3-49 is shown in Figure 3-51. The **comments** in VHDL follow two dashes (--). Typing two successive dashes allows the designer to write information from that point to the end of the line. The information following the two successive dashes will be ignored by computer programs using this file, but can be read by any person trying to decipher the code.

VHDL

FIGURE 3-51
Intermediate signals in
VHDL described in Figure
3-49.

```
1    -- Intermediate variables in VHDL (Figure 3-49)
2    -- Digital Systems 9th ed
3    -- NS Widmer
4    -- MAY 22, 2002
5
6    ENTITY  fig3_51 IS
7    PORT( a, b, c   :IN BIT;     -- define inputs to block
8    y                :OUT BIT);  -- define block output
9    END fig3_51;
10
11   ARCHITECTURE ckt OF fig3_51 IS
12
13      SIGNAL m     :BIT;        -- name an intermediate signal
14
15   BEGIN
16       m <= a AND b;            -- generate buried product term
17       y <= m OR c;             -- generate sum on output
18   END ckt;
```

The intermediate signal m is defined on line 13 following the keyword
SIGNAL. The keyword BIT designates the type of the signal. Notice that a
colon separates the signal name from its type designation. In the hardware
description on line 16, the intermediate signal is assigned (connected to) a
value ($m <=a$ AND b;) and then m is used in the statement on line 17 to assign (connect) a value to y ($y <=m$ OR c;). Remember that the assignment
statements are concurrent and, thus, the order in which they are given does
not matter. For human readability, it may seem more logical to assign values
to intermediate signals before they are utilized in other assignment statements, as shown here.

REVIEW QUESTIONS

1. What is the designation used for intermediate signals?
2. Where are these signals declared?
3. Does it matter whether the m or y equation comes first?
4. What characters are used to comment a single line?

SUMMARY

1. Boolean algebra is a mathematical tool used in the analysis and design of
   digital circuits.
2. The basic Boolean operations are the OR, AND, and NOT operations.
3. An OR gate produces a HIGH output when any input is HIGH. An AND
   gate produces a HIGH output only when all inputs are HIGH. A NOT circuit (INVERTER) produces an output that is the opposite logic level compared to the input.

4. A NOR gate is the same as an OR gate with its output connected to an INVERTER. A NAND gate is the same as an AND gate with its output connected to an INVERTER.

5. Boolean theorems and rules can be used to simplify the expression of a logic circuit and can lead to a simpler way of implementing the circuit.

6. NAND gates can be used to implement any of the basic Boolean operations. NOR gates can be used likewise.

7. Either standard or alternate symbols can be used for each logic gate, depending on whether the output is to be active-HIGH or active-LOW.

8. The IEEE/ANSI standard for logic symbols uses rectangular symbols for each logic device, with special notations inside the rectangles to show how the outputs depend on the inputs.

9. Hardware description languages have become an important method of describing digital circuits.

10. HDL code should always contain comments that document its vital characteristics so a person reading it later can understand what it does.

11. Every HDL circuit description contains a definition of the inputs and outputs, followed by a section that describes the circuit's operation.

12. In addition to inputs and outputs, intermediate connections that are buried within the circuit can be defined. These intermediate connections are called nodes or signals.

IMPORTANT TERMS

logic level
Boolean algebra
truth table
OR operation
OR gate
AND operation
AND gate
NOT operation
inversion
    (complementation)
NOT circuit
    (INVERTER)
NOR gate
NAND gate
Boolean theorems
DeMorgan's theorems
alternate logic
    symbols

active logic levels
active-HIGH
active-LOW
asserted
unasserted
IEEE/ANSI
hardware description
    languages (HDLs)
Altera hardware
    description
    language (AHDL)
very high speed
    integrated circuit
    (VHSIC) hardware
    description
    language (VHDL)
programmable logic
    devices (PLDs)

concurrent
compiler
syntax
mode
type
SUBDESIGN
concurrent
    assignment
    statement
ENTITY
BIT
ARCHITECTURE
buried nodes (local
    signals)
comments
VARIABLE
NODE

PROBLEMS

The **blue** letters preceding some of the problems are used to indicate the nature or type of problem as follows:

B basic problem

T troubleshooting problem

D design or circuit-modification problem

N new concept or technique not covered in text

C challenging problem

H HDL problem

SECTION 3-3

B   3-1. Draw the output waveform for the OR gate of Figure 3-52.

FIGURE 3-52

3-2. Suppose that the A input in Figure 3-52 is unintentionally shorted to ground (i.e., $A = 0$). Draw the resulting output waveform.

3-3. Suppose that the A input in Figure 3-52 is unintentionally shorted to the $+5$ V supply line (i.e., $A = 1$). Draw the resulting output waveform.

3-4. Read the statements below concerning an OR gate. At first, they may appear to be valid, but after some thought you should realize that neither one is *always* true. Prove this by showing a specific example to refute each statement.

(a) If the output waveform from an OR gate is the same as the waveform at one of its inputs, the other input is being held permanently LOW.

(b) If the output waveform from an OR gate is always HIGH, one of its inputs is being held permanently HIGH.

3-5. How many different sets of input conditions will produce a HIGH output from a five-input OR gate?

SECTION 3-4

B   3-6. Change the OR gate in Figure 3-52 to an AND gate.

(a) Draw the output waveform.

(b) Draw the output waveform if the A input is permanently shorted to ground.

(c) Draw the output waveform if A is permanently shorted to $+5$ V.

D   3-7. Refer to Figure 3-4. Modify the circuit so that the alarm is to be activated only when the pressure and the temperature exceed their maximum limits at the same time.

B   3-8. Change the OR gate in Figure 3-6 to an AND gate and draw the output waveform.

3-9. Suppose that you have an unknown two-input gate that is either an OR gate or an AND gate. What combination of input levels should you apply to the gate's inputs to determine which type of gate it is?

3-10. *True or false:* No matter how many inputs it has, an AND gate will produce a HIGH output for only one combination of input levels.

SECTIONS 3-5 TO 3-7

B   3-11. Apply the A waveform from Figure 3-23 to the input of an INVERTER. Draw the output waveform. Repeat for waveform B.

B 3-12. (a) Write the Boolean expression for output x in Figure 3-53(a). Determine the value of x for all possible input conditions, and list the values in a truth table.

(b) Repeat for the circuit in Figure 3-53(b).

FIGURE 3-53

(a)

(b)

3-13. Determine the complete truth table for the circuit of Figure 3-15(b) by finding the logic levels present at each gate output for each of the 32 possible input combinations.

3-14. Change each OR to an AND, and each AND to an OR, in Figure 3-15(b). Then write the expression for the output.

3-15. Determine the complete truth table for the circuit of Figure 3-16 by finding the logic levels present at each gate output for each of the 16 possible combinations of input levels.

SECTION 3-8

B 3-16. For each of the following expressions, construct the corresponding logic circuit, using AND and OR gates and INVERTERs.

(a) $x = \overline{AB(C + D)}$

(b) $z = \overline{A + B + \overline{CD\overline{E}}) + \overline{BC\overline{D}}$

(c) $y = (\overline{M + N} = \overline{P}Q)$

(d) $x = \overline{W + P\overline{Q}}$

(e) $z = MN(P + \overline{N})$

(f) $x = (A + B)(\overline{A} + \overline{B})$

SECTION 3-9

B   3-17. (a) Apply the input waveforms of Figure 3-54 to a NOR gate, and draw the output waveform.

(b) Repeat with C held permanently LOW.

(c) Repeat with C held HIGH.

FIGURE 3-54

B   3-18. Repeat Problem 3-17 for a NAND gate.

C   3-19. Write the expression for the output of Figure 3-55, and use it to determine the complete truth table. Then apply the waveforms of Figure 3-54 to the circuit inputs, and draw the resulting output waveform.

FIGURE 3-55

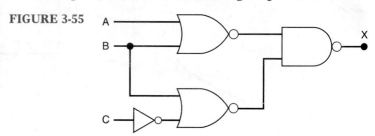

3-20. Determine the truth table for the circuit of Figure 3-24.

3-21. Modify the circuits that were constructed in Problem 3-16 so that NAND gates and NOR gates are used wherever appropriate.

SECTION 3-10

B   3-22. DRILL QUESTION

Complete each expression.

(a) $A + 1 =$ _____   (f) $D \cdot 1 =$ _____

(b) $A \cdot A =$ _____   (g) $D + 0 =$ _____

(c) $B \cdot \overline{B} =$ _____   (h) $C + \overline{C} =$ _____

(d) $C + C =$ _____   (i) $G + GF =$ _____

(e) $x \cdot 0 =$ _____   (j) $y + \overline{w}y =$ _____

3-23. Prove theorems (15a) and (15b) by trying all possible cases.

C   3-24. (a) Simplify the following expression using theorems (13b), (3), and (4):

$$x = (M + N)(\overline{M} + P)(\overline{N} + \overline{P})$$

(b) Simplify the following expression using theorems (13a), (8), and (6):

$$z = \overline{A}B\overline{C} + AB\overline{C} + B\overline{C}D$$

SECTIONS 3-11 AND 3-12

3-25. Prove DeMorgan's theorems by trying all possible cases.

B 3-26. Simplify each of the following expressions using DeMorgan's theorems.

(a) $\overline{\overline{AB}\,\overline{C}}$ (d) $\overline{A + \overline{\overline{B}}}$ (g) $\overline{A(\overline{B + \overline{C}})D}$

(b) $\overline{A + \overline{BC}}$ (e) $\overline{\overline{\overline{AB}}}$ (h) $\overline{(M + \overline{N})(\overline{M} + N)}$

(c) $\overline{\overline{A}\overline{B}\overline{C}D}$ (f) $\overline{\overline{A} + \overline{C} + \overline{D}}$ (i) $\overline{\overline{A}\,\overline{BCD}}$

3-27. Use DeMorgan's theorems to simplify the expression for the output of Figure 3-55.

C 3-28. Convert the circuit of Figure 3-53(b) to one using only NAND gates. Then write the output expression for the new circuit, simplify it using DeMorgan's theorems, and compare it with the expression for the original circuit.

3-29. Convert the circuit of Figure 3-53(a) to one using only NOR gates. Then write the expression for the new circuit, simplify it using DeMorgan's theorems, and compare it with the expression for the original circuit.

B 3-30. Show how a two-input NAND gate can be constructed from two-input NOR gates.

B 3-31. Show how a two-input NOR gate can be constructed from two-input NAND gates.

3-32. A jet aircraft employs a system for monitoring the rpm, pressure, and temperature values of its engines using sensors that operate as follows:

RPM sensor output = 0 only when speed < 4800 rpm
P sensor output = 0 only when pressure < 220 psi
T sensor output = 0 only when temperature < 200° F

Figure 3-56 shows the logic circuit that controls a cockpit warning light for certain combinations of engine conditions. Assume that a HIGH at output W activates the warning light.

(a) Determine what engine conditions will give a warning to the pilot.

(b) Change this circuit to one using all NAND gates.

FIGURE 3-56

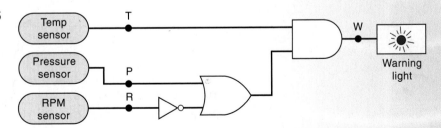

SECTIONS 3-13 AND 3-14

B   3-33. Draw the standard representations for each of the basic logic gates. Then draw the alternate representations.

3-34. For each statement below, draw the appropriate logic-gate symbol—standard or alternate—for the given operation.

(a) A HIGH output occurs only when all three inputs are LOW.

(b) A LOW output occurs when any of the four inputs is LOW.

(c) A LOW output occurs only when all eight inputs are HIGH.

3-35. The circuit of Figure 3-55 is supposed to be a simple digital combination lock whose output will generate an active-LOW $\overline{UNLOCK}$ signal for only one combination of inputs.

(a) Modify the circuit diagram so that it represents more effectively the circuit operation.

(b) Use the new circuit diagram to determine the input combination that will activate the output. Do this by working back from the output using the information given by the gate symbols, as was done in Examples 3-22 and 3-23. Compare the results with the truth table obtained in Problem 3-19.

3-36. (a) Determine the input conditions needed to activate output Z in Figure 3-37(b). Do this by working back from the output, as was done in Examples 3-22 and 3-23.

(b) Assume that it is the LOW state of Z that is to activate the alarm. Change the circuit diagram to reflect this, and then use the revised diagram to determine the input conditions needed to activate the alarm.

D   3-37. Modify the circuit of Figure 3-40 so that $A_1 = 0$ is needed to produce $DRIVE = 1$ instead of $A_1 = 1$.

3-38. Determine the input conditions needed to cause the output in Figure 3-57 to go to its active state.

FIGURE 3-57

3-39. Use the results of Problem 3-38 to obtain the complete truth table for the circuit of Figure 3-57.

B   3-40. What is the asserted state for the output of Figure 3-57? For the output of Figure 3-36(c)?

3-41. Figure 3-58 shows an application of logic gates that simulates a two-way switch like the ones used in our homes to turn a light on or off from two different switches. Here the light is an LED that will be ON (conducting) when the NOR gate output is LOW. Note that this output is labeled $\overline{LIGHT}$ to indicate that it is active-LOW. Determine the input conditions needed to turn on the LED. Then verify that the circuit operates as a two-way switch using switches *A* and *B*. (In Chapter 4, you will learn how to design circuits like this one to produce a given relationship between inputs and outputs.)

FIGURE 3-58

SECTION 3-15

3-42. Redraw the circuits of (a) Figure 3-57 and (b) Figure 3-58 using the IEEE/ANSI symbols.

SECTION 3-17

HDL DRILL QUESTIONS

H   3-43. *True or false:*

   (a) VHDL is a computer programming language.
   (b) VHDL can accomplish the same thing as AHDL.
   (c) AHDL is an IEEE standard language.
   (d) Each intersection in a switch matrix can be programmed as an open or short circuit between a row and column wire.
   (e) The first item that appears at the top of an HDL listing is the functional description.
   (f) The type of an object indicates if it is an input or an output.
   (g) The mode of an object determines if it is an input or an output.
   (h) Buried nodes are nodes that have been deleted and will never be used again.
   (i) Local signals are another name for intermediate variables.
   (j) The header is a block of comments that document vital information about the project.

SECTION 3-18

B   3-44. Redraw the programmable connection matrix from Figure 3-44. Label the output signals (horizontal lines) from the connection matrix (from top row to bottom row) as follows: AAABADHE. Draw an X in the appropriate intersections to short-circuit a row to a column and create these connections to the logic circuit.

H   3-45. Write the HDL code in the language of your choice that will produce the following output functions:

$$X = A + B$$
$$Y = AB$$
$$Z = A + B + C$$

H 3-46. Write the HDL code in the language of your choice that will implement the logic circuit of Figure 3-39.

(a) Use a single Boolean equation.

(b) Use the intermediate variables V, W, X, and Y.

MICROCOMPUTER APPLICATION

C 3-47. Refer to Figure 3-40 in Example 3-23. Inputs A_7 through A_0 are *address* inputs that are supplied to this circuit from outputs of the microprocessor chip in a microcomputer. The eight-bit address code A_7 to A_0 selects which device the microprocessor wants to activate. In Example 3-23, the required address code to activate the disk drive was A_7 through $A_0 = 11111110_2 = FE_{16}$.

Modify the circuit so that the microprocessor must supply an address code of $4A_{16}$ to activate the disk drive.

CHALLENGING EXERCISES

C 3-48. Show how $x = AB\overline{C}$ can be implemented with one two-input NOR and one two-input NAND gate.

C 3-49. Implement $y = ABCD$ using only two-input NAND gates.

ANSWERS TO SECTION REVIEW QUESTIONS

SECTION 3-2

1. $x = 1$ 2. $x = 0$ 3. 32

SECTION 3-3

1. All inputs LOW   2. $x = A + B + C + D + E + F$ 3. Constant HIGH

SECTION 3-4

1. All five inputs $= 1$ 2. A LOW input will keep the output LOW.   3. False; see truth table of each gate.

SECTION 3-5

1. Output of second INVERTER will be same as input A.   2. y will be LOW only for $A = B = 1$.

SECTION 3-6

1. $x = \overline{A} + B + C + \overline{AD}$ 2. $x = D(\overline{AB + C}) + E$

SECTION 3-7

1. $x = 1$ 2. $x = 1$ 3. $x = 1$ for both.

SECTION 3-8

1. See Figure 3-15(a).   2. See Figure 3-17(b).   3. See Figure 3-15(b).

SECTION 3-9

1. All inputs LOW.   2. $x = 0$ 3. $x = \overline{A + B + \overline{CD}}$

SECTION 3-10

1. $y = A\overline{C}$ 2. $y = \overline{A}\,\overline{B}D$ 3. $y = \overline{A}D + BD$

SECTION 3-11

1. $z = \overline{AB} + C$ 2. $y = (\overline{R} + S + \overline{T})Q$ 3. Same as Figure 3-28 except NAND is replaced by NOR.   4. $y = \overline{A}B(C + \overline{D})$

SECTION 3-12

1. Three.   2. NOR circuit is more efficient because it can be implemented with one 74LS02 IC.   3. $x = \overline{(\overline{AB})\,(\overline{CD})} = AB + \overline{(\overline{CD})} + AB + CD$

SECTION 3-13

1. Output goes LOW when any input is HIGH.   2. Output goes HIGH only when all inputs are LOW.   3. Output goes LOW when any input is LOW.   4. Output goes HIGH only when all inputs are HIGH.

SECTION 3-14

1. Z will go HIGH when $A = B = 0$ and $C = D = 1$.   2. Z will go LOW when $A = B = 0$, $E = 1$, and either C or D or both are 0.   3. Two   4. Two   5. LOW
6. $A = B = 0$, $C = D = 1$ 7. $\overline{W}$

SECTION 3-15

1. See Figure 3-41.   2. Rectangle with & inside, and triangles on inputs.

SECTION 3-16

1. Boolean equation, truth table, logic diagram, timing diagram, language.

SECTION 3-17

1. Hardware description language   2. To describe a digital circuit and its operation.   3. To give a computer a sequential list of tasks.   4. HDL describes concurrent hardware circuits; computer instructions execute one at a time.

SECTION 3-18

1. Programmable logic device   2. By making and breaking connections in a switching matrix   3. It translates HDL code into a pattern of bits to configure the switching matix.

SECTION 3-19

AHDL

1. The input and output definitions.   2. The description of how it operates.

VHDL

1. To give a name to the circuit and define its inputs and outputs.   2. The ARCHITECTURE description.   3. <=

SECTION 3-20

AHDL

1. NODE   2. After the I/O definition and before BEGIN.   3. No   4. %   5. --

VHDL

1. SIGNAL   2. Inside ARCHITECTURE before BEGIN.   3. No   4. --

C H A P T E R   4

COMBINATIONAL LOGIC CIRCUITS

■ OUTLINE

■ OBJECTIVES

Upon completion of this chapter, you will be able to:

■ Convert a logic expression into a sum-of-products expression.

■ Perform the necessary steps to reduce a sum-of-products expression to its simplest form.

■ Use Boolean algebra and the Karnaugh map as tools to simplify and design logic circuits.

■ Explain the operation of both exclusive-OR and exclusive-NOR circuits.

■ Design simple logic circuits without the help of a truth table.

■ Implement enable circuits.

■ Cite the basic characteristics of TTL and CMOS digital ICs.

■ Use the basic troubleshooting rules of digital systems.

■ Deduce from observed results the faults of malfunctioning combinational logic circuits.

■ Describe the fundamental idea of programmable logic devices (PLDs).

■ Outline the steps involved in programming a PLD to perform a simple combinational logic function.

■ Go to the Altera user manuals to acquire the information needed to do a simple programming experiment in the lab.

■ Describe hierarchical design methods.

■ Identify proper data types for single-bit, bit array, and numeric value variables.

■ Describe logic circuits using HDL control structures IF/ELSE, IF/ELSIF, and CASE.

■ Select the appropriate control structure for a given problem.

■ INTRODUCTION

In Chapter 3, we studied the operation of all the basic logic gates, and we used Boolean algebra to describe and analyze circuits that were made up of combinations of logic gates. These circuits can be classified as *combinational* logic circuits because, at any time, the logic level at the output depends on the combination of logic levels present at the inputs. A combinational circuit has no *memory* characteristic, so its output depends *only* on the current value of its inputs.

In this chapter, we will continue our study of combinational circuits. To start, we will go further into the simplification of logic circuits. Two methods will be used: one uses Boolean algebra theorems; the other uses a *mapping* technique. In addition, we will study simple techniques for designing combinational logic circuits to satisfy a given set of requirements. A complete study of logic-circuit design is not one of our objectives, but the methods we introduce will provide a good introduction to logic design.

A good portion of the chapter is devoted to the troubleshooting of combinational circuits. This first exposure to troubleshooting should begin to develop the type of analytical skills needed for successful troubleshooting. To make this material as practical as possible, we will first present some of the basic characteristics of logic-gate ICs in the TTL and CMOS logic families along with a description of the most common types of faults encountered in digital IC circuits.

In the last sections of this chapter, we will extend our knowledge of programmable logic devices and hardware description languages. The concept of programmable hardware connections will be reinforced, and we will provide more details regarding the role of the development system. You will learn the steps followed in the design and development of digital systems today. Enough information will be provided to allow you to choose the correct types of data objects for use in simple projects to be presented later in this text. Finally, several control structures will be explained, along with some instruction regarding their appropriate use.

4-1   SUM-OF-PRODUCTS FORM

The methods of logic-circuit simplification and design that we will study require the logic expression to be in a **sum-of-products (SOP)** form. Some examples of this form are:

1. $ABC + \overline{A}B\overline{C}$
2. $AB + \overline{A}B\overline{C} + \overline{C}\overline{D} + D$
3. $\overline{A}B + C\overline{D} + EF + GK + H\overline{L}$

Each of these sum-of-products expressions consists of two or more AND terms (products) that are ORed together. Each AND term consists of one or more variables *individually* appearing in either complemented or uncomplemented form. For example, in the sum-of-products expression $ABC + \overline{A}B\overline{C}$, the first AND product contains the variables A, B, and C in their uncomplemented (not inverted) form. The second AND term contains A and C in their complemented (inverted) form. Note that in a sum-of-products expression, one inversion sign *cannot* cover more than one variable in a term (e.g., we cannot have $\overline{ABC}$ or $\overline{RST}$).

Product-of-Sums

Another general form for logic expressions is sometimes used in logic-circuit design. Called the **product-of-sums (POS)** form, it consists of two or more OR terms (sums) that are ANDed together. Each OR term contains one or more variables in complemented or uncomplemented form. Here are some product-of-sum expressions:

1. $(A + \overline{B} + C)(A + C)$
2. $(A + \overline{B})(\overline{C} + D)F$
3. $(A + C)(B + \overline{D})(\overline{B} + C)(A + \overline{D} + \overline{E})$

The methods of circuit simplification and design that we will be using are based on the sum-of-products (SOP) form, so we will not be doing much with the product-of-sums (POS) form. It will, however, occur from time to time in some logic circuits that have a particular structure.

REVIEW QUESTIONS

1. Which of the following expressions is in SOP form?
   (a) $AB + CD + E$
   (b) $AB(C + D)$
   (c) $(A + B)(C + D + F)$
   (d) $\overline{MN} + PQ$
2. Repeat question 1 for the POS form.

4-2 SIMPLIFYING LOGIC CIRCUITS

Once the expression for a logic circuit has been obtained, we may be able to reduce it to a simpler form containing fewer terms or fewer variables in one or more terms. The new expression can then be used to implement a circuit that is equivalent to the original circuit but that contains fewer gates and connections.

To illustrate, the circuit of Figure 4-1(a) can be simplified to produce the circuit of Figure 4-1(b). Both circuits perform the same logic, so it should be obvious that the simpler circuit is more desirable because it contains fewer gates and will therefore be smaller and cheaper than the original. Furthermore, the circuit reliability will improve because there are fewer interconnections that can be potential circuit faults.

In subsequent sections, we will study two methods for simplifying logic circuits. One method will utilize the Boolean algebra theorems and, as we shall see,

(a)

(b)

FIGURE 4-1 It is often possible to simplify a logic circuit such as that in part (a) to produce a more efficient implementation, shown in (b).

is greatly dependent on inspiration and experience. The other method (Karnaugh mapping) is a systematic, step-by-step approach. Some instructors may wish to skip over this latter method because it is somewhat mechanical and probably does not contribute to a better understanding of Boolean algebra. This can be done without affecting the continuity or clarity of the rest of the text.

4-3 ALGEBRAIC SIMPLIFICATION

We can use the Boolean algebra theorems that we studied in Chapter 3 to help us simplify the expression for a logic circuit. Unfortunately, it is not always obvious which theorems should be applied to produce the simplest result. Furthermore, there is no easy way to tell whether the simplified expression is in its simplest form or whether it could have been simplified further. Thus, algebraic simplification often becomes a process of trial and error. With experience, however, one can become adept at obtaining reasonably good results.

The examples that follow will illustrate many of the ways in which the Boolean theorems can be applied in trying to simplify an expression. You should notice that these examples contain two essential steps:

1. The original expression is put into SOP form by repeated application of DeMorgan's theorems and multiplication of terms.

2. Once the original expression is in SOP form, the product terms are checked for common factors, and factoring is performed wherever possible. The factoring should result in the elimination of one or more terms.

EXAMPLE 4-1

Simplify the logic circuit shown in Figure 4-2(a).

FIGURE 4-2 Example 4-1.

Solution

The first step is to determine the expression for the output using the method presented in Section 3-6. The result is

$$z = ABC + A\overline{B} \cdot (\overline{\overline{A}\,\overline{C}})$$

Once the expression is determined, it is usually a good idea to break down all large inverter signs using DeMorgan's theorems and then multiply out all terms.

$$
\begin{aligned}
z &= ABC + A\overline{B}(\overline{\overline{A}} + \overline{\overline{C}}) && [\text{theorem (17)}] \\
&= ABC + A\overline{B}(A + C) && [\text{cancel double inversions}] \\
&= ABC + A\overline{B}A + A\overline{B}C && [\text{multiply out}] \\
&= ABC + A\overline{B} + A\overline{B}C && [A \cdot A = A]
\end{aligned}
$$

With the expression now in SOP form, we should look for common variables among the various terms with the intention of factoring. The first and third terms above have AC in common, which can be factored out:

$$z = AC(B + \overline{B}) + A\overline{B}$$

Since $B + \overline{B} = 1$, then

$$
\begin{aligned}
z &= AC(1) + A\overline{B} \\
&= AC + A\overline{B}
\end{aligned}
$$

We can now factor out A, which results in

$$z = A(C + \overline{B})$$

This result can be simplified no further. Its circuit implementation is shown in Figure 4-2(b). It is obvious that the circuit in Figure 4-2(b) is a great deal simpler than the original circuit in Figure 4-2(a).

EXAMPLE 4-2

Simplify the expression $z = A\overline{B}\,\overline{C} + A\overline{B}C + ABC$.

Solution

The expression is already in SOP form.

Method 1: The first two terms in the expression have the product $A\overline{B}$ in common. Thus,

$$
\begin{aligned}
z &= A\overline{B}(\overline{C} + C) + ABC \\
&= A\overline{B}(1) + ABC \\
&= A\overline{B} + ABC
\end{aligned}
$$

We can factor the variable A from both terms:

$$z = A(\overline{B} + BC)$$

Invoking theorem (15b):

$$z = A(\bar{B} + C)$$

Method 2: The original expression is $z = AB\bar{C} + A\bar{B}C + ABC$. The first two terms have $A\bar{B}$ in common. The last two terms have AC in common. How do we know whether to factor $A\bar{B}$ from the first two terms or AC from the last two terms? Actually, we can do both by using the ABC term *twice.* In other words, we can rewrite the expression as:

$$z = AB\bar{C} + A\bar{B}C + A\bar{B}C + ABC$$

where we have added an extra term $A\bar{B}C$. This is valid and will not change the value of the expression because $A\bar{B}C + A\bar{B}C = A\bar{B}C$ [theorem (7)]. Now we can factor $A\bar{B}$ from the first two terms and AC from the last two terms:

$$\begin{aligned} z &= A\bar{B}(C + \bar{C}) + AC(\bar{B} + B) \\ &= A\bar{B} \cdot 1 + AC \cdot 1 \\ &= A\bar{B} + AC = A(\bar{B} + C) \end{aligned}$$

Of course, this is the same result obtained with method 1. This trick of using the same term twice can always be used. In fact, the same term can be used more than twice if necessary.

EXAMPLE 4-3

Simplify $z = \overline{AC(\overline{ABD})} + \overline{ABC}D + A\bar{B}C$.

Solution

First, use DeMorgan's theorem on the first term:

$$z = \bar{A}C(A + \bar{B} + \bar{D}) + \bar{A}B\bar{C}D + A\bar{B}C \qquad \text{(step 1)}$$

Multiplying out yields

$$z = \bar{A}CA + \bar{A}C\bar{B} + \bar{A}C\bar{D} + \bar{A}B\bar{C}D + A\bar{B}C \qquad (2)$$

Because $\bar{A} \cdot A = 0$, the first term is eliminated:

$$z = \bar{A}\bar{B}C + \bar{A}C\bar{D} + \bar{A}B\bar{C}D + A\bar{B}C \qquad (3)$$

This is the desired SOP form. Now we must look for common factors among the various product terms. The idea is to check for the largest common factor between any two or more product terms. For example, the first and last terms have the common factor $\bar{B}C$, and the second and third terms share the common factor $\bar{A}D$. We can factor these out as follows:

$$z = \bar{B}C(\bar{A} + A) + \bar{A}D(C + B\bar{C}) \qquad (4)$$

Now, because $\bar{A} + A = 1$, and $C + B\bar{C} = C + B$ [theorem (15a)], we have

$$z = \bar{B}C + \bar{A}D(B + C) \qquad (5)$$

This same result could have been reached with other choices for the factoring. For example, we could have factored C from the first, second, and fourth product terms in step 3 to obtain

$$z = C(\overline{A}\,\overline{B} + \overline{A}\,\overline{D} + A\overline{B}) + \overline{A}B\overline{C}\overline{D}$$

The expression inside the parentheses can be factored further:

$$z = C(\overline{B}[\overline{A} + A] + \overline{A}\,\overline{D}) + \overline{A}B\overline{C}\overline{D}$$

Because $\overline{A} + A = 1$, this becomes

$$z = C(\overline{B} + \overline{A}\,\overline{D}) + \overline{A}B\overline{C}\overline{D}$$

Multiplying out yields

$$z = \overline{B}C + \overline{A}C\overline{D} + \overline{A}B\overline{C}\overline{D}$$

Now we can factor $\overline{A}\,\overline{D}$ from the second and third terms to get

$$z = \overline{B}C + \overline{A}\,\overline{D}(C + B\overline{C})$$

If we use theorem (15a), the expression in parentheses becomes $B + C$. Thus, we finally have

$$z = \overline{B}C + \overline{A}\,\overline{D}(B + C)$$

This is the same result that we obtained earlier, but it took us many more steps. This illustrates why you should look for the largest common factors: it will generally lead to the final expression in the fewest steps.

Example 4-3 illustrates the frustration often encountered in Boolean simplification. Because we have arrived at the same equation (which appears irreducible) by two different methods, it might seem reasonable to conclude that this final equation is the simplest form. In fact, the simplest form of this equation is

$$z = \overline{A}B\overline{D} + \overline{B}C$$

But there is no apparent way to reduce step (5) to reach this simpler version. In this case, we missed an operation earlier in the process that could have led to the simpler form. The question is, "How could we have known that we missed a step?" Later in this chapter, we will examine a mapping technique that will always lead to the simplest SOP form.

EXAMPLE 4-4

Simplify the expression $x = (\overline{A} + B)(A + B + D)\overline{D}$.

Solution

The expression can be put into sum-of-products form by multiplying out all the terms. The result is

$$x = \overline{A}A\overline{D} + \overline{A}B\overline{D} + \overline{A}D\overline{D} + BA\overline{D} + BB\overline{D} + BD\overline{D}$$

The first term can be eliminated because $\overline{A}A = 0$. Likewise, the third and sixth terms can be eliminated because $D\overline{D} = 0$. The fifth term can be simplified to $B\overline{D}$ because $BB = B$. This gives us

$$x = \overline{A}B\overline{D} + AB\overline{D} + B\overline{D}$$

We can factor $B\overline{D}$ from each term to obtain

$$x = B\overline{D}(\overline{A} + A + 1)$$

Clearly, the term inside the parentheses is always 1, so we finally have

$$x = B\overline{D}$$

EXAMPLE 4-5

Simplify the circuit of Figure 4-3(a).

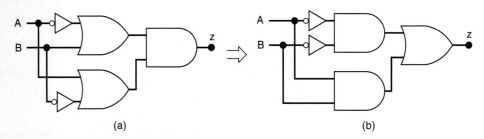

(a)   (b)

FIGURE 4-3   Example 4-5.

Solution

The expression for output z is

$$z = (\overline{A} + B)(A + \overline{B})$$

Multiplying out to get the sum-of-products form, we obtain

$$z = \overline{A}A + \overline{A}\,\overline{B} + BA + B\overline{B}$$

We can eliminate $\overline{A}A = 0$ and $B\overline{B} = 0$ to end up with

$$z = \overline{A}\,\overline{B} + AB$$

This expression is implemented in Figure 4-3(b), and if we compare it with the original circuit, we see that both circuits contain the same number of gates and connections. In this case, the simplification process produced an equivalent, but not simpler, circuit.

EXAMPLE 4-6

Simplify $x = A\overline{B}C + \overline{A}BD + \overline{C}\overline{D}$.

Solution

You can try, but you will not be able to simplify this expression any further.

1. State which of the following expressions are *not* in the sum-of-products form:
   (a) $R S \overline{T} + \overline{R} S \overline{T} + \overline{T}$
   (b) $A \overline{D} C + \overline{A} D C$
   (c) $M N \overline{P} + (M + \overline{N}) P$
   (d) $A B + \overline{A} B \overline{C} + A \overline{B} \, \overline{C} D$

2. Simplify the circuit in Figure 4-1(a) to arrive at the circuit of Figure 4-1(b).

3. Change each AND gate in Figure 4-1(a) to a NAND gate. Determine the new expression for x and simplify it.

4-4   DESIGNING COMBINATIONAL LOGIC CIRCUITS

When the desired output level of a logic circuit is given for all possible input conditions, the results can be conveniently displayed in a truth table. The Boolean expression for the required circuit can then be derived from the truth table. For example, consider Figure 4-4(a), where a truth table is shown for a circuit that has two inputs, A and B, and output x. The table shows that output x is to be at the 1 level *only* for the case where $A = 0$ and $B = 1$. It now remains to determine what logic circuit will produce this desired operation. It should be apparent that one possible solution is that shown in Figure 4-4(b). Here an AND gate is used with inputs $\overline{A}$ and B, so that $x = \overline{A} \cdot B$. Obviously x will be 1 *only if* both inputs to the AND gate are 1, namely, $\overline{A} = 1$ (which means that $A = 0$) and $B = 1$. For all other values of A and B, the output x will be 0.

FIGURE 4-4 Circuit that produces a 1 output only for the $A = 0, B = 1$ condition.

A	B	x
0	0	0
0	1	1
1	0	0
1	1	0

(a)   (b)

A similar approach can be used for the other input conditions. For instance, if x were to be high only for the $A = 1, B = 0$ condition, the resulting circuit would be an AND gate with inputs A and $\overline{B}$. In other words, for any of the four possible input conditions, we can generate a high x output by using an AND gate with appropriate inputs to generate the required AND product. The four different cases are shown in Figure 4-5. Each of the AND gates shown generates an output that is 1 *only* for one given input condition and the output is 0 for all other conditions. It should be noted that the AND inputs are inverted or not inverted depending on the values that the variables have for the given condition. If the variable is 0 for the given condition, it is inverted before entering the AND gate.

Let us now consider the case shown in Figure 4-6(a), where we have a truth table that indicates that the output x is to be 1 for two different cases: $A = 0, B = 1$ and $A = 1, B = 0$. How can this be implemented? We know that the AND term $\overline{A} \cdot B$ will generate a 1 only for the $A = 0, B = 1$ condition, and the AND term $A \cdot \overline{B}$ will generate a 1 for the $A = 1, B = 0$ condition. Because

FIGURE 4-5 An AND gate with appropriate inputs can be used to produce a 1 output for a specific set of input levels.

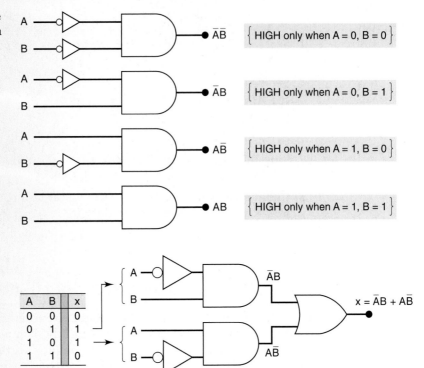

$\overline{A}\,\overline{B}$ { HIGH only when A = 0, B = 0 }

$\overline{A}B$ { HIGH only when A = 0, B = 1 }

$A\overline{B}$ { HIGH only when A = 1, B = 0 }

AB { HIGH only when A = 1, B = 1 }

FIGURE 4-6 Each set of input conditions that is to produce a HIGH output is implemented by a separate AND gate. The AND outputs are ORed to produce the final output.

A	B	x
0	0	0
0	1	1
1	0	1
1	1	0

(a)

$\overline{A}B$

$A\overline{B}$

$x = \overline{A}B + A\overline{B}$

(b)

x must be HIGH for *either* condition, it should be clear that these terms should be ORed together to produce the desired output, x. This implementation is shown in Figure 4-6(b), where the resulting expression for the output is $x = \overline{A}B + A\overline{B}$.

In this example, an AND term is generated for each case in the table where the output x is to be a 1. The AND gate outputs are then ORed together to produce the total output x, which will be 1 when either AND term is 1. This same procedure can be extended to examples with more than two inputs. Consider the truth table for a three-input circuit (Table 4-1). Here there are three cases where the output x is to be 1. The required AND term for each of these cases is shown. Again, note that for each case where a variable is 0, it appears inverted in the AND term. The sum-of-products expression for x is obtained by ORing the three AND terms.

$$x = \overline{A}\,\overline{B}\,\overline{C} + \overline{A}BC + ABC$$

Complete Design Procedure

Once the output expression has been determined from the truth table in sum-of-products form, it can easily be implemented using AND and OR gates and INVERTERs. Usually, however, the expression can be simplified, thereby resulting in a more efficient circuit. The following example illustrates the complete design procedure.

TABLE 4-1

A	B	C	x	
0	0	0	0	
0	0	1	0	
0	1	0	1	$\rightarrow \overline{A}B\overline{C}$
0	1	1	1	$\rightarrow \overline{A}BC$
1	0	0	0	
1	0	1	0	
1	1	0	0	
1	1	1	1	$\rightarrow ABC$

EXAMPLE 4-7

Design a logic circuit that has three inputs, A, B, and C, and whose output will be HIGH only when a majority of the inputs are HIGH.

Solution

Step 1.   Set up the truth table.

On the basis of the problem statement, the output x should be 1 whenever two or more inputs are 1; for all other cases, the output should be 0 (Table 4-2).

TABLE 4-2

A	B	C	x	
0	0	0	0	
0	0	1	0	
0	1	0	0	
0	1	1	1	$\rightarrow \overline{A}BC$
1	0	0	0	
1	0	1	1	$\rightarrow A\overline{B}C$
1	1	0	1	$\rightarrow AB\overline{C}$
1	1	1	1	$\rightarrow ABC$

Step 2.   Write the AND term for each case where the output is a 1.

There are four such cases. The AND terms are shown next to the truth table (Table 4-2). Again note that each AND term contains each input variable in either inverted or noninverted form.

Step 3.   Write the sum-of-products expression for the output.

$$x = \overline{A}BC + A\overline{B}C + AB\overline{C} + ABC$$

Step 4.   Simplify the output expression.

This expression can be simplified in several ways. Perhaps the quickest way is to realize that the last term ABC has two variables in common with each of the other terms. Thus, we can use the ABC term to factor with each of the other terms. The expression is rewritten with the ABC term occurring three times (recall from Example 4-2 that this is legal in Boolean algebra):

$$x = \overline{A}BC + ABC + A\overline{B}C + ABC + AB\overline{C} + ABC$$

Factoring the appropriate pairs of terms, we have

$$x = BC(\overline{A} + A) + AC(\overline{B} + B) + AB(\overline{C} + C)$$

Each term in parentheses is equal to 1, so we have

$$x = BC + AC + AB$$

Step 5.   Implement the circuit for the final expression.

This expression is implemented in Figure 4-7. Since the expression is in SOP form, the circuit consists of a group of AND gates working into a single OR gate.

FIGURE 4-7   Example 4-7.

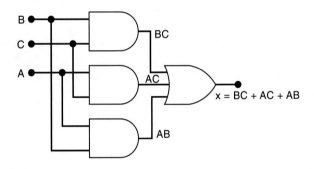

EXAMPLE 4-8

Refer to Figure 4-8(a), where an analog-to-digital converter is monitoring the dc voltage of a 12-V storage battery on an orbiting spaceship. The converter's output is a four-bit binary number, $ABCD$, corresponding to the battery voltage in steps of 1 V, with A as the MSB. The converter's binary outputs are fed to a logic circuit that is to produce a HIGH output as long as the binary value is greater than $0110_2 = 6_{10}$; that is, the battery voltage is greater than 6 V. Design this logic circuit.

Solution

The truth table is shown in Figure 4-8(b). For each case in the truth table, we have indicated the decimal equivalent of the binary number represented by the $ABCD$ combination.

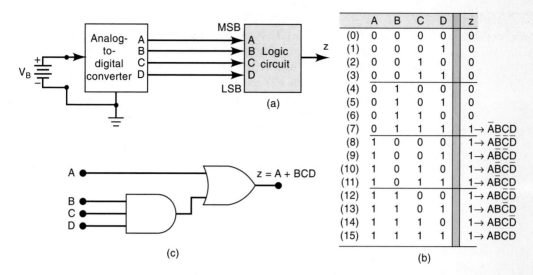

		A	B	C	D	z
(0)		0	0	0	0	0
(1)		0	0	0	1	0
(2)		0	0	1	0	0
(3)		0	0	1	1	0
(4)		0	1	0	0	0
(5)		0	1	0	1	0
(6)		0	1	1	0	0
(7)		0	1	1	1	$1 \rightarrow \overline{A}BCD$
(8)		1	0	0	0	$1 \rightarrow A\overline{B}\,\overline{C}\,\overline{D}$
(9)		1	0	0	1	$1 \rightarrow A\overline{B}\,\overline{C}D$
(10)		1	0	1	0	$1 \rightarrow A\overline{B}C\overline{D}$
(11)		1	0	1	1	$1 \rightarrow A\overline{B}CD$
(12)		1	1	0	0	$1 \rightarrow AB\overline{C}\,\overline{D}$
(13)		1	1	0	1	$1 \rightarrow AB\overline{C}D$
(14)		1	1	1	0	$1 \rightarrow ABC\overline{D}$
(15)		1	1	1	1	$1 \rightarrow ABCD$

(a)

(b)

(c)

FIGURE 4-8 Example 4-8.

The output z is set equal to 1 for all those cases where the binary number is greater than 0110. For all other cases, z is set equal to 0. This truth table gives us the following sum-of-products expression:

$$z = \overline{A}BCD + A\overline{B}\,\overline{C}\,\overline{D} + A\overline{B}\,\overline{C}D + A\overline{B}C\overline{D} + A\overline{B}CD + AB\overline{C}\,\overline{D} \\ + AB\overline{C}D + ABC\overline{D} + ABCD$$

Simplification of this expression will be a formidable task, but with a little care it can be accomplished. The step-by-step process involves factoring and eliminating terms of the form $A + \overline{A}$:

$$z = \overline{A}BCD + A\overline{B}\,\overline{C}(\overline{D} + D) + A\overline{B}C(\overline{D} + D) + AB\overline{C}(\overline{D} + D) + ABC(\overline{D} + D)$$
$$= \overline{A}BCD + A\overline{B}\,\overline{C} + A\overline{B}C + AB\overline{C} + ABC$$
$$= \overline{A}BCD + A\overline{B}(\overline{C} + C) + AB(\overline{C} + C)$$
$$= \overline{A}BCD + A\overline{B} + AB$$
$$= \overline{A}BCD + A(\overline{B} + B)$$
$$= \overline{A}BCD + A$$

This can be reduced further by invoking theorem (15a), which says that $x + \overline{x}y = x + y$. In this case $x = A$ and $y = BCD$. Thus,

$$z = \overline{A}BCD + A = BCD + A$$

This final expression is implemented in Figure 4-8(c).

As this example demonstrates, the algebraic simplification method can be quite lengthy when the original expression contains a large number of terms. This is a limitation that is not shared by the Karnaugh mapping method, as we will see later.

EXAMPLE 4-9

Refer to Figure 4-9(a). In a simple copy machine, a stop signal, S, is to be generated to stop the machine operation and energize an indicator light whenever either of the following conditions exists: (1) there is no paper in the paper feeder tray; or (2) the two microswitches in the paper path are activated, indicating a jam in the paper path. The presence of paper in the feeder tray is indicated by a HIGH at logic signal P. Each of the microswitches produces a logic signal (Q and R) that goes HIGH whenever paper is passing over the switch to activate it. Design the logic circuit to produce a HIGH at output signal S for the stated conditions, and implement it using the 74HC00 CMOS quad two-input NAND chip.

FIGURE 4-9   Example 4-9.

Solution

We will use the five-step process used in Example 4-7. The truth table is shown in Table 4-3. The S output will be a logic 1 whenever $P = 0$ because this indicates no paper in the feeder tray. S will also be a 1 for the two cases where Q and R are both 1, indicating a paper jam. As the table shows, there are five different input conditions that produce a HIGH output. (Step 1)

TABLE 4-3

P	Q	R	S	
0	0	0	1	$\overline{P}\,\overline{Q}\,\overline{R}$
0	0	1	1	$\overline{P}\,\overline{Q}R$
0	1	0	1	$\overline{P}Q\overline{R}$
0	1	1	1	$\overline{P}QR$
1	0	0	0	
1	0	1	0	
1	1	0	0	
1	1	1	1	PQR

The AND terms for each of these cases are shown. (Step 2)
The sum-of-products expression becomes

$$S = \overline{P}\,\overline{Q}\,\overline{R} + \overline{P}\,\overline{Q}R + \overline{P}Q\overline{R} + \overline{P}QR + PQR \qquad \text{(Step 3)}$$

We can begin the simplification by factoring out $\overline{P}\,\overline{Q}$ from terms 1 and 2 and by factoring out $\overline{P}Q$ from terms 3 and 4:

$$S = \overline{P}\,\overline{Q}(\overline{R} + R) + \overline{P}Q(\overline{R} + R) + PQR \qquad \text{(Step 4)}$$

Now we can eliminate the $\overline{R} + R$ terms because they equal 1:

$$S = \overline{P}\,\overline{Q} + \overline{P}Q + PQR$$

Factoring $\overline{P}$ from terms 1 and 2 allows us to eliminate Q from these terms:

$$S = \overline{P} + PQR$$

Here we can apply theorem (15b) ($\overline{x} + xy = \overline{x} + y$) to obtain

$$S = \overline{P} + QR$$

The AND/OR implementation for this circuit is shown in Figure 4-9(b). (Step 5)
To implement this circuit using the 74HC00 quad two-input NAND chip, we must convert each gate and the INVERTER by their NAND-gate equivalents (per Section 3-12). This is shown in Figure 4-9(c). Clearly, we can eliminate the double inverters to produce the NAND-gate implementation shown in Figure 4-9(d).

The final wired-up circuit is obtained by connecting two of the NAND gates on the 74HC00 chip. This CMOS chip has the same gate configuration and pin numbers as the TTL 74LS00 chip of Figure 3-31. Figure 4-10 shows the wired-up circuit with pin numbers, including the +5 V and GROUND pins. It also includes an output driver transistor and LED to indicate the state of output S.

FIGURE 4-10   Circuit of
Figure 4-9(d) implemented
using 74HC00 NAND chip.

(e)

Note: The other two
       gates on the chip
       are not connected.

1. Write the sum-of-products expression for a circuit with four inputs and
   an output that is to be HIGH only when input A is LOW at the same time
   that exactly two other inputs are LOW.

2. Implement the expression of question 1 using all four-input NAND gates.
   How many are required?

4-5   KARNAUGH MAP METHOD

The **Karnaugh map (K map)** is a graphical tool used to simplify a logic equation or to convert a truth table to its corresponding logic circuit in a simple,
orderly process. Although a K map can be used for problems involving any
number of input variables, its practical usefulness is limited to five or six
variables. The following discussion will be limited to problems with up to
four inputs because even five- and six-input problems are too involved and
are best done by a computer program.

Karnaugh Map Format

The K map, like a truth table, is a means for showing the relationship between logic inputs and the desired output. Figure 4-11 shows three examples
of K maps for two, three, and four variables, together with the corresponding
truth tables. These examples illustrate the following important points:

1. The truth table gives the value of output X for each combination of input
   values. The K map gives the same information in a different format. Each
   case in the truth table corresponds to a square in the K map. For example,
   in Figure 4-11(a), the $A = 0, B = 0$ condition in the truth table corresponds
   to the $\overline{A}\,\overline{B}$ square in the K map. Because the truth table shows $X = 1$ for this
   case, a 1 is placed in the $\overline{A}\,\overline{B}$ square in the K map. Similarly, the $A = 1, B = 1$
   condition in the truth table corresponds to the AB square of the K map. Be-
   cause $X = 1$ for this case, a 1 is placed in the AB square. All other squares
   are filled with 0s. This same idea is used in the three- and four-variable
   maps shown in the figure.

2. The K-map squares are labeled so that horizontally adjacent squares differ
   only in one variable. For example, the upper left-hand square in the four-
   variable map is $\overline{A}\,\overline{B}\,\overline{C}\,\overline{D}$, while the square immediately to its right is $\overline{A}\,\overline{B}\,C\overline{D}$

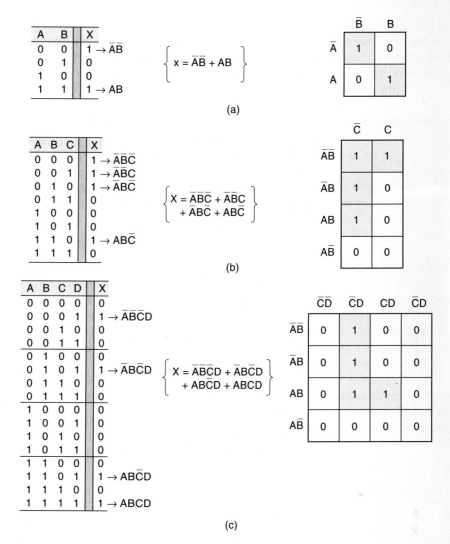

FIGURE 4-11   Karnaugh maps and truth tables for (a) two, (b) three, and (c) four variables.

(only the D variable is different). Similarly, vertically adjacent squares differ only in one variable. For example, the upper left-hand square is $\overline{A}\,\overline{B}\,\overline{C}D$, while the square directly below it is $\overline{A}B\overline{C}\overline{D}$ (only the B variable is different).

Note that each square in the top row is considered to be adjacent to a corresponding square in the bottom row. For example, the $\overline{A}BCD$ square in the top row is adjacent to the $A\overline{B}CD$ square in the bottom row because they differ only in the A variable. You can think of the top of the map as being wrapped around to touch the bottom of the map. Similarly, squares in the leftmost column are adjacent to corresponding squares in the rightmost column.

3. In order for vertically and horizontally adjacent squares to differ in only one variable, the top-to-bottom labeling must be done in the order shown: $\overline{A}\,\overline{B}$, $\overline{A}B$, AB, $A\overline{B}$. The same is true of the left-to-right labeling: $\overline{C}\overline{D}$, $\overline{C}D$, CD, $C\overline{D}$.

4. Once a K map has been filled with 0s and 1s, the sum-of-products expression for the output X can be obtained by ORing together those squares that contain a 1. In the three-variable map of Figure 4-11(b), the $\overline{A}\,\overline{B}\,\overline{C}$, $\overline{A}\,\overline{B}C$, $\overline{A}B\overline{C}$, and $AB\overline{C}$ squares contain a 1, so that $X = \overline{A}\,\overline{B}\,\overline{C} + \overline{A}\,\overline{B}C + \overline{A}B\overline{C} + AB\overline{C}$.

Also notice in the figure that AND and NOR gates produce a constant LOW output when they are in the disabled condition. Conversely, the NAND and OR gates produce a constant HIGH output in the disabled condition.

There will be many situations in digital-circuit design where the passage of a logic signal is to be enabled or disabled, depending on conditions present at one or more control inputs. Several are shown in the following examples.

EXAMPLE 4-21

Design a logic circuit that will allow a signal to pass to the output only when control inputs B and C are both HIGH; otherwise, the output will stay LOW.

Solution

An AND gate should be used because the signal is to be passed without inversion, and the disable output condition is a LOW. Because the enable condition must occur only when $B = C = 1$, a three-input AND gate is used, as shown in Figure 4-27(a).

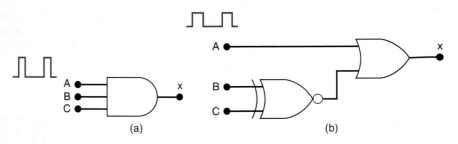

FIGURE 4-27 Examples 4-21 and 4-22.

EXAMPLE 4-22

Design a logic circuit that allows a signal to pass to the output only when one, but not both, of the control inputs are HIGH; otherwise, the output will stay HIGH.

Solution

The result is drawn in Figure 4-27(b). An OR gate is used because we want the output disable condition to be a HIGH, and we do not want to invert the signal. Control inputs B and C are combined in an XNOR gate. When B and C are different, the XNOR sends a LOW to enable the OR gate. When B and C are the same, the XNOR sends a HIGH to disable the OR gate.

EXAMPLE 4-23

Design a logic circuit with input signal A, control input B, and outputs X and Y to operate as follows:

1. When $B = 1$, output X will follow input A, and output Y will be 0.
2. When $B = 0$, output X will be 0, and output Y will follow input A.

Solution

The two outputs will be 0 when they are disabled and will follow the input signal when they are enabled. Thus, an AND gate should be used for each output.

Because X is to be enabled when $B = 1$, its AND gate must be controlled by B, as shown in Figure 4-28. Because Y is to be enabled when $B = 0$, its AND gate is controlled by $\overline{B}$. The circuit in Figure 4-28 is called a *pulse-steering circuit* because it steers the input pulse to one output or the other, depending on B.

FIGURE 4-28   Example 4-23.

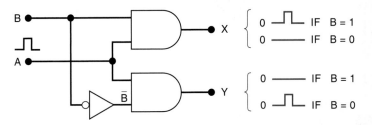

REVIEW QUESTIONS

1. Design a logic circuit with three inputs A, B, C and an output that goes LOW only when A is HIGH while B and C are different.
2. Which logic gates produce a 1 output in the disabled state?
3. Which logic gates pass the inverse of the input signal when they are enabled?

4-9   BASIC CHARACTERISTICS OF DIGITAL ICs

Digital ICs are a collection of resistors, diodes, and transistors fabricated on a single piece of semiconductor material (usually silicon) called a *substrate*, which is commonly referred to as a *chip*. The chip is enclosed in a protective plastic or ceramic package from which pins extend for connecting the IC to other devices. One of the more common types of package is the **dual-in-line package (DIP)**, shown in Figure 4-29(a), so called because it contains two parallel rows of pins. The pins are numbered counterclockwise when viewed from the top of the package with respect to an identifying notch or dot at one end of the package [see Figure 4-29(b)]. The DIP shown here is a 14-pin package that measures 0.75 in. by 0.25 in.; 16-, 20-, 24-, 28-, 40-, and 64-pin packages are also used.

Figure 4-29(c) shows that the actual silicon chip is much smaller than the DIP; typically, it might be a 0.05-in. square. The silicon chip is connected to the pins of the DIP by very fine wires (1-mil diameter).

The DIP is probably the most common digital IC package found in older digital equipment, but other types are becoming more and more popular. The IC shown in Figure 4-29(d) is only one of the many packages common to modern digital circuits. This particular package uses J-shaped leads that curl under the IC. We will take a look at some of these other types of IC packages in Chapter 8.

Digital ICs are often categorized according to their circuit complexity as measured by the number of equivalent logic gates on the substrate. There are currently six levels of complexity that are commonly defined as shown in Table 4-5.

All of the specific ICs referred to in Chapter 3 and this chapter are **SSI** chips having a small number of gates. In modern digital systems, medium-scale integration **(MSI)** and large-scale integration devices **(LSI, VLSI, ULSI, GSI)** perform most of the functions that once required several circuit boards full of SSI devices. However, SSI chips are still used as the "interface," or "glue," between these more complex chips. The small scale ICs also offer an excellent way to learn the basic building blocks of digital systems. Consequently, many laboratory-based courses use these ICs to build and test small projects.

TABLE 4-6 Various series within the TTL logic family.

TTL Series	Prefix	Example IC
Standard TTL	74	7404 (hex INVERTER)
Schottky TTL	74S	74S04 (hex INVERTER)
Low-power Schottky TTL	74LS	74LS04 (hex INVERTER)
Advanced Schottky TTL	74AS	74AS04 (hex INVERTER)
Advanced low-power Schottky TTL	74ALS	74ALS04 (hex INVERTER)

that are part of the standard TTL series have an identification number that starts with 74. The 7402, 7438, and 74123 are all ICs in this series. Likewise, ICs that are part of the low-power Schottky TTL series have an identification number that starts with 74LS. The 74LS02, 74LS38, and 74LS123 are examples of devices in the 74LS series.

The principal differences in the various TTL series have to do with their electrical characteristics such as power dissipation and switching speed. They do not differ in the pin layout or logic operations performed by the circuits on the chip. For example, the 7404, 74S04, 74LS04, 74AS04, and 74ALS04 are all hex-INVERTER ICs, each containing *six* INVERTERs on a single chip.

CMOS Family

Several CMOS series are available and some of these are listed in Table 4-7. The 4000 series is the oldest CMOS series. This series contains many of the same logic functions as the TTL family but was not designed to be *pin-compatible* with TTL devices. For example, the 4001 quad NOR chip contains four two-input NOR gates, as does the TTL 7402 chip, but the gate inputs and outputs on the CMOS chip will not have the same pin numbers as the corresponding signals on the TTL chip.

The 74C, 74HC, 74HCT, 74AC, and 74ACT series are newer CMOS series. The first three are pin-compatible with correspondingly numbered TTL devices. For example, the 74C02, 74HC02, and 74HCT02 have the same pin layout as the 7402, 74LS02, and so on. The 74HC and 74HCT series operate at a higher speed than 74C devices. The 74HCT series is designed to be *electrically compatible* with TTL devices; that is, a 74HCT integrated circuit can be connected directly to TTL devices without any interfacing circuitry. The 74AC and 74ACT series are advanced-performance ICs. Neither is pin-compatible

TABLE 4-7 Various series within the CMOS logic family.

CMOS Series	Prefix	Example IC
Metal-gate CMOS	40	4001 (quad NOR gates)
Metal-gate, pin-compatible with TTL	74C	74C02 (quad NOR gates)
Silicon-gate, pin-compatible with TTL, high-speed	74HC	74HC02 (quad NOR gates)
Silicon-gate, high-speed, pin-compatible and electrically compatible with TTL	74HCT	74HCT02 (quad NOR gates)
Advanced-performance CMOS, not pin-compatible or electrically compatible with TTL	74AC	74AC02 (quad NOR)
Advanced-performance CMOS, not pin-compatible with TTL, but electrically compatible with TTL	74ACT	74ACT02 (quad NOR)

with TTL. 74ACT devices are electrically compatible with TTL. We explore the various TTL and CMOS series in greater detail in Chapter 8.

Power and Ground

To use digital ICs, it is necessary to make the proper connections to the IC pins. The most important connections are *dc power* and *ground*. These are required for the circuits on the chip to operate correctly. In Figure 4-30, you can see that both the TTL and the CMOS circuits have a dc power supply voltage connected to one of their pins, and ground to another. The power supply pin is labeled V_{CC} for the TTL circuit, and V_{DD} for the CMOS circuit. Many of the newer CMOS integrated circuits that are designed to be compatible with TTL integrated circuits also use the V_{CC} designation as their power pin.

If either the power or the ground connection is not made to the IC, the logic gates on the chip will not respond properly to the logic inputs, and the gates will not produce the expected output logic levels.

Logic-Level Voltage Ranges

For TTL devices, V_{CC} is nominally $+5$ V. For CMOS integrated circuits, V_{DD} can range from $+3$ to $+18$ V, although $+5$ V is most often used when CMOS integrated circuits are used in the same circuit with TTL integrated circuits.

For standard TTL devices, the acceptable input voltage ranges for the logic 0 and logic 1 levels are defined as shown in Figure 4-31(a). A logic 0 is any voltage in the range from 0 to 0.8 V; a logic 1 is any voltage from 2 to 5 V. Voltages that are not in either of these ranges are said to be **indeterminate** and should not be used as inputs to any TTL device. The IC manufacturers cannot guarantee how a TTL circuit will respond to input levels that are in the indeterminate range (between 0.8 and 2.0 V).

The logic input voltage ranges for CMOS integrated circuits operating with $V_{DD} = +5$ V are shown in Figure 4-31(b). Voltages between 0 and 1.5 V are defined as a logic 0, and voltages from 3.5 to 5 V are defined as a logic 1. The indeterminate range includes voltages between 1.5 and 3.5 V.

FIGURE 4-31 Logic-level input voltage ranges for (a) TTL and (b) CMOS digital ICs.

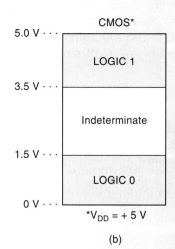

Unconnected (Floating) Inputs

What happens when the input to a digital IC is left unconnected? An unconnected input is often called a **floating** input. The answer to this question will be different for TTL and CMOS.

1. What is the most common type of digital IC package?
2. Name the six common categories of digital ICs according to complexity.
3. *True or false:* A 74S74 chip will contain the same logic and pin layout as the 74LS74.
4. *True or false:* A 74HC74 chip will contain the same logic and pin layout as the 74AS74.
5. Which CMOS series are not pin-compatible with TTL?
6. What is the acceptable input voltage range of a logic 0 for TTL? What is it for a logic 1?
7. Repeat question 6 for CMOS operating at $V_{DD} = 5$ V.
8. How does a TTL integrated circuit respond to a floating input?
9. How does a CMOS integrated circuit respond to a floating input?
10. Which CMOS series can be connected directly to TTL with no interfacing circuitry?
11. What is the purpose of pin numbers on a logic circuit connection diagram?
12. What are the key similarities of graphic design files used for programmable logic and traditional logic circuit connection diagrams?

4-10 TROUBLESHOOTING DIGITAL SYSTEMS

There are three basic steps in fixing a digital circuit or system that has a fault (failure):

1. **Fault detection.** Observe the circuit/system operation and compare it with the expected correct operation.
2. **Fault isolation.** Perform tests and make measurements to isolate the fault.
3. **Fault correction.** Replace the faulty component, repair the faulty connection, remove the short, and so on.

Although these steps may seem relatively apparent and straightforward, the actual troubleshooting procedure that is followed is highly dependent on the type and complexity of the circuitry, and on the kinds of troubleshooting tools and documentation that are available.

Good troubleshooting techniques can be learned only in a laboratory environment through experimentation and actual troubleshooting of faulty circuits and systems. There is absolutely no better way to become an effective troubleshooter than to do as much troubleshooting as possible, and no amount of textbook reading can provide that kind of experience. We can, however, help you to develop the analytical skills that are the most essential part of effective troubleshooting. We will describe the types of faults that are common to systems that are made primarily from digital ICs and we will tell you how to recognize them. We will then present typical case studies to

	LEDs		
Red	Green	Yellow	Logic Condition
OFF	ON	OFF	LOW
ON	OFF	OFF	HIGH
OFF	OFF	OFF	INDETERMINATE*
X	X	FLASHING	PULSING

* Includes open or floating condition

FIGURE 4-34 A logic probe is used to monitor the logic level activity at an IC pin or any other accessible point in a logic circuit.

illustrate the analytical processes involved in troubleshooting. In addition, there will be end-of-chapter problems to provide you with the opportunity to go through these analytical processes to reach conclusions about faulty digital circuits.

For the troubleshooting discussions and exercises we will be doing in this book, it will be assumed that the troubleshooting technician has the basic troubleshooting tools available: *logic probe, oscilloscope, logic pulser.* Of course, the most important and effective troubleshooting tool is the technician's brain, and that's the tool we are hoping to develop by presenting troubleshooting principles and techniques, examples and problems, here and in the following chapters.

In the next three sections on troubleshooting, we will use only our brain and a **logic probe** such as the one illustrated in Figure 4-34. The logic probe has a pointy metal tip that is touched to the specific point you want to test. Here, it is shown probing (touching) pin 3 of an IC. It can also be touched to a printed circuit board trace, an uninsulated wire, a connector pin, a lead on a discrete component such as a transistor, or any other conducting point in a circuit. The logic level that is present at the probe tip will be indicated by the status of the indicator LEDs in the probe. The four possibilities are given in the table of Figure 4-34. Note that an *indeterminate* logic level produces no indicator light. This includes the condition where the probe tip is touched to a point in a circuit that is open or floating—that is, not connected to any source of voltage. This type of probe also offers a yellow LED to indicate the presence of a pulse train. Any transitions (LOW to HIGH or HIGH to LOW) will cause the yellow LED to flash on for a fraction of a second and then turn off. If the transitions are occurring frequently, the LED will continue to flash at around 3 Hz. By observing the green and red LEDs along with the flashing yellow, you can tell whether the signal is mostly HIGH or mostly LOW.

range (i.e., close to 0.8 V), although it may also be in the indeterminate range. For CMOS devices, it will often be a voltage in the indeterminate range.

    Whenever you see a waveform like the Z1-2, Z1-4 signal in Figure 4-39 with three different levels, you should suspect that two output signals may be shorted together.

REVIEW QUESTIONS

1. List the different internal digital IC faults.
2. Which internal IC fault can produce signals that show three different voltage levels?
3. What would a logic probe indicate at Z1-2 and Z1-4 of Figure 4-39 if A = 0 and B = 1?
4. What is signal contention?

4-12   EXTERNAL FAULTS

We have seen how to recognize the effects of various faults internal to digital ICs. Many more things can go wrong external to the ICs; we will describe the most common ones in this section.

Open Signal Lines

This category includes any fault that produces a break or discontinuity in the conducting path such that a voltage level or signal is prevented from going from one point to another. Some of the causes of open signal lines are:

1. Broken wire
2. Poor solder connection; loose wire-wrap connection
3. Crack or cut trace on a printed circuit board (some of these are hairline cracks that are difficult to see without a magnifying glass)
4. Bent or broken pin on an IC
5. Faulty IC socket such that the IC pin does not make good contact with the socket

This type of circuit fault can often be detected by a careful visual inspection and then verified by disconnecting power from the circuit and checking for continuity (i.e., a low-resistance path) with an ohmmeter between the two points in question.

EXAMPLE 4-27

Consider the CMOS circuit of Figure 4-40 and the accompanying logic probe indications. What is the most probable circuit fault?

Solution

The indeterminate level at the NOR gate output is probably due to the indeterminate input at pin 2. Because there is a LOW at Z1-6, this LOW should also be at Z2-2. Clearly, the LOW from Z1-6 is not reaching Z2-2, and there

Pin	Condition
Z1-1	Pulsing
Z1-2	HIGH
Z1-3	Pulsing
Z1-4	LOW
Z1-5	Pulsing
Z1-6	LOW
Z2-3	Pulsing
Z2-2	Indeterminate
Z2-1	Indeterminate

All ICs
are CMOS
Z1: 74HC08
Z2: 74HC02

FIGURE 4-40   Example 4-27.

must be an open circuit in the signal path between these two points. The location of this open circuit can be determined by starting at Z1-6 with the logic probe and tracing the LOW level along the signal path toward Z2-2 until it changes into an indeterminate level.

Shorted Signal Lines

This type of fault has the same effect as an internal short between IC pins. It causes two signals to be exactly the same (signal contention). A signal line may be shorted to ground or V_{CC} rather than to another signal line. In those cases, the signal will be forced to the LOW or the HIGH state. The main causes for unexpected shorts between two points in a circuit are as follows:

1. **Sloppy wiring.** An example of this is stripping too much insulation from ends of wires that are in close proximity.
2. **Solder bridges.** These are splashes of solder that short two or more points together. They commonly occur between points that are very close together, such as adjacent pins on a chip.
3. **Incomplete etching.** The copper between adjacent conducting paths on a printed circuit board is not completely etched away.

Again, a careful visual inspection can very often uncover this type of fault, and an ohmmeter check can verify that the two points in the circuit are shorted together.

Faulty Power Supply

All digital systems have one or more dc power supplies that supply the V_{CC} and V_{DD} voltages required by the chips. A faulty power supply or one that is overloaded (supplying more than its rated amount of current) will provide poorly regulated supply voltages to the ICs, and the ICs either will not operate or will operate erratically.

A power supply may go out of regulation because of a fault in its internal circuitry, or because the circuits that it is powering are drawing more current than the supply is designed for. This can happen if a chip or a component has a fault that causes it to draw much more current than normal.

It is good troubleshooting practice to check the voltage levels at each power supply in the system to see that they are within their specified ranges. It is also a good idea to check them on an oscilloscope to verify that there is no significant amount of ac ripple on the dc levels and to verify that the voltage levels stay regulated during the system operation.

One of the most common signs of a faulty power supply is one or more chips operating erratically or not at all. Some ICs are more tolerant of power supply variations and may operate properly, while others do not. You should always check the power and ground levels at each IC that appears to be operating incorrectly.

Output Loading

When a digital IC has its output connected to too many IC inputs, its output current rating will be exceeded, and the output voltage can fall into the indeterminate range. This effect is called *loading* the output signal (actually it's overloading the output signal) and is usually the result of poor design or an incorrect connection.

REVIEW QUESTIONS

1. What are the most common types of external faults?
2. List some of the causes of signal-path open circuits.
3. What symptoms are caused by a faulty power supply?
4. How might loading affect an IC output voltage level?

4-13 TROUBLESHOOTING CASE STUDY

The following example will illustrate the analytical processes involved in troubleshooting digital circuits. Although the example is a fairly simple combinational logic circuit, the reasoning and the troubleshooting procedures used can be applied to the more complex digital circuits that we encounter in subsequent chapters.

EXAMPLE 4-28

Consider the circuit of Figure 4-41. Output Y is supposed to go HIGH for either of the following conditions:

1. $A = 1, B = 0$ regardless of the level on C
2. $A = 0, B = 1, C = 1$

You may wish to verify these results for yourself.

When the circuit is tested, the technician observes that output Y goes HIGH whenever A is HIGH or C is HIGH, regardless of the level at B. She takes logic probe measurements for the condition where $A = B = 0, C = 1$ and comes up with the indications recorded in Figure 4-41.

Examine the recorded levels and list the possible causes for the malfunction. Then develop a step-by-step procedure to determine the exact fault.

Pin	Condition
Z1-1	LOW
Z1-2	LOW
Z1-3	HIGH
Z2-4	LOW
Z2-5	HIGH
Z2-6,10	HIGH
Z2-13	HIGH
Z2-12	HIGH
Z2-9,11	LOW
Z2-8	HIGH

ICs are TTL
Z1: 74LS86
Z2: 74LS00

FIGURE 4-41   Example 4-28.

Solution

All of the NAND gate outputs are correct for the levels present at their inputs. The XOR gate, however, should be producing a LOW at output pin 3 because both of its inputs are at the same LOW level. It appears that Z1-3 is stuck HIGH, even though its inputs should produce a LOW. There are several possible causes for this:

1. An internal component failure in Z1 that prevents its output from going LOW
2. An external short to V_{CC} from any point along the conductors connected to node X (shaded in the diagram of the figure)
3. Pin 3 of Z1 internally shorted to V_{CC}
4. Pin 5 of Z2 internally shorted to V_{CC}
5. Pin 13 of Z2 internally shorted to V_{CC}

All of these possibilities except for the first one will short node X (and every IC pin connected to it) directly to V_{CC}.

The following procedure can be used to isolate the fault. This procedure is not the only approach that can be used and, as we stated earlier, the actual troubleshooting procedure that a technician uses is very dependent on what test equipment is available.

1. Check the V_{CC} and ground levels at the appropriate pins of Z1. Although it is unlikely that the absence of either of these might cause Z1-3 to stay HIGH, it is a good idea to make this check on any IC that is producing an incorrect output.
2. Turn off power to the circuit and use an ohmmeter to check for a short (resistance less than 1 Ω) between node X and any point connected to V_{CC} (such as Z1-14 or Z2-14). If no short is indicated, the last four possibilities in our list can be eliminated. This means that it is very likely that Z1 has an internal failure and should be replaced.

3. If step 2 shows that there is a short from node X to V_{CC}, perform a thorough visual examination of the circuit board and look for solder bridges, unetched copper slivers, uninsulated wires touching each other, and any other possible cause of an external short to V_{CC}. A likely spot for a solder bridge would be between adjacent pins 13 and 14 of Z2. Pin 14 is connected to V_{CC}, and pin 13 to node X. If an external short is found, remove it and perform an ohmmeter check to verify that node X is no longer shorted to V_{CC}.

4. If step 3 does not reveal an external short, the three possibilities that remain are internal shorts to V_{CC} at Z1-3, Z2-13, or Z2-5. One of these is shorting node X to V_{CC}.

To determine which of these IC pins is the culprit, we should disconnect each of them from node X *one at a time* and recheck for a short to V_{CC} after each disconnection. When the pin that is internally shorted to V_{CC} is disconnected, node X will no longer be shorted to V_{CC}.

The process of disconnecting each suspected pin from node X can be easy or difficult depending on how the circuit is constructed. If the ICs are in sockets, all you need to do is to pull the IC from its socket, bend out the suspected pin, and reinsert the IC into its socket. If the ICs are soldered into a printed circuit board, you will have to cut the trace that is connected to the pin and repair the cut trace when you are finished.

Example 4-28, although fairly simple, shows you the kinds of thinking that a troubleshooter must employ to isolate a fault. You will have the opportunity to begin developing your own troubleshooting skills by working on many end-of-chapter problems that have been designated with a **T** for troubleshooting.

4-14   PROGRAMMABLE LOGIC DEVICES*

In the previous sections, we briefly considered the class of ICs known as programmable logic devices. In Chapter 3, we introduced the concept of describing a circuit's operation using a hardware description language. In this section, we will explore these topics further and become prepared to use the tools of the trade to develop and implement digital systems using PLDs. Of course, it is impossible to understand all the complex details of how a PLD works before grasping the fundamentals of digital circuits. As we examine new fundamental concepts, we will expand our knowledge of the PLDs and the programming methods. The material is presented in such a way that anyone who is not interested in PLDs can easily skip over these sections without loss of continuity in the coverage of the basic principles.

Let's review the process we covered earlier of designing combinational digital circuits. The input devices are identified and assigned an algebraic name like *A, B, C,* or *LOAD, SHIFT, CLOCK.* Likewise, output devices are given names like *X, Z,* or *CLOCK_OUT, SHIFT_OUT.* Then a truth table is constructed that lists all the possible input combinations and identifies the required state of the outputs under each input condition. The truth table is one way of describing how the circuit is to operate. Another way to describe the circuit's operation is the Boolean expression. From this point the designer must find the simplest algebraic relationship and select digital ICs that can be wired together to implement the circuit. You have probably experienced that these last steps are the most tedious, time consuming, and prone to errors.

---

*All sections covering PLDs may be skipped without loss of continuity with the balance of Chapters 1–11.

Programmable logic devices allow most of these tedious steps to be automated by a computer and PLD *development software*. Using programmable logic improves the efficiency of the design and development process. Consequently, most modern digital systems are implemented in this way. The job of the circuit designer is to identify inputs and outputs, specify the logical relationship in the most convenient manner, and select a programmable device that is capable of implementing the circuit at the lowest cost. The concept behind programmable logic devices is simple: put lots of logic gates in a single IC and control the interconnection of these gates electronically.

PLD Hardware

Recall from Chapter 3 that many digital circuits today are implemented using programmable logic devices (PLDs). These devices are configured electronically and their internal circuits are "wired" together electronically to form a logic circuit. This programmable wiring can be thought of as thousands of connections that are either connected (1) or not connected (0). It is very tedious to try to configure these devices by manually placing 1s and 0s in a grid, so the next logical question is, "How do we control the interconnection of gates in a PLD electronically?" This section describes the general architecture of several types of PLDs and provides an introduction to the way all PLDs are constructed internally.

A common method of connecting one of many signals entering a network to one of many signal lines exiting the network is a switching matrix. Refer back to Figure 3-44, where this concept was introduced. A matrix is simply a grid of conductors (wires) arranged in rows and columns. Input signals are connected to the columns of the matrix, and the outputs are connected to the rows of the matrix. At each intersection of a row and a column is a switch that can electrically connect that row to that column. The switches that connect rows to columns can be mechanical switches, fusible links, electromagnetic switches (relays), or transistors. This is the general structure used in many applications and will be explored further when we study memory devices in Chapter 12.

PLDs also use a switch matrix that is often referred to as a programmable array. By deciding which intersections are connected and which ones are not, we can "program" the way the inputs are connected to the outputs of the array. In Figure 4-42, a programmable array is used to select the inputs for each AND gate. Notice that in this simple matrix, we can produce any logical product combination of variables A, B at any of the AND gate outputs. A matrix or programmable array such as the one shown in the figure can also be used to connect the AND outputs to OR gates. The details of various PLD architectures will be covered thoroughly in Chapter 13.

Programming a PLD

There are two ways to "program" a PLD IC. Programming means making the actual connections in the array. In other words, it means determining which of those connections are supposed to be open (0) and which are supposed to be closed (1). The first method involves removing the PLD IC chip from its circuit board. The chip is then placed in a special fixture called a **programmer**, shown in Figure 4-43. Most modern programmers are connected to a personal computer that is running software containing libraries of information about the many types of programmable devices available.

The programming software is invoked (called up and executed) on the PC to establish communication with the programmer. This software allows the user to set up the programmer for the type of device that is to be programmed,

FIGURE 4-42   A programmable array for selecting inputs as product terms.

FIGURE 4-43   A PLD
development system.

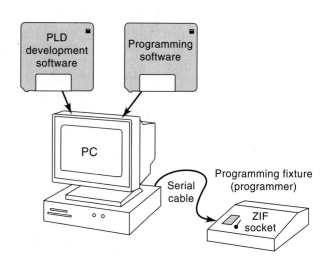

check if the device is blank, read the state of any fuse in the device, and provide instructions for the user to program a chip. Ultimately, the part is placed into a special socket that allows you to drop the chip in and then clamp the contacts onto the pins. This is called a **zero insertion force (ZIF)** socket. *Universal programmers* that can program any type of programmable device are available from numerous manufacturers.

Fortunately, as programmable parts began to proliferate, manufacturers saw the need to standardize pin assignments and programming methods. As a result, the Joint Electronic Device Engineering Council (**JEDEC**) was formed. One of the results was JEDEC standard 3, a format for transferring programming data for PLDs, independent of the PLD manufacturer or programming software. Pin assignments for various IC packages were also standardized, making universal programmers less complicated. Consequently, programming fixtures can program numerous types of PLDs. The software that allows the designer to specify a configuration for a PLD simply needs to produce an output file that conforms to the JEDEC standards. Then this JEDEC file can be loaded into any JEDEC-compatible PLD programmer that is capable of programming the desired type of PLD.

The second method is referred to as in system programming (ISP). As its name implies, the chip does not need to be removed from its circuit for storage of the programming information. A standard interface has been developed by the Joint Test Action Group (**JTAG**). The interface was developed to allow ICs to be tested without actually connecting test equipment to every pin of the IC. It also allows for internal programming. Four pins on the IC are used as a portal to store data and retrieve information about the inner condition of the IC. Many ICs, including PLDs and microcontrollers, are manufactured today to include the JTAG interface. An interface cable connects the four JTAG pins on the IC to an output port (like the printer port) of a personal computer. Software running on the PC establishes contact with the IC and loads the information in the proper format.

Development Software

We have examined several methods of describing logic circuits now, including schematic capture, logic equations, truth tables, and HDL. We also described the fundamental methods of storing 1s and 0s into a PLD IC to connect the logic circuits in the desired way. The biggest challenge in getting a PLD programmed is converting from any form of description into the array of 1s and 0s. Fortunately, this task is accomplished quite easily by a computer running the development software. The development software that we will be referring to and using for examples is produced by Altera and is called MAX+PLUS II. This software allows the designer to enter a circuit description in any one of the many ways we have been discussing: graphic design files (schematics), AHDL, and VHDL. It also allows the use of another HDL, called Verilog, and the option of describing the circuit with timing diagrams. Circuit blocks described by any of these methods can also be "connected" together to implement a much larger digital system, as shown in Figure 4-44.

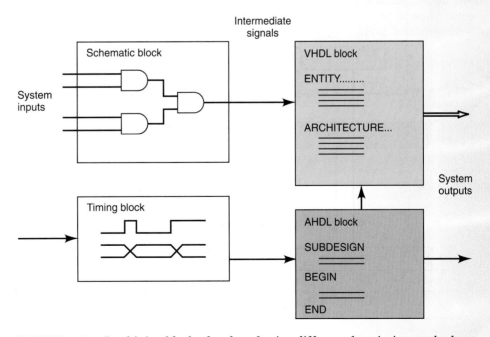

FIGURE 4-44 Combining blocks developed using different description methods.

FIGURE 4-47 A timing simulation of a circuit described in HDL.

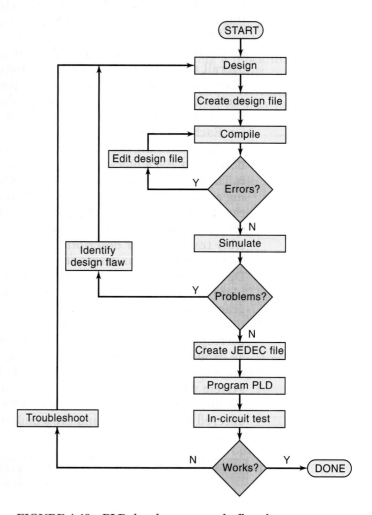

FIGURE 4-48 PLD development cycle flowchart.

When the designer is satisfied that the design works, the design can be verified by actually programming a chip and testing. For a complex PLD, the designer can either let the development system assign pins and then lay out the final circuit board accordingly, or specify the pins for each signal using the software features. If the compiler assigns the pins, the assignments can be found in the report file (.rpt), which provides many details about the implementation of the design. If the designer specifies the pins, it is important to know the constraints and limitations of the chip's architecture. These details will be covered in Chapter 13. The flowchart of Figure 4-48 summarizes the design process for designing each block.

After each circuit in a subsection has been tested, all can be combined and the subsection can be tested following the same process that was used for the small circuits. Then the subsections are combined and the system is tested. This approach lends itself very well to a typical project environment, where a team of people are working together, each responsible for his or her own circuits and sections that will ultimately come together to make up the system.

REVIEW QUESTIONS

1. What is actually being "programmed" in a PLD?
2. What bits (column, row) in Figure 4-42 must be connected to make Product 1 = AB?
3. What bits (column, row) in Figure 4-42 must be connected to make Product 3 = $A\bar{B}$?

4-15 REPRESENTING DATA IN HDL

Numeric data can be represented in various ways. We have studied the use of the hexadecimal number system as a convenient way to communicate bit patterns. We naturally prefer to use the decimal number system for numeric data, but computers and digital systems can operate only on binary information, as we studied in previous chapters. When we write in HDL, we often need to use these various number formats, and the computer must be able to understand which number system we are using. So far in this text, we have used a subscript to indicate the number system. For example, 101_2 was binary, 101_{16} was hexadecimal, and 101_{10} was decimal. Every programming language and HDL has its own unique way of identifying the various number systems, generally done with a prefix to indicate the number system. In most languages, a number with no prefix is assumed to be decimal. When we read one of these number designations, we must think of it as a symbol that represents a binary bit pattern. These numeric values are referred to as scalars or **literals**. Table 4-8 summarizes the methods of specifying values in binary, hex, and decimal for AHDL and VHDL.

TABLE 4-8 Designating number systems in HDL

Number System	AHDL	VHDL	Bit Pattern	Decimal Equivalent
Binary	B"101"	B"101"	101	5
Hexadecimal	H"101"	X"101"	100000001	257
Decimal	101	101"	1100101	101

EXAMPLE 4-29

Express the following bit pattern's numeric value in binary, hex, and decimal using AHDL and VHDL notation:

11001

Solution

Binary is designated the same in both AHDL and VHDL: **B "11001"**.

Converting the binary to hex, we have 19_{16}.

VHDL

Intermediate signals can also be declared as an array of bits. As with single bits, they are declared just inside the ARCHITECTURE definition. As an example, the eight-bit temperature on port *p1* can be assigned (connected) to a signal named *temp*, as follows:

```
SIGNAL      temp  :BIT_VECTOR (7 DOWNTO 0);
BEGIN
    temp <= p1;
END;
```

Notice that the input port *p1* has the data applied to it, and it is driving the signal wires named *temp*. No elements in the bit vector are specified, which means that all the bits are being connected. Individual bits can also be "connected" using signal assignments and by specifying the bit numbers inside parentheses. For example, to connect only the least significant bit of *p1* to the LSB of *temp*, the statement would be temp(0) <= p1(0);.

VHDL is very particular regarding the definitions of each type of the data. The type "bit_vector" describes an array of individual bits. This is interpreted differently than an eight-bit binary number (called a scaler quantity), which has the type **integer.** Unfortunately, VHDL does not allow us to assign an integer value to a BIT_VECTOR signal directly. Data can be represented by any of the types shown in Table 4-9, but data assignments and other operations must be done between objects of the same type. For example, the compiler will not allow you to take a number from a keypad declared as an integer and display it on four LEDs that are declared as BIT_VECTOR outputs. Notice in Table 4-9, under Possible Values, that individual BIT and STD_LOGIC data **objects** (e.g., signals, variables, inputs, and outputs) are designated by single quotes, whereas values assigned to BIT_VECTOR and STD_LOGIC_VECTOR types are strings of valid bit values enclosed in double quotes.

VHDL also offers some standardized data types that are necessary when using logic functions that are contained in the **libraries.** As you might have guessed, libraries are simply collections of little pieces of VHDL code that can be used in your hardware descriptions without reinventing the wheel. These libraries offer convenient functions, called **macrofunctions,** like many of the standard TTL devices that are described throughout this text. Rather than writing a new description of a familiar TTL device, we can simply pull its macrofunction out of the library and use it in our system. Of course, you need to get signals into and out of these macrofunctions, and the types of the signals in your code must match the types in the functions (which someone else wrote). This means that everyone must use the same standard data types.

TABLE 4-9   Common VHDL data types.

Data Type	Sample Declaration	Possible Values	Use
BIT	y :OUT BIT;	'0' '1'	y <= '0';
STD_LOGIC	driver :STD_LOGIC	'0' '1' 'z' 'x' '-'	driver <= 'z'
BIT_VECTOR	bcd_data :BIT_VECTOR (3 DOWNTO 0);	"0101" "1001" "0000"	digit <= bcd_data
STD_LOGIC_VECTOR	dbus :STD_LOGIC_VECTOR (3 DOWNTO 0);	"0Z1X"	IF rd THEN dbus <= "zzzz";
INTEGER	SIGNAL z:INTEGER (−32 TO 31);	−32..−2,−1,0,1,2 . . . 31	IF z > 5 THEN . . .

TABLE 4-10   STD_LOGIC
                 values.

'1'	Logic 1 (just like BIT type)
'0'	Logic 0 (just like BIT type)
'z'	High impedance*
'-'	don't care (just like you used in your K maps)
'U'	Uninitialized
'X'	Unknown
'W'	Weak unknown
'L'	Weak '0'
'H'	Weak '1'

*We will study tristate logic in Chapter 8.

When VHDL was standardized through the IEEE, many data types were created at the same time. The two that we will use in this text are **STD_LOGIC**, which is equivalent to BIT type, and **STD_LOGIC_VECTOR**, which is equivalent to BIT_VECTOR. As you recall, BIT type can have values of only '0' and '1'. The standard logic types are defined in the IEEE library and have a broader range of possible values than their built-in counterparts. The possible values for a **STD_LOGIC** type or for any element in a STD_LOGIC_VECTOR are given in Table 4-10. The names of these categories will make much more sense after we study the characteristics of logic circuits in Chapter 8. For now, we will show examples using values of only '1' and '0'.

REVIEW QUESTIONS

1. How would you declare a six-bit input array named push_buttons in (a) AHDL or (b) VHDL?
2. What statement would you use to take the MSB from the array in question 1 and put it on a single-bit output port named z? Use (a) AHDL or (b) VHDL.
3. In VHDL, what is the IEEE standard type that is equivalent to the BIT type?
4. In VHDL, what is the IEEE standard type that is equivalent to the BIT_VECTOR type?

4-16   TRUTH TABLES USING HDL

We have learned that a truth table is another way of expressing the operation of a circuit block. It relates the output of the circuit to every possible combination of its inputs. As we saw in Section 4-4, a truth table is the starting point for a designer to define how the circuit should operate. Then a Boolean expression is derived from the truth table and simplified using K maps or Boolean algebra. Finally the circuit is implemented from the final Boolean equation. Wouldn't it be great if we could go from the truth table directly to the final circuit without all those steps? We can do exactly that by entering the truth table using HDL.

AHDL

TRUTH TABLES USING AHDL

The code in Figure 4-50 uses AHDL to implement a circuit and uses a truth table to describe its operation. The truth table for this design was presented in Example 4-7. The key point of this example is the use of the TABLE keyword in AHDL. It allows the designer to specify the operation of the circuit just like you would fill out a truth table. On the first line after TABLE, the input variables (a,b,c) are listed exactly like you would create a column heading on a truth table. By including the three binary variables in parentheses, we tell the compiler that we want to use these three bits as a group and to refer to them as a three-bit binary number or bit pattern. The specific values for this bit pattern are listed below the group and are referred to as binary **literals**. The special operator (=>) is used in truth tables to separate the inputs from the output (y).

The TABLE in Figure 4-50 is intended to show the relationship between the HDL code and a truth table. A more common way of representing the input data heading is to use a variable bit array to represent the value on a, b, c. This method involves a declaration of the bit array on the line before BEGIN, such as:

```
VARIABLE in_bits[2..0]     :NODE;
```

Just before the TABLE keyword, the input bits can be assigned to the array, *inbits[ ]*:

```
in_bits[] = (a,b,c);
```

Grouping three independent bits in order like this is referred to as **concatenating**, and it is often done to connect individual bits to a bit array. The table

FIGURE 4-50   AHDL design file for Figure 4-7.

```
%       Figure 4-7 in AHDL
        Digital Systems 9th ed
        Neal Widmer
        MAY 22, 2002                        %
SUBDESIGN FIG4_50
(
        a,b,c :INPUT;              --define inputs to block
        y     :OUTPUT;             --define block output
)
BEGIN
        TABLE
                (a,b,c)            =>    y;    --column headings
                (0,0,0)            =>    0;
                (0,0,1)            =>    0;
                (0,1,0)            =>    0;
                (0,1,1)            =>    1;
                (1,0,0)            =>    0;
                (1,0,1)            =>    1;
                (1,1,0)            =>    1;
                (1,1,1)            =>    1;
        END TABLE;
END;
```

heading on the input bit sets can be represented by *in_bits[ ]*, in this case. Note that as we list the possible combinations of the inputs, we have several options. We can make up a group of 1s and 0s in parentheses, as shown in Figure 4-50, or we can represent the same bit pattern using the equivalent binary, hex, or decimal number. It is up to the designer to decide which format is most appropriate depending on what the input variables represent.

TRUTH TABLES USING VHDL: SELECTED SIGNAL ASSIGNMENT

The code in Figure 4-51 uses VHDL to implement a circuit using a **selected signal assignment** to describe its operation. It allows the designer to specify the operation of the circuit, just like you would fill out a truth table. The truth table for this design was presented in Example 4-7. The primary point of this example is the use of the WITH signal_name SELECT statement in VHDL. A secondary point presented here shows how to put the data into a format that can be used conveniently with the selected signal assignment. Notice that the inputs are defined in the ENTITY declaration as three independent bits *a, b,* and *c.* Nothing in this declaration makes one of these more significant than another. The order in which they are listed does not matter. We want to compare the current value of these bits with each of the possible combinations that could be present. If we drew out a truth table, we would decide which bit to place on the left (MSB) and which to place on the right (LSB). This is accomplished in VHDL by **concatenating** (connecting in order) the bit variables to form a bit vector. The concatenation operator is "&". A signal is

```
--      Figure 4-7 in VHDL
--      Digital Systems 9th ed
--      Neal Widmer
--      MAY 22, 2002
ENTITY fig4_51 IS
PORT(
        a,b,c :IN BIT;              --declare individual input bits
        y     :OUT BIT);
END fig4_51;

ARCHITECTURE truth OF fig4_51 IS
        SIGNAL in_bits :BIT_VECTOR(2 DOWNTO 0);
        BEGIN
        in_bits <= a & b & c;      --concatenate input bits into bit_vector
            WITH in_bits SELECT
                y    <=   '0'  WHEN "000",    --Truth Table
                          '0'  WHEN "001",
                          '0'  WHEN "010",
                          '1'  WHEN "011",
                          '0'  WHEN "100",
                          '1'  WHEN "101",
                          '1'  WHEN "110",
                          '1'  WHEN "111";
        END truth;
```

FIGURE 4-51   VHDL design file for Figure 4–7.

a bit pattern representing the three conditions of *too_cold, just_right,* and *too_hot.* The sequential section of the code uses the IF, ELSIF, ELSE to identify the range in which the temperature lies and assigns the correct bit pattern to status. After the decisions are made, the value of *status* is assigned to the actual output port bits. These bits have been ordered in a group that relates to the bit patterns assigned to *status[ ].* This could also have been written as three concurrent statements: too_cold = status[2]; just_right = status[1]; too_hot = status[0];.

ELSIF USING VHDL

The VHDL code in Figure 4-59 defines the inputs as a four-bit integer. The outputs are three individual bits that drive the three range indicators. This example uses an intermediate signal (*status*) that allows us to assign a bit pattern representing all three conditions of *too_cold, just_right,* and *too_hot.* The process section of the code uses the IF, ELSIF, and ELSE to identify the range in which the temperature lies and assigns the correct bit pattern to status. After the decisions are made, the value of each bit of *status* is assigned to the correct output port bit.

```
ENTITY fig4_59  IS
PORT(digital_value:IN INTEGER RANGE 0 TO 15;     -- declare 4-bit input
     too_cold, just_right, too_hot :OUT BIT);
END fig4_59 ;

ARCHITECTURE howhot OF fig4_59  IS
SIGNAL status    :BIT_VECTOR (2 downto 0);
BEGIN
    PROCESS (digital_value)
      BEGIN
        IF (digital_value <= 8) THEN status <= "100";
        ELSIF (digital_value > 8 AND digital_value < 11) THEN
              status <= "010";
        ELSE  status <= "001";
        END IF;
      END PROCESS ;
    too_cold   <= status(2);     -- assign status bits to output
    just_right <= status(1);
    too_hot    <= status(0);
END howhot;
```

FIGURE 4-59   Temperature range example in VHDL using ELSIF.

CASE

One more important control structure is useful for choosing actions based on current conditions. It is called by various names, depending on the programming language, but it nearly always involves the word **CASE.** This construct determines the value of an expression or object and then goes through a list of possible values (cases) for the expression or object being evaluated. Each case has a list of actions that should take place. A CASE construct is different from an IF/ELSIF because a case correlates one unique value of an object with a set of actions. Recall that an IF/ELSIF correlates a set of actions with a true statement. There can be only one match for a CASE statement. An IF/ELSIF can have more than one statement that is true, but will THEN perform the action associated with the first true statement it evaluates.

Another important point in the examples that follow is the need to combine several independent variables into a set of bits, called a bit vector. Recall that this action of linking several bits in a particular order is called *concatenation.* It allows us to consider the bit pattern as an ordered group.

CASE USING AHDL

The AHDL example in Figure 4-60 demonstrates a case construct implementing the circuit of Figure 4-9 (see also Table 4-3). It uses individual bits as its inputs. In the first statement after BEGIN, these bits are concatenated and assigned to the intermediate variable called *status.* The CASE statement evaluates the variable *status* and finds the bit pattern (following the keyword WHEN) that matches the value of *status.* It then performs the action described following =>. In this example, it simply assigns logic 0 to the output for each of the three specified cases. All *other* cases result in a logic 1 on the output.

```
SUBDESIGN fig4_60
(
    p, q, r          :INPUT;      -- define inputs to block
    s                :OUTPUT;     -- define outputs
)
VARIABLE
  status[2..0]    :NODE;
BEGIN
    status[]= (p, q, r);  -- link input bits in order
    CASE status[] IS
        WHEN b"100"      => s = GND;
        WHEN b"101"      => s = GND;
        WHEN b"110"      => s = GND;
        WHEN OTHERS      => s = VCC;
    END CASE;
END;
```

FIGURE 4-60   Figure 4-9 represented in AHDL.

CASE USING VHDL

The VHDL example in Figure 4-61 demonstrates the case construct implementing the circuit of Figure 4-9 (see also Table 4-3). It uses individual bits as its inputs. In the first statement after BEGIN, these bits are concatenated and assigned to the intermediate variable called *status* using the & operator. The CASE statement evaluates the variable status and finds the bit pattern (following the keyword WHEN) that matches the value of *status.* It then performs the action described following =>. In this simple example, it merely

```
ENTITY fig4_61 IS
PORT( p, q, r     :IN bit;                  --declare 3 bits input
      s           :OUT BIT);
END fig4_61;


ARCHITECTURE copy OF fig4_61 IS
SIGNAL status     :BIT_VECTOR (2 downto 0);
BEGIN
   status <= p & q & r;                     --link bits in order.
   PROCESS (status)
      BEGIN
         CASE status IS
            WHEN "100" =>  s <= '0';
            WHEN "101" =>  s <= '0';
            WHEN "110" =>  s <= '0';
            WHEN OTHERS => s <= '1';
         END CASE;
      END PROCESS ;
END copy;
```

FIGURE 4-61   Figure 4-9 represented in VHDL.

assigns logic 0 to the output for each of the three specified cases. All *other* cases result in a logic 1 on the output.

REVIEW QUESTIONS

1. Which control structure decides to do or not to do?
2. Which control structure decides to do this or to do that?
3. Which control structure(s) decides which one of several different actions to take?
4. Declare an input named *count* that can represent a numeric quantity as big as 205. Use AHDL or VHDL.

SUMMARY

1. The two general forms for logic expressions are the sum-of-products form and the product-of-sums form.
2. One approach to the design of a combinatorial logic circuit is to (1) construct its truth table, (2) convert the truth table to a sum-of-products expression, (3) simplify the expression using Boolean algebra or K mapping, (4) implement the final expression.
3. The K map is a graphical method for representing a circuit's truth table and generating a simplified expression for the circuit output.
4. An exclusive-OR circuit has the expression $x = A\overline{B} + \overline{A}B$. Its output x will be HIGH only when inputs A and B are at opposite logic levels.
5. An exclusive-NOR circuit has the expression $x = \overline{A}\,\overline{B} + AB$. Its output x will be HIGH only when inputs A and B are at the same logic level.

6. Each of the basic gates (AND, OR, NAND, NOR) can be used to enable or disable the passage of an input signal to its output.

7. The main digital IC families are the TTL and CMOS families. Digital ICs are available in a wide range of complexities (gates per chip), from the basic to the high-complexity logic functions.

8. To perform basic troubleshooting requires—at a minimum—an understanding of circuit operation, a knowledge of the types of possible faults, a complete logic-circuit connection diagram, and a logic probe.

9. A programmable logic device (PLD) is an IC that contains a large number of logic gates whose interconnections can be programmed by the user to generate the desired logic relationship between inputs and outputs.

10. To program a PLD, you need a development system that consists of a computer, PLD development software, and a programmer fixture that does the actual programming of the PLD chip.

11. The Altera system allows convenient hierarchical design techniques using any form of hardware description.

12. The type of data objects must be specified so that the HDL compiler knows the range of numbers to be represented.

13. Truth tables can be entered directly into the source file using the features of HDL.

14. Logical control structures such as IF, ELSE, and CASE can be used to describe the operation of a logic circuit, making the code and the problem's solution much more straightforward.

IMPORTANT TERMS

sum-of-products (SOP)	indeterminate	libraries
product-of-sums (POS)	floating	macrofunction
Karnaugh map (K map)	logic probe	STD_LOGIC
looping	contention	STD_LOGIC_VECTOR
don't-care condition	programmer	selected signal
exclusive-OR (XOR)	ZIF socket	assignment
exclusive-NOR (XNOR)	JEDEC	concatenate
parity generator	JTAG	decision control
parity checker	hierarchical design	structure
enable/disable	top-down	concurrent
dual-in-line package (DIP)	test vectors	sequential
SSI, MSI, LSI, VLSI,	literals	PROCESS
ULSI, GSI	bit array	sensitivity list
transistor-transistor logic	BIT_VECTOR	IF/THEN
(TTL)	index	ELSE
complementary metal-	integer	ELSIF
oxide-semiconductor	objects	CASE
(CMOS)		

PROBLEMS

SECTIONS 4-2 AND 4-3

B   4-1. Simplify the following expressions using Boolean algebra.

(a) $x = ABC + \overline{A}C$

(b) $y = (Q + R)(\overline{Q} + \overline{R})$

(c) $w = ABC + A\bar{B}C + \bar{A}$

(d) $q = \overline{RST(R + S + T)}$

(e) $x = \bar{A}\bar{B}\bar{C} + \bar{A}BC + ABC + AB\bar{C} + A\bar{B}C$

(f) $z = (B + \bar{C})(\bar{B} + C) + \overline{\bar{A} + B + \bar{C}}$

(g) $y = \overline{(C + D)} + \bar{A}C\bar{D} + A\bar{B}\bar{C} + \bar{A}BCD + AC\bar{D}$

(h) $x = AB(\overline{CD}) + \bar{A}BD + \overline{B}\,\overline{C}\,\overline{D}$

B   4-2. Simplify the circuit of Figure 4-62 using Boolean algebra.

FIGURE 4-62   Problems 4-2 and 4-3.

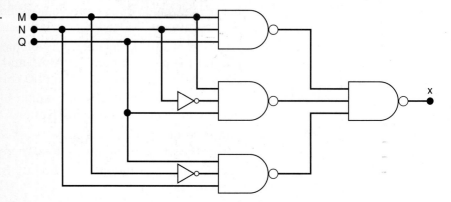

B   4-3. Change each gate in Problem 4-2 to a NOR gate, and simplify the circuit using Boolean algebra.

SECTION 4-4

B, D   4-4. Design the logic circuit corresponding to the truth table shown in Table 4-11.

TABLE 4-11

A	B	C	x
0	0	0	1
0	0	1	0
0	1	0	1
0	1	1	1
1	0	0	1
1	0	1	0
1	1	0	0
1	1	1	1

B, D   4-5. Design a logic circuit whose output is HIGH *only* when a majority of inputs A, B, and C are LOW.

D   4-6. A manufacturing plant needs to have a horn sound to signal quitting time. The horn should be activated when either of the following conditions is met:

1. It's after 5 o'clock and all machines are shut down.

2. It's Friday, the production run for the day is complete, and all machines are shut down.

Design a logic circuit that will control the horn. (*Hint:* Use four logic input variables to represent the various conditions; for example, input A will be HIGH only when the time of day is 5 o'clock or later.)

D   4-7. A four-bit binary number is represented as $A_3A_2A_1A_0$, where A_3, A_2, A_1, and A_0 represent the individual bits and A_0 is equal to the LSB. Design a logic circuit that will produce a HIGH output whenever the binary number is greater than 0010 and less than 1000.

D   4-8. Figure 4-63 shows a diagram for an automobile alarm circuit used to detect certain undesirable conditions. The three switches are used to indicate the status of the door by the driver's seat, the ignition, and the headlights, respectively. Design the logic circuit with these three switches as inputs so that the alarm will be activated whenever either of the following conditions exists:

- The headlights are on while the ignition is off.
- The door is open while the ignition is on.

FIGURE 4-63   Problem 4-8.

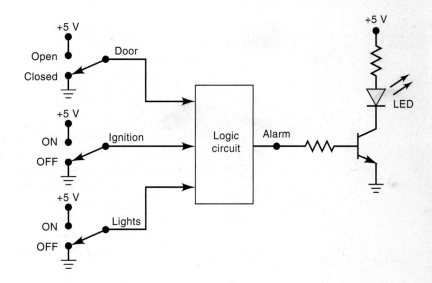

4-9. Implement the circuit of Problem 4-4 using all NAND gates.

4-10. Implement the circuit of Problem 4-5 using all NAND gates.

SECTION 4-5

B   4-11. Determine the minimum expression for each K map in Figure 4-64. Pay particular attention to step 5 for the map in (a).

	$\overline{C}\overline{D}$	$\overline{C}D$	CD	$C\overline{D}$
$\overline{A}\overline{B}$	1	1	1	1
$\overline{A}B$	1	1	0	0
AB	0	0	0	1
$A\overline{B}$	0	0	1	1

(a)

	$\overline{C}\overline{D}$	$\overline{C}D$	CD	$C\overline{D}$
$\overline{A}\overline{B}$	1	0	1	1
$\overline{A}B$	1	0	0	1
AB	0	0	0	0
$A\overline{B}$	1	0	1	1

(b)

	$\overline{C}$	C
$\overline{A}\overline{B}$	0	1
$\overline{A}B$	0	0
AB	1	0
$A\overline{B}$	1	X

(c)

FIGURE 4-64   Problem 4-11.

B 4-12. Simplify the expression in Problem 4-1(e) using a K map.

B 4-13. Simplify the expression in Problem 4-1(g) using a K map.

B 4-14. Simplify the expression in Problem 4-1(h) using a K map.

B 4-15. Obtain the output expression for Problem 4-7 using a K map.

C, D 4-16. Figure 4-65 shows a *BCD counter* that produces a four-bit output representing the BCD code for the number of pulses that have been applied to the counter input. For example, after four pulses have occurred, the counter outputs are $DCBA = 0100_2 = 4_{10}$. The counter resets to 0000 on the tenth pulse and starts counting over again. In other words, the *DCBA* outputs will never represent a number greater than $1001_2 = 9_{10}$. Design the logic circuit that produces a HIGH output whenever the count is 2, 3, or 9. Use K mapping and take advantage of the don't-care conditions.

FIGURE 4-65 Problem 4-16.

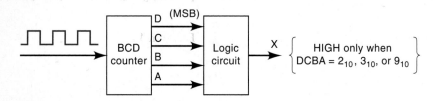

D 4-17. Figure 4-66 shows four switches that are part of the control circuitry in a copy machine. The switches are at various points along the path of the copy paper as the paper passes through the machine. Each switch is normally open, and as the paper passes over a switch, the switch closes. It is impossible for switches SW1 and SW4 to be closed at the same time. Design the logic circuit to produce a HIGH output whenever *two or more* switches are closed at the same time. Use K mapping and take advantage of the don't-care conditions.

FIGURE 4-66 Problem 4-17.

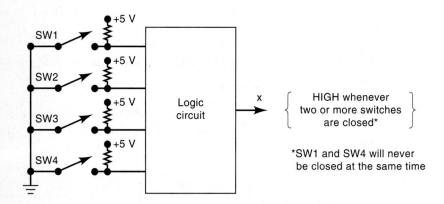

B 4-18. Example 4-3 demonstrated algebraic simplification. Step 3 resulted in the SOP equation $z = \overline{A}\,\overline{B}C + \overline{A}C\overline{D} + \overline{A}B\overline{C}\overline{D} + A\overline{B}C$. Use a K map to prove that this equation can be simplified further than the answer shown in the example.

C 4-19. Use Boolean algebra to arrive at the same result obtained by the K map method of Problem 4-18.

SECTION 4-6

B   4-20. (a) Determine the output waveform for the circuit of Figure 4-67.
            (b) Repeat with the *B* input held LOW.
            (c) Repeat with *B* held HIGH.

FIGURE 4-67   Problem 4-20.

B   4-21. Determine the input conditions needed to produce *x* = 1 in Figure 4-68.

FIGURE 4-68   Problem 4-21.

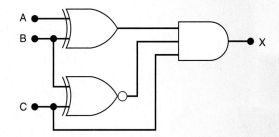

B   4-22. A 7486 chip contains four XOR gates. Show how to make an XNOR
            gate using only a 7486 chip. *Hint:* See Example 4-16.

B   4-23. Modify the circuit of Figure 4-23 to compare two 4-bit numbers and
            produce a HIGH output when the two numbers match exactly.

C   4-24. Figure 4-69 represents a *relative-magnitude detector* that takes two
            three-bit binary numbers, $x_2x_1x_0$ and $y_2y_1y_0$, and determines whether
            they are equal and, if not, which one is larger. There are three outputs,
            defined as follows:

            1. *M* = 1 only if the two input numbers are equal.
            2. *N* = 1 only if $x_2x_1x_0$ is greater than $y_2y_1y_0$.
            3. *P* = 1 only if $y_2y_1y_0$ is greater than $x_2x_1x_0$.

            Design the logic circuitry for this detector. The circuit has *six* inputs
            and *three* outputs and is therefore much too complex to handle using
            the truth-table approach. Refer to Example 4-17 as a hint about how
            you might start to solve this problem.

FIGURE 4-69   Problem 4-24.

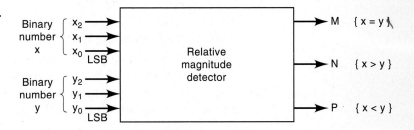

MORE DESIGN PROBLEMS

C, D   4-25. Figure 4-70 represents a multiplier circuit that takes two-bit binary numbers, x_1x_0 and y_1y_0, and produces an output binary number $z_3z_2z_1z_0$ that is equal to the arithmetic product of the two input numbers. Design the logic circuit for the multiplier. (*Hint:* The logic circuit will have four inputs and four outputs.)

FIGURE 4-70   Problem 4-25.

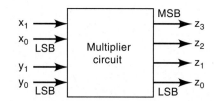

D   4-26. A BCD code is being transmitted to a remote receiver. The bits are A_3, A_2, A_1, and A_0, with A_3 as the MSB. The receiver circuitry includes a *BCD error detector* circuit that examines the received code to see if it is a legal BCD code (i.e., ≤1001). Design this circuit to produce a HIGH for any error condition.

D   4-27. Design a logic circuit whose output is HIGH whenever A and B are both HIGH as long as C and D are either both LOW or both HIGH. Try to do this without using a truth table. Then check your result by constructing a truth table from your circuit to see if it agrees with the problem statement.

D   4-28. Four large tanks at a chemical plant contain different liquids being heated. Liquid-level sensors are being used to detect whenever the level in tank A or tank B rises above a predetermined level. Temperature sensors in tanks C and D detect when the temperature in either of these tanks drops below a prescribed temperature limit. Assume that the liquid-level sensor outputs A and B are LOW when the level is satisfactory and HIGH when the level is too high. Also, the temperature-sensor outputs C and D are LOW when the temperature is satisfactory and HIGH when the temperature is too low. Design a logic circuit that will detect whenever the level in tank A or tank B is too high at the same time that the temperature in either tank C or tank D is too low.

C, D   4-29. Figure 4-71 shows the intersection of a main highway with a secondary access road. Vehicle-detection sensors are placed along lanes C and D (main road) and lanes A and B (access road). These sensor outputs are LOW (0) when no vehicle is present and HIGH (1) when a vehicle is present. The intersection traffic light is to be controlled according to the following logic:

1. The east-west (E-W) traffic light will be green whenever *both* lanes C and D are occupied.

2. The E-W light will be green whenever *either* C or D is occupied but lanes A and B are not *both* occupied.

3. The north-south (N-S) light will be green whenever *both* lanes A and B are occupied but C and D are not *both* occupied.

4. The N-S light will also be green when *either* A or B is occupied while C and D are *both* vacant.

5. The E-W light will be green when *no* vehicles are present.

Using the sensor outputs *A, B, C,* and *D* as inputs, design a logic circuit to control the traffic light. There should be two outputs, N-S and E-W, that go HIGH when the corresponding light is to be *green*. Simplify the circuit as much as possible and show *all* steps.

FIGURE 4-71   Problem 4-29.

SECTION 4-7

D   4-30. Redesign the parity generator and checker of Figure 4-25 to operate using odd parity. (*Hint:* What is the relationship between an odd-parity bit and an even-parity bit for the same set of data bits?)

D   4-31. Redesign the parity generator and checker of Figure 4-25 to operate on eight data bits.

SECTION 4-8

B   4-32. (a) Under what conditions will an OR gate allow a logic signal to pass through to its output unchanged?

(b) Repeat (a) for an AND gate.

(c) Repeat for a NAND gate.

(d) Repeat for a NOR gate.

B   4-33. (a) Can an INVERTER be used as an enable/disable circuit? Explain.

(b) Can an XOR gate be used as an enable/disable circuit? Explain.

D   4-34. Design a logic circuit that will allow input signal *A* to pass through to the output only when control input *B* is LOW while control input *C* is HIGH; otherwise, the output is LOW.

D   4-35. Design a circuit that will *disable* the passage of an input signal only when control inputs *B, C,* and *D* are all HIGH; the output is to be HIGH in the disabled condition.

D   4-36. Design a logic circuit that controls the passage of a signal A according to the following requirements:

    1. Output X will equal A when control inputs B and C are the same.

    2. X will remain HIGH when B and C are different.

D   4-37. Design a logic circuit that has two signal inputs, A_1 and A_0, and a control input S so that it functions according to the requirements given in Figure 4-72. (This type of circuit is called a *multiplexer* and will be covered in Chapter 9.)

FIGURE 4-72   Problem 4-37.

D   4-38. Use K mapping to design a circuit to meet the requirements of Example 4-17. Compare this circuit with the solution in Figure 4-23. This points out that the K-map method cannot take advantage of the XOR and XNOR gate logic. The designer must be able to determine when these gates are applicable.

SECTIONS 4-9 TO 4-13

T*   4-39. (a) A technician testing a logic circuit sees that the output of a particular INVERTER is stuck LOW while its input is pulsing. List as many possible reasons as you can for this faulty operation.

    (b) Repeat part (a) for the case where the INVERTER output is stuck at an indeterminate logic level.

T   4-40. The signals shown in Figure 4-73 are applied to the inputs of the circuit of Figure 4-32. Suppose that there is an internal open circuit at Z1-4.

    (a) What will a logic probe indicate at Z1-4?

    (b) What dc voltage reading would you expect a VOM to register at Z1-4? (Remember that the ICs are TTL.)

    (c) Sketch what you think the $\overline{CLKOUT}$ and $\overline{SHIFTOUT}$ signals will look like.

FIGURE 4-73   Problem 4-40.

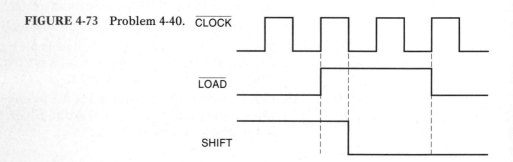

---

*Recall that T indicates a troubleshooting exercise.

(d) Instead of the open at Z1-4, suppose that pins 9 and 10 of Z2 are internally shorted. Sketch the probable signals at Z2-10, $\overline{CLOCKOUT}$, and $\overline{SHIFTOUT}$.

T 4-41. Assume that the ICs in Figure 4-32 are CMOS. Describe how the circuit operation would be affected by an open circuit in the conductor connecting Z2-2 and Z2-10.

T 4-42. In Example 4-24, we listed three possible faults for the situation of Figure 4-36. What procedure would you follow to determine which of the faults is the actual one?

T 4-43. Refer to the circuit of Figure 4-38. Assume that the devices are CMOS. Also assume that the logic probe indication at Z2-3 is "indeterminate" rather than "pulsing." List the possible faults, and write a procedure to follow to determine the actual fault.

T 4-44. Refer to the logic circuit of Figure 4-41. Recall that output Y is supposed to be HIGH for either of the following conditions:

1. $A = 1, B = 0$, regardless of C
2. $A = 0, B = 1, C = 1$

When testing the circuit, the technician observes that Y goes HIGH only for the first condition but stays LOW for all other input conditions. Consider the following list of possible faults. For each one, write yes or no to indicate whether or not it could be the actual fault. Explain your reasoning for each no response.

(a) An internal short to ground at Z2-13

(b) An open circuit in the connection to Z2-13

(c) An internal short to V_{CC} at Z2-11

(d) An open circuit in the V_{CC} connection to Z2

(e) An internal open circuit at Z2-9

(f) An open in the connection from Z2-11 to Z2-9

(g) A solder bridge between pins 6 and 7 of Z2

T 4-45. Develop a procedure for isolating the fault that is causing the malfunction described in Problem 4-44.

T 4-46. Assume that the gates in Figure 4-41 are all CMOS. When the technician tests the circuit, he finds that it operates correctly except for the following conditions:

1. $A = 1, B = 0, C = 0$
2. $A = 0, B = 1, C = 1$

For these conditions, the logic probe indicates indeterminate levels at Z2-6, Z2-11, and Z2-8. What do you think is the probable fault in the circuit? Explain your reasoning.

T 4-47. Figure 4-74 is a combinational logic circuit that operates an alarm in a car whenever the driver and/or passenger seats are occupied and the seatbelts are not fastened when the car is started. The active-HIGH signals *DRIV* and *PASS* indicate the presence of the driver and passenger, respectively, and are taken from pressure-actuated switches in the seats. The signal *IGN* is active-HIGH when the ignition switch is on. The signal $\overline{BELTD}$ is active-LOW and indicates that the driver's seatbelt is *unfastened*; $\overline{BELTP}$ is the corresponding signal for the passenger seatbelt. The alarm will be activated (LOW) whenever the car is started and either of the front seats is occupied and its seatbelt is not fastened.

FIGURE 4-74   Problems 4-47, 4-48, and 4-49.

(a) Verify that the circuit will function as described.

(b) Describe how this alarm system would operate if Z1-2 were internally shorted to ground.

(c) Describe how it would operate if there were an open connection from Z2-6 to Z2-10.

T   4-48. Suppose that the system of Figure 4-74 is functioning so that the alarm is activated as soon as the driver and/or passenger are seated and the car is started, regardless of the status of the seatbelts. What are the possible faults? What procedure would you follow to find the actual fault?

T   4-49. Suppose that the alarm system of Figure 4-74 is operating so that the alarm goes on continuously as soon as the car is started, regardless of the state of the other inputs. List the possible faults and write a procedure to isolate the fault.

DRILL QUESTIONS ON PLDs (50 THROUGH 55)

4-50. *True or false*

(a) Top-down design begins with an overall description of the entire system and it specifications.

(b) A JEDEC file can be used as the input file for a programmer.

(c) If an input file compiles with no errors, it means the PLD circuit will work correctly.

(d) A compiler can interpret code in spite of syntax errors.

(e) Test vectors are used to simulate and test a device.

H, B   4-51. What are the % characters used for in the AHDL design file?

H, B   4-52. How are comments indicated in a VHDL design file?

B   4-53. What is a ZIF socket?

B   4-54. Name three entry modes used to input a circuit description into PLD development software.

B   4-55. What do JEDEC and HDL stand for?

SECTION 4-15

H, B   4-56. Declare the following data objects in AHDL or VHDL.

(a) An array of eight output bits named *gadgets*.

(b) A single-output bit named *buzzer.*

(c) A 16-bit numeric input port named *altitude*.

(d) A single, intermediate bit within a hardware description file named *wire2*.

H, B   4-57. Express the following literal numbers in hex, binary, and decimal using the syntax of AHDL or VHDL.

(a) 152_{10}

(b) 1001010100_2

(c) $3C4_{16}$

H, B   4-58. The following similar I/O definition is given for AHDL and VHDL. Write four concurrent assignment statements that will connect the inputs to the outputs as shown in Figure 4-75.

```
SUBDESIGN hw
(
    inbits[3..0]    :INPUT;
    outbits[3..0]   :OUTPUT;
)
```

```
ENTITY hw IS
Port (
    inbits      :IN BIT_VECTOR (3 downto 0);
    outbits     :OUT BIT_VECTOR (3 downto 0)
    );
END hw;
```

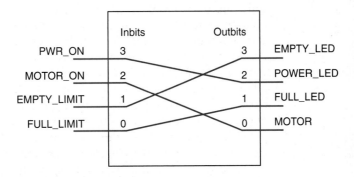

FIGURE 4-75   Problem 4-58.

SECTION 4-16

H, D   4-59. Modify the AHDL truth table of Figure 4-50 to implement $A B + A \overline{C} + \overline{A} B$.

H, D   4-60. Modify the VHDL truth table of Figure 4-51 to implement $A B + A \overline{C} + \overline{A} B$.

H, D 4-61. Modify the AHDL design in Figure 4-54 so that $z = 1$ only when the digital value is less than 1010_2.

H, D 4-62. Modify the VHDL design in Figure 4-55 so that $z = 1$ only when the digital value is less than 1010_2.

H, D 4-63. Modify the AHDL design in Figure 4-60 to implement Table 4-1.

H, D 4-64. Modify the VHDL design in Figure 4-61 to implement Table 4-1.

H, D 4-65. Write the hardware description design file Boolean equation to implement Example 4-9.

4-66. Write the hardware description design file Boolean equation to implement a four-bit parity generator as shown in Figure 4-25(a).

DRILL QUESTION

B 4-67. Define each of the following terms.

(a) Karnaugh map

(b) Sum-of-products form

(c) Parity generator

(d) Octet

(e) Enable circuit

(f) Don't-care condition

(g) Floating input

(h) Indeterminate voltage level

(i) Contention

(j) PLD

(k) TTL

(l) CMOS

MICROCOMPUTER APPLICATIONS

C 4-68. In a microcomputer, the microprocessor unit (MPU) is always communicating with one of the following: (1) random-access memory (RAM), which stores programs and data that can be readily changed; (2) read-only memory (ROM), which stores programs and data that never change; and (3) external input/output (I/O) devices such as keyboards, video displays, printers, and disk drives. As it is executing a program, the MPU will generate an address code that selects which type of device (RAM, ROM, or I/O) it wants to communicate with. Figure 4-76 shows a typical arrangement where the MPU outputs an eight-bit address code A_{15} through A_8. Actually, the MPU outputs a 16-bit address code, but the low-order bits A_7 through A_0 are not used in the device selection process. The address code is applied to a logic circuit that uses it to generate the device select signals: $\overline{RAM}$, $\overline{ROM}$, and $\overline{I/O}$.

Analyze this circuit and determine the following.

(a) The range of addresses A_{15} through A_8 that will activate $\overline{RAM}$

(b) The range of addresses that activate $\overline{I/O}$

(c) The range of addresses that activate $\overline{ROM}$

Express the addresses in binary and hexadecimal. For example, the answer to (a) is A_{15} to $A_8 = 00000000_2$ to $11101111_2 = 00_{16}$ to EF_{16}.

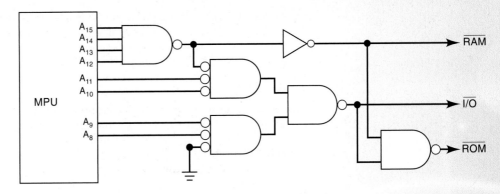

FIGURE 4-76 Problem 4-68.

C, D **4-69.** In some microcomputers, the MPU can be *disabled* for short periods of time while another device controls the RAM, ROM, and I/O. During these intervals, a special control signal ($\overline{DMA}$) is activated by the MPU and is used to disable (deactivate) the device select logic so that the $\overline{RAM}$, $\overline{ROM}$, and $\overline{I/O}$ are all in their inactive state. Modify the circuit of Figure 4-76 so that $\overline{RAM}$, $\overline{ROM}$, and $\overline{I/O}$ will be deactivated whenever the signal $\overline{DMA}$ is active, regardless of the state of the address code.

ANSWERS TO SECTION REVIEW QUESTIONS

SECTION 4-1

1. Only (a)   2. Only (c)

SECTION 4-3

1. Expression (b) is not in sum-of-products form because of the inversion sign over both the C and D variables (i.e., the $\overline{ACD}$ term). Expression (c) is not in sum-of-products form because of the $(M + \overline{N})P$ term.   3. $x = \overline{A} + \overline{B} + \overline{C}$

SECTION 4-4

1. $x = \overline{A}\,\overline{B}\,\overline{C}D + \overline{A}\,\overline{B}C\overline{D} + \overline{A}BC\overline{D}$ 2. Eight

SECTION 4-5

1. $x = AB + AC + BC$ 2. $x = A + BCD$ 3. $S = \overline{P} + QR$ 4. An input condition for which there is no specific required output condition; i.e., we are free to make it 0 or 1.

SECTION 4-6

2. A constant LOW   3. No; the available XOR gate can be used as an INVERTER by connecting one of its inputs to a constant HIGH (see Example 4-16).

SECTION 4-8

1. $x = \overline{A(B \oplus C)}$ 2. OR, NAND   3. NAND, NOR

SECTION 4-9

1. DIP   2. SSI, MSI, LSI, VLSI, ULSI, GSI   3. True   4. True   5. 40, 74AC, 74ACT series   6. 0 to 0.8 V; 2.0 to 5.0 V   7. 0 to 1.5 V; 3.5 to 5.0 V   8. As if the

input were HIGH   9. Unpredictably; it may overheat and be destroyed.
10. 74HCT and 74ACT   11. They describe exactly how to interconnect the chips for laying out the circuit and troubleshooting.   12. Inputs and outputs are defined, and logical relationships are described.

SECTION 4-11

1. Open inputs or outputs; inputs or outputs shorted to V_{CC}; inputs or outputs shorted to ground; pins shorted together; internal circuit failures   2. Pins shorted together   3. For TTL, a LOW; for CMOS, indeterminate   4. Two or more outputs connected together

SECTION 4-12

1. Open signal lines; shorted signal lines; faulty power supply; output loading
2. Broken wires; poor solder connections; cracks or cuts in PC board; bent or broken IC pins; faulty IC sockets   3. ICs operating erratically or not at all   4. Logic level indeterminate

SECTION 4-14

1. Electrically controlled connections are being programmed as open or closed.
2. (4, 1) (2, 2) or (2, 1) (4, 2)   3. (4, 5) (1, 6) or (4, 6) (1, 5)

SECTION 4-15

1. (a) push_buttons[5..0] :INPUT;   (b) push_buttons :IN BIT_VECTOR (5 DOWNTO 0),
2. (a) z = push_buttons[5]; (b) z <= push_buttons(5);   3. STD_LOGIC
4. STD_LOGIC_VECTOR

SECTION 4-16

1. (AHDL) omega[] = (x, y, z); (VHDL) omega <= x & y & z;   2. Using the keyword TABLE   3. Using selected signal assignments

SECTION 4-17

1. IF/THEN   2. IF/THEN/ELSE   3. CASE or IF/ELSIF   4. (AHDL) count[7..0] :INPUT; (VHDL) count :IN INTEGER RANGE 0 TO 205

CHAPTER 5

FLIP-FLOPS AND RELATED DEVICES

■ OUTLINE

■ OBJECTIVES

Upon completion of this chapter, you will be able to:

- ■ Construct and analyze the operation of a latch flip-flop made from NAND or NOR gates.
- ■ Describe the difference between synchronous and asynchronous systems.
- ■ Understand the operation of edge-triggered flip-flops.
- ■ Analyze and apply the various flip-flop timing parameters specified by the manufacturers.
- ■ Understand the major differences between parallel and serial data transfers.
- ■ Draw the output timing waveforms of several types of flip-flops in response to a set of input signals.
- ■ Recognize the various IEEE/ANSI flip-flop symbols.
- ■ Use state transition diagrams to describe counter operation.
- ■ Use flip-flops in synchronization circuits.
- ■ Connect shift registers as data transfer circuits.
- ■ Employ flip-flops as frequency-division and counting circuits.
- ■ Understand the typical characteristics of Schmitt triggers.
- ■ Apply two different types of one-shots in circuit design.
- ■ Design a free-running oscillator using a 555 timer.
- ■ Recognize and predict the effects of clock skew on synchronous circuits.
- ■ Troubleshoot various types of flip-flop circuits.
- ■ Write HDL code for latches.
- ■ Use logic primitives, components, and libraries in HDL code.
- ■ Build structural level circuits from components.

■ INTRODUCTION

The logic circuits considered thus far have been combinational circuits whose output levels at any instant of time are dependent on the levels present at the inputs at that time. Any prior input-level conditions have no

FIGURE 5-1 General digital system diagram.

effect on the present outputs because combinational logic circuits have no memory. Most digital systems consist of both combinational circuits and memory elements.

Figure 5-1 shows a block diagram of a general digital system that combines combinational logic gates with memory devices. The combinational portion accepts logic signals from external inputs and from the outputs of the memory elements. The combinational circuit operates on these inputs to produce various outputs, some of which are used to determine the binary values to be stored in the memory elements. The outputs of some of the memory elements, in turn, go to the inputs of logic gates in the combinational circuits. This process indicates that the external outputs of a digital system are functions of both its external inputs and the information stored in its memory elements.

The most important memory element is the **flip-flop**, which is made up of an assembly of logic gates. Even though a logic gate, by itself, has no storage capability, several can be connected together in ways that permit information to be stored. Several different gate arrangements are used to produce these flip-flops (abbreviated FF).

Figure 5-2(a) is the general type of symbol used for a flip-flop. It shows two outputs, labeled Q and $\overline{Q}$, that are the inverse of each other. $Q/\overline{Q}$ are the most common designations used for a FF's outputs. From time to time, we will use other designations such as $X/\overline{X}$ and $A/\overline{A}$ for convenience in identifying different FFs in a logic circuit.

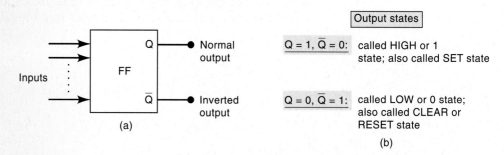

FIGURE 5-2 General flip-flop symbol and definition of its two possible output states.

The Q output is called the *normal* FF output, and $\overline{Q}$ is the *inverted* FF output. Whenever we refer to the state of a FF, we are referring to the state of its normal (Q) output; it is understood that its inverted output ($\overline{Q}$) is in the opposite state. For example, if we say that a FF is in the HIGH (1) state, we mean that $Q = 1$; if we say that a FF is in the LOW (0) state, we mean that $Q = 0$. Of course, the $\overline{Q}$ state will always be the inverse of Q.

The two possible operating states for a FF are summarized in Figure 5-2(b). Note that the HIGH or 1 state ($Q = 1/\overline{Q} = 0$) is also referred to as the **SET** state. Whenever the inputs to a FF cause it to go to the $Q = 1$ state, we call this *setting* the FF; the FF has been set. In a similar way, the LOW or 0 state ($Q = 0/\overline{Q} = 1$) is also referred to as the **CLEAR** or **RESET** state. Whenever the inputs to a FF cause it to go to the $Q = 0$ state, we call this *clearing* or *resetting* the FF; the FF has been cleared (reset). As we shall see, many FFs will have a **SET** input and/or a **CLEAR (RESET)** input that is used to drive the FF into a specific output state.

As the symbol in Figure 5-2(a) implies, a FF can have one or more inputs. These inputs are used to cause the FF to switch back and forth ("flip-flop") between its possible output states. We will find out that most FF inputs need only to be momentarily activated (pulsed) in order to cause a change in the FF output state, and the output will remain in that new state even after the input pulse is over. This is the FF's *memory* characteristic.

The flip-flop is known by other names, including *latch* and *bistable multivibrator*. The term *latch* is used for certain types of flip-flops that we will describe. The term *bistable multivibrator* is the more technical name for a flip-flop, but it is too much of a mouthful to be used regularly.

5-1 NAND GATE LATCH

The most basic FF circuit can be constructed from either two NAND gates or two NOR gates. The NAND gate version, called a **NAND gate latch** or simply a **latch**, is shown in Figure 5-3(a). The two NAND gates are cross-coupled so that the output of NAND-1 is connected to one of the inputs of NAND-2, and vice versa. The gate outputs, labeled Q and $\overline{Q}$, respectively, are the latch outputs. Under normal conditions, these outputs will always be the inverse of each other. There are two latch inputs: the SET input is the input that *sets* Q to the 1 state; the CLEAR input is the input that *clears* Q to the 0 state.

The SET and CLEAR inputs are both normally resting in the HIGH state, and one of them will be pulsed LOW whenever we want to change the latch outputs. We begin our analysis by showing that there are two equally likely output states when SET = CLEAR = 1. One possibility is shown in Figure 5-3(a), where we have $Q = 0$ and $\overline{Q} = 1$. With $Q = 0$, the inputs to NAND-2 are 0 and 1, which produce $\overline{Q} = 1$. The 1 from $\overline{Q}$ causes NAND-1 to have a 1 at both inputs to produce a 0 output at Q. In effect, what we have is the LOW at the NAND-1 output producing a HIGH at the NAND-2 output, which, in turn, keeps the NAND-1 output LOW.

The second possibility is shown in Figure 5-3(b), where $Q = 1$ and $\overline{Q} = 0$. The HIGH from NAND-1 produces a LOW at the NAND-2 output, which, in

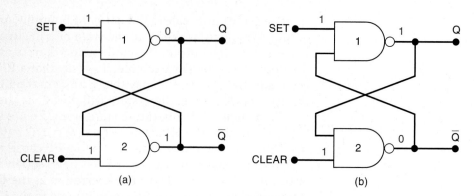

FIGURE 5-3 A NAND latch has two possible resting states when SET = CLEAR = 1.

turn, keeps the NAND-1 output HIGH. Thus, there are two possible output states when SET = CLEAR = 1; as we shall soon see, the one that actually exists will depend on what has occurred previously at the inputs.

Setting the Latch (FF)

Now let's investigate what happens when the SET input is momentarily pulsed LOW while CLEAR is kept HIGH. Figure 5-4(a) shows what happens when $Q = 0$ prior to the occurrence of the pulse. As SET is pulsed LOW at time t_0, Q will go HIGH, and this HIGH will force $\overline{Q}$ to go LOW so that NAND-1 now has two LOW inputs. Thus, when SET returns to the 1 state at t_1, the NAND-1 output *remains* HIGH, which, in turn, keeps the NAND-2 output LOW.

Figure 5-4(b) shows what happens when $Q = 1$ and $\overline{Q} = 0$ prior to the application of the SET pulse. Since $\overline{Q} = 0$ is already keeping the NAND-1 output HIGH, the LOW pulse at SET will not change anything. Thus, when SET returns HIGH, the latch outputs are still in the $Q = 1$, $\overline{Q} = 0$ state.

We can summarize Figure 5-4 by stating that a LOW pulse on the SET input will always cause the latch to end up in the $Q = 1$ state. This operation is called *setting* the latch or FF.

FIGURE 5-4 Pulsing the SET input to the 0 state when (a) $Q = 0$ prior to SET pulse; (b) $Q = 1$ prior to SET pulse. Note that, in both cases, Q ends up HIGH.

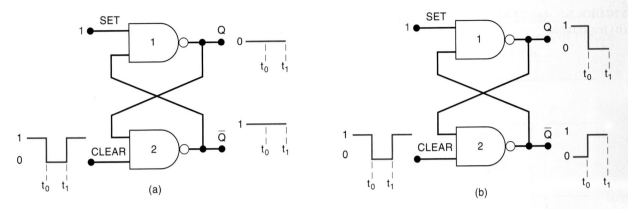

FIGURE 5-5 Pulsing the CLEAR input to the LOW state when (a) $Q = 0$ prior to CLEAR pulse; (b) $Q = 1$ prior to CLEAR pulse. In each case, Q ends up LOW.

Clearing the Latch (FF)

Now let's consider what occurs when the CLEAR input is pulsed LOW while SET is kept HIGH. Figure 5-5(a) shows what happens when $Q = 0$ and $\overline{Q} = 1$ prior to the application of the pulse. Since $Q = 0$ is already keeping the NAND-2 output HIGH, the LOW pulse at CLEAR will not have any effect. When CLEAR returns HIGH, the latch outputs are still $Q = 0$ and $\overline{Q} = 1$.

Figure 5-5(b) shows the situation where $Q = 1$ prior to the occurrence of the CLEAR pulse. As CLEAR is pulsed LOW at t_0, $\overline{Q}$ will go HIGH, and this HIGH forces Q to go LOW so that NAND-2 now has two LOW inputs. Thus, when CLEAR returns HIGH at t_1, the NAND-2 output *remains* HIGH, which, in turn, keeps the NAND-1 output LOW.

Figure 5-5 can be summarized by stating that a LOW pulse on the CLEAR input will always cause the latch to end up in the $Q = 0$ state. This operation is called *clearing* or *resetting* the latch.

Simultaneous Setting and Clearing

The last case to consider is the case where the SET and CLEAR inputs are simultaneously pulsed LOW. This will produce HIGH levels at both NAND outputs so that $Q = \overline{Q} = 1$. Clearly, this is an undesired condition because the two outputs are supposed to be inverses of each other. Furthermore, when the SET and CLEAR inputs return HIGH, the resulting output state will depend on which input returns HIGH first. Simultaneous transitions back to the 1 state will produce unpredictable results. For these reasons the SET = CLEAR = 0 condition is normally not used for the NAND latch.

Summary of NAND Latch

The operation described above can be conveniently placed in a truth table (Figure 5-6) and is summarized as follows:

1. SET = CLEAR = 1. This condition is the normal resting state, and it has no effect on the output state. The Q and $\overline{Q}$ outputs will remain in whatever state they were in prior to this input condition.

2. SET = 0, CLEAR = 1. This will always cause the output to go to the $Q = 1$ state, where it will remain even after SET returns HIGH. This is called *setting* the latch.

Solution

Initially, SET = CLEAR = 0, which has no effect on Q, and Q stays LOW. When SET goes HIGH at time T_1, Q will be set to 1 and will remain there even after SET returns to 0 at T_2.

At T_3 the CLEAR input goes HIGH and clears Q to the 0 state, where it remains even after CLEAR returns LOW at T_4.

The CLEAR pulse at T_5 has no effect on Q because Q is already LOW. The SET pulse at T_6 again sets Q back to 1, where it will stay.

Example 5-3 shows that the FF "remembers" the last input that was activated, and it will not change states until the opposite input is activated.

EXAMPLE 5-4

Figure 5-12 shows a simple circuit that can be used to detect the interruption of a light beam. The light is focused on a phototransistor that is connected in the common-emitter configuration to operate as a switch. Assume that the latch has previously been cleared to the 0 state by momentarily opening switch SW1, and describe what happens if the light beam is momentarily interrupted.

FIGURE 5-12   Example 5-4.

Solution

With light on the phototransistor, we can assume that it is fully conducting so that the resistance between the collector and the emitter is very small. Thus, v_0 will be close to 0 V. This places a LOW on the SET input of the latch so that SET = CLEAR = 0.

When the light beam is interrupted, the phototransistor turns off, and its collector-emitter resistance becomes very high (i.e., essentially an open circuit). This causes v_0 to rise to approximately 5 V; this activates the SET input, which sets Q HIGH and turns on the alarm.

Q will remain HIGH and the alarm will remain on even if v_0 returns to 0 V (i.e., the light beam was interrupted only momentarily) because SET and CLEAR will both be LOW, which will produce no change in Q.

In this application, the latch's memory characteristic is used to convert a momentary occurrence (beam interruption) into a constant output.

Flip-Flop State on Power-Up

When power is applied to a circuit, it is not possible to predict the starting state of a flip-flop's output if its SET and CLEAR inputs are in their inactive state (e.g., $S = C = 1$ for a NAND latch, $S = C = 0$ for a NOR latch). There is just as much chance that the starting state will be $Q = 0$ as $Q = 1$. It will depend on factors such as internal propagation delays, parasitic capacitance, and external loading. If a latch or FF must start off in a particular state to ensure the proper operation of a circuit, then it must be placed in that state by momentarily activating the SET or CLEAR input at the start of the circuit's operation. This is often achieved by application of a pulse to the appropriate input.

REVIEW QUESTIONS

1. What is the normal resting state of the NOR latch inputs? What is the active state?
2. When a FF is set, what are the states of Q and $\overline{Q}$?
3. What is the only way to cause the Q output of a NOR latch to change from 1 to 0?
4. If the NOR latch in Figure 5-12 were replaced by a NAND latch, why wouldn't the circuit work properly?

5-3 TROUBLESHOOTING CASE STUDY

The following two examples present an illustration of the kinds of reasoning used in troubleshooting a circuit containing a latch.

EXAMPLE 5-5

Analyze and describe the operation of the circuit in Figure 5-13.

Solution

The switch is used to set or clear the NAND latch to produce clean, bounce-free signals at Q and $\overline{Q}$. These latch outputs control the passage of the 1-kHz pulse signal through to the AND outputs X_A and X_B.

When the switch moves to position A, the latch is set to $Q = 1$. This enables the 1-kHz pulses to pass through to X_A, while the LOW at $\overline{Q}$ keeps $X_B = 0$. When the switch moves to position B, the latch is cleared to $Q = 0$, which keeps $X_A = 0$, while the HIGH at $\overline{Q}$ enables the pulses to pass through to X_B.

Switch position	X_A	X_B
A	Pulses	LOW
B	LOW	Pulses

FIGURE 5-13   Examples 5-5 and 5-6.

EXAMPLE 5-6

A technician tests the circuit of Figure 5-13 and records the observations shown in Table 5-1. He notices that when the switch is in position B, the circuit functions correctly, but in position A the latch does not set to the $Q = 1$ state. What are the possible faults that could produce this malfunction?

TABLE 5-1

Switch Position	SET (Z1-1)	CLEAR (Z1-5)	Q (Z1-3)	$\overline{Q}$ (Z1-6)	X_A (Z2-3)	X_B (Z2-6)
A	LOW	HIGH	LOW	HIGH	LOW	Pulses
B	HIGH	LOW	LOW	HIGH	LOW	Pulses

Solution

There are several possibilities:

1. An internal open connection at Z1-1, which would prevent Q from responding to the $\overline{SET}$ input.

2. An internal component failure in NAND gate Z1 that prevents it from responding properly.

3. The Q output is stuck LOW, which could be caused by:
   (a) Z1-3 internally shorted to ground
   (b) Z1-4 internally shorted to ground
   (c) Z2-2 internally shorted to ground
   (d) The Q node externally shorted to ground

An ohmmeter check from Q to ground will determine if any of these conditions are present. A visual check should reveal any external short.

What about $\overline{Q}$ internally or externally shorted to V_{CC}? A little thought will lead to the conclusion that this could not be the fault. If $\overline{Q}$ were shorted to V_{CC}, this would not prevent the Q output from going HIGH when $\overline{SET}$ goes LOW. Because Q *does not* go HIGH, this cannot be the fault. The reason that $\overline{Q}$ looks as if it is stuck HIGH is that Q is stuck LOW, and that keeps $\overline{Q}$ HIGH through the bottom NAND gate.

5-4 CLOCK SIGNALS AND CLOCKED FLIP-FLOPS

Digital systems can operate either *asynchronously* or *synchronously*. In asynchronous systems, the outputs of logic circuits can change state any time one or more of the inputs change. An asynchronous system is generally more difficult to design and troubleshoot than a synchronous system.

In synchronous systems, the exact times at which any output can change states are determined by a signal commonly called the **clock**. This clock signal is generally a rectangular pulse train or a square wave, as shown in Figure 5-14. The clock signal is distributed to all parts of the system, and most (if not all) of the system outputs can change state only when the clock makes a transition. The transitions (also called *edges*) are pointed out in Figure 5-14. When the clock changes from a 0 to a 1, this is called the **positive-going transition (PGT)**; when the clock goes from 1 to 0, this is the **negative-going transition (NGT)**. We will use the abbreviations PGT and NGT because these terms appear so often throughout the text.

Most digital systems are principally synchronous (although there are always some asynchronous parts) because synchronous circuits are easier to design and troubleshoot. They are easier to troubleshoot because the circuit outputs can change only at specific instants of time. In other words, almost everything is synchronized to the clock-signal transitions.

The synchronizing action of the clock signals is accomplished through the use of **clocked flip-flops** that are designed to change states on one or the other of the clock's transitions.

FIGURE 5-14   Clock signals.

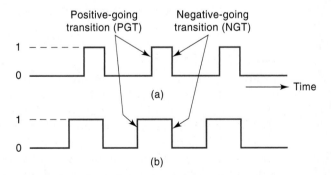

Clocked Flip-Flops

Several types of clocked FFs are used in a wide range of applications. Before we begin our study of the different clocked FFs, we will describe the principal ideas that are common to all of them.

FIGURE 5-15 Clocked FFs have a clock input (*CLK*) that is active on either (a) the PGT or (b) the NGT. The control inputs determine the effect of the active clock transition.

CLK is activated by a PGT

(a)

CLK is activated by an NGT

(b)

1. Clocked FFs have a clock input that is typically labeled *CLK, CK,* or *CP*. We will normally use *CLK*, as shown in Figure 5-15. In most clocked FFs, the *CLK* input is **edge-triggered**, which means that it is activated by a signal transition; this is indicated by the presence of a small triangle on the *CLK* input. This contrasts with the latches, which are level-triggered.

    Figure 5-15(a) is a FF with a small triangle on its *CLK* input to indicate that this input is activated *only* when a positive-going transition (PGT) occurs; no other part of the input pulse will have an effect on the *CLK* input. In Figure 5-15(b), the FF symbol has a bubble as well as a triangle on its *CLK* input. This signifies that the *CLK* input is activated *only* when a negative-going transition occurs; no other part of the input pulse will have an effect on the *CLK* input.

2. Clocked FFs also have one or more **control inputs** that can have various names, depending on their operation. The control inputs will have no effect on *Q* until the active clock transition occurs. In other words, their effect is synchronized with the signal applied to *CLK*. For this reason they are called **synchronous control inputs**.

    For example, the control inputs of the FF in Figure 5-15(a) will have no effect on *Q* until the PGT of the clock signal occurs. Likewise, the control inputs in Figure 5-15(b) will have no effect until the NGT of the clock signal occurs.

3. In summary, we can say that the control inputs get the FF outputs ready to change, while the active transition at the *CLK* input actually *triggers* the change. The control inputs control the WHAT (i.e., what state the output will go to); the *CLK* input determines the WHEN.

Setup and Hold Times

Two timing requirements must be met if a clocked FF is to respond reliably to its control inputs when the active *CLK* transition occurs. These requirements are illustrated in Figure 5-16 for a FF that triggers on a PGT.

The **setup time, t_S**, is the time interval immediately preceding the active transition of the *CLK* signal during which the control input must be maintained at the proper level. IC manufacturers usually specify the minimum allowable setup time $t_S(\text{min})$. If this time requirement is not met, the FF may not respond reliably when the clock edge occurs.

The **hold time, t_H**, is the time interval immediately following the active transition of the *CLK* signal during which the synchronous control input must be maintained at the proper level. IC manufacturers usually specify the minimum acceptable value of hold time $t_H(\text{min})$. If this requirement is not met, the FF will not trigger reliably.

FIGURE 5-16 Control inputs must be held stable for (a) a time t_S prior to active clock transition and for (b) a time t_H after the active block transition.

Thus, to ensure that a clocked FF will respond properly when the active clock transition occurs, the control inputs must be stable (unchanging) for at least a time interval equal to t_S(min) *prior* to the clock transition, and for at least a time interval equal to t_H(min) *after* the clock transition.

IC flip-flops will have minimum allowable t_S and t_H values in the nanosecond range. Setup times are usually in the range of 5 to 50 ns, whereas hold times are generally from 0 to 10 ns. Notice that these times are measured between the 50 percent points on the transitions.

These timing requirements are very important in synchronous systems because, as we shall see, there will be many situations where the synchronous control inputs to a FF are changing at approximately the same time as the *CLK* input.

1. What two types of inputs does a clocked FF have?
2. What is meant by the term *edge-triggered*?
3. *True or false:* The *CLK* input will affect the FF output only when the active transition of the control input occurs.
4. Define the setup time and hold time requirements of a clocked FF.

5-5 CLOCKED S-C FLIP-FLOP

Figure 5-17(a) shows the logic symbol for a **clocked S-C flip-flop** that is triggered by the positive-going edge of the clock signal. This means that the FF can change states *only* when a signal applied to its clock input makes a transition from 0 to 1. The S and C inputs control the state of the FF in the same manner as described earlier for the NOR gate latch, but the FF does not respond to these inputs until the occurrence of the PGT of the clock signal.

The truth table in Figure 5-17(b) shows how the FF output will respond to the PGT at the *CLK* input for the various combinations of S and C inputs. This truth table uses some new nomenclature. The up arrow ($\uparrow$) indicates that a PGT is required at *CLK*; the label Q_0 indicates the level at Q prior to the PGT. This nomenclature is often used by IC manufacturers in their IC data manuals.

Inputs			Output
S	C	CLK	Q
0	0	↑	Q_0 (no change)
1	0	↑	1
0	1	↑	0
1	1	↑	Ambiguous

Q_0 is output level prior to ↑ of CLK.
↓ of CLK produces no change in Q.

(b)

(a)

FF triggers on positive transition

(c)

FIGURE 5-17 (a) Clocked S-C flip-flop that responds only to the positive-going edge of a clock pulse; (b) truth table; (c) typical waveforms.

The waveforms in Figure 5-17(c) illustrate the operation of the clocked S-C flip-flop. If we assume that the setup and hold time requirements are being met in all cases, we can analyze these waveforms as follows:

1. Initially all inputs are 0 and the Q output is assumed to be 0; that is, $Q_0 = 0$.
2. When the PGT of the first clock pulse occurs (point *a*), the S and C inputs are both 0, so the FF is not affected and remains in the $Q = 0$ state (i.e., $Q = Q_0$).
3. At the occurrence of the PGT of the second clock pulse (point *c*), the S input is now high, with C still low. Thus, the FF sets to the 1 state at the rising edge of this clock pulse.
4. When the third clock pulse makes its positive transition (point *e*), it finds that $S = 0$ and $C = 1$, which causes the FF to clear to the 0 state.

5. The fourth pulse sets the FF once again to the $Q = 1$ state (point g) because $S = 1$ and $C = 0$ when the positive edge occurs.

6. The fifth pulse also finds that $S = 1$ and $C = 0$ when it makes its positive-going transition. However, Q is already high, so it remains in that state.

7. The $S = C = 1$ condition should not be used because it results in an ambiguous condition.

It should be noted from these waveforms that the FF is not affected by the negative-going transitions of the clock pulses. Also, note that the S and C levels have no effect on the FF, except upon the occurrence of a positive-going transition of the clock signal. The S and C inputs are synchronous *control* inputs; they control which state the FF will go to when the clock pulse occurs. The *CLK* input is the **trigger** input that causes the FF to change states according to what the S and C inputs are when the active clock transition occurs.

Figure 5-18 shows the symbol and the truth table for a clocked S-C flip-flop that triggers on the *negative*-going transition at its *CLK* input. The small circle and triangle on the *CLK* input indicates that this FF will trigger only when the *CLK* input goes from 1 to 0. This FF operates in the same manner as the positive-edge FF except that the output can change states only on the falling edge of the clock pulses (points b, d, f, h, and j in Figure 5-17). Both positive-edge and negative-edge triggering FFs are used in digital systems.

FIGURE 5-18 Clocked S-C flip-flop that triggers only on negative-going transitions.

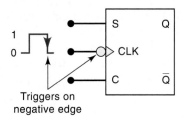

Inputs			Output
S	C	CLK	Q
0	0	↓	Q_0 (no change)
1	0	↓	1
0	1	↓	0
1	1	↓	Ambiguous

Internal Circuitry of the Edge-Triggered S-C Flip-Flop

A detailed analysis of the internal circuitry of a clocked FF is not necessary because all types are readily available as ICs. Although our main interest is in the FF's external operation, our understanding of this external operation can be aided by taking a look at a simplified version of the FF's internal circuitry. Figure 5-19 shows this for an edge-triggered S-C flip-flop.

The circuit contains three sections:

1. A basic NAND gate latch formed by NAND-3 and NAND-4

2. A **pulse-steering circuit** formed by NAND-1 and NAND-2

3. An **edge-detector circuit**

As shown in Figure 5-19, the edge detector produces a narrow positive-going spike (*CLK**) that occurs coincident with the active transition of the *CLK* input pulse. The pulse-steering circuit "steers" the spike through to the SET or the CLEAR input of the latch in accordance with the levels present at S and C. For example, with $S = 1$ and $C = 0$, the *CLK** signal is inverted and passed through NAND-1 to produce a LOW pulse at the SET input of the latch that sets $Q = 1$. With $S = 0$, $C = 1$, the *CLK** signal is inverted and passed

FIGURE 5-19 Simplified version of the internal circuitry for an edge-triggered S-C flip-flop.

through NAND-2 to produce a low pulse at the CLEAR input of the latch that resets $Q = 0$.

Figure 5-20(a) shows how the *CLK*★ signal is generated for edge-triggered FFs that trigger on a PGT. The INVERTER produces a delay of a few nanoseconds so that the transitions of $\overline{CLK}$ occur a little bit after those of *CLK*. The AND gate produces an output spike that is HIGH only for the few nanoseconds when *CLK* and $\overline{CLK}$ are both HIGH. The result is a narrow pulse at *CLK*★, which occurs on the PGT of *CLK*. The arrangement of Figure 5-20(b) likewise produces *CLK*★ on the NGT of *CLK* for FFs that are to trigger on a NGT.

Because the *CLK*★ signal is HIGH for only a few nanoseconds, Q is affected by the levels at S and C only for a short time during and after the occurrence of the active edge of *CLK*. This is what gives the FF its edge-triggered property.

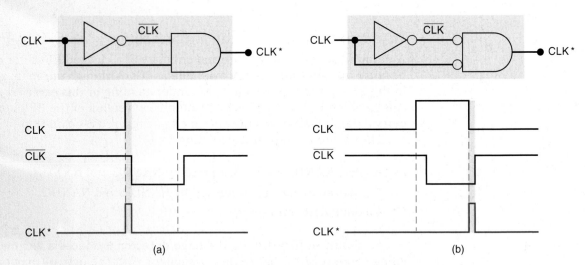

FIGURE 5-20 Implementation of edge-detector circuits used in edge-triggered flip-flops: (a) PGT; (b) NGT. The duration of the *CLK*★ pulses is typically 2–5 nanoseconds.

REVIEW QUESTIONS

1. Suppose that the waveforms of Figure 5-17(c) are applied to the inputs of the FF of Figure 5-18. What will happen to Q at point b? At point f? At point h?

2. Explain why the S and C inputs affect Q only during the active transition of CLK.

5-6   CLOCKED J-K FLIP-FLOP

Figure 5-21(a) shows a **clocked J-K flip-flop** that is triggered by the positive-going edge of the clock signal. The J and K inputs control the state of the FF in the same ways as the S and C inputs do for the clocked S-C flip-flop except for one major difference: *the $J = K = 1$ condition does not result in an ambiguous output.* For this 1, 1 condition, the FF will always go to its *opposite* state upon the positive transition of the clock signal. This is called the **toggle mode** of operation. In this mode, if both J and K are left HIGH, the FF will change states (toggle) for each PGT of the clock.

The truth table in Figure 5-21(a) summarizes how the J-K flip-flop responds to the PGT for each combination of J and K. Notice that the truth table is the same as for the clocked S-C flip-flop (Figure 5-17) except for the $J = K = 1$ condition. This condition results in $Q = \overline{Q}_0$, which means that the

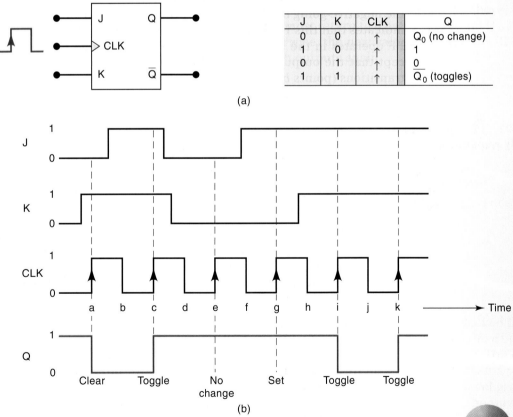

J	K	CLK	Q
0	0	↑	Q_0 (no change)
1	0	↑	1
0	1	↑	0
1	1	↑	$\overline{Q}_0$ (toggles)

(a)

(b)

FIGURE 5-21   (a) Clocked J-K flip-flop that responds only to the positive edge of the clock; (b) waveforms.

FIGURE 5-35 Q_2 will respond properly to the level present at Q_1 prior to the NGT of *CLK*, provided that Q_2's hold time requirement, t_H, is less than Q_1's propagation delay.

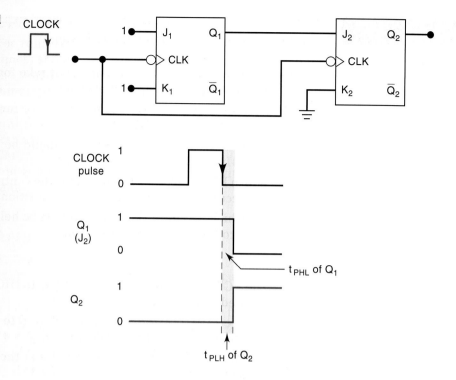

Fortunately, all modern edge-triggered FFs have hold time requirements that are 5 ns or less; most have $t_H = 0$, which means that they have no hold time requirement. For these FFs, situations like that in Figure 5-35 will not be a problem.

Unless stated otherwise, in all of the FF circuits that we encounter throughout the text, we will assume that the FF's hold time requirement is short enough to respond reliably according to the following rule:

The FF output will go to a state determined by the logic levels present at its synchronous control inputs just prior to the active clock transition.

If we apply this rule to Figure 5-35, it says that Q_2 will go to a state determined by the $J_2 = 1, K_2 = 0$ condition that is present just prior to the NGT of the clock pulse. The fact that J_2 is changing in response to the same NGT has no effect.

EXAMPLE 5-10

Determine the Q output for a negative-edge-triggered J-K flip-flop for the input waveforms shown in Figure 5-36. Assume that $t_H = 0$ and that $Q = 0$ initially.

Solution

The FF will respond only at times T_2, T_4, T_6, and T_8. At T_2, Q will respond to the $J = K = 0$ condition present just prior to T_2. At T_4, Q will respond to the $J = 1$, $K = 0$ condition present just prior to T_4. At T_6, Q will respond to the $J = 0$, $K = 1$ condition present just prior to T_6. At T_8, Q responds to $J = K = 1$.

FIGURE 5-36   Example 5-10.

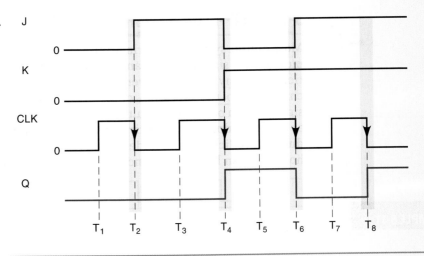

5-13   MASTER/SLAVE FLIP-FLOPS

Before the development of edge-triggered FFs with little or no hold-time re-quirements, timing problems such as that in Figure 5-35 were often handled by using a class of FFs called **master/slave FFs.** A master/slave FF actually contains two FFs—a master and a slave. On the rising edge of the *CLK* sig-nal, the levels on the control inputs (*D, J, K*) are used to determine the out-put of the master. When the *CLK* signal goes LOW, the state of the master is transferred to the slave, whose outputs are Q and $\overline{Q}$. Thus, Q and $\overline{Q}$ change just after the NGT of the clock. These master/slave FFs function very much like the negative-edge-triggered FFs except for one major disadvantage: the control inputs must be held stable while *CLK* is HIGH, or unpredictable operation may occur. This problem with master/slave FFs has been over-come with an improved master/slave version called a *master/slave with data lockout.*

The master/slave FF has become obsolete, although you may encounter it in older equipment. Examples of this type are the standard TTL 7473, 7476, and 74107, and the data lockout versions, 74110 and 74111. The newer IC technologies (74LS, 74AS, 74ALS, 74HC, 74HCT) do not include any mas-ter/slave FFs in their series. In fact, the 74LS76 and 74LS107 have been man-ufactured as edge-triggered FFs even though their standard 74 series counterparts are master/slave.

For most purposes, if you encounter a master/slave FF in a piece of equip-ment, you can analyze its operation like a negative-edge-triggered FF.

5-14   FLIP-FLOP APPLICATIONS

Edge-triggered (clocked) flip-flops are versatile devices that can be used in a wide variety of applications including counting, storing of binary data, transferring binary data from one location to another, and many more. Al-most all of these applications utilize the FF's clocked operation. Many of them fall into the category of **sequential circuits.** A sequential circuit is one in which the outputs follow a predetermined sequence of states, with a new state occurring each time a clock pulse occurs. We will introduce some of the basic applications in the following sections, and we will expand on them in subsequent chapters.

Once you are clear about how these Boolean equations have been developed for the NAND latch and *D* latch, you may wish to try them out in the lab by programming them into a PLD using an HDL.

1. Figure 5-6(a) is the NAND gate latch drawn using the standard representation for NAND gates. Write the Boolean equations in an HDL format for this latch. Do you expect this latch to operate in the same way as the file in Figures 5-61 *and* 5-62?
2. What is the distinguishing hardware characteristic of latching logic circuits?
3. What is the major characteristic of sequential circuits?

5-27   EDGE-TRIGGERED DEVICES

Earlier in this chapter, we introduced edge-triggered devices whose outputs respond to the inputs when the clock input sees an "edge." An edge simply means a transition from HIGH to LOW, or vice versa, and is often referred to as an **event**. If we are writing statements in the code that are concurrent, how can outputs change only when a clock input detects an edge event? The answer to this question differs substantially, depending on the HDL you use. In this section, we want to concentrate on creating clocked logic circuits in their simplest form using HDL. We will use JK flip-flops to correlate with many of the examples found earlier in this chapter.

The JK flip-flop is a standard building block of clocked (sequential) logic circuits known as a **logic primitive**. In its most common form, it has five inputs and two outputs, as shown in Figure 5-63. The input/output names can be standardized to allow us to refer to the connections of this primitive or fundamental circuit. The actual operation of the primitive circuit is defined in a library of components that is available to the HDL compiler as it generates a circuit from our description. AHDL uses logic primitives to describe flip-flop operation. VHDL offers something similar, but it also allows the designer to describe the clocked logic circuit's operation explicitly in the code.

FIGURE 5-63   JK flip-flop logic primitive.

AHDL FLIP-FLOPS

A flip-flop can be used in AHDL by declaring a register (even one flip-flop is called a register). Several different types of register primitives are available for use in AHDL, for example, JK, D, SR, and latch register primitives. Each different type of register primitive has its own official identifiers (according to Altera software) for the ports of these primitives. These can be found by using the HELP menu in the MAX+PLUS software and looking under Primitives. Table 5-3 lists some of these identifiers. Registers that use these primitives are declared in the VARIABLE section of the code. The register is given an instance name, just as we have named intermediate variables or buried nodes in previous examples. Instead of declaring it as a node, however, it is declared by the type of the register primitive. For example, a JK flip-flop can be declared as:

```
VARIABLE
    ff1 :JKFF;
```

TABLE 5-3 Altera primitive port identifiers.

Standard Part Function	Primitive Identifier
Clock input	clk
Asynchronous preset (active-LOW)	prn
Asynchronous clear (active-LOW)	clrn
J, K, S, R, D inputs	j, k, s, r, d
Level triggered ENABLE input	ena
Q output	q

The instance name is ff1 (which you can make up) and the register primitive type is JKFF (which Altera requires you to use). Once you have declared a register, it is connected to the other logic in the design using its standard port identifiers. The ports (or pins) on the flip-flop are referred to using the instance name, with a dot extension that designates the particular input or output. An example for a JK flip-flop in AHDL is shown in Figure 5-64. Notice that we have made up our own input/output names for this SUBDESIGN in order to distinguish them from the primitive port identifiers. The single flip-flop is declared on line 8, as previously described. The J input or port for this device is then labeled *ff1.j*, the K input is *ff1.k*, the clock input is *ff1.clk*, and so on. Each of the given port assignment statements will make the needed wiring connections for this design block. The *prn* and *clrn* ports are both active-LOW, asynchronous controls such as those commonly found on a standard flip-flop. These asynchronous controls are optional in AHDL and will default to a disabled condition (at a logic 1) if they are omitted from the logic section. In other words, if lines 10 and 11 were deleted, the *prn* and *clrn* ports of ff1 would automatically be tied to V_{CC}.

```
1    %      JK flip-flop circuit      %
2    SUBDESIGN fig5_64
3    (
4         jin, kin, clkin, preset, clear   :INPUT;
5         qout                             :OUTPUT;
6    )
7    VARIABLE
8    ff1         :JKFF;          -- define this flip-flop as a JKFF type
9    BEGIN
10        ff1.prn = preset;  -- these are optional and default to vcc
11        ff1.clrn = clear;
12        ff1.j = jin;       -- connect primitive to the input signal
13        ff1.k = kin;
14        ff1.clk = clkin;
15        qout = ff1.q;      -- connect the output pin to the primitive
16   END;
```

FIGURE 5-64   Single JK flip-flop using AHDL.

D   5-26. In the circuit of Figure 5-83, inputs *A*, *B*, and *C* are all initially LOW. Output *Y* is supposed to go HIGH only when *A*, *B*, and *C* go HIGH in a certain sequence.

         (a) Determine the sequence that will make *Y* go HIGH.

         (b) Explain why the START pulse is needed.

         (c) Modify this circuit to use D FFs.

FIGURE 5-83   Problem 5-26.

SECTIONS 5-17 AND 5-18

D   5-27. (a) Draw a circuit diagram for the synchronous parallel transfer of data from one three-bit register to another using J-K flip-flops.

         (b) Repeat for asynchronous parallel transfer.

     5-28. A *recirculating* shift register is a shift register that keeps the binary information circulating through the register as clock pulses are applied. The shift register of Figure 5-43 can be made into a circulating register by connecting X_0 to the DATA IN line. No external inputs are used. Assume that this circulating register starts out with 1011 stored in it (i.e., $X_3 = 1$, $X_2 = 0$, $X_1 = 1$, and $X_0 = 1$). List the sequence of states that the register FFs go through as eight shift pulses are applied.

D   5-29. Refer to Figure 5-44, where a three-bit number stored in register *X* is serially shifted into register *Y*. How can the circuit be modified so that, at the end of the transfer operation, the original number stored in *X* is present in both registers? (*Hint:* See Problem 5-28.)

SECTION 5-19

B   5-30. Refer to the counter circuit of Figure 5-45 and answer the following:

         (a) If the counter starts at 000, what will be the count after 13 clock pulses? After 99 pulses? After 256 pulses?

         (b) If the counter starts at 100, what will be the count after 13 pulses? After 99 pulses? After 256 pulses?

         (c) Connect a fourth J-K FF (X_3) to this counter and draw the state transition diagram for this 4-bit counter. If the input clock frequency is 80 MHz, what will the waveform at X_3 look like?

     5-31. Refer to the binary counter of Figure 5-45. Change it by connecting $\overline{X_0}$ to the *CLK* of flip-flop X_1, and $\overline{X_1}$ to the *CLK* of flip-flop X_2. Start with all FFs in the 1 state, and draw the various FF output waveforms (X_0, X_1, X_2) for 16 input pulses. Then list the sequence of FF states as was done in Figure 5-46. This counter is called a *down counter*. Why?

     5-32. Draw the state transition diagram for this down counter, and compare it with the diagram of Figure 5-47. How are they different?

B   5-33. (a) How many FFs are required to build a binary counter that counts from 0 to 1023?

(b) Determine the frequency at the output of the last FF of this counter for an input clock frequency of 2 MHz.

(c) What is the counter's MOD number?

(d) If the counter is initially at zero, what count will it hold after 2060 pulses?

5-34. A binary counter is being pulsed by a 256-kHz clock signal. The output frequency from the last FF is 2 kHz.

(a) Determine the MOD number.

(b) Determine the counting range.

5-35. A photodetector circuit is being used to generate a pulse each time a customer walks into a certain establishment. The pulses are fed to an eight-bit counter. The counter is used to count these pulses as a means for determining how many customers have entered the store. After closing the store, the proprietor checks the counter and finds that it shows a count of $00001001_2 = 9_{10}$. He knows that this is incorrect because there were many more than nine people in his store. Assuming that the counter circuit is working properly, what could be the reason for the discrepancy?

SECTION 5-20

D   5-36. Modify the circuit of Figure 5-48 so that only the presence of address code 10110110 will allow data to be transferred to register X.

T   5-37. Suppose that the circuit of Figure 5-48 is malfunctioning so that data are being transferred to X for either of the address codes 11111110 or 11111111. What are some circuit faults that could be causing this?

D   5-38. Modify the circuit of Figure 5-48 so that the MPU has eight data output lines connected to transfer eight bits of data to an eight-bit register made up of two 74HC175 ICs [Figure 5-32(b)]. Show all circuit connections.

SECTION 5-22

B   5-39. Refer to the waveforms in Figure 5-51(a). Change the OS pulse duration to 0.5 ms and determine the Q output for both types of OS. Then repeat using a OS pulse duration of 1.5 ms.

N   5-40. Figure 5-84 shows three nonretriggerable one-shots connected in a timing chain that produces three sequential output pulses. Note the "1" in front of the pulse on each OS symbol to indicate nonretriggerable operation. Draw a timing diagram showing the relationship between the input pulse and the three OS outputs. Assume an input pulse duration of 10 ms.

FIGURE 5-84   Problem 5-40.

2. If the sum is 15 or less, it can be directly expressed as a hex digit.

3. If the sum is greater than or equal to 16, subtract 16 and carry a 1 to the next digit position.

The following examples will illustrate the procedure.

EXAMPLE 6-6

Add the hex numbers 58 and 24.

Solution

```
 58
+24
 7C
```

Adding the LSDs (8 and 4) produces 12, which is C in hex. There is no carry into the next digit position. Adding 5 and 2 produces 7.

EXAMPLE 6-7

Add the hex numbers 58 and 4B.

Solution

```
 58
+4B
 A3
```

Start by adding 8 and B, mentally substituting decimal 11 for B. This produces a sum of 19. Because 19 is greater than 16, subtract 16 to get 3; write down the 3 and carry a 1 into the next position. This carry is added to the 5 and 4 to produce a sum of 10_{10}, which is then converted to hexadecimal A.

EXAMPLE 6-8

Add 3AF to 23C.

Solution

```
 3AF
+23C
 5EB
```

The sum of F and C is considered as $15 + 12 = 27_{10}$. Because this is greater than 16, subtract 16 to get 11_{10}, which is hexadecimal B, and carry a 1 into the second position. Add this carry to A and 3 to obtain E. There is no carry into the MSD position.

Hex Subtraction

Remember that hex numbers are just an efficient way to represent binary numbers. Thus, we can subtract hex numbers using the same method we used for binary numbers. The 2's complement of the hex subtrahend will be taken and then *added* to the minuend, and any carry out of the MSD position will be disregarded.

How do we find the 2's complement of a hex number? One way is to convert it to binary, take the 2's complement of the binary equivalent, and then convert it back to hex. This process is illustrated below.

$$
\begin{array}{ccc}
 & 73A & \leftarrow \text{hex number} \\
0111 \quad 0011 \quad 1010 & & \leftarrow \text{convert to binary} \\
1000 \quad 1100 \quad 0110 & & \leftarrow \text{take 2's complement} \\
 & 8C6 & \leftarrow \text{convert back to hex}
\end{array}
$$

There is a quicker procedure: subtract *each* hex digit from F; then add 1. Let's try this for the same hex number from the example above.

$$
\begin{array}{ccc}
F & F & F \\
-7 & -3 & -A \\
\hline
8 & C & 5
\end{array} \leftarrow \text{subtract each digit from F}
$$

$$+1 \leftarrow \text{add 1}$$

$$8 \quad C \quad 6 \leftarrow \text{hex equivalent of 2's complement}$$

Try either of the procedures above on the hex number E63. The correct result for the 2's complement is 19D.

EXAMPLE 6-9

Subtract $3A5_{16}$ from 592_{16}.

Solution

First, convert the subtrahend (3A5) to its 2's-complement form by using either method presented above. The result is C5B. Then add this to the minuend (592):

$$
\begin{array}{r}
592 \\
+ \underline{C5B} \\
11ED
\end{array}
$$

Disregard carry.

Ignore the carry out of the MSD addition; the result is 1ED. We can prove that this is correct by adding 1ED to 3A5 and checking to see that it equals 592_{16}.

Hex Representation of Signed Numbers

The data stored in a microcomputer's internal working memory or on a hard disk or CD ROM are typically stored in bytes (groups of eight bits). The data byte stored in a particular memory location is often expressed in hexadecimal because it is more efficient and less error-prone than expressing it in binary. When the data consist of *signed* numbers, it is helpful to be able to recognize whether a hex value represents a positive or a negative number. For example, Table 6-2 lists the data stored in a small segment of memory starting at address 4000.

Each memory location stores a single byte (eight bits), which is the binary equivalent of a signed decimal number. The table also shows the hex equivalent

TABLE 6-2

Hex Address	Stored Binary Data	Hex Value	Decimal Value
4000	00111010	3A	+58
4001	11100101	E5	−29
4002	01010111	57	+87
4003	10000000	80	−128

of each byte. For a negative data value, the sign bit (MSB) of the binary number will be a 1; this will always make the MSD of the hex number 8 or greater. When the data value is positive, the sign bit will be a 0, and the MSD of the hex number will be 7 or less. The same holds true no matter how many digits are in the hex number. *When the MSD is 8 or greater, the number being represented is negative; when the MSD is 7 or less, the number is positive.*

REVIEW QUESTIONS

1. Add 67F + 2A4.
2. Subtract 67F − 2A4.
3. Which of the following hex numbers represent positive values: 2F, 77EC, C000, 6D, FFFF?

6-9   ARITHMETIC CIRCUITS

One essential function of most computers and calculators is the performance of arithmetic operations. These operations are all performed in the arithmetic/logic unit of a computer, where logic gates and flip-flops are combined so that they can add, subtract, multiply, and divide binary numbers. These circuits perform arithmetic operations at speeds that are not humanly possible. Typically, an addition operation will take less than 100 ns.

We will now study some of the basic arithmetic circuits that are used to perform the arithmetic operations discussed earlier. In some cases, we will go through the actual design process, even though the circuits may be commercially available in integrated-circuit form, to provide more practice in the use of the techniques learned in Chapter 4.

Arithmetic/Logic Unit

All arithmetic operations take place in the **arithmetic/logic unit (ALU)** of a computer. Figure 6-3 is a block diagram showing the major elements included in a typical ALU. The main purpose of the ALU is to accept binary data that are stored in the memory and to execute arithmetic and logic operations on these data according to instructions from the control unit.

The arithmetic/logic unit contains at least two flip-flop registers: the *B register* and the **accumulator register.** It also contains combinational logic, which performs the arithmetic and logic operations on the binary numbers that are stored in the *B* register and the accumulator. A typical sequence of operations may occur as follows:

FIGURE 6-3 Functional parts of an ALU.

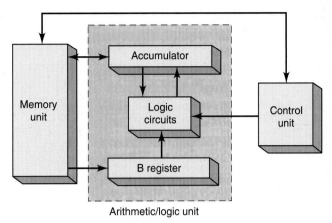

Arithmetic/logic unit

1. The control unit receives an instruction (from the memory unit) specifying that a number stored in a particular memory location (address) is to be added to the number presently stored in the accumulator register.
2. The number to be added is transferred from memory to the *B* register.
3. The number in the *B* register and the number in the accumulator register are added together in the logic circuits (upon command from the control unit). The resulting sum is then sent to the accumulator to be stored.
4. The new number in the accumulator can remain there so that another number can be added to it or, if the particular arithmetic process is finished, it can be transferred to memory for storage.

These steps should make it apparent how the accumulator register derives its name. This register "accumulates" the sums that occur when performing successive additions between new numbers acquired from memory and the previously accumulated sum. In fact, for any arithmetic problem containing several steps, the accumulator usually contains the results of the intermediate steps as they are completed as well as the final result when the problem is finished.

6-10   PARALLEL BINARY ADDER

Computers and calculators perform the addition operation on two binary numbers at a time, where each binary number can have several binary digits. Figure 6-4 illustrates the addition of two five-bit numbers. The **augend** is stored in the accumulator register; that is, the accumulator contains five FFs, storing the values 10101 in successive FFs. Similarly, the **addend**, the number that is to be added to the augend, is stored in the *B* register (in this case, 00111).

The addition process starts by adding the least significant bits (LSBs) of the augend and addend. Thus, $1 + 1 = 10$, which means that the *sum* for that position is 0, with a *carry* of 1.

This carry must be added to the next position along with the augend and addend bits in that position. Thus, in the second position, $1 + 0 + 1 = 10$, which is again a sum of 0 and a carry of 1. This carry is added to the next position together with the augend and addend bits in that position, and so on, for the remaining positions, as shown in Figure 6-4.

FIGURE 6-4   Typical binary addition process.

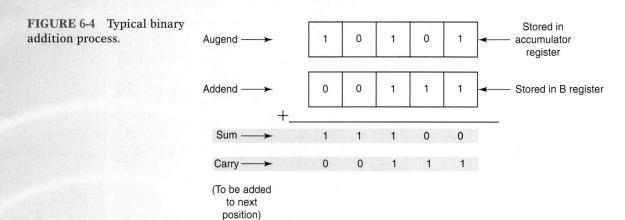

| | 1 | 0 | 1 | 0 | 1 | Augend → Stored in accumulator register |

| | 0 | 0 | 1 | 1 | 1 | Addend → Stored in B register |

| Sum → | 1 | 1 | 1 | 0 | 0 |

| Carry → | 0 | 0 | 1 | 1 | 1 |

(To be added to next position)

At each step in this addition process, we are performing the addition of three bits: the augend bit, the addend bit, and a carry bit from the previous position. The result of the addition of these three bits produces two bits: a *sum* bit, and a *carry* bit that is to be added to the next position. It should be clear that the same process is followed for each bit position. Thus, if we can design a logic circuit that can duplicate this process, then all we have to do is to use the identical circuit for each of the bit positions. This is illustrated in Figure 6-5.

In this diagram, variables A_4, A_3, A_2, A_1, and A_0 represent the bits of the augend that are stored in the accumulator (which is also called the A register). Variables B_4, B_3, B_2, B_1, and B_0 represent the bits of the addend stored in the B register. Variables C_4, C_3, C_2, C_1, and C_0 represent the carry bits into the corresponding positions. Variables S_4, S_3, S_2, S_1, S_0 are the sum output bits for each position. Corresponding bits of the augend and addend are fed to a logic circuit called a **full adder (FA)**, along with a carry bit from the previous position. For example, bits A_1 and B_1 are fed into full adder 1 along with C_1, which is the carry bit produced by the addition of the A_0 and B_0 bits. Bits A_0 and B_0 are fed into full adder 0 along with C_0. A_0 and B_0 are the LSBs of the augend and addend, so it appears that C_0 would always have to be 0 because

Sum appears at S_4, S_3, S_2, S_1, S_0 outputs.

FIGURE 6-5   Block diagram of a five-bit parallel adder circuit using full adders.

there can be no carry into that position. We shall see, however, that there will be situations when C_0 can also be 1.

The full-adder circuit used in each position has three inputs: an A bit, a B bit, and a C bit. It also produces two outputs: a sum bit and a carry bit. For example, full adder 0 has inputs A_0, B_0, and C_0, and it produces outputs S_0 and C_1. Full adder 1 had A_1, B_1, and C_1 as inputs and S_1 and C_2 as outputs, and so on. This arrangement is repeated for as many positions as there are in the augend and addend. Although this illustration is for five-bit numbers, in modern computers the numbers usually range from 8 to 64 bits.

The arrangement in Figure 6-5 is called a **parallel adder** because all of the bits of the augend and addend are present and are fed into the adder circuits *simultaneously*. This means that the additions in each position are taking place at the same time. This is different from how we add on paper, taking each position one at a time starting with the LSB. Clearly, parallel addition is extremely fast. More will be said about this later.

1. How many inputs does a full adder have? How many outputs?
2. Assume the following input levels in Figure 6-5: $A_4A_3A_2A_1A_0 = 01001$; $B_4B_3B_2B_1B_0 = 00111$; $C_0 = 0$.
   (a) What are the logic levels at the outputs of FA #2?
   (b) What is the logic level at the C_5 output?

6-11 DESIGN OF A FULL ADDER

Now that we know the function of the full adder, we can design a logic circuit that will perform this function. First, we must construct a truth table showing the various input and output values for all possible cases. Figure 6-6 shows the truth table having three inputs, A, B, and C_{IN}, and two outputs, S and C_{OUT}. There are eight possible cases for the three inputs, and for each case the desired output values are listed. For example, consider the case $A = 1$, $B = 0$, and $C_{IN} = 1$. The full adder (FA) must add these bits to produce a sum (S) of 0 and a carry (C_{OUT}) of 1. The reader should check the other cases to be sure they are understood.

FIGURE 6-6 Truth table for a full-adder circuit.

Augend bit input	Addend bit input	Carry bit input	Sum bit output	Carry bit output
A	B	C_{IN}	S	C_{OUT}
0	0	0	0	0
0	0	1	1	0
0	1	0	1	0
0	1	1	0	1
1	0	0	1	0
1	0	1	0	1
1	1	0	0	1
1	1	1	1	1

Because there are two outputs, we will design the circuitry for each output individually, starting with the S output. The truth table shows that there are four cases where S is to be a 1. Using the sum-of-products method, we can write the expression for S as

$$S = \overline{A}\,\overline{B}C_{IN} + \overline{A}B\overline{C}_{IN} + A\overline{B}\,\overline{C}_{IN} + ABC_{IN} \qquad (6\text{-}1)$$

We can now try to simplify this expression by factoring. Unfortunately, none of the terms in the expression has two variables in common with any of the other terms. However, $\overline{A}$ can be factored from the first two terms, and A can be factored from the last two terms:

$$S = \overline{A}(\overline{B}C_{IN} + B\overline{C}_{IN}) + A(\overline{B}\,\overline{C}_{IN} + BC_{IN})$$

The first term in parentheses should be recognized as the exclusive-OR combination of B and C_{IN}, which can be written as $B \oplus C_{IN}$. The second term in parentheses should be recognized as the exclusive-NOR of B and C_{IN}, which can be written as $\overline{B \oplus C_{IN}}$. Thus, the expression for S becomes

$$S = \overline{A}(B \oplus C_{IN}) + A(\overline{B \oplus C_{IN}})$$

If we let $X = B \oplus C_{IN}$, this can be written as

$$S = \overline{A} \cdot X + A \cdot \overline{X} = A \oplus X$$

which is simply the exclusive-OR of A and X. Replacing the expression for X, we have

$$S = A \oplus [B \oplus C_{IN}] \qquad (6\text{-}2)$$

Consider now the output C_{OUT} in the truth table of Figure 6-6. We can write the sum-of-products expression for C_{OUT} as follows:

$$C_{OUT} = \overline{A}BC_{IN} + A\overline{B}C_{IN} + AB\overline{C}_{IN} + ABC_{IN}$$

This expression can be simplified by factoring. We will employ the trick introduced in Chapter 4, whereby we will use the ABC_{IN} term *three* times because it has common factors with each of the other terms. Hence,

$$C_{OUT} = BC_{IN}(\overline{A} + A) + AC_{IN}(\overline{B} + B) + AB(\overline{C}_{IN} + C_{IN}) \qquad (6\text{-}3)$$
$$= BC_{IN} + AC_{IN} + AB$$

This expression cannot be simplified further.

Expressions (6-2) and (6-3) can be implemented as shown in Figure 6-7. Several other implementations can be used to produce the same expressions for S and C_{OUT}, none of which has any particular advantage over those shown. The complete circuit with inputs A, B, and C_{IN} and outputs S and C_{OUT} represents the full adder. Each of the FAs in Figure 6-5 contains this same circuitry (or its equivalent).

FIGURE 6-7 Complete circuitry for a full adder.

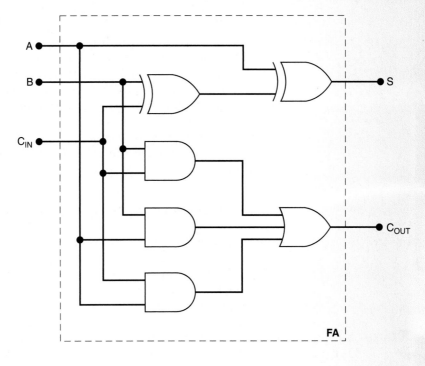

K-Map Simplification

We simplified the expressions for S and C_{OUT} using algebraic methods. The K-map method can also be used. Figure 6-8(a) shows the K map for the S output. This map has no adjacent 1s, and so there are no pairs or quads to loop. Thus, the expression for S cannot be simplified using the K map. This points out a limitation of the K-map method compared with the algebraic method. We were able to simplify the expression for S through factoring and the use of XOR and XNOR operations.

The K map for the C_{OUT} output is shown in Figure 6-8(b). The three pairs that are looped will produce the same expression obtained from the algebraic method.

FIGURE 6-8 K mappings for the full-adder outputs.

K map for S

$$S = \overline{A}\overline{B}C_{IN} + \overline{A}B\overline{C}_{IN} + ABC_{IN} + A\overline{B}\overline{C}_{IN}$$

(a)

K map for C_{OUT}

$$C_{OUT} = BC_{IN} + AC_{IN} + AB$$

(b)

Half Adder

The FA operates on three inputs to produce a sum and carry output. In some cases, a circuit is needed that will add only two input bits, to produce a sum and carry output. An example would be the addition of the LSB position of two binary numbers where there is no carry input to be added. A special logic circuit can be designed to take *two* input bits, A and B, and to produce sum (S) and carry (C_{OUT}) outputs. This circuit is called a **half adder (HA)**. Its operation is similar to that of an FA except that it operates on only two bits. We shall leave the design of the HA as an exercise at the end of the chapter.

6-12 COMPLETE PARALLEL ADDER WITH REGISTERS

In a computer, the numbers that are to be added are stored in FF registers. Figure 6-9 shows the diagram of a four-bit parallel adder, including the storage registers. The augend bits A_3 through A_0 are stored in the accumulator (A register); the addend bits B_3 through B_0 are stored in the B register. Each of these registers is made up of D flip-flops for easy transfer of data.

The contents of the A register (i.e., the binary number stored in A_3 through A_0) is added to the contents of the B register by the four FAs, and the sum is produced at outputs S_3 through S_0. C_4 is the carry out of the fourth FA, and it can be used as the carry input to a fifth FA, or as an *overflow* bit to indicate that the sum exceeds 1111.

Note that the sum outputs are connected to the D inputs of the A register. This will allow the sum to be parallel-transferred into the A register on the positive-going transition (PGT) of the TRANSFER pulse. In this way, the sum can be stored in the A register.

Also note that the D inputs of the B register are coming from the computer's memory, so that binary numbers from memory will be parallel-transferred into the B register on the PGT of the LOAD pulse. In most computers, there is also provision for parallel-transferring binary numbers from memory into the accumulator (A register). For simplicity, the circuitry necessary for performing this transfer is not shown in this diagram; it will be addressed in an end-of-chapter exercise.

Finally, note that the A register outputs are available for transfer to other locations such as another computer register or the computer's memory. This will make the adder circuit available for a new set of numbers.

Register Notation

Before we go through the complete process of how this circuit adds two binary numbers, it will be helpful to introduce some notation that makes it easy to describe the contents of a register and data transfer operations.

Whenever we want to give the levels that are present at each FF in a register or at each output of a group of outputs, we will use brackets, as illustrated below:

$$[A] = 1011$$

This is the same as saying that $A_3 = 1, A_2 = 0, A_1 = 1, A_0 = 1$. In other words, think of $[A]$ as representing "the contents of register A."

Whenever we want to indicate the transfer of data to or from a register, we will use an arrow, as illustrated below:

$$[B] \rightarrow [A]$$

FIGURE 6-9 (a) Complete four-bit parallel adder with registers; (b) signals used to add binary numbers from memory and store their sum in the accumulator.

This means that the contents of the *B* register have been transferred to the *A* register. The old contents of the *A* register will be lost as a result of this operation, and the *B* register will be unchanged. This type of notation is very common, especially in data books describing microprocessor and microcontroller operations. In many ways, it is very similar to the notation used to refer to bit-array data objects using hardware description languages.

Sequence of Operations

We will now describe the process by which the circuit of Figure 6-9 will add the binary numbers 1001 and 0101. Assume that $C_0 = 0$; that is, there is no carry into the LSB position.

1. $[A] = 0000$. A $\overline{\text{CLEAR}}$ pulse is applied to the asynchronous inputs ($\overline{CLR}$) of each FF in register A. This occurs at time t_1.
2. $[M] \rightarrow [B]$. This first binary number is transferred from memory (M) to the B register. In this case, the binary number 1001 is loaded into register B on the PGT of the LOAD pulse at t_2.
3. $[S]^\star \rightarrow [A]$. With $[B] = 1001$ and $[A] = 0000$, the full adders produce a sum of 1001; that is, $[S] = 1001$. These sum outputs are transferred into the A register on the PGT of the TRANSFER pulse at t_3. This makes $[A] = 1001$.
4. $[M] \rightarrow [B]$. The second binary number, 0101, is transferred from memory into the B register on the PGT of the second LOAD pulse at t_4. This makes $[B] = 0101$.
5. $[S] \rightarrow [A]$. With $[B] = 0101$ and $[A] = 1001$, the FAs produce $[S] = 1110$. These sum outputs are transferred into the A register when the second TRANSFER pulse occurs at t_5. Thus, $[A] = 1110$.
6. At this point, the sum of the two binary numbers is present in the accumulator. In most computers, the contents of the accumulator, $[A]$, will usually be transferred to the computer's memory so that the adder circuit can be used for a new set of numbers. The circuitry that performs this $[A] \rightarrow [M]$ transfer is not shown in Figure 6-9.

REVIEW QUESTIONS

1. Suppose that four different four-bit numbers are to be taken from memory and added by the circuit of Figure 6-9. How many $\overline{\text{CLEAR}}$ pulses will be needed? How many TRANSFER pulses? How many LOAD pulses?
2. Determine the contents of the A register after the following sequence of operations: $[A] = 0000$, $[0110] \rightarrow [B]$, $[S] \rightarrow [A]$, $[1110] \rightarrow [B]$, $[S] \rightarrow [A]$.

6-13 CARRY PROPAGATION

The parallel adder of Figure 6-9 performs additions at a relatively high speed because it adds the bits from each position simultaneously. However, its speed is limited by an effect called **carry propagation** or **carry ripple**, which can best be explained by considering the following addition:

$$\begin{array}{r} 0111 \\ + \ 0001 \\ \hline 1000 \end{array}$$

Addition of the LSB position produces a carry into the second position. This carry, when added to the bits of the second position, produces a carry into the third position. The latter carry, when added to the bits of the third position, produces a carry into the last position. The key point to notice in this example is that the sum bit generated in the *last* position (MSB) de-

---
*Even though S is not a register, we will use $[S]$ to represent the group of S outputs.

pended on the carry that was generated by the addition in the *first* position (LSB).

Looking at this from the viewpoint of the circuit of Figure 6-9, S_3 out of the last full adder depends on C_1 out of the first full adder. But the C_1 signal must pass through three FAs before it produces S_3. What this means is that the S_3 output will not reach its correct value until C_1 has propagated through the intermediate FAs. This represents a time delay that depends on the propagation delay produced in each FA. For example, if each FA has a propagation delay of 40 ns, then S_3 will not reach its correct level until 120 ns after C_1 is generated. This means that the add command pulse cannot be applied until 160 ns after the augend and addend numbers are present in the FF registers (the extra 40 ns is due to the delay of the LSB full adder, which generates C_1).

Obviously, the situation becomes much worse if we extend the adder circuitry to add a greater number of bits. If the adder were handling 32-bit numbers, the carry propagation delay could be 1280 ns = 1.28 μs. The add pulse could not be applied until at least 1.28 μs after the numbers were present in the registers.

This magnitude of delay is prohibitive for high-speed computers. Fortunately, logic designers have come up with several ingenious schemes for reducing this delay. One of the schemes, called **look-ahead carry**, utilizes logic gates to look at the lower-order bits of the augend and addend to see if a higher-order carry is to be generated. For example, it is possible to build a logic circuit with B_2, B_1, B_0, A_2, A_1, and A_0 as inputs and C_3 as an output. This logic circuit would have a shorter delay than is obtained by the carry propagation through the FAs. This scheme requires a large amount of extra circuitry but is necessary to produce high-speed adders. The extra circuitry is not a significant consideration with the present use of integrated circuits. Many high-speed adders available in integrated-circuit form utilize the look-ahead carry or a similar technique for reducing overall propagation delays.

6-14   INTEGRATED-CIRCUIT PARALLEL ADDER

Several parallel adders are available as ICs. The most common is a four-bit parallel adder IC that contains four interconnected FAs and the look-ahead carry circuitry needed for high-speed operation. The 7483A, 74LS83A, 74283, and 74LS283 are all TTL four-bit parallel-adder chips. The 283s are identical to the 83s except that they have V_{CC} and ground on pins 16 and 8, respectively; it has become standard on all new chips to have the power and ground pins at the corners of the chip. The 74HC283 is the high-speed CMOS version of the same four-bit parallel adder.

Figure 6-10(a) shows the functional symbol for the 74HC283 four-bit parallel adder (and its equivalents). The inputs to this IC are two four-bit numbers, $A_3A_2A_1A_0$ and $B_3B_2B_1B_0$, and the carry, C_0, into the LSB position. The outputs are the sum bits and the carry, C_4, out of the MSB position. The sum bits are labeled $\Sigma_3\Sigma_2\Sigma_1\Sigma_0$, where Σ is the Greek capital letter *sigma*. The Σ label is just a common alternative to the S label for a sum bit.

Cascading Parallel Adders

Two or more IC adders can be connected together (cascaded) to accomplish the addition of larger binary numbers. Figure 6-10(b) shows two 74HC283 adders connected to add two 8-bit numbers $A_7\,A_6\,A_5\,A_4\,A_3\,A_2\,A_1\,A_0$ and $B_7\,B_6\,B_5\,B_4\,B_3\,B_2\,B_1\,B_0$. The adder on the right adds the lower-order bits of the numbers. The adder on the left adds the higher-order bits *plus* the C_4 carry

FIGURE 6-10 (a) Block symbol for the 74HC283 four-bit parallel adder; (b) cascading two 74HC283s.

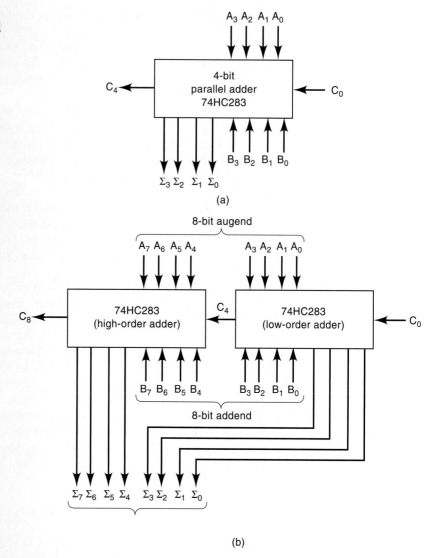

out of the lower-order adder. The eight sum outputs are the resultant sum of the two 8-bit numbers. C_8 is the carry out of the MSB position. It can be used as the carry input to a third adder stage if larger binary numbers are to be added.

The look-ahead carry feature of the 74HC283 speeds up the operation of this two-stage adder because the logic level at C_4, the carry out of the lower-order stage, is generated more rapidly than it would be if there were no look-ahead carry circuitry on the 74HC283 chip. This allows the higher-order stage to produce its sum outputs more quickly.

EXAMPLE 6-10

Determine the logic levels at the inputs and outputs of the eight-bit adder in Figure 6-10(b) when 72_{10} is added to 137_{10}.

Solution

First, convert each number to an eight-bit binary number:

$$137 = 10001001$$
$$72 = 01001000$$

These two binary values will be applied to the A and B inputs; that is, the A inputs will be 10001001 from left to right, and the B inputs will be 01001000 from left to right. The adder will produce the binary sum of the two numbers:

$$[A] = 10001001$$
$$[B] = \underline{01001000}$$
$$[\Sigma] = 11010001$$

The sum outputs will read 11010001 from left to right. There is no overflow into the C_8 bit, and so it will be a 0.

1. How many 74HC283 chips are needed to add two 20-bit numbers?
2. If a 74HC283 has a maximum propagation delay of 30 ns from C_0 to C_4, what will be the total propagation delay of a 32-bit adder constructed from 74HC283s?
3. What will be the logic level at C_4 in Example 6-10?

6-15 2's-COMPLEMENT SYSTEM

Most modern computers use the 2's-complement system to represent negative numbers and to perform subtraction. The operations of addition and subtraction of signed numbers can be performed using only the addition operation if we use the 2's-complement form to represent negative numbers.

Addition

Positive and negative numbers, including the sign bits, can be added together in the basic parallel-adder circuit when the negative numbers are in 2's-complement form. This is illustrated in Figure 6-11 for the addition of -3 and $+6$. The -3 is represented in its 2's-complement form as 1101, where the first 1 is the sign bit; the $+6$ is represented as 0110, with the first zero as the sign bit. These numbers are stored in their corresponding registers. The four-bit parallel adder produces sum outputs of 0011, which represents $+3$. The C_4 output is 1, but remember that it is disregarded in the 2's-complement method.

Subtraction

When the 2's-complement system is used, the number to be subtracted (the subtrahend) is changed to its 2's complement and then *added* to the minuend (the number the subtrahend is being subtracted from). For example, we can assume that the minuend is already stored in the accumulator (A register). The subtrahend is then placed in the B register (in a computer it would be transferred here from memory) and is changed to its 2's-complement form before it is added to the number in the A register. The sum outputs of the adder circuit now represent the *difference* between the minuend and the subtrahend.

The parallel-adder circuit that we have been discussing can be adapted to perform the subtraction described above if we provide a means for taking

$\Sigma_7\Sigma_6\Sigma_5\Sigma_4$, the BCD code for the second digit of the sum, and so on. This arrangement can, of course, be extended to decimal numbers of any size by simply adding more FFs to the registers and including a BCD adder for each digit position.

EXAMPLE 6-11

Determine the inputs and outputs when the circuit of Figure 6-15 is used to add 538_{10} to 247_{10}. Assume CARRY IN = 0.

Solution

First, the decimal numbers are represented in BCD.

$$247 = 0010 \quad 0100 \quad 0111 \quad \text{(BCD)}$$
$$538 = 0101 \quad 0011 \quad 1000 \quad \text{(BCD)}$$

These BCD numbers will be placed in the A and B registers, respectively, so that

$$[A] = 0010 \quad 0100 \quad 0111$$
$$[B] = 0101 \quad 0011 \quad 1000$$

The CARRY IN to the LSD adder will be a 0.

Once the data are in the registers, the BCD adders will begin to produce the correct BCD sums at their outputs. The LSD adder will add the 0111 (7) and 1000 (8) to produce a sum of 0101 (5) and a CARRY of 1 into the middle adder. The middle adder will add the 0100 (4) and 0011 (3) and the CARRY of 1 to produce a sum of 1000 (8) and a CARRY of 0 into the MSD adder. The MSD adder will add 0010 (2) and 0101 (5) for a sum of 0111 (7) and no CARRY OUT. Thus, at the sum outputs we have

$$[\Sigma] = 0111 \quad 1000 \quad 0101$$

and there is a CARRY output of 0 from the MSD adder. This result is, of course, the BCD representation of the decimal sum 785_{10}.

REVIEW QUESTIONS

1. What are the three basic parts of a BCD adder circuit?
2. Describe how the BCD adder circuit detects the need for a correction and executes it.

6-17 ALU INTEGRATED CIRCUITS

Several integrated circuits are called arithmetic/logic units (ALUs), even though they do not have the full capabilities of a computer's arithmetic/logic unit. These ALU chips are capable of performing several different arithmetic and logic operations on binary data inputs. The specific operation that an ALU IC is to perform is determined by a specific binary code applied to its function-select inputs. Some of the ALU ICs are fairly complex, and it would require a great amount of time and space to explain and illustrate their operation. In this section, we will use a relatively simple, yet useful, ALU chip to show the basic concepts behind all ALU chips. The ideas presented here can then be extended to the more complex devices.

The 74LS382/HC382 ALU

Figure 6-16(a) shows the block symbol for an ALU that is available as a 74LS382 (TTL) and as a 74HC382 (CMOS). This 20-pin IC operates on two four-bit input numbers, $A_3A_2A_1A_0$ and $B_3B_2B_1B_0$, to produce a four-bit output result $F_3F_2F_1F_0$. This ALU can perform *eight* different operations. At any given time, the operation that it is performing depends on the input code applied to the function-select inputs $S_2S_1S_0$. The table in Figure 6-16(b) shows the eight available operations. We will now describe each of these operations.

CLEAR OPERATION   With $S_2S_1S_0 = 000$, the ALU will *clear* all of the bits of the F output so that $F_3F_2F_1F_0 = 0000$.

ADD OPERATION   With $S_2S_1S_0 = 011$, the ALU will add $A_3A_2A_1A_0$ to $B_3B_2B_1B_0$ to produce their sum at $F_3F_2F_1F_0$. For this operation, C_N is the carry into the LSB position, and it must be made a 0. C_{N+4} is the carry output from the MSB position. *OVR* is the overflow indicator output; it detects overflow when signed numbers are being used. *OVR* will be a 1 when an add or a subtract operation produces a result that is too large to fit into four bits (including the sign bit).

SUBTRACT OPERATIONS   With $S_2S_1S_0 = 001$, the ALU will subtract the A input number from the B input number. With $S_2S_1S_0 = 010$, the ALU will subtract B from A. In either case, the difference appears at $F_3F_2F_1F_0$. Note that the subtract operations require that the C_N input be a 1.

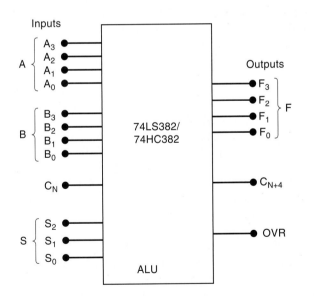

S_2	S_1	S_0	Operation	Comments
0	0	0	CLEAR	$F_3F_2F_1F_0 = 0000$
0	0	1	B minus A	Needs $C_N = 1$
0	1	0	A minus B	Needs $C_N = 1$
0	1	1	A plus B	Needs $C_N = 0$
1	0	0	A ⊕ B	Exclusive-OR
1	0	1	A + B	OR
1	1	0	AB	AND
1	1	1	PRESET	$F_3F_2F_1F_0 = 1111$

Function Table

Notes: S inputs select operation.
         OVR = 1 for signed-number overflow.

(b)

A = 4-bit input number
B = 4-bit input number
C_N = carry into LSB position
S = 3-bit operation select inputs

F = 4-bit output number
C_{N+4} = carry out of MSB position
OVR = overflow indicator

(a)

FIGURE 6-16   (a) Block symbol for 74LS382/HC382 ALU chip; (b) function table showing how select inputs (S) determine what operation is to be performed on A and B inputs.

XOR OPERATION With $S_2S_1S_0 = 100$, the ALU will perform a bit-by-bit XOR operation on the A and B inputs. This is illustrated below for $A_3A_2A_1A_0 = 0110$ and $B_3B_2B_1B_0 = 1100$.

$$A_3 \oplus B_3 = 0 \oplus 1 = 1 = F_3$$
$$A_2 \oplus B_2 = 1 \oplus 1 = 0 = F_2$$
$$A_1 \oplus B_1 = 1 \oplus 0 = 1 = F_1$$
$$A_0 \oplus B_0 = 0 \oplus 0 = 0 = F_0$$

The result is $F_3F_2F_1F_0 = 1010$.

OR OPERATION With $S_2S_1S_0 = 101$, the ALU will perform a bit-by-bit OR operation on the A and B inputs. For example, with $A_3A_2A_1A_0 = 0110$ and $B_3B_2B_1B_0 = 1100$, the ALU will generate a result of $F_3F_2F_1F_0 = 1110$.

AND OPERATION With $S_2S_1S_0 = 110$, the ALU will perform a bit-by-bit AND operation on the A and B inputs. For example, with $A_3A_2A_1A_0 = 0110$ and $B_3B_2B_1B_0 = 1100$, the ALU will generate a result of $F_3F_2F_1F_0 = 0100$.

PRESET OPERATIONS With $S_2S_1S_0 = 111$, the ALU will *set* all of the bits of the output so that $F_3F_2F_1F_0 = 1111$.

EXAMPLE 6-12

(a) Determine the 74HC382 outputs for the following inputs: $S_2S_1S_0 = 010$, $A_3A_2A_1A_0 = 0100$, $B_3B_2B_1B_0 = 0001$, and $C_N = 1$.

(b) Change the select code to 011 and repeat.

Solution

(a) From the function table in Figure 6-16(b), 010 selects the $(A - B)$ operation. The ALU will perform the 2's-complement subtraction by complementing B and adding it to A and C_N. Note that $C_N = 1$ is needed to complete the 2's complement of B effectively.

$$
\begin{array}{r}
1 \quad \leftarrow C_N \\
0100 \quad \leftarrow A \\
+ \quad 1110 \quad \leftarrow \overline{B} \\
\hline
10011
\end{array}
$$
$$C_{N+4} \quad\uparrow \quad \uparrow\!\!-\!\!- F_3\, F_2\, F_1\, F_0$$

As always in 2's-complement subtraction, the CARRY OUT of the MSB is discarded. The correct result of the $(A - B)$ operation appears at the F outputs.

The *OVR* output is determined by considering the input numbers to be signed numbers. Thus, we have $A_3A_2A_1A_0 = 0100 = +4_{10}$ and $B_3B_2B_1B_0 = 0001 = +1_{10}$. The result of the subtract operation is $F_3F_2F_1F_0 = 0011 = +3_{10}$, which is correct. Therefore, no overflow has occurred, and $OVR = 0$. If the result had been negative, it would have been in 2's-complement form.

(b) A select code of 011 will produce the sum of the A and B inputs. However, because $C_N = 1$, there will be a carry of 1 added into the LSB position. This will produce a result of $F_3F_2F_1F_0 = 0110$, which is 1 greater than $(A + B)$. The C_{N+4} and *OVR* outputs will both be 0. For the correct sum to appear at F, the C_N input must be at 0.

Expanding the ALU

A single 74LS382 or 74HC382 operates on four-bit numbers. Two or more of these chips can be connected together to operate on larger numbers. Figure 6-17 shows how two four-bit ALUs can be combined to add two eight-bit numbers, $B_7B_6B_5B_4B_3B_2B_1B_0$ and $A_7A_6A_5A_4A_3A_2A_1A_0$, to produce the output sum $\Sigma_7\Sigma_6\Sigma_5\Sigma_4\Sigma_3\Sigma_2\Sigma_1\Sigma_0$. Study the circuit diagram and note the following points:

1. Chip Z1 operates on the four lower-order bits of the two input numbers. Chip Z2 operates on the four higher-order bits.
2. The sum appears at the F outputs of Z1 and Z2. The lower-order bits appear at Z1, and the higher-order bits appear at Z2.
3. The C_N input of Z1 is the carry into the LSB position. For addition, it is made a 0.
4. The carry output $[C_{N+4}]$ of Z1 is connected to the carry input $[C_N]$ of Z2.
5. The *OVR* output of Z2 is the overflow indicator when signed eight-bit numbers are being used.
6. The corresponding select inputs of the two chips are connected together so that Z1 and Z2 are always performing the same operation. For addition, the select inputs are shown as 011.

EXAMPLE 6-13

How would the arrangement of Figure 6-17 have to be changed in order to perform the subtraction $(B - A)$?

Notes: Z1 adds lower-order bits.
Z2 adds higher-order bits.
Σ_7–Σ_0 = 8-bit sum.
OVR of Z2 is 8-bit overflow indicator.

FIGURE 6-17 Two 74HC382 ALU chips connected as an eight-bit adder.

Solution

The select input code [see the table in Figure 6-16(b)] must be changed to 001, and the C_N input of Z1 must be made a 1.

Other ALUs

The 74LS181/HC181 is another four-bit ALU. It has four select inputs that can select any of 16 different operations. It also has a mode input bit that can switch between logic operations and arithmetic operations (add and subtract). This ALU has an $A = B$ output that is used to compare the magnitudes of the A and B inputs. When the two input numbers are exactly equal, the $A = B$ output will be a 1; otherwise, it is a 0.

The 74LS881/HC881 is similar to the 181 chip, but it has the capability of performing some additional logic operations.

1. Apply the following inputs to the ALU of Figure 6-16, and determine the outputs: $S_2S_1S_0 = 001$, $A_3A_2A_1A_0 = 1110$, $B_3B_2B_1B_0 = 1001$, $C_N = 1$.
2. Change the select code to 011 and C_N to 0, and repeat review question 1.
3. Change the select code to 110, and repeat review question 1.
4. Apply the following inputs to the circuit of Figure 6-17, and determine the outputs: $B = 01010011$, $A = 00011000$.
5. Change the select code to 111, and repeat review question 4.
6. How many 74HC382s are needed to add two 32-bit numbers?

6-18   IEEE/ANSI SYMBOLS

Figure 6-18(a) shows the IEEE/ANSI symbol for a one-bit adder (full adder). Note that the symbol Σ is used to indicate the addition operation. Figure 6-18(b) is the IEEE/ANSI symbol for the 7483/74283 four-bit parallel adder. Note that the letters P and Q are used to represent the two four-bit inputs, and Σ is used for the four-bit output sum. The P, Q, and Σ are specified by the IEEE/ANSI standard and must be used inside the symbol outline. The letters used for the inputs and outputs external to the symbol outline are not specified by the standard.

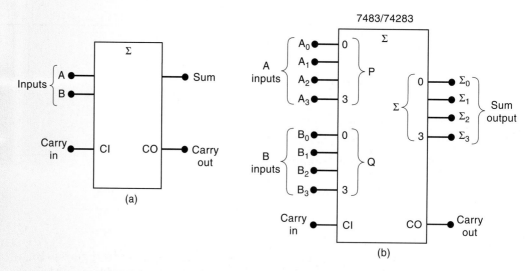

FIGURE 6-18   IEEE/ANSI symbols for (a) a full adder and (b) a four-bit parallel adder IC (7483/74283).

6-19 TROUBLESHOOTING CASE STUDY

A technician is testing the adder/subtractor redrawn in Figure 6-19 and records the following test results for the various operating modes:

- **Mode 1: ADD = 0, SUB = 0.** The sum outputs are always equal to the number in the A register *plus one*. For example, when $[A] = 0110$, the sum is $[\Sigma] = 0111$. This is incorrect because the OR outputs and C_0 should all be 0 in this mode to produce $[\Sigma] = [A]$.

- **Mode 2: Add = 1, SUB = 0.** The sum is always 1 more than it should be. For example, with $[A] = 0010$ and $[B] = 0100$, the sum output is 0111 instead of 0110.

- **Mode 3: Add = 0, SUB = 1.** The Σ outputs are always equal to $[A] - [B]$, as expected.

When she examines these test results, the technician sees that the sum outputs exceed the expected results by 1 for the first two modes of operation. At first, she suspects a possible fault in one of the LSB inputs to the adder, but she dismisses this because such a fault would also affect the subtraction operation, which is working correctly. Eventually, she realizes that there is another fault that could add an extra 1 to the results for the first two modes without causing an error in the subtraction mode.

Recall that C_0 is made a 1 in the subtraction mode as part of the 2's-complement operation on $[B]$. For the other modes, C_0 is to be a 0. The technician checks the connection between the SUB signal and the C_0 input to the adder and finds that it is open due to a bad solder connection. This open connection explains the observed results because the TTL adder responds as if

FIGURE 6-19 Parallel adder/subtractor circuit.

C_0 were a constant logic 1, causing an extra 1 to be added to the result in modes 1 and 2. The open connection would have no effect on mode 3 because C_0 is supposed to be a 1 anyway.

EXAMPLE 6-14

Consider again the adder/subtractor circuit. Suppose that there is a break in the connection path between the SUB input and the AND gates at point X in Figure 6-19. Describe the effects of this open on the circuit operation for each mode.

Solution

First, realize that this fault will produce a logic 1 at the affected input of AND gates 2, 4, 6, and 8, which will permanently enable each of these gates to pass its $\overline{B}$ input to the following OR gate as shown.

- **Mode 1: ADD = 0, SUB = 0.** The fault will cause the circuit to perform subtraction—almost. The 1's complement of [B] will reach the OR gate outputs and be applied to the adder along with [A]. With $C_0 = 0$, the 2's complement of [B] will not be complete; it will be short by 1. Thus, the adder will produce [A] − [B] − 1. To illustrate, let's try [A] = +6 = 0110 and [B] = +3 = 0011. The adder will add as follows:

$$
\begin{array}{rl}
\text{1's complement of } [B] = & 1100 \\
[A] = & 0110 \\
\hline
\text{result} = & \cancel{1}0010 \\
& \quad\uparrow \\
& \quad\llcorner\!\!-\!\!-\text{ Disregard carry.}
\end{array}
$$

  The result is 0010 = +2 instead of 0011 = +3, as it would be for normal subtraction.

- **Mode 2: ADD = 1, SUB = 0.** With ADD = 1, AND gates 1, 3, 5, and 7 will pass the B inputs to the following OR gate. Thus, each OR gate will have a $\overline{B}$ and a B at its inputs, thereby producing a 1 output. For example, the inputs to OR gate 9 will be $\overline{B}_0$ coming from AND gate 2 (because of the fault), and B_0 coming from AND gate 1 (because ADD = 1). Thus, OR gate 9 will produce an output of $\overline{B}_0 + B_0$, which will always be a logic 1.

  The adder will add the 1111 from the OR gates to the [A] to produce a sum that is 1 less than [A]. Why? Because $1111_2 = -1_{10}$.

- **Mode 3: ADD = 0, SUB = 1.** This mode will work correctly because SUB = 1 is supposed to enable AND gates 2, 4, 6, and 8 anyway.

6-20   USING TTL LIBRARY FUNCTIONS WITH HDL

The adder and ALU ICs that we have looked at in this chapter are just a few of the many MSI chips that have served as the building blocks of digital systems for decades. Whenever a technology serves such a long and useful lifetime, it has a lasting impact on the field and the people who use it. TTL integrated circuits certainly fall into this category and continue in various forms today. Experienced engineers and technicians (we want to avoid the word *old*) are familiar with the standard parts. Existing designs can be remanufactured and upgraded using the same basic circuits if they can be implemented in a VLSI PLD. Data sheets for these devices are readily available, and studying these old TTL parts is still an excellent way to learn the fundamentals of any digital system.

For all of these reasons, the Altera development system offers what they refer to as old-style macrofunctions. A **macrofunction** is a self-contained description of a logic circuit with all its inputs, outputs, and operational characteristics defined. In other words, they have gone to the trouble of writing the code necessary to get a PLD to emulate the operation of many conventional TTL MSI devices. All the designer needs to know is how to hook it into the rest of the system. In this section, we will expand on the concepts of logic primitives and libraries presented in Chapter 5 to see how we can use the HDL equivalent of standard MSI parts in our designs.

The 74382 arithmetic logic unit (ALU) is a fairly sophisticated IC. The task of describing its operation using HDL code is challenging but certainly within our reach. Refer again to the examples of this IC and its operation, which were covered in Section 6-17. Specifically, look at Figure 6-17, which shows how to cascade two four-bit ALU chips to make an eight-bit ALU that could serve as the heart of a microcontroller's central processing unit (CPU). Figure 6-20 shows the graphic method of describing the eight-bit circuit using Altera's graphic description file and macrofunction blocks from its library of components. The 74382 symbols are simply chosen from the list in

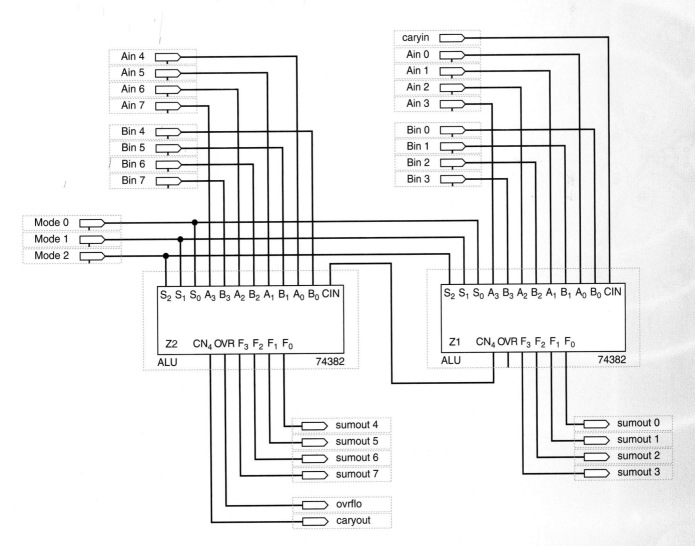

FIGURE 6-20 An Altera graphic description file tf an eight-bit ALU.

the macrofunction library and placed on the screen. Wiring these chips together is simple and intuitive, but it can be tedious and time consuming for positioning, labeling, and drawing connections. This section will present the techniques needed to implement this same circuit using only text-based HDL and macrofunctions.

The first thing you need if you are going to use a macrofunction is good documentation. If you want to write a macro, a function, a subroutine, or any piece of code that someone else is going to use, what is the most important thing you *must* provide? If your answer is "Good documentation," then you have gotten the point. Documentation that is understood by the person writing the code is not necessarily good documentation. Good documentation must be understood easily by the person trying to *use* the code. Altera provides some good documentation through the HELP menu. The documentation tells us what the macrofunction is, what it does, and the names of the inputs and outputs. The proper syntax is also given to define and use the functions in each of the HDLs supported by the compiler.

For the macrofunction of the 74382, the key item of interest is the **function prototype**, shown below.

AHDL Function Prototype (port name and order also apply to Verilog HDL):

```
FUNCTION 74382 (s[2..0], a[3..0], b[3..0], cin)
RETURNS (ovr, cn4, f[3..0]);
```

It is like a text-based, block-level model of the component, and it serves the same purpose as a logic symbol in a graphic design file. A function prototype is needed in the HDL source code to provide the compiler with all the vital information it needs to interpret the code that refers to the function. In this sense, it is very much like a declaration of a variable or signal. For example, if you try to connect an eight-bit signal to a four-bit input (defined in the prototype), the compiler can inform you of your mistake. The compiler knows that the operational description of the circuit is in the library, and it will "look it up" at the appropriate time to generate the actual circuit for the PLD.

AHDL MACROFUNCTIONS

The first item that should go into any source file is a field of comments that documents your code. In AHDL, the function prototypes should come next, along with any constant definitions and other universal (often called global) definitions. When many different functions are used in a design, the function prototypes can occupy a large area at the top of the file, and typing all this information for each design file gets rather burdensome. As a convenience for the designer, AHDL allows the use of the INCLUDE directive, followed by a file name with a .inc extension. The INCLUDE directive causes the specified file to be opened and essentially inserted into the code at that point. Consequently, a single line can take the place of many prototypes and make it easier to use functions. Include files made up for the library functions come with the MAX+PLUS software (see the maxplus2\max2inc directory). For the following example, we will not use include files but rather show the actual prototype. In Figure 6-21, the function prototype (lines 4 and 5) has been cut and pasted directly from the help menu documentation for this macrofunction. From this prototype, we know the part number of the MSI chip and the names of the inputs and outputs. The input *s* is a three-bit number that selects the mode of operation (add, subtract, XOR, etc.), the inputs *a* and *b* are four-bit

```
1    %     8-bit ALU cascaded 74382 ALU in AHDL
2          Digital Systems 9th ed
3          MAY 22, 2002                    %
4    FUNCTION 74382 (s[2..0], a[3..0], b[3..0], cin)
5    RETURNS (ovr, cn4, f[3..0]);
6
7    SUBDESIGN fig6_21
8    (
9          caryin              :INPUT;
10         mode[2..0]          :INPUT;              -- add/subtract controls
11         bin[7..0]           :INPUT;
12         ain[7..0]           :INPUT;
13         sumout[7..0]        :OUTPUT;
14         caryout             :OUTPUT;
15         ovrflo              :OUTPUT;
16    )
17    VARIABLE
18         z1    :74382;
19         z2    :74382;
20    BEGIN
21         z1.s[] = mode[];          -- both ALUs in same mode
22         z1.a[] = ain[3..0];       -- low nibble A
23         z1.b[] = bin[3..0];       -- low nibble B
24         z1.cin = caryin;
25         z2.s[] = mode[];          -- both ALUs in same mode
26         z2.a[] = ain[7..4];       -- high nibble A
27         z2.b[] = bin[7..4];       -- high nibble B
28         z2.cin = z1.cn4;          -- low Cout to high Cin
29
30         caryout = z2.cn4;         -- 9th bit out
31         sumout[3..0] = z1.f[];    -- low nibble result
32         sumout[7..4] = z2.f[];    -- high nibble result
33         ovrflo = z2.ovr;
34    END;
```

FIGURE 6-21   AHDL text definition file of an eight-bit ALU.

numbers that are logically or arithmetically combined, and *cin* is a single-bit input that allows us to cascade these devices. The *ovr* is a bit that indicates if 2's complement overflow has occurred, as described in Section 6-4. The output *cn4* is the carry out of the fourth bit position used for cascading chips, and *f* is the four-bit numeric result of the function.

In this subdesign, two of these chips are cascaded to make an eight-bit ALU. Consequently, all of the numeric inputs (lines 11–12) are eight-bit binary arrays with bit positions 0 through 7. Notice that the *mode* input (line 10) to the overall circuit still has only three bits. As Figure 6-17 indicates, there are only three control lines selecting one of eight different modes (functions) for these chips. By cascading, we are simply expanding the number of bits in the operands, not the number of operations. Consequently, the *mode* inputs of the two chips are connected in parallel (see lines 21 and 25).

The components to be cascaded must be given names (z1 and z2) and identified with the macrofunction in the library. This is accomplished in the instance declarations (lines 18–19) given in the VARIABLE section. Between

AHDL

BEGIN and END, all the concurrent statements define the circuit's interconnections. Every connection to the component is referred to by using the component instance name (defined under VARIABLE) and a dot extension. Notice that the dot extensions use the same designation that was given in the function prototype. For example, in line 21, to connect the *mode* inputs to the *s* inputs of the ALU chip *z1,* we use z1.s[] = mode[]. Each component input appears on the left side of the assignment operator (=), and the signal in the system that is driving it appears on the right. For the outputs (lines 30–33) of each component, the name of the output is on the right of the assignment operator. On the left is the signal in the system to which this output is assigned. In AHDL, the destination of the signal is written on the left and the source of the signal is on the right of the assignment operator. Refer to the graphic design file of Figure 6-20 to compare the way that this HDL description parallels the graphic diagram.

VHDL MACROFUNCTIONS

The VHDL compiler requires libraries to be specified at the beginning of the code if components from those libraries are being used. Of course, the first item that should go into any design file is a field of comments that documents your code. The code in Figure 6-22 has had the comments removed to save space. This circuit requires the Altera MAXPLUS2 library (lines 3, 4), which includes the macrofunction for a 74382 ALU chip. The macrofunction uses objects defined according to the IEEE 1164 standard (lines 1, 2), which includes definitions for STD_LOGIC and STD_LOGIC_VECTOR.

In this architecture, we cascade two of the 74382 chips to make an eight-bit ALU. Consequently, all the numeric inputs defined in the ENTITY declaration are eight-bit binary arrays with bit positions 0 through 7 (see lines 9–10). Notice that the *mode* input to the overall circuit (line 8) still has only three bits. As Figure 6-20 indicates, there are only three control lines selecting one of eight different modes (functions) for these chips. By cascading, we are simply expanding the number of bits in the operands, not the number of operations. Consequently, the *mode* inputs of the two chips are connected in parallel (lines 32 and 40). Remember, VHDL is very particular about matching object types when components are connected in a circuit. Because the library components use std_logic and std_logic_vector types, the inputs and outputs must also use them.

The architecture declaration always begins with the declaration of any variables, signals, or components (lines 17–26) that will be used in the concurrent description between BEGIN and END. Refer to Figure 6-20. Notice that there is really only one signal wire that originates from one of the components and terminates at the other (the *cn4* to *cin* connection). In VHDL, we need to declare a SIGNAL to serve as the wire connecting these two points. In this file, it is named *cascade_wire* on line 17. The equivalent of a function prototype in VHDL is the component declaration, as shown on lines 18–26. It is very important to use the naming convention for components required by the Altera library. It is simply the MSI chip number preceded by *a_*. In this case, the macrofunction name is *a_74382*. Notice that the names for the inputs and outputs used here are exactly the same as those given in the function prototype from the HELP menu. Using the component name and given object names in this manner makes any supported macrofunction component available to the designer as a building block in the system that he or she is designing.

```
1    LIBRARY ieee;
2    USE ieee.std_logic_1164.all;
3    LIBRARY altera;
4    USE altera.maxplus2.all;
5
6    ENTITY fig6_22 IS
7    PORT( caryin       :IN STD_LOGIC;
8          mode         :IN STD_LOGIC_VECTOR(2 DOWNTO 0);
9          ain          :IN STD_LOGIC_VECTOR(7 DOWNTO 0);
10         bin          :IN STD_LOGIC_VECTOR(7 DOWNTO 0);
11         sumout       :OUT STD_LOGIC_VECTOR(7 DOWNTO 0);
12         caryout      :OUT STD_LOGIC;
13         ovrflo       :OUT STD_LOGIC);
14   END fig6_22;
15
16   ARCHITECTURE a OF fig6_22 IS
17   SIGNAL cascade_wire: STD_LOGIC; --to connect carry
18   COMPONENT   a_74382
19       PORT(        s     :IN STD_LOGIC_VECTOR (2 DOWNTO 0);
20                    a     :IN STD_LOGIC_VECTOR (3 DOWNTO 0);
21                    b     :IN STD_LOGIC_VECTOR (3 DOWNTO 0);
22                    cin   :IN STD_LOGIC;
23                    f     :OUT STD_LOGIC_VECTOR (3 DOWNTO 0);
24                    cn4   :OUT STD_LOGIC;
25                    ovr   :OUT STD_LOGIC);
26   END COMPONENT;
27
28   BEGIN
29       z1: a_74382 --least significant nibble
30             PORT MAP
31             (    cin    => caryin,
32                  s      => mode,
33                  a      => ain(3 DOWNTO 0),
34                  b      => bin(3 DOWNTO 0),
35                  f      => sumout(3 DOWNTO 0),
36                  cn4    => cascade_wire);
37       z2: a_74382 --most significant nibble
38             PORT MAP
39             (    cin    => cascade_wire,
40                  s      => mode,
41                  a      => ain(7 DOWNTO 4),
42                  b      => bin(7 DOWNTO 4),
43                  f      => sumout(7 DOWNTO 4),
44                  cn4    => caryout,
45                  ovr    => ovrflo);
46   END a;
```

FIGURE 6-22   VHDL source file for an eight-bit ALU.

The concurrent section of the hardware description is where the components are interconnected. Immediately following the BEGIN keyword is the first instantiation of the 74382 chip (line 29). This instance of the component is given a name (z1) to correlate with the graphic description in Figure 6-20. This is followed by a colon and the name of the component described (a_74382). The PORT MAP section from lines 30–36 allows us to specify, at the structural level, exactly how the components are interconnected. Notice that in VHDL, the signal name associated with the macrofunction is on the left of the => operator, and the system signal name is on the right, regardless of the source or destination of the signal. Line 36 connects the carry out to the cascade_wire signal defined earlier.

The second instantiation (z2) of the 74382 chip (starting at line 37) connects the upper four bits of each eight-bit input and output port name to their respective component port names. The cascade_wire (line 39) is connected to the carry in of z2, which processes the most significant nibble (four bits) of the eight-bit operands.

Note that the PORT MAP definition can be abbreviated by simply listing the system signals (i.e., those on the right side of =>), separated by commas, in the same order that their corresponding component signals are listed in the COMPONENT declaration (lines 18–26). The assignments are written explicitly here for the sake of clarity.

REVIEW QUESTIONS

1. Where can you find information about using a 74283 full adder in your HDL design?
2. What is a macrofunction?
3. What do we call the block-level text description that defines a macrofunction in an AHDL source file?
4. What do we call the block-level text description that defines a macrofunction in a VHDL source file?

6-21 LOGICAL OPERATIONS ON BIT ARRAYS

In the previous section, we examined the use of macrofunctions to build systems from standard parts. Now we need to practice writing HDL code rather than using a macrofunction to make an adder similar to Figure 6-5. In this section, we will expand our understanding of the HDL techniques in two main areas: specifying groups of bits in an array and using logical operations to combine arrays of bits using Boolean expressions.

In Section 6-12, we discussed register notation, which makes it easy to describe the contents of registers and signals consisting of multiple bits. The HDLs use arrays of bits in a similar notation to describe signals, as we discussed in Chapter 4. For example, in AHDL, the four-bit signal named d is defined as:

```
VARIABLE d [3..0] :NODE.
```

In VHDL, the same data format is expressed as:

```
SIGNAL d :BIT_VECTOR (3 DOWNTO 0).
```

Each bit in these data types is designated by an element number. In this example of a bit array named d, the bits can be referred to as d3, d2, d1, d0. Bits can also be grouped into **sets**. For example, if we want to refer to the three most significant bits of d as a set, we can use the expression d[3..1] in AHDL and the expression d (3 DOWNTO 1) in VHDL. Once a value is assigned to the array and the desired set of bits is identified, logical operations can be performed on the entire set of bits. As long as the sets are the same size (the same number of bits), two sets can be combined in a logical expression, just like you would combine single variables in a Boolean equation. Each of the corresponding pairs of bits in the two sets is combined as stated in the logic equation. This allows one equation to describe the logical operation performed on each bit in a set.

EXAMPLE 6-15

Assume D_3, D_2, D_1, D_0 has the value 1011 and G_3, G_2, G_1, G_0 has the value 1100. Let's define Dnum = $[D_3, D_2, D_1, D_0]$ and Gnum = $[G_3, G_2, G_1, G_0]$. Let's also define X = $[X_3, X_2, X_1, X_0]$ where X is related to Dnum and Gnum as follows:

$$X = Dnum \bullet Gnum;$$

What is the value of X after this operation?

Solution

$$
\begin{array}{ll}
\begin{array}{ll}
D_3, D_2, D_1, D_0 & 1\ 0\ 1\ 1 \\
\updownarrow\ \updownarrow\ \updownarrow\ \updownarrow & \updownarrow\ \updownarrow\ \updownarrow\ \updownarrow \\
G_3, G_2, G_1, G_0 & 1\ 1\ 0\ 0 \\
\hline
X_3, X_2, X_1, X_0 & 1\ 0\ 0\ 0
\end{array}
& \text{AND each bit position together}
\end{array}
$$

Thus, X is a set of four bits with value 1000.

EXAMPLE 6-16

For the register values described in Example 6-15, declare each d, g, and x. Then write an expression using your favorite HDL that performs the ANDing operation on all bits.

Solution

```
SUBDESIGN bitwise_and
(    d[3..0], g[3..0]        :INPUT;
     xx[3..0]               :OUTPUT;)
BEGIN
     xx[] = d[] & g[];
END; -- Note: x is a reserved identifier in AHDL: use xx
```

```
ENTITY bitwise_and IS
PORT(d, g              :IN  BIT_VECTOR (3 DOWNTO 0);
        x              :OUT BIT_VECTOR (3 DOWNTO 0));
END bitwise_and;
ARCHITECTURE a OF bitwise_and IS
BEGIN
     x <= d AND g;
END a;
```

6-22   HDL ADDERS

In this section, we will see how to create a parallel adder circuit that can be used to add bit arrays using the logic equation for a single-bit full adder. Figure 6-23 shows the basic block diagram with signals labeled to create a four-bit adder. Notice that each bit of the augend [A], addend [B], carry [C], and sum [S] are bit array variables and have index numbers associated with them. The equation for the sum using register notation is:

$$[S] = [A] \oplus [B] \oplus [C];$$

Notice that the carry signals between stages are not inputs or outputs of this overall circuit, but rather intermediate variables. A strategy must be developed for labeling the carry bits so that they can be used in an array. We have chosen to let each carry bit serve as an input to its corresponding adder stage, as shown in Figure 6-23. For example, C0 is an input to the bit 0 stage, C1 is the carry input to the bit 1 stage, and so on. The bits of the carry array can be thought of as the "wires" that connect the adders. This array must have a carry in for each stage and also a carry out for the most significant state. In this example, there will be five bits, labeled C4 through C0, in the carry array. The set of data that represents the carry outputs would be C4 through C1, and the set of data that represents the carry inputs would be C3 through C0.

AUGEND	A_3	A_2	A_1	A_0	A	
ADDEND	B_3	B_2	B_1	B_0	B	
CARRYin	C_3	C_2	C_1	C_0	Cin	Generate sum
SUM	S_3	S_2	S_1	S_0	S	$S = A \oplus [B \oplus Cin]$

AUGEND	A_3	A_2	A_1	A_0	A	
ADDEND	B_3	B_2	B_1	B_0	B	Generate carry bits
CARRYin	C_3	C_2	C_1	C_0	Cin	$Cout = A{\cdot}B + A{\cdot}Cin + B{\cdot}Cin$
CARRYout	C_4	C_3	C_2	C_1	Cout	

FIGURE 6-23   Four-bit parallel adder.

$$Cin = C[3..0]$$
$$Cout = C[4..1]$$

The HDLs allow us to specify which sets of bits out of the entire array we want to use in an equation. To ensure that all variables being combined in a logic equation contain the same number of bits, we can start with the general equation for the carry out of a one-bit adder as follows:

$$Cout = AB + ACin + BCin$$

Substituting our definition of Cin and Cout (above), we have the following equation for the four-bit adder carry out:

$$C[4..1] = A[3..0] \text{ \& } B[3..0] + A[3..0] \text{ \& } C[3..0] + B[3..0] \text{ \& } C[3..0]$$

The graphic symbol for this device is shown in Figure 6-24. Notice that it does not show the carry array. It is a variable or signal *inside* the block. Altera allows any SUBDESIGN in AHDL or ENTITY in VHDL, even those that you create, to be represented by a graphic block diagram symbol such as this. This is all part of the hierarchical design scheme of the Altera development system (described in Chapter 4).

To summarize this information, Figure 6-23 shows the "insides" of the block diagram in Figure 6-24 and summarizes the operation described by the two equations. Now let's look at text-based files that can be used to generate a block symbol like the one in Figure 6-24 using AHDL and VHDL.

FIGURE 6-24   Block symbol generated by Altera MAX+PLUS.

AHDL FOUR-BIT ADDER

In lines 14 and 15 of the AHDL code of Figure 6-25, notice the syntax for referring to bit arrays in their entirety. The name is given, followed by []. If no bits are designated inside the square brackets, it means that all the bits that were declared are included in the operations. Lines 14 and 15 describe fully all four adder circuits and come up with the sum. In order to choose a specific set of elements from the array (i.e., a subset of the array), the name is followed by the range of element numbers in square brackets. For example, the carry equation (line 15) in AHDL syntax is:

$$c[4..1] = a[] \text{ \& } b[] \text{ \# } a[] \text{ \& } c[3..0] \text{ \# } b[] \text{ \& } c[3..0];$$

Notice that only four bits of the carry array *c[ ]* are being assigned, even though the array is five bits long. In this way, the carry out of each single-bit adder is assigned as the carry in of the next stage.

AHDL

```
1     SUBDESIGN fig6_25
2     (
3          cin          :INPUT;      -- carry in
4          a[3..0]      :INPUT;      -- augend
5          b[3..0]      :INPUT;      -- addend
6          s[3..0]      :OUTPUT;     -- sum
7          cout         :OUTPUT;     -- carry OUT
8     )
9     VARIABLE
10         c[4..0]      :NODE;       -- carry array is 5 bits long!
11
12    BEGIN
13         c[0] = cin;
14         s[] = a[] $ b[] $ c[3..0];   -- generate sum
15         c[4..1] = (a[] & b[]) # (a[] & c[3..0]) # (b[] & c[3..0]);
16         cout = c[4];                   -- carry out
17    END;
```

FIGURE 6-25   AHDL adder.

VHDL FOUR-BIT ADDER

In the VHDL code of Figure 6-26, notice the syntax for referring to bit arrays in their entirety. The name is simply used with no bit designations. Lines 15 and 16 describe fully all four adder circuits that will come up with the sum. In order to choose a specific set of elements from the array (i.e., a subset of the array), the name is followed by the range of element numbers in parentheses. The carry equation (line 16) in VHDL syntax is:

```
1     ENTITY fig6_26 IS
2     PORT(
3        cin    :IN BIT;
4        a      :IN BIT_VECTOR(3 DOWNTO 0);
5        b      :IN BIT_VECTOR(3 DOWNTO 0);
6        s      :OUT BIT_VECTOR(3 DOWNTO 0);
7        cout   :OUT BIT);
8     END fig6_26;
9
10    ARCHITECTURE a OF fig6_26 IS
11    SIGNAL c :BIT_VECTOR (4 DOWNTO 0);   -- carries require 5 bit array
12
13    BEGIN
14       c(0) <= cin;          -- Read the carry in to bit array
15       s <= a XOR b  XOR c(3 DOWNTO 0); -- Generate the sum bits
16       c(4 DOWNTO 1) <=      (a AND b)
17                        OR   (a AND c(3 DOWNTO 0))
18                        OR   (b AND c(3 DOWNTO 0));
19       cout <= c(4);         -- output the carry of the MSB.
20    END a;
```

FIGURE 6-26   VHDL adder.

c(4 DOWNTO 1) <= (a AND b) OR (a AND c(3 DOWNTO 0)) OR (b AND c(3 DOWNTO 0));

Notice that only four bits of the carry array *c* are being assigned, even though the array is five bits long. In this way, the carry out of each single-bit adder is assigned as the carry in of the next stage.

REVIEW QUESTIONS

1. If [*A*] = 1001 and [*B*] = 0011, what is the value of (a) [*A*] · [*B*]? (b) [*A*] + [*B*]? (Note that · means AND; + means OR.)
2. If $A[7..0]$ = 1010 1100, what is the value of (a) $A[7..4]$? (b) $A[5..2]$?
3. In AHDL, the following object is declared: toggles[7..0] :INPUT. Give an expression for the least significant four bits using AHDL syntax.
4. In VHDL, the following object is declared: toggles :IN BIT_VECTOR (7 DOWNTO 0). Give an expression for the least significant four bits using VHDL syntax.
5. What would be the result of ORing the two registers of Example 6-15?
6. Write an HDL statement that would OR the two objects *d* and *g* together. Use your favorite HDL.
7. Write an HDL statement that would XOR the two most significant bits of *d* with the two least significant bits of *g* and put the result in the middle two bits of *x*.

6-23 EXPANDING THE BIT CAPACITY OF A CIRCUIT

One way we have learned to expand the capacity of a circuit is to cascade stages, like we did with the 74382 ALU chip in the previous section. This can be done using the Altera graphic design file approach (like Figure 6-20) or the structural text-based HDL approach (like Figures 6-21 and 6-22). With either of these methods, we need to specify all the inputs, outputs, and interconnections between blocks. In the case of this adder circuit, it would be much easier to start with the HDL file for a four-bit adder and simply increase the size of each operand variable in the equation. For example, if we wanted an eight-bit adder, we simply need to expand *a*, *b*, and *s* to eight bits. The code would remain almost identical to the four-bit adder shown above. This is just a glimpse of some of the efficiency improvements that HDL offers. The way this code is written, however, the indices of each signal, and each *bit array* specification in the equation, would also have to be redefined. In other words, the designer would need to examine the code carefully and change all the 3s to 7s, all the 4s to 8s, and so on.

An important principle in software engineering is symbolic representation of the **constants** that are used throughout the code. Constants are simply fixed numbers represented by a name (symbol). If we can define a symbol (i.e., make up a name) at the top of the source code that is assigned the value for the total number of bits and then use this symbol (name) throughout the code, it is much easier to modify the circuit. Only one line of code needs to be changed to expand the capacity of the circuit. The examples that follow add this feature to the code and also upgrade the code to implement the adder/subtractor circuit like the one in Figure 6-13. It should be noted here that expanding the capacity of an adder circuit such as this one will also reduce the speed of the circuit because of carry propagation (described in Section 6-13). In order to keep these examples simple, we have not added any logic to generate a look-ahead carry.

AHDL ADDER/SUBTRACTOR

In AHDL, using constants is very simple, as shown on lines 1 and 2 of Figure 6-27. The keyword CONSTANT is followed by the symbolic name and the value it is to be assigned. Notice that we can allow the compiler to do some simple math calculations to establish a value for one constant based on another. We can also use this feature as we refer to the constant in the code, as shown on lines 14, 20, and 21. For example, we can refer to $c[7]$ as $c[n]$ and $c[8]$ as $c[n+1]$. The size of this adder/subtractor can be expanded by simply changing the value of *number_of_bits* to the desired number of bits and then recompiling.

As mentioned, this code has been upgraded from the previous example to make it an adder/subtractor like the one in Figure 6-13. The *add* and *sub* inputs have been included on lines 6 and 7, and a new intermediate variable named $b[\ ]$ has been included on line 15. The first concurrent statement on line 17 describes all the SOP logic that drives the b inputs to the adder in Figure 6-13. First, it describes a logical AND operation between every bit of *bin[ ]* and the logic level on *add*. This result is ORed (bit for bit) with the result of ANDing the complement of every bit of *bin[ ]* with *sub*. In other words, it creates the following Boolean function for each bit: $b = bin \cdot add + \overline{bin} \cdot sub$. The signal $b[\ ]$ is then used in the adder equations instead of *bin[ ]*, as was used in the adder examples. Notice on line 18 that *sub* is also used to connect the carry array LSB (carry into bit 0) with the value on *sub*, which needs to be 0 when adding and 1 when subtracting.

```
1    CONSTANT number_of_bits = 8;                -- set total number of bits
2    CONSTANT n = number_of_bits - 1;            -- n is highest bit index
3
4    SUBDESIGN fig6_27
5    (
6        add          :INPUT;      -- add control
7        sub          :INPUT;      -- subtract control and LSB Carry in
8        a[n..0]      :INPUT;      -- Augend bits
9        bin[n..0]    :INPUT;      -- Addend bits
10       s[n..0]      :OUTPUT;     -- Sum bits
11       caryout      :OUTPUT;     -- MSB carry OUT
12   )
13   VARIABLE
14       c[n+1..0]    :NODE;       -- intermediate carry vector
15       b[n..0]      :NODE;       -- intermediate operand vector
16   BEGIN
17       b[] = bin[] & add # NOT bin[] & sub;
18       c[0] = sub;                          --Read the carry in to group variable
19       s[] = a[] $ b[] $ c[n..0];           --Generate the sums
20       c[n+1..1] = (a[] & b[]) # (a[] & c[n..0]) # (b[] & c[n..0]);
21       caryout = c[n+1];                    -- output the carry of the MSB.
22   END;
```

FIGURE 6-27   An *n*-bit adder/subtractor description in AHDL.

VHDL ADDER/SUBTRACTOR

In VHDL, using constants is a little bit more involved. Constants must be included in a **PACKAGE**, as shown in Figure 6-28, lines 1–4. Packages are also used to contain component definitions and other information that must be available to all entities in the design file. Notice on line 6 that the keyword USE tells the compiler to use the definitions in this package throughout this design file. Inside the package, the keyword CONSTANT is followed by the symbolic name, its type, and the value it is to be assigned using the := operator. Notice on line 3 that we can allow the compiler to do some simple math

```
1   PACKAGE const IS
2     CONSTANT number_of_bits :INTEGER:=8;  -- set total number of bits
3     CONSTANT n :INTEGER:= number_of_bits - 1;  -- MSB index number
4   END const;
5
6   USE work.const.all;
7
8   ENTITY fig6_28 IS
9   PORT(
10      add      :IN BIT; -- add control
11      sub      :IN BIT; -- subtract control and LSB carry in
12      a        :IN BIT_VECTOR(n DOWNTO 0);
13      bin      :IN BIT_VECTOR(n DOWNTO 0);
14      s        :OUT BIT_VECTOR(n DOWNTO 0);
15      carryout :OUT BIT);
16  END fig6_28;
17
18  ARCHITECTURE a OF fig6_28 IS
19  SIGNAL c :BIT_VECTOR (n+1 DOWNTO 0);   -- define intermediate carries
20  SIGNAL b :BIT_VECTOR (n DOWNTO 0);      -- define intermediate operand
21  SIGNAL bnot :BIT_VECTOR (n DOWNTO 0);
22  SIGNAL mode :BIT_VECTOR (1 DOWNTO 0);
23  BEGIN
24     bnot <= NOT bin;
25     mode <= add & sub;
26
27     WITH mode SELECT
28        b <= bin       WHEN "10",    -- add
29             bnot      WHEN "01",    -- sub
30             "0000"    WHEN OTHERS;
31
32     c(0) <= sub;                           -- read the carry_in to bit array
33     s <= a XOR b XOR c(n DOWNTO 0); -- generate the sum bits
34     c(n+1 DOWNTO 1) <= (a AND b) OR
35                         (a AND c(n DOWNTO 0)) OR
36                         (b AND c(n DOWNTO 0)); --generate carries
37     carryout <= c(n+1);          -- output the carry of the MSB.
38
39  END a;
```

FIGURE 6-28 An *n*-bit adder/subtractor description in VHDL.

VHDL

calculations to establish a value of one constant based on another. We can also use this feature as we refer to the constant in the code, as shown on lines 34 and 37. For example, we can refer to $c(7)$ as $c(n)$ and $c(8)$ as $c(n+1)$. The size of this adder/subtractor can be expanded by simply changing the value of *number_of_bits* to the desired number of bits and then recompiling.

As mentioned, this code has been upgraded from the previous example to make it an adder/subtractor like the one in Figure 6-13. The *add* and *sub* inputs have been included on lines 10 and 11, and a new signal named *b* has been included on line 20, *bnot* has been included on line 21, and *mode* has been included on line 22. The first concurrent statement on line 24 serves to create the 1's complement of *bin*. The SOP circuits in Figure 6-13 that drive the *b* inputs to the adder select the *bin* inputs if *add* = 1 or the 1's complement (*bnot*) if *sub* = 1. This is an excellent application of the VHDL selected signal assignment, as shown on lines 27–30. When *add* is 1, *bin* is channeled to *b*. When *sub* is 1, *bnot* is channeled to *b*. The signal *b* is then used in the adder equations instead of *bin*, as was used in the previous adder examples. Notice on line 32 that *sub* is also used to connect the carry array LSB (carry into bit 0) with the value on *sub*, which needs to be 0 when adding and 1 when subtracting.

VHDL GENERATE Statement

Another way to make circuits that handle more bits is the VHDL **GENERATE** statement. It is a very concise way of telling the compiler to replicate several components that are cascaded together. As we have shown, there are many other ways to accomplish the same thing and if the abstract nature of this method seems difficult, use another method. The GENERATE statement is offered here for the sake of completeness. The adder circuits we have been discussing are cascaded chains of single-bit full adder modules. The VHDL code for a single-bit full adder module is shown in Figure 6-29. Multiple instances of this module need to be connected to each other to form an *n*-bit adder circuit. Of course, you can do it using the same component techniques that we have discussed previously, but it would result in very lengthy code.

To make the code more concise and easier to modify, a strategy is needed for the way we label the inputs and outputs for each module. As we mentioned previously, the bit-0 adder has inputs that have an index of 0 (e.g., *a0*,

```
1    ENTITY add1 IS
2    PORT(
3          cin    :IN BIT;
4          a      :IN BIT;
5          b      :IN BIT;
6          s      :OUT BIT;
7          cout   :OUT BIT);
8    END add1;
9
10   ARCHITECTURE a OF add1 IS
11   BEGIN
12
13       s      <= a XOR b XOR cin;
14       cout   <= (a AND b) OR (a AND cin) OR (b AND cin);
15   END a;
```

FIGURE 6-29 Single-bit full adder in VHDL.

b0, c0, s0). The carry out of bit 0 is labeled *c1*, and it becomes the carry input for the bit 1 adder module. Each time we instantiate another component for the next bit of the multibit adder, the index number of all connections goes up by 1 (*a1, b1, c1, s1*). The GENERATE statement allows us to repeat an instantiation of a component *n* times, increasing the index number by 1 for each instantiation up to *n*. In line 27 of Figure 6-30, the GENERATE keyword is used in an **iterative loop (FOR loop)**, which means that a set of descriptive actions (PORT MAP) will be repeated a certain number of times. The variable *i* represents an index number that starts at 0 (for the first iteration) and ends at *n* (the last iteration). The advantages of this method are code compactness and the ease with which the number of bits can be expanded. The code in Figure 6-30 shows how to use a single-bit adder (Figure 6-29) as a component to generate an eight-bit adder circuit. Remember that the file for the single-bit adder (add1.vhd in Figure 6-29) must be saved in the same folder as the design file that uses it to generate multiple instances of the adder (fig6_30.vhd).

```
1    PACKAGE const IS
2       CONSTANT number_of_bits :INTEGER:=8;    -- Specify number of bits
3       CONSTANT n :INTEGER:=number_of_bits - 1;    --n is MSB bit number
4    END const;
5    USE work.const.all;
6    ENTITY fig6_30 IS
7    PORT(
8       caryin   :IN bit;
9       ain      :IN BIT_VECTOR (n DOWNTO 0);
10      bin      :IN BIT_VECTOR (n DOWNTO 0);
11      sout     :OUT BIT_VECTOR (n DOWNTO 0);
12     carryout  :OUT bit);
13   END fig6_30;
14
15   ARCHITECTURE a OF fig6_30 IS
16
17   COMPONENT add1        -- declare single bit full adder
18      PORT  (
19             cin   :IN BIT;
20             a,b   :IN BIT;
21             s     :OUT BIT;
22             cout  :OUT BIT);
23   END COMPONENT;
24   SIGNAL c :BIT_VECTOR (n+1 DOWNTO 0);    -- declare bit array for carries
25   BEGIN
26      c(0)  <= caryin;                      -- put LSB in array (carry in)
27      repeat:FOR i IN 0 TO n GENERATE       -- instantiate n+1 adders
28           -- cascade them
29      casc:add1 PORT MAP (cin=> c(i), a=> ain(i), b=> bin(i),
30                          s=> sout(i), cout=> c(i+1));
31      END GENERATE;
32      carryout    <= c(n+1);                -- out the carry from nth bit stage
33   END a;
```

FIGURE 6-30   Use of the VHDL GENERATE statement.

The single-bit adder component is defined on lines 17–23 of Figure 6-30. For the first iteration of lines 29 and 30, the value of i is 0, creating an adder stage for bit 0. The second iteration, i, has been increased to 1 to form adder stage 1. This continues until i is equal to n, generating each stage of the $(n+1)$-bit adder. Notice that VHDL allows us the option of placing labels at the beginning of a line of code to help describe its purpose. For example, on line 27, the label *repeat* is used and on line 29, the label *casc* is used. Labels are optional but they must always end with a colon.

Libraries of Parameterized Modules

Using HDL techniques clearly makes it easy to alter the bit capacity of a generic circuit. In this chapter, we can see that it is easy to change from a four-bit adder to an eight-, 12-, or 16-bit adder. When Altera was creating its library of useful functions, they also took advantage of these techniques and created what they refer to as **megafunctions**, which include a **library of parameterized modules (LPMs)**. These functions do not attempt to imitate a particular standard IC like the old-style macrofunctions; instead, they offer a generic solution for the various types of logic circuits that are useful in digital systems. Examples of these generic circuits that we have covered so far are logic gates (AND, OR, XOR), latches, counters, shift registers, and adders. The term *parameterized* means that when you instantiate a function from the library, you also specify some parameters that define certain attributes (bit capacity, for example) for the circuit you are describing. Altera recommends the use of LPMs instead of writing HDL code to perform these generic functions because they have optimized the use of the Altera PLD resources for each LPM. You will understand this issue better after reading Chapter 13, but for now, this simply means you can put more digital circuitry in a given PLD by using LPMs in your code rather than writing each module yourself. For most people who are new to HDL, the most endearing quality of LPMs is not the efficiency but rather their ease of use.

The various LPMs that are available can be found through the HELP menu under megafunctions/LPM. This documentation describes the parameters that the user can specify as well as the port features of the device. For example, the LPM_ADD_SUB megafunction has a parameter, LPM_WIDTH, that determines the number of bits being added or subtracted. The ports for the LPM include two multibit operand inputs, one multibit result output, a carry in, a carry out, a function control bit (to choose between the add or subtract mode), and a signed integer overflow indicator. The simple examples that follow demonstrate the proper syntax for creating an eight-bit adder/subtractor using LPMs.

AHDL LPM ADDER/SUBTRACTOR

The function prototype for this AHDL megafunction is included in the source file using the syntax on line 1 of Figure 6-31. The name of the function with an extension *(.inc)* specifies the file containing the appropriate definitions. The port names on lines 5 and 6 are specified by the designer. Notice that the input and output arrays are eight bits wide. This is because we intend to use this megafunction to make an eight-bit adder/subtractor. On line 9, the name *eightbit* was also made up by the designer to serve as a label for this particular instance of an eight-bit adder/subtractor. The name

```
1    INCLUDE "LPM_ADD_SUB.INC";
2
3    SUBDESIGN fig6_31
4    (
5         a[7..0], b[7..0], caryin, functn    :INPUT;
6         s[7..0], caryout, ovr               :OUTPUT;
7    )
8    VARIABLE
9         eightbit    :LPM_ADD_SUB WITH (LPM_WIDTH = 8);
10
11   BEGIN
12        eightbit.dataa[]  = a[];
13        eightbit.datab[]  = b[];
14        eightbit.cin      = caryin;
15        eightbit.add_sub  = functn;
16        s[] = eightbit.result[];
17        caryout = eightbit.cout;
18        ovr = eightbit.overflow;
19   END;
```

FIGURE 6-31   The adder/subtractor LPM used in AHDL.

of the megafunction is followed by its parameter *LPM_WIDTH*, which is set to 8. It is just that easy to make this circuit have as many bits as we desire. In the logic section (after BEGIN), the names that follow the dot extensions (e.g., *dataa*[] on line 12) are specific names from the function prototype and cannot be altered. These official port names are listed in the HELP files. Lines 12–18 simply connect the ports of the eight-bit adder/subtractor to the other signals in the system, which in this case are simply inputs and outputs for this module.

VHDL LPM ADDER/SUBTRACTOR

Because the megafunctions use VHDL data objects that are defined by IEEE 1164, all the signals, inputs, and outputs must be of the STD_LOGIC and STD_LOGIC_VECTOR type. In Figure 6-32, lines 1–4 are used to include the appropriate definitions from the libraries for the signals and components. The port names on lines 7 through 11 are specified by the designer. Notice that the input and output vectors are eight bits wide. This is because we intend to use this megafunction to make an eight-bit adder/subtractor. On line 16, the name *eightbit* was also made up by the designer. It serves as a label for this particular instance of an eight-bit adder/subtractor, which is specified by the name of the megafunction, LPM_ADD_SUB. The GENERIC MAP keywords are used on line 17 to specify the parameter *LPM_WIDTH*, which is assigned a value of 8 in this case. It is just that easy to make this circuit have as many bits as we desire. In the PORT MAP section, the names to the left of => (e.g., *dataa*[] on line 18) are specified in the function prototype and cannot be altered. These official port names are listed in the HELP files. Lines 18–21 simply connect these ports to the other signals in the system, which in this case are simply inputs and outputs for this module.

VHDL

```
1    LIBRARY ieee;
2    USE ieee.std_logic_1164.all;
3    LIBRARY lpm;
4    USE lpm.lpm_components.all;
5
6    ENTITY fig6_32 IS
7    PORT (   a,b              : IN STD_LOGIC_VECTOR (7 DOWNTO 0);
8             caryin, functn   : IN STD_LOGIC;
9             output           : OUT STD_LOGIC_VECTOR (7 DOWNTO 0);
10            caryout          : OUT STD_LOGIC;
11            ovr              : OUT STD_LOGIC);
12   END fig6_32;
13
14   ARCHITECTURE lpm OF fig6_32 IS
15   BEGIN
16      eightbit  : LPM_ADD_SUB
17         GENERIC MAP (LPM_WIDTH => 8)
18         PORT MAP (  dataa => a (7 DOWNTO 0), datab => b (7 DOWNTO 0),
19                     cin => caryin, add_sub => functn,
20                     result => output (7 DOWNTO 0),
21                     cout => caryout, overflow => ovr);
22   END lpm;
```

FIGURE 6-32   The adder/subtractor LPM used in VHDL.

REVIEW QUESTIONS

1. What keyword is used to assign a symbolic name to a fixed number?
2. In AHDL, where are constants defined? Where are they defined in VHDL?
3. Why are constants useful?
4. If the constant max_val has a value of 127, how will a compiler interpret the expression max_val – 5?
5. What is the GENERATE statement used for in VHDL?

SUMMARY

1. To represent signed numbers in binary, a sign bit is attached as the MSB. A + sign is a 0, and a − sign is a 1.
2. The 2's complement of a binary number is obtained by complementing each bit and then adding 1 to the result.
3. In the 2's-complement method of representing signed binary numbers, positive numbers are represented by a sign bit of 0 followed by the magnitude in its true binary form. Negative numbers are represented by a sign bit of 1 followed by the magnitude in 2's-complement form.
4. A signed binary number is negated (changed to a number of equal value but opposite sign) by taking the 2's complement of the number, including the sign bit.
5. Subtraction can be performed on signed binary numbers by negating (2's complementing) the subtrahend and adding it to the minuend.
6. In BCD addition, a special correction step is needed whenever the sum of a digit position exceeds 9 (1001).

7. When signed binary numbers are represented in hexadecimal, the MSD of the hex number will be 8 or greater when the number is negative; it will be 7 or less when the number is positive.

8. The arithmetic/logic unit (ALU) of a computer contains the circuitry needed to perform arithmetic and logic operations on binary numbers stored in memory.

9. The accumulator is a register in the ALU. It holds one of the numbers being operated upon, and it also is where the result of the operation is stored in the ALU.

10. A full adder performs the addition on two bits plus a carry input. A parallel binary adder is made up of cascaded full adders.

11. The problem of excessive delays caused by carry propagation can be reduced by a look-ahead carry logic circuit.

12. IC adders such as the 74LS83/HC83 and the 74LS283/HC283 can be used to construct high-speed parallel adders and subtractors.

13. A BCD adder circuit requires special correction circuitry.

14. Integrated-circuit ALUs are available that can be commanded to perform a wide range of arithmetic and logic operations on two input numbers.

15. Prefabricated functions are available in the Altera libraries.

16. These library parts and the HDL circuits you create can be interconnected using either graphic or structural HDL techniques.

17. Logical operations can be performed on all the bits in a set using Boolean equations.

18. Practicing good software engineering techniques, specifically the use of symbols to represent constants, allows for easy code modification and expansion of the bit capacity of circuits such as full adders.

19. Libraries of parameterized modules (LPMs) offer a flexible, easily modified or expanded solution for many types of digital circuits.

IMPORTANT TERMS

carry	accumulator register	constants
sign bit	full adder (FA)	PACKAGE
sign-magnitude system	parallel adder	GENERATE
2's-complement system	half adder (HA)	iterative loop
negation	carry propagation (carry	FOR loop
augend	ripple)	megafunctions
addend	look-ahead carry	library of parameterized
subtrahend	adder/subtractor	functions (LPMs)
minuend	BCD adder	
overflow	macrofunction	
arithmetic/logic	function prototype	
unit (ALU)	sets	

PROBLEMS

SECTION 6-1

B   6-1. Add the following in binary. Check your results by doing the addition in decimal.

6-20. A full adder can be implemented in many different ways. Figure 6-33 shows how one may be constructed from two half adders. Construct a truth table for this arrangement, and verify that it operates as a FA.

FIGURE 6-33   Problem 6-20.

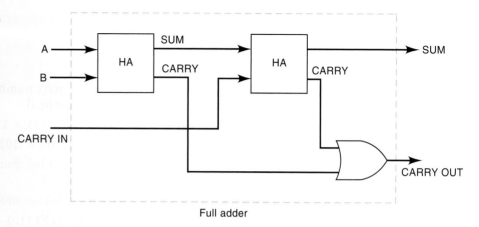

Full adder

SECTION 6-12

6-21. Refer to Figure 6-9. Determine the contents of the A register after the following sequence of operations: $[A] = 0000$, $[0100] \rightarrow [B]$, $[S] \rightarrow [A]$, $[1011] \rightarrow [B]$, $[S] \rightarrow [A]$.

6-22. Refer to Figure 6-9. Assume that each FF has $t_{PLH} = t_{PHL} = 30$ ns and a setup time of 10 ns, and that each FA has a propagation delay of 40 ns. What is the minimum time allowed between the PGT of the LOAD pulse and the PGT of the TRANSFER pulse for proper operation?

D   6-23. In the adder and subtractor circuits discussed in this chapter, we gave no consideration to the possibility of *overflow*. Overflow occurs when the two numbers being added or subtracted produce a result that contains more bits than the capacity of the accumulator. For example, using four-bit registers, including a sign bit, numbers ranging from +7 to −8 (in 2's complement) can be stored. Therefore, if the result of an addition or subtraction exceeds +7 or −8, we would say that an overflow has occurred. When an overflow occurs, the results are useless because they cannot be stored correctly in the accumulator register. To illustrate, add +5 (0101) and +4 (0100), which results in 1001. This 1001 would be interpreted incorrectly as a negative number because there is a 1 in the sign-bit position.

In computers and calculators, there are usually circuits that are used to detect an overflow condition. There are several ways to do this. One method that can be used for the adder that operates in the 2's-complement system works as follows:

1. Examine the sign bits of the two numbers being added.
2. Examine the sign bit of the result.
3. Overflow occurs whenever the numbers being added are *both positive* and the sign bit of the result is 1 *or* when the numbers are *both negative* and the sign bit of the result is 0.

This method can be verified by trying several examples. Readers should try the following cases for their own clarification: (1) 5 + 4; (2) − 4 + (−6); (3) 3 + 2. Cases 1 and 2 will produce an overflow, and case 3 will not. Thus, by examining the sign bits, one can design a logic circuit that will produce a 1 output whenever the overflow condition occurs. Design this overflow circuit for the adder of Figure 6-9.

C, D 6-24. Add the necessary logic circuitry to Figure 6-9 to accommodate the transfer of data from memory into the *A* register. The data values from memory are to enter the *A* register through its *D* inputs on the PGT of the *first* TRANSFER pulse; the data from the sum outputs of the FAs will be loaded into *A* on the PGT of the *second* TRANSFER. In other words, a LOAD pulse followed by two TRANSFER pulses is required to perform the complete sequence of loading the *B* register from memory, loading the *A* register from memory, and then transferring their sum into the *A* register. (*Hint:* Use a flip-flop *X* to control which source of data gets loaded into the *D* inputs of the accumulator.)

SECTION 6-13

C, D 6-25. Design a look-ahead carry circuit for the adder of Figure 6-9 that generates the carry C_3 to be fed to the FA of the MSB position based on the values of $A_0, B_0, C_0, A_1, B_1, A_2,$ and B_2. In other words, derive an expression for C_3 in terms of $A_0, B_0, C_0, A_1, B_1, A_2,$ and B_2. (*Hint:* Begin by writing the expression for C_1 in terms of $A_0, B_0,$ and C_0. Then write the expression for C_2 in terms of $A_1, B_1,$ and C_1. Substitute the expression for C_1 into the expression for C_2. Then write the expression for C_3 in terms of $A_2, B_2,$ and C_2. Substitute the expression for C_2 into the expression for C_3. Simplify the final expression for C_3 and put it in sum-of-products form. Implement the circuit.)

SECTION 6-14

6-26. Show the logic levels at each input and output of Figure 6-10(a) when 354_8 is added to 103_8.

SECTION 6-15

6-27. For the circuit of Figure 6-13, determine the sum outputs for the following cases.
(a) *A* register = 0101 (+5), *B* register = 1110 (−2); SUB = 1, ADD = 0
(b) *A* register = 1100 (−4), *B* register = 1110 (−2); SUB = 0, ADD = 1
(c) Repeat (b) with ADD = SUB = 0.

D 6-28. Show how the gates of Figure 6-13 can be implemented using three 74HC00 chips.

D 6-29. Modify the circuit of Figure 6-13 so that a single control input, *X*, is used in place of ADD and SUB. The circuit is to function as an adder when $X = 0$, and as a subtractor when $X = 1$. Then simplify each set of gates. (*Hint:* Note that now each set of gates is functioning as a controlled inverter.)

SECTION 6-16

6-30. Assume the following inputs in Figure 6-14: [*A*] = 0101, [*B*] = 1001, $C_0 = 0$. Determine the logic levels at [*S*], *X*, [Σ], and CARRY.

C 6-31. Would it make any difference in the BCD adder of Figure 6-14 if the C_0 of the upper adder was held LOW while the C_0 of the lower adder was used as the carry input? Explain.

6-32. Assume that the *A* register in Figure 6-15 holds the BCD code for 376 and that the *B* register holds the BCD code for 469. Determine the outputs.

counter states and that the arrows indicate how one state changes to another in response to an input clock pulse.

If we assume a starting count of 000, the diagram shows that the states of the counter change normally up until the count of 101. When the next clock pulse occurs, the counter temporarily goes to the 110 count before going to the stable 000 count. The dotted lines indicate the temporary nature of the 110 state. As stated earlier, the duration of this temporary state is so short that for most purposes we can consider that the counter goes directly from 101 to 000 (solid arrow).

Note that there is no arrow into the 111 state because the counter can never advance to that state. However, the 111 state can occur on power-up when the FFs come up in random states. If that happens, the 111 condition will produce a LOW at the NAND gate output and immediately clear the counter to 000. Thus, the 111 state is also a temporary condition that ends up at 000.

Displaying Counter States

Sometimes during normal operation, and very often during testing, it is necessary to have a visible display of how a counter is changing states in response to the input pulses. We will take a detailed look at several ways of doing this later in the text. For now, Figure 7-5(b) shows one of the simplest methods using individual indicator LEDs for each FF output. Each FF output is connected to an INVERTER whose output provides the current path for the LED. For example, when output A is HIGH, the INVERTER output goes LOW and the LED turns ON. The bright LED indicates $A = 1$. When output A is LOW, the INVERTER output is HIGH and the LED turns OFF. The dark LED indicates $A = 0$.

EXAMPLE 7-4

(a) What will be the status of the LEDs when the counter is holding the count of five?

(b) What will the LEDs display as the counter is clocked by a 1-kHz input?

(c) Will the 110 state be visible on the LEDs?

Solution

(a) Because $5_{10} = 101_2$, the 2^0 and 2^2 LEDs will be ON, and the 2^1 LED will be OFF.

(b) At 1 kHz, the LEDs will be switching ON and OFF so rapidly that they will appear to the human eye to be ON all the time at about half the normal brightness.

(c) No; the 110 state will persist for only a few nanoseconds as the counter recycles to 000.

Changing the MOD Number

The counter of Figures 7-4 and 7-5 is a MOD-6 counter because of the choice of inputs to the NAND gate. Any desired MOD number can be obtained by changing these inputs. For example, using a three-input NAND gate with inputs A, B, and C, the counter would function normally until the 111 condition was reached, at which point it would immediately reset to the 000 state. Ignoring the very temporary excursion into the 111 state, the counter would go from 000 through 110 and then recycle back to 000, resulting in a MOD-7 counter (seven states).

EXAMPLE 7-5

Determine the MOD number of the counter in Figure 7-6(a). Also determine the frequency at the *D* output.

*All J, K inputs are HIGH

(a)

*All J, K inputs are HIGH

(b)

FIGURE 7-6 (a) MOD-14 ripple counter; (b) MOD-10 (decade) ripple counter.

Solution

This is a four-bit counter, which would normally count from 0000 through 1111. The NAND inputs are *D*, *C*, and *B*, which means that the counter will immediately recycle to 0000 when the 1110 (decimal 14) count is reached. Thus, the counter actually has 14 stable states 0000 through 1101 and is therefore a *MOD-14* counter. Because the input frequency is 30 kHz, the frequency at output *D* will be

$$\frac{30 \text{ kHz}}{14} = 2.14 \text{ kHz}$$

General Procedure

To construct a counter that starts counting from all 0s and has a MOD number of *X*:

1. Find the smallest number of FFs such that $2^N \geq X$, and connect them as a counter. If $2^N = X$, do not do steps 2 and 3.

2. Connect a NAND gate to the asynchronous CLEAR inputs of all the FFs.
3. Determine which FFs will be in the HIGH state at a count = X; then connect the normal outputs of these FFs to the NAND gate inputs.

EXAMPLE 7-6

Construct a MOD-10 counter that will count from 0000 (zero) through 1001 (decimal 9).

Solution

$2^3 = 8$ and $2^4 = 16$; thus, four FFs are required. Because the counter is to have stable operation up to the count of 1001, it must be reset to zero when the count of 1010 is reached. Therefore, FF outputs D and B must be connected as the NAND gate inputs. Figure 7-6(b) shows the arrangement.

Decade Counters/BCD Counters

The MOD-10 counter of Example 7-6 is also referred to as a **decade counter**. In fact, a decade counter is any counter that has 10 distinct states, no matter what the sequence. A decade counter such as the one in Figure 7-6(b), which counts in sequence from 0000 (zero) through 1001 (decimal 9), is also commonly called a **BCD counter** because it uses only the 10 BCD code groups 0000, 0001, . . . , 1000, and 1001. To reiterate, any MOD-10 counter is a decade counter; and any decade counter that counts in binary from 0000 to 1001 is a BCD counter.

Decade counters, especially the BCD type, find widespread use in applications where pulses or events are to be counted and the results displayed on some type of decimal numerical readout. We shall examine this later in more detail. A decade counter is also often used for dividing a pulse frequency *exactly* by 10. The input pulses are applied to flip-flop A, and the output pulses are taken from the output of flip-flop D, which has one-tenth the frequency of the input.

EXAMPLE 7-7

In Example 7-3, a MOD-60 counter was needed to divide the 60-Hz line frequency down to 1 Hz. Construct an appropriate MOD-60 counter.

Solution

$2^5 = 32$ and $2^6 = 64$, and so we need six FFs, as shown in Figure 7-7. The counter is to be cleared when it reaches the count of 60 (111100). Thus, the outputs of flip-flops Q_5, Q_4, Q_3, and Q_2 must be connected to the NAND gate. The output of flip-flop Q_5 will have a frequency of 1 Hz.

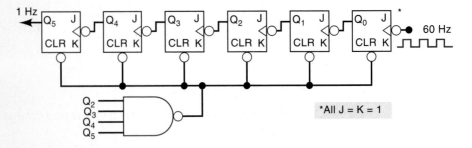

FIGURE 7-7   MOD-60 counter.

1. What FF outputs should be connected to the clearing NAND gate to form a MOD-13 counter?
2. *True or false:* All BCD counters are decade counters.
3. What is the output frequency of a decade counter that is clocked from a 50-kHz signal?

7-3   IC ASYNCHRONOUS COUNTERS

There are several TTL and CMOS asynchronous counter ICs. One of them is the TTL 74LS293. Figure 7-8(a) shows the logic diagram for the 74LS293 as it would appear in the manufacturer's TTL data book. Some of the nomenclature is different from what we have been using, but it should be easy to figure out. Note the following points:

FIGURE 7-8   (a) Logic diagram for 74LS293 asynchronous counter IC; (b) block symbol, with pin numbers in parentheses.

1. The 74LS293 has four J-K flip-flops with outputs Q_0, Q_1, Q_2, Q_3, where Q_0 is the LSB and Q_3 is the MSB. The FFs are shown arranged with the LSB on the left. This is done to satisfy the convention that the circuit input signals appear on the left. We have been drawing our counters with the LSB on the right, so you will have to get used to this arrangement.

2. Each FF has a *CP* (clock pulse) input, which is just another name for the *CLK* input. The clock inputs to Q_0 and Q_1, labeled $\overline{CP_0}$ and $\overline{CP_1}$, respectively, are externally accessible. The inversion bars over these inputs indicate that they are activated by a NGT.

3. Each FF has an asynchronous active-LOW CLEAR input, C_D. These are connected together to the output of a two-input NAND gate with inputs MR_1 and MR_2, where *MR* stands for *master reset*. Both *MR* inputs must be HIGH to clear the counter to 0000.

4. Flip-flops Q_1, Q_2, and Q_3 are already connected as a three-bit ripple counter. Flip-flop Q_0 is not connected to anything internally. This allows the user the option of either connecting Q_0 to Q_1 to form a four-bit counter or using Q_0 separately if desired.

The following examples will illustrate some of the ways the 74LS293 can be wired to produce different counters. In these examples, we will use the simplified logic symbol shown in Figure 7-8(b).

EXAMPLE 7-8

Show how the 74LS293 should be connected to operate as a MOD-16 counter with a 10-kHz clock input. Determine the frequency at Q_3.

Solution

A MOD-16 counter requires four FFs, and so we must connect the Q_0 output to $\overline{CP_1}$, the clock input of flip-flop Q_1 (see Figure 7-9). The 10-kHz pulses are applied to $\overline{CP_0}$, the clock input of Q_0. The output at Q_3 will have a frequency $\frac{1}{16}$ of the clock input frequency.

FIGURE 7-9 74LS293 wired as a MOD-16 counter.

EXAMPLE 7-9

Show how to wire the 74LS293 as a MOD-10 counter.

Solution

A MOD-10 counter requires four FFs, and so again we need to connect Q_0 to $\overline{CP_1}$. This time, however, we want the counter to recycle back to 0000 when it tries to go to the count of 1010 (10). Thus, the Q_3 and Q_1 outputs must be connected to the master reset inputs; when they both go HIGH at the count of 1010, the NAND output will immediately reset the counter to 0000.

The circuit wiring is shown in Figure 7-10. The state transition diagram is also shown. Note that the temporary 1010 state is not shown. We also have not shown the 1011, 1100, 1101, 1110, and 1111 states because they are not part of the normal sequence of states for this counter. However, there is always a chance that the counter could come up in one of these states at power-up, so it is important to know how the counter will operate in the event that this happens. We will explore this in one of the problems at the end of the chapter.

FIGURE 7-10 74LS293 wired as a MOD-10 counter.

EXAMPLE 7-10

Show how to wire a 74LS293 as a MOD-14 counter.

Solution

When the counter reaches the count of 1110 (14), the Q_3, Q_2, and Q_1 outputs are all HIGH. Unfortunately, the 74LS293's built-in reset NAND gate has only two inputs. Thus, we must add some extra logic to ensure that the counter will reset to 0000 when $Q_3 = Q_2 = Q_1 = 1$. In fact, all we need is a two-input AND gate as shown in Figure 7-11. You should verify that this arrangement operates as a MOD-14 counter.

FIGURE 7-11 An external AND gate is needed to wire the 74LS293 as a MOD-14 counter.

EXAMPLE 7-11

In Example 7-7, we divided the input frequency by 60 with a MOD-60 counter using six J-K flip-flops and a NAND gate. Another way to get a MOD-60 counter is shown in Figure 7-12. Explain how this circuit works.

FIGURE 7-12 Two 74LS293s combined to provide a frequency division of 60 by successive divisions by 10 (MOD-10) and 6 (MOD-6). Note that the MOD-6 does not use $\overline{CP}_0$ or Q_0; it uses only Q_3, Q_2, and Q_1.

Solution

This circuit divides the input frequency by 60 in two steps. The 74LS293 counter on the left is wired as a MOD-10 counter so that its output Q_3 has a frequency of $f_{in}/10$. This signal is connected to the $\overline{CP}_1$ input of the second 74LS293 counter, which is wired as a MOD-6 counter (note that Q_0 is not being used). Thus, the Q_3 output of the second counter will have the following frequency:

$$f_{out} = \frac{f_{in}/10}{6} = \frac{f_{in}}{60}$$

Example 7-11 shows that two (or more) counters can be cascaded to produce an overall MOD number equal to the *product* of their individual MOD numbers. This can be very useful in applications where a large amount of frequency division is required.

IEEE/ANSI Symbol for 74LS293 Counter

Figure 7-13 shows the IEEE/ANSI symbol for the 74LS293. This symbol contains several new aspects of the IEEE/ANSI standard. As we describe these, you should continue to appreciate how the new IEEE/ANSI symbology is designed to tell us a lot about the IC's operation.

The symbol contains three distinct blocks. The top block (with the notches) is the common-control block. The notation "CTR" defines this IC as a counter. Recall from our discussion in Chapter 5 that the common-control block is used whenever an IC has one or more inputs that are common to more than one of the circuits on the chip. For the 74LS293, the MR_1 and MR_2 inputs are common to all the FFs in the counter.

MR_1 and MR_2 are shown as active-HIGH inputs that are internally combined using the AND operation as indicated by the "&" notation. This indi-

FIGURE 7-13 IEEE/ANSI symbol for the 74LS293 IC.

cates that *both MR₁ and MR₂* must be in their active states in order to clear the counter. The notation "CT = 0" tells us that the action of the *MR* inputs is to make the count equal zero.

The middle block is labeled "DIV2" to indicate that it is a MOD-2 counter, which, of course, is a single FF. DIV2 means that the counter will divide its clock input frequency by 2. The bottom block is labeled "DIV8" to indicate that it is a MOD-8 counter. The clock inputs to each of these blocks are shown as activated by negative-going transitions. The "+" notation on each clock input indicates that the NGT of the clock will cause the count to be *incremented by 1.* In other words, this is an **up counter,** meaning that it *counts up* on each NGT. A "−" notation would be used for a **down counter** (see Section 7-4).

While we will continue to use the traditional symbol for the 74LS293 and other counters, we will use some of the notation from the IEEE/ANSI symbology. For example, we will indicate a counter's MOD number using the DIV*n* notation, where *n* is the MOD number.

CMOS Asynchronous Counters

There are several asynchronous counters in the CMOS family. Most of them are equivalents to the TTL versions. However, some CMOS asynchronous counter ICs do not have a TTL counterpart. One of these is the 74HC4024; its logic symbol is shown in Figure 7-14. It is a seven-bit counter with one asynchronous master reset input. The seven FFs are internally connected as a MOD-128 ripple counter. The *MR* input is active-HIGH and can be used to reset all of the FFs to the 0 state. Note that we have used the notation "CTR DIV128" to signify that this is a MOD-128 counter.

FIGURE 7-14 Logic symbol for the CMOS 74HC4024 ripple counter. The + sign indicates count-up operation.

The MOD number of this counter can be changed to less than 128 by using the *MR* input, as we did for the 74LS293. For example, if we connect outputs Q_4 and Q_5 to an AND gate, and connect the gate output to *MR*, as soon as the counter reaches 0110000 (48_{10}), it will immediately clear to 0000000. Thus, the usable count sequence will be 0000000 to 0101111 (zero to 47), and it is a MOD-48 counter.

Another CMOS ripple counter that has no TTL counterpart is the 74HC4040, which is a 12-bit counter with a single active-HIGH master reset input. The clock input to this counter is a Schmitt-trigger type of input that permits the use of slow-changing signals without producing erratic counting.

1. A 2-kHz clock signal is applied to $\overline{CP}_1$ of a 74LS293. What is the frequency at Q_3?
2. What would be the final output frequency if the order of the counters were reversed in Figure 7-12?
3. What is the MOD number of a 74HC4040 counter?
4. What would the notation "DIV64" mean on a counter symbol?
5. Which outputs would you connect to an AND gate to convert the 74HC4024 to a MOD-120 counter?

7-4   ASYNCHRONOUS DOWN COUNTER

All of the counters we have looked at thus far have counted *upward* from zero; that is, they were **up counters.** It is a relatively simple matter to construct asynchronous (ripple) **down counters,** which will count downward from a maximum count to zero. Before looking at the circuit for a ripple down counter, let us examine the count-down sequence for a three-bit down counter:

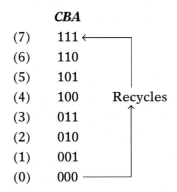

A, B, and C represent the FF output states as the counter goes through its sequence. You can see that the A flip-flop (LSB) changes states (toggles) at each step in the sequence, just as it does in the up counter. The B flip-flop changes states each time A goes from LOW to HIGH; C changes states each time B goes from LOW to HIGH. Thus, in a down counter, each FF, except the first, must toggle when the preceding FF goes from LOW to HIGH. If the FFs have *CLK* inputs that respond to negative transitions (HIGH to LOW), then an inverter can be placed in front of each *CLK* input; however, the same effect can be accomplished by driving each FF *CLK* input from the *inverted* output of the preceding FF. This is illustrated in Figure 7-15 for a MOD-8 down counter.

FIGURE 7-15 MOD-8 down counter.

The input pulses are applied to the *A* flip-flop. The $\overline{A}$ output serves as the *CLK* input for the *B* flip-flop; the $\overline{B}$ output serves as the *CLK* input for the *C* flip-flop. The waveforms at *A, B,* and *C* show that *B* toggles whenever *A* goes LOW to HIGH (so that $\overline{A}$ goes HIGH to LOW) and *C* toggles whenever *B* goes LOW to HIGH. This results in the desired down-counting sequence at the *C, B,* and *A* outputs. The state transition diagram shows the sequence. Compare it with the diagram for the MOD-8 up counter in Figure 5-47.

Down counters are not as widely used as up counters. Their major application is in situations where it must be known when a desired number of input pulses have occurred. In these situations, the down counter is *preset* to the desired number and then allowed to count down as the pulses are applied. When the counter reaches the *zero* state, it is detected by a logic gate whose output then indicates that the preset number of pulses has occurred. We shall discuss presettable counters in Section 7-8.

1. What is the difference between the counting sequence of an up counter and a down counter?
2. Describe how an asynchronous down-counter circuit differs from an up-counter circuit.

7-5 PROPAGATION DELAY IN RIPPLE COUNTERS

Ripple counters are the simplest type of binary counters because they require the fewest components to produce a given counting operation. They do, however, have one major drawback, which is caused by their basic principle of operation: each FF is triggered by the transition at the output of the preceding FF. Because of the inherent propagation delay time (t_{pd}) of each FF, this means that the second FF will not respond until a time t_{pd} after the first FF receives an active clock transition; the third FF will not respond until a time equal to $2 \times t_{pd}$ after that clock transition; and so on. In other words, the propagation delays of the FFs accumulate so that the Nth FF cannot change states until a time equal to $N \times t_{pd}$ after the clock transition occurs. This is illustrated in Figure 7-16, where the waveforms for a three-bit ripple counter are shown.

The first set of waveforms in Figure 7-16(a) shows a situation where an input pulse occurs every 1000 ns (the clock period $T = 1000$ ns) and it is assumed that each FF has a propagation delay of 50 ns ($t_{pd} = 50$ ns). Notice that the A flip-flop output toggles 50 ns after the NGT of each input pulse. Similarly, B toggles 50 ns after A goes from 1 to 0, and C toggles 50 ns after B goes from 1 to 0. As a result, when the fourth input NGT occurs, the C output goes HIGH after a delay of 150 ns. In this situation, the counter does operate properly in the sense that the FFs do eventually get to their correct states, representing the binary count. However, the situation worsens if the input pulses are applied at a much higher frequency.

The waveforms in Figure 7-16(b) show what happens if the input pulses occur once every 100 ns. Again, each FF output responds 50 ns after the 1-to-0 transition at its *CLK* input (note the change in the relative time scale). Of particular interest is the situation after the falling edge of the *fourth* input pulse, where the C output does not go HIGH until 150 ns later, which is the same time that the A output goes HIGH in response to the *fifth* input pulse. In other words, the condition $C = 1$, $B = A = 0$ (count of 100) never appears because the input frequency is too high. This could cause a serious problem if this condition were supposed to be used to control some other operation in a digital system. Problems such as this can be avoided if the period between input pulses is made longer than the total propagation delay of the counter. That is, for proper counter operation we need

$$T_{\text{clock}} \geq N \times t_{pd} \qquad (7\text{-}2)$$

where N = the number of FFs. Stated in terms of input-clock frequency, the maximum frequency that can be used is given by

$$f_{\text{max}} = \frac{1}{N \times t_{pd}} \qquad (7\text{-}3)$$

For example, suppose that a four-bit ripple counter is constructed using the 74LS112 J-K flip-flop. Table 5-2 shows that the 74LS112 has $t_{PLH} = 16$ ns and

FIGURE 7-16 Waveforms of a three-bit ripple counter illustrating the effects of FF propagation delays for different input pulse frequencies.

$t_{PHL} = 24$ ns as the propagation delays from *CLK* to *Q*. To calculate f_{max}, we will assume the "worst case"; that is, we will use $t_{pd} = t_{PHL} = 24$ ns, so that

$$f_{max} = \frac{1}{4 \times 24 \text{ ns}} = 10.4 \text{ MHz}$$

Clearly, as the number of FFs in the counter increases, the total propagation delay increases and f_{max} decreases. For example, a ripple counter that uses six 74LS112 FFs will have

$$f_{max} = \frac{1}{6 \times 24 \text{ ns}} = 6.9 \text{ MHz}$$

Thus, asynchronous counters are not useful at very high frequencies, especially for large numbers of bits. Another problem caused by propagation delays in asynchronous counters occurs when the counter outputs are *decoded*. This problem is discussed in Section 7-12. Despite these problems, the simplicity of asynchronous counters makes them useful for applications where their frequency limitation is not critical.

1. Explain why a ripple counter's maximum frequency limitation decreases as more FFs are added to the counter.
2. A certain J-K flip-flop has $t_{pd} = 12$ ns. What is the largest MOD counter that can be constructed from these FFs and still operate up to 10 MHz?

7-6 SYNCHRONOUS (PARALLEL) COUNTERS

The problems encountered with ripple counters are caused by the accumulated FF propagation delays; stated another way, the FFs do not all change states simultaneously in synchronism with the input pulses. These limitations can be overcome with the use of **synchronous** or **parallel counters** in which all of the FFs are triggered simultaneously (in parallel) by the clock input pulses. Because the input pulses are applied to all the FFs, some means must be used to control when an FF is to toggle and when it is to remain unaffected by a clock pulse. This is accomplished by using the J and K inputs and is illustrated in Figure 7-17 for a four-bit, MOD-16 synchronous counter.

If we compare the circuit arrangement for this synchronous counter with its asynchronous counterpart in Figure 7-1, we can see the following notable differences:

- The *CLK* inputs of all of the FFs are connected together so that the input clock signal is applied to each FF simultaneously.

- Only flip-flop A, the LSB, has its J and K inputs permanently at the HIGH level. The J, K inputs of the other FFs are driven by some combination of FF outputs.

- The synchronous counter requires more circuitry than does the asynchronous counter.

Circuit Operation

For this circuit to count properly, on a given NGT of the clock, only those FFs that are supposed to toggle on that NGT should have $J = K = 1$ when that NGT occurs. Let's look at the counting sequence in Figure 7-17(b) to see what this means for each FF.

The counting sequence shows that the A flip-flop must change states at each NGT. For this reason, its J and K inputs are permanently HIGH so that it will toggle on each NGT of the clock input.

The counting sequence shows that flip-flop B must change states on each NGT that occurs while $A = 1$. For example, when the count is 0001, the next NGT must toggle B to the 1 state; when the count is 0011, the next NGT must toggle B to the 0 state; and so on. This operation is accomplished by connecting output A to the J and K inputs of flip-flop B so that $J = K = 1$ only when $A = 1$.

The counting sequence shows that flip-flop C must change states on each NGT that occurs while $A = B = 1$. For example, when the count is 0011, the next NGT must toggle C to the 1 state; when the count is 0111, the next NGT must toggle C to the 0 state; and so on. By connecting the logic signal AB to FF C's J and K inputs, this FF will toggle only when $A = B = 1$.

In a like manner, we can see that flip-flop D must toggle on each NGT that occurs while $A = B = C = 1$. When the count is 0111, the next NGT must tog-

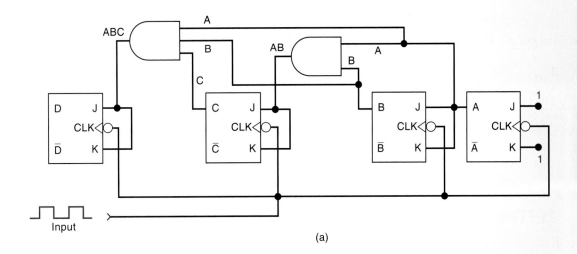

(a)

Count	D	C	B	A
0	0	0	0	0
1	0	0	0	1
2	0	0	1	0
3	0	0	1	1
4	0	1	0	0
5	0	1	0	1
6	0	1	1	0
7	0	1	1	1
8	1	0	0	0
9	1	0	0	1
10	1	0	1	0
11	1	0	1	1
12	1	1	0	0
13	1	1	0	1
14	1	1	1	0
15	1	1	1	1
0	0	0	0	0
.	.	.	.	.
.	.	.	.	.
.	.	etc.	.	.

(b)

FIGURE 7-17 Synchronous MOD-16 counter. Each FF is clocked by the NGT of the clock input signal so that all FF transitions occur at the same time.

gle *D* to the 1 state; when the count is 1111, the next NGT must toggle *D* to the 0 state. By connecting the logic signal *ABC* to FF *D*'s *J* and *K* inputs, this FF will toggle only when $A = B = C = 1$.

The basic principle for constructing a synchronous counter can therefore be stated as follows:

> **Each FF should have its *J* and *K* inputs connected so that they are HIGH only when the outputs of all lower-order FFs are in the HIGH state.**

Advantage of Synchronous Counters over Asynchronous

In a parallel counter, all of the FFs will change states simultaneously; that is, they are all synchronized to the NGTs of the input clock pulses. Thus,

CLOCK

Up/Down

(a)

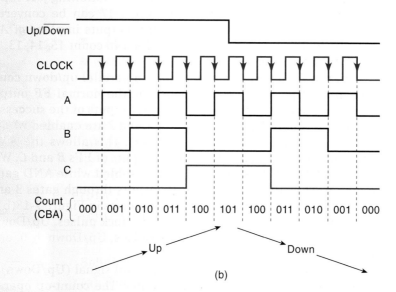

(b)

FIGURE 7-18   (a) MOD-8 synchronous up/down counter. (b) The counter counts up when the control input Up/$\overline{\text{Down}}$ = 1; it counts down when the control input = 0.

7-8   PRESETTABLE COUNTERS

Many synchronous (parallel) counters that are available as ICs are designed to be **presettable**; in other words, they can be preset to any desired starting count either asynchronously (independent of the clock signal) or synchronously (on the active transition of the clock signal). This presetting operation is also referred to as **parallel loading** the counter.

Figure 7-19 shows the logic circuit for a three-bit presettable parallel up counter. The *J*, *K*, and *CLK* inputs are wired for operation as a parallel up counter. The asynchronous PRESET and CLEAR inputs are wired to perform asynchronous presetting. The counter is loaded with any desired count at any time by doing the following:

1. Apply the desired count to the parallel data inputs, P_2, P_1, and P_0.
2. Apply a LOW pulse to the PARALLEL LOAD input, $\overline{PL}$.

FIGURE 7-19   Presettable parallel counter with asynchronous preset.

This procedure will perform an asynchronous transfer of the P_2, P_1, and P_0 levels into flip-flops Q_2, Q_1, and Q_0, respectively (Section 5-17). This *jam transfer* occurs independently of the *J, K,* and *CLK* inputs. The effect of the *CLK* input will be disabled as long as $\overline{PL}$ is in its active-LOW state because each FF will have one of its asynchronous inputs activated while $\overline{PL} = 0$. Once $\overline{PL}$ returns HIGH, the FFs can respond to their *CLK* inputs and can resume the counting-up operation starting from the count that was loaded into the counter.

For example, let's say that $P_2 = 1$, $P_1 = 0$, and $P_0 = 1$. While $\overline{PL}$ is HIGH, these parallel data inputs have no effect. If clock pulses are present, the counter will perform the normal count-up operation. Now let's say that $\overline{PL}$ is pulsed LOW when the counter is at the 010 count (i.e., $Q_2 = 0$, $Q_1 = 1$, and $Q_0 = 0$). This LOW at $\overline{PL}$ will produce LOWs at the *CLR* input of Q_1 and at the *PRE* inputs of Q_2 and Q_0 so that the counter will go to the 101 count *regardless of what is occurring at the CLK input*. The count will hold at 101 until $\overline{PL}$ is deactivated (returned HIGH); at that time the counter will resume counting up clock pulses from the count of 101.

This asynchronous presetting is used by several IC counters, such as the TTL 74ALS190, 74ALS191, 74ALS192, and 74ALS193 and the CMOS equivalents, 74HC190, 74HC191, 74HC192, and 74HC193.

will be LOW only when the counter is in the 1111 state and CP_U is LOW. Thus, $\overline{TC}_U$ will remain HIGH as the counter counts up from 0000 to 1110. On the next PGT of CP_U, the count goes to 1111, but $\overline{TC}_U$ does not go LOW until CP_U returns LOW. The next PGT at CP_U recycles the count to 0000 and also causes $\overline{TC}_U$ to return HIGH. This PGT at $\overline{TC}_U$ occurs when the counter recycles from 1111 to 0000, and it can be used to clock a second 74ALS193 up counter to its next higher count.

$\overline{TC}_D$ is the *terminal count-down* (also called the *borrow*) output. It is generated as shown in Figure 7-21(b). It is normally HIGH and does not go LOW until the counter has counted down to the 0000 state and CP_D is LOW. When the next PGT at CP_D recycles the counter to 1111, it causes $\overline{TC}_D$ to return HIGH. This PGT at $\overline{TC}_D$ can be used to clock a second 74ALS193 down counter to its next lower count.

EXAMPLE 7-14

Refer to Figure 7-22(a), where a 74HC193 is wired as an *up counter*. The parallel data inputs are permanently connected as 1011, and the CP_U, $\overline{PL}$, and MR input waveforms are shown in Figure 7-22(b). Assume that the counter is initially in the 0000 state, and determine the counter output waveforms.

(a)

(b)

FIGURE 7-22   Example 7-14.

Solution

Initially (at t_0) the counter FFs are all LOW. This causes $\overline{TC}_U$ to be HIGH. Just prior to time t_1 the $\overline{PL}$ input is pulsed LOW. This immediately loads the counter with 1011 to produce $Q_3 = 1$, $Q_2 = 0$, $Q_1 = 1$, and $Q_0 = 1$. At t_1 the CP_U input makes a PGT, but the counter cannot respond to this because $\overline{PL}$ is still active at that time. At times t_2, t_3, t_4, and t_5 the counter counts up on each PGT at CP_U. After the PGT at t_5, the counter is in the 1111 state, but $\overline{TC}_U$ does not go LOW until CP_U goes LOW at t_6. When the next PGT occurs at t_7, the counter recycles to 0000, and $\overline{TC}_U$ returns HIGH.

The counter will count up in response to the PGTs at t_8 and t_9. The PGT at t_{10} will have no effect because the MR goes HIGH prior to t_{10} and remains active at t_{10}. This will reset all FFs to 0 and overrides the CP_U signal.

EXAMPLE 7-15

Figure 7-23(a) shows the 74HC193 wired as a *down counter*. The parallel data inputs are permanently wired as 0111, and the CP_D and $\overline{PL}$ waveforms are shown in Figure 7-23(b). Assume that the counter is initially in the 0000 state, and determine the output waveforms.

(a)

(b)

FIGURE 7-23 Example 7-15.

Solution

At t_0, all of the FF outputs are LOW and CP_D is LOW. These are the conditions that produce $\overline{TC}_D = 0$. Prior to t_1 the $\overline{PL}$ input is pulsed LOW. This immediately presets the counter to 0111 and therefore causes $\overline{TC}_D$ to go HIGH. The PGT of CP_D at t_1 will have no effect because $\overline{PL}$ is still active. The counter will respond to the PGTs at t_2 to t_8 and counts down to 0000 at t_8. $\overline{TC}_D$ does not go LOW until t_9, when CP_D goes LOW. At t_{10} the PGT of CP_D causes the counter to recycle to 1111 and also drives $\overline{TC}_D$ back HIGH.

Variable MOD Number Using the 74ALS193/HC193

Presettable counters can be easily wired for different MOD numbers without the need for additional logic circuitry. We will demonstrate this for the 74HC193 using the circuit of Figure 7-24(a). Here, the 74HC193 is used as a down counter with its parallel load inputs permanently connected at 0101 (5_{10}). Note that the $\overline{TC}_D$ output is connected back to the $\overline{PL}$ input.

    We will begin our analysis by assuming that the counter is in the 0101 state at time t_0. Refer to Figure 7-24(b) for the counter waveforms.

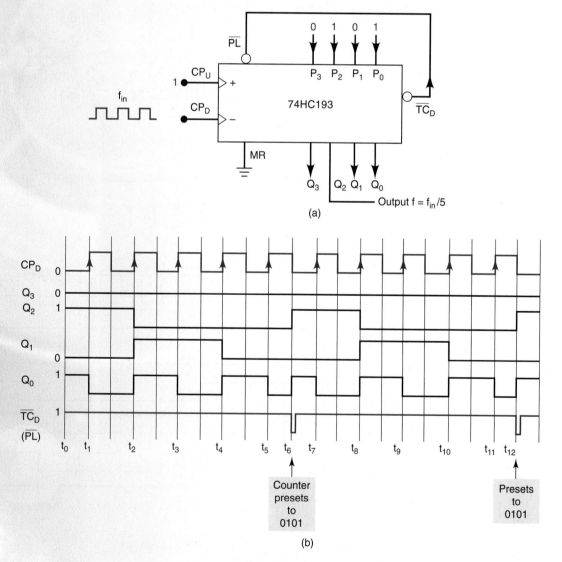

(a)

(b)

FIGURE 7-24   (a) 74HC193 wired as a MOD-5 down counter; (b) waveforms.

The counter will decrement (count down) on the PGTs of CP_D at times t_1 to t_5. At t_5, the counter is in the 0000 state. When CP_D goes LOW at t_6, it drives $\overline{TC}_D$ LOW. This immediately activates the $\overline{PL}$ input and presets the counter back to the 0101 state. Note that $\overline{TC}_D$ stays LOW for only a short interval because once the counter outputs go to 0101 in response to $\overline{PL} = 0$, the condition needed to keep $\overline{TC}_D = 0$ is removed. Thus, there is only a narrow glitch at $\overline{TC}_D$.

This same sequence is repeated at times t_7 to t_{12} and at equal intervals thereafter. If we examine the Q_2 waveform, we can see that it goes through one complete cycle for every *five* cycles of CP_D. For example, there are *five* clock cycles between the PGT of Q_2 at t_6 and the PGT of Q_2 at t_{12}. Thus, the frequency of the Q_2 waveform is one-fifth of the clock frequency.

This arrangement does have a peculiarity which you may have noticed: it counts through *six* different states (5, 4, 3, 2, 1, 0), yet it divides the frequency by *five*. This is due to the unusual way the count gets preset back to 5 in the middle of a clock cycle. Thus, this counter's operation violates our general rule that the number of states and the frequency-division ratio are the same. This type of arrangement is used principally for frequency division, so we will ignore the counting sequence and call it a MOD-5 counter because it divides the clock frequency by 5.

It is no coincidence that the frequency-division ratio (5) is the same as the number applied to the parallel data inputs (0101 = 5). In fact, we can vary the frequency division by changing the logic levels applied to the parallel data inputs.

A *variable frequency-divider* circuit can be easily implemented by connecting switches to the parallel data inputs of the circuit in Figure 7-24. The switches can be set to a value equal to the desired frequency-division ratio. Notice that care must be taken to choose the appropriate Q output, depending on the frequency-division ratio that is selected.

Multistage Arrangement

As stated earlier, the $\overline{TC}_D$ and $\overline{TC}_U$ outputs are used when two or more 74HC193 counters are connected in a **multistage** arrangement. In Figure 7-25, two 74HC193s are connected in a two-stage up/down counter arrangement that

Note: Reset input has priority over $\overline{Load}$ and clock inputs.
$\overline{Load}$ input has priority over clock inputs.

FIGURE 7-25 Two 74HC193s connected in a two-stage arrangement to extend maximum counting range.

EXAMPLE 7-16

How many AND gates are required to decode completely all of the states of a MOD-32 binary counter? What are the inputs to the gate that decodes for the count of 21?

Solution

A MOD-32 counter has 32 possible states. One AND gate is needed to decode for each state; therefore, the decoder requires 32 AND gates. Because $32 = 2^5$, the counter contains five FFs. Thus, each gate will have five inputs, one from each FF. Decoding for the count of 21 (that is, 10101_2) requires AND gate inputs of E, $\overline{D}$, C, $\overline{B}$, and A, where E is the MSB flip-flop.

Active-LOW Decoding

If NAND gates are used in place of AND gates, the decoder outputs produce a normally HIGH signal, which goes LOW only when the number being decoded occurs. Both types of decoders are used, depending on the type of circuits being driven by the decoder outputs.

EXAMPLE 7-17

Figure 7-28 shows a common situation in which a counter is used to generate a control waveform, which could be used to control devices such as a motor, solenoid valve, or heater. The MOD-16 counter cycles and recycles through its counting sequence. Each time it goes to the count of 8 (1000), the upper NAND gate will produce a LOW output, which sets flip-flop X to the 1 state. Flip-flop X stays HIGH until the counter reaches the count of 14 (1110), at which time the lower NAND gate decodes it and produces a LOW output to clear X to the 0 state. Thus, the X output is HIGH between the counts of 8 and 14 for each cycle of the counter.

FIGURE 7-28   Example 7-17.

BCD Counter Decoding

A BCD counter has 10 states that can be decoded using the techniques described previously. BCD decoders provide 10 outputs corresponding to the decimal digits 0 through 9 and represented by the states of the counter FFs. These 10 outputs can be used to control 10 individual indicator LEDs for a visual display. More often, instead of using 10 separate LEDs, a single display device is used to display the decimal numbers 0 through 9. One class of decimal displays contains seven small segments made of a material (usually LEDs or liquid-crystal displays) that either emits light or reflects ambient light. The BCD decoder outputs control which segments are illuminated in order to produce a pattern representing one of the decimal digits.

We will go into more detail concerning these types of decoders and displays in Chapter 9. However, because BCD counters and their associated decoders and displays are very commonplace, we will use the decoder/display unit (see Figure 7-29) to represent the complete circuitry used to display visually the contents of a BCD counter as a decimal digit.

FIGURE 7-29 BCD counters usually have their count displayed on a single display device.

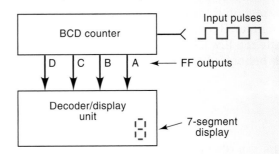

REVIEW QUESTIONS

1. How many gates are needed to decode a six-bit counter fully?
2. Describe the decoding gate needed to produce a LOW output when a MOD-64 counter is at the count of 23.

7-12 DECODING GLITCHES

In Section 7-5, we discussed the effects of FF propagation delays in ripple counters. As we saw then, the accumulated propagation delays serve essentially to limit the frequency response of ripple counters. The delays between FF transitions can also cause problems when decoding a ripple counter. The problem occurs in the form of **decoding glitches** or spikes at the outputs of some of the decoding gates. This is illustrated in Figure 7-30 for a MOD-4 ripple counter.

The waveforms at the outputs of each FF and decoding gate are shown in the figure. Notice the propagation delay between the clock waveform and the A output waveform and between the A waveform and the B waveform. The glitches in the X_0 and X_2 decoding waveforms are caused by the delay between the A and B waveforms. X_0 is the output of the AND gate decoding for the normal 00 count. The 00 condition also occurs momentarily as the counter goes from the 01 to the 10 count, as shown by the waveforms. This is because B cannot change states until A goes LOW. This momentary 00 state lasts only for several nanoseconds (depending on t_{pd} of flip-flop B) but can be detected to enable the decoding gates some time t_D *after* the clock pulse goes LOW. t_D is chosen to be greater than the total time it takes the counter to reach a stable count, and it depends, of course, on the FF delays and the number of FFs in the counter. In this way, the decoding gate outputs will not contain any spikes because they are disabled during the time the FFs are in transition.

The strobe method is not used if a counter is used only for display purposes because the decoding spikes are too narrow to affect the display. The strobe signal is used when the counter is used in control applications like that of Figure 7-28, where the spikes could cause erroneous operation.

REVIEW QUESTIONS

1. Explain why the decoding gates for an asynchronous counter may have glitches on their outputs.
2. How does strobing eliminate decoding glitches?

7-13 CASCADING BCD COUNTERS

BCD counters are often used whenever pulses are to be counted and the results displayed in decimal. A single BCD counter counts from 0 through 9 and then recycles to 0. To count to larger decimal values, we can **cascade** BCD counter stages as shown in Figure 7-32. This multistage arrangement operates as follows:

1. Initially, all counters are cleared to the 0 state. Thus, the decimal display is 000.
2. As input pulses arrive, the units BCD counter advances one count per pulse. After nine pulses have occurred, the hundreds and tens BCD counters are still at 0, and the units counter is at 9 (binary 1001). Thus, the decimal display reads 009.
3. On the tenth input pulse, the units counter recycles to 0, causing its flip-flop D output to go from 1 to 0. This 1-to-0 transition acts as the clock input for the tens counter and causes it to advance one count. Thus, after 10 input pulses, the decimal readout is 010.
4. As additional pulses occur, the units counter advances one count per pulse, and each time the units counter recycles to 0, it advances the tens

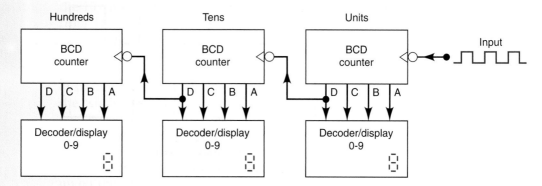

FIGURE 7-32 Cascading BCD counters to count and display numbers from 000 to 999.

counter one count. Thus, after 99 input pulses have occurred, the tens counter is at 9, as is the units counter. The decimal readout is thus 099.

5. On the hundredth input pulse, the units counter recycles to 0, which in turn causes the tens counter to recycle to 0. The flip-flop D output of the tens counter thus makes a 1-to-0 transition, which acts as the clock input for the hundreds counter and causes it to advance one count. Thus, after 100 pulses, the decimal readout is 100.

6. This process continues up until 999 pulses. On the 1000th pulse, all of the counters recycle back to 0.

It should be obvious that this arrangement can be expanded to any desired number of decimal digits simply by adding more stages. For example, to count up to 999,999 will require six BCD counters and associated decoders and displays. In general, then, we need one BCD counter per decimal digit.

Each BCD counter in a cascaded arrangement, such as in Figure 7-32, could be a variable-MOD counter such as the 74LS293 wired as a MOD-10 counter; or it could be an IC that is internally wired as a BCD counter, such as the 74LS90 and 74HC192.

7-14   SYNCHRONOUS COUNTER DESIGN*

Many different counter arrangements are available as ICs—asynchronous, synchronous, and combined asynchronous/synchronous. Most of these count in a normal binary sequence, although their counting sequences can be somewhat altered using the methods we demonstrated for the 74293 and 74193 ICs. There are situations, however, where a counter is required that follows a sequence that is not counting in normal binary, for example, 000, 010, 101, 001, 110, 000, ...

Several methods exist for designing counters that follow arbitrary sequences. We will present the details for one common method that uses J-K flip-flops in a synchronous counter configuration. The same method can be used in designs with D flip-flops. The technique is one of several design procedures that are part of an area of digital circuit design called **sequential circuit design**, which is normally part of an advanced course.

Basic Idea

In synchronous counters, all of the FFs are clocked at the same time. Before each clock pulse, the J and K input of each FF in the counter must be at the correct level to ensure that the FF goes to the correct state. For example, consider the situation shown in Table 7-1. When the next clock pulse occurs, the J and K inputs of the FFs must be at the correct levels that will cause flip-flop C to change from 1 to 0, flip-flop B from 0 to 1, and flip-flop A from 1 to 1 (i.e., no change).

The process of designing a synchronous counter thus becomes one of designing the logic circuits that *decode* the various states of the counter to sup-

TABLE 7-1

PRESENT State			NEXT State		
C	B	A	C	B	A
1	0	1	0	1	1

*This topic may be omitted without affecting the continuity of the remainder of the book.

ply the logic levels to each J and K input. The inputs to these decoder circuits will come from the outputs of one or more of the FFs. To illustrate, for the synchronous counter of Figure 7-17, the AND gate that feeds the J and K inputs of flip-flop C decodes the states of flip-flops A and B. Likewise, the AND gate that feeds the J and K inputs of flip-flop D decodes the states of A, B, and C.

J-K Excitation Table

Before we begin the process of designing the decoder circuits for each J and K input, we must first review the operation of the J-K flip-flop using a different approach, one called an *excitation table* (Table 7-2). The leftmost column of this table lists each possible FF output transition. The second and third columns list the FF's PRESENT state, symbolized as $Q(N)$, and the NEXT state, symbolized as $Q(N+1)$, for each transition. The last two columns list the J and K levels required to produce each transition. Let's examine each case.

$0 \rightarrow 0$ **TRANSITION** The FF PRESENT state is at 0 and is to remain at 0 when a clock pulse is applied. From our understanding of how a J-K flip-flop works, this can happen when either $J = K = 0$ (no-change condition) or $J = 0$ and $K = 1$ (clear condition). Thus, J must be at 0, but K can be at either level. The table indicates this with a "0" under J and an "x" under K. Recall that "x" means the don't-care condition.

$0 \rightarrow 1$ **TRANSITION** The PRESENT state is 0 and is to change to a 1, which can happen when either $J = 1$ and $K = 0$ (set condition) or $J = K = 1$ (toggle condition). Thus, J must be a 1, but K can be at either level for this transition to occur.

$1 \rightarrow 0$ **TRANSITION** The PRESENT state is 1 and is to change to a 0, which can happen when either $J = 0$ and $K = 1$ or $J = K = 1$. Thus, K must be a 1, but J can be at either level.

$1 \rightarrow 1$ **TRANSITION** The PRESENT state is a 1 and is to remain a 1, which can happen when either $J = K = 0$ or $J = 1$ and $K = 0$. Thus, K must be a 0 while J can be at either level.

The use of this **J-K excitation table** (Table 7-2) is a principal part of the synchronous counter design procedure.

TABLE 7-2
J-K flip-flop
excitation table.

Transition at Output	PRESENT State $Q(N)$	NEXT State $Q(N+1)$	J	K
$0 \rightarrow 0$	0	0	0	x
$0 \rightarrow 1$	0	1	1	x
$1 \rightarrow 0$	1	0	x	1
$1 \rightarrow 1$	1	1	x	0

Design Procedure

We will now go through a complete synchronous counter design procedure. Although we will do it for a specific counting sequence, the same steps can be followed for any desired sequence.

Step 1. Determine the desired number of bits (FFs) and the desired counting sequence.

TABLE 7-3

C	B	A
0	0	0
0	0	1
0	1	0
0	1	1
1	0	0
0	0	0
0	0	1
	etc.	

For our example, we will design a three-bit counter that goes through the sequence shown in Table 7-3. Notice that this sequence does not include the 101, 110, and 111 states. We will refer to these states as *undesired states*.

Step 2. Draw the state transition diagram showing *all* possible states, including those that are not part of the desired counting sequence.

For our example, the state transition diagram appears as shown in Figure 7-33. The 000 through 100 states are connected in the expected sequence. The new idea used in this diagram is the inclusion of the undesired states. They must be included in our design in case the counter accidentally gets into one of these states upon power-up or due to noise. The circuit designer can choose to have each of these undesired states go to any state upon the application of the next clock pulse. We choose to have them all go to the 000 state, from which the correct sequence will be generated.

FIGURE 7-33 State transition diagram for the synchronous counter design example.

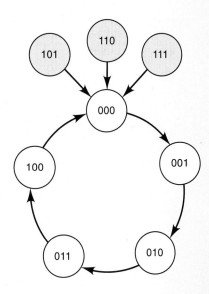

Step 3. Use the state transition diagram to set up a table that lists *all* PRESENT states and their NEXT states.

For our example, the information is shown in Table 7-4. The left-hand portion of the table lists *every* possible state, even those that are not part of the sequence. We label these as the PRESENT states. The right-hand portion lists the NEXT state for each PRESENT state. These are obtained from the state transition diagram in Figure 7-33. For instance, line 1 shows that the PRESENT state of 000 has the NEXT state of 001, and line 5 shows that the PRESENT state of 100 has the NEXT state of 000. Lines 6, 7, and 8 show that the undesired PRESENT states 101, 110, and 111 all have the NEXT state of 000.

TABLE 7-4

	PRESENT State			NEXT State		
	C	B	A	C	B	A
Line 1	0	0	0	0	0	1
2	0	0	1	0	1	0
3	0	1	0	0	1	1
4	0	1	1	1	0	0
5	1	0	0	0	0	0
6	1	0	1	0	0	0
7	1	1	0	0	0	0
8	1	1	1	0	0	0

Step 4. Add a column to this table for each J and K input. For each PRESENT state, indicate the levels required at each J and K input in order to produce the transition to the NEXT state.

Our design example uses three FFs—C, B, and A—and each one has a J and a K input. Therefore, we must add six new columns as shown in Table 7-5. This completed table is called the **circuit excitation table**. The six new columns are the J and K inputs of each FF. The entries under each J and K are obtained from Table 7-2, the J-K flip-flop excitation table that we developed earlier. We will demonstrate this for several of the cases, and you can verify the rest.

Let's look at line 1 in Table 7-5. The PRESENT state of 000 is to go to the NEXT state of 001 on the occurrence of a clock pulse. For this state transition, the C flip-flop goes from 0 to 0. From the J-K excitation table, we see that J_C must be at 0 and K_C at "x" for this transition to occur. The B flip-flop also goes from 0 to 0, and so $J_B = 0$ and $K_B = x$. The A flip-flop goes from 0 to 1. Also from Table 7-2, we see that $J_A = 1$ and $K_A = x$ for this transition.

In line 4 in Table 7-5, the PRESENT state of 011 has a NEXT state of 100. For this state transition, flip-flop C goes from 0 to 1, which requires $J_C = 1$ and $K_C = x$. Flip-flops B and A are both going from 1 to 0. The J-K excitation table indicates that these two FFs need $J = x$ and $K = 1$ for this to occur.

The required J and K levels for all other lines in Table 7-5 can be determined in the same manner.

TABLE 7-5
Circuit excitation table.

	PRESENT State			NEXT State			J_C	K_C	J_B	K_B	J_A	K_A
	C	B	A	C	B	A						
Line 1	0	0	0	0	0	1	0	x	0	x	1	x
2	0	0	1	0	1	0	0	x	1	x	x	1
3	0	1	0	0	1	1	0	x	x	0	1	x
4	0	1	1	1	0	0	1	x	x	1	x	1
5	1	0	0	0	0	0	x	1	0	x	0	x
6	1	0	1	0	0	0	x	1	0	x	x	1
7	1	1	0	0	0	0	x	1	x	1	0	x
8	1	1	1	0	0	0	x	1	x	1	x	1

Step 5. Design the logic circuits needed to generate the levels required at each J and K input.

Table 7-5, the circuit excitation table, lists six J, K inputs—J_C, K_C, J_B, K_B, J_A, and K_A. We must consider each of these as an output from its own logic circuit with inputs from flip-flops C, B, and A. Then we must design the circuit for each one. Let's design the circuit for J_A.

To do this, we need to look at the PRESENT states of C, B, and A and the desired levels at J_A for each case. This information has been extracted from Table 7-5 and presented in Figure 7-34(a). This truth table shows the desired levels at J_A for each PRESENT state. Of course, for some of the cases, J_A is a don't-care. To develop the logic circuit for J_A, we must first determine its expression in terms of C, B, and A. We will do this by transferring the truth-table information to a three-variable Karnaugh map and performing the K-map simplification, as in Figure 7-34(b).

FIGURE 7-34   (a) Portion of circuit excitation table showing J_A for each PRESENT state; (b) K map used to obtain the simplified expression for J_A.

There are only two 1s in this K map, and they can be looped to obtain the term $\overline{A}\,\overline{C}$, but if we use the don't-care conditions at $A\,\overline{B}\,\overline{C}$ and $AB\overline{C}$ as 1s, we can loop a quad to obtain the simpler term $\overline{C}$. Thus, the final expression is

$$J_A = \overline{C}$$

Now let's consider K_A. We can follow the same steps as we did for J_A. However, a look at the entries under K_A in the circuit excitation table shows only 1s and don't-cares. If we change all the don't-cares to 1s, then K_A is always a 1. Thus, the final expression is

$$K_A = 1$$

In a similar manner, we can derive the expressions for J_C, K_C, J_B, and K_B. The K maps for these expressions are given in Figure 7-35. You might want to confirm their correctness by checking them against the circuit excitation table.

Step 6. Implement the final expressions.

The logic circuits for each J and K input are implemented from the expressions obtained from the K map. The complete synchronous counter design is implemented in Figure 7-36. Note that all FFs are clocked in parallel. You might want to verify that the logic for the J and K inputs agrees with Figures 7-34 and 7-35.

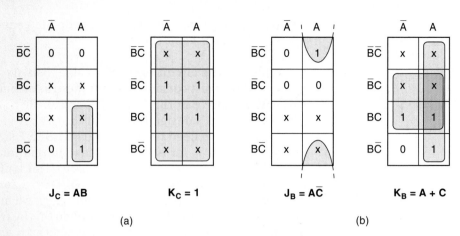

$$J_C = AB \qquad K_C = 1 \qquad J_B = A\overline{C} \qquad K_B = A + C$$

(a)   (b)

FIGURE 7-35 (a) K maps for the J_C and K_C logic circuits; (b) K maps for the J_B and K_B logic circuits.

FIGURE 7-36 Final implementation of the synchronous counter design example.

Stepper Motor Control

We will now apply this design procedure to a practical situation—driving a *stepper motor*. A stepper motor is a motor that rotates in steps, typically 15° per step, rather than in a continuous motion. Magnetic coils or windings within the motor must be energized and deenergized in a specific sequence in order to produce this stepping action. Digital signals are normally used to control the current in each of the motor's coils. Stepper motors are used extensively in situations where precise position control is needed, such as in positioning of read/write heads on magnetic disks, in controlling print heads in printers, and in robots.

Figure 7-37(a) is a diagram of a typical stepper motor with four coils. For the motor to rotate properly, coils 1 and 2 must always be in opposite states; that is, when coil 1 is energized, coil 2 is not, and vice versa. Likewise, coil 3 and coil 4 must always be in opposite states. The outputs of a two-bit synchronous counter are used to control the current in the four coils; A and $\overline{A}$ control coils 1 and 2, and B and $\overline{B}$ control coils 3 and 4. The current amplifiers are needed because the FF outputs cannot supply the amount of current that the coils require.

(a)

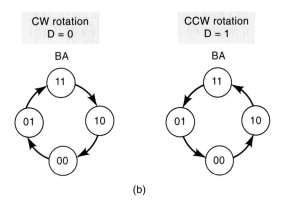

(b)

FIGURE 7-37 (a) A synchronous counter supplies the appropriate sequential outputs to drive a stepper motor; (b) state transition diagrams for both states of Direction input, D.

Because this stepper motor can rotate either clockwise (CW) or counter-clockwise (CCW), we have a Direction input, D, which is used to control the direction of rotation. The state diagrams in Figure 7-37(b) show the two cases. For CW rotation to occur, we must have $D = 0$, and the state of the counter, BA, must follow the sequence 11, 10, 00, 01, 11, 10, . . . , and so on, as it is clocked by the Step input signal. For CCW rotation, $D = 1$, and the counter must follow the sequence 11, 01, 00, 10, 11, 01, . . . , and so on.

We are now ready to follow the six steps of the synchronous counter de-sign procedure. Steps 1 and 2 have already been done, so we can proceed with steps 3 and 4. Table 7-6 shows each possible PRESENT state of D, B, and A and the desired NEXT state, along with the levels at each J and K input

TABLE 7-6
Circuit excitation table for
Figure 7-37(b).

PRESENT State			NEXT State						
D	**B**	**A**	**D**	**B**	**A**	J_B	K_B	J_A	K_A
0	0	0	0	0	1	0	x	1	x
0	0	1	0	1	1	1	x	x	0
0	1	0	0	0	0	x	1	0	x
0	1	1	0	1	0	x	0	x	1
1	0	0	1	1	0	1	x	0	x
1	0	1	1	0	0	0	x	x	1
1	1	0	1	1	1	x	0	1	x
1	1	1	1	0	1	x	1	x	0

K inputs are usually simpler than the logic circuits needed to drive a similar counter made from D flip-flops. When designing counters that will be implemented in PLDs and when abundant gates are available, it makes sense to use D flip-flops instead of J-Ks. In this section, we will examine another way to design synchronous counters using D flip-flops.

Designing counter circuits using D flip-flops is even easier than using J-K flip-flops. The first three steps of synchronous counter design are identical to those used to demonstrate the design of a MOD-5 counter in Section 7-14. After the first three steps, you should have a PRESENT state/NEXT state table, as shown in Table 7-7. Step 4 for the J-K design process involved listing all the input levels necessary to make the flip-flops go to the desired NEXT state when the clock edge occurs. With D flip-flops, this step is essentially unnecessary because the NEXT state of the flip-flop is the same as its PRESENT D input values. Notice that the NEXT state column is identical to the D input column. Also notice that the D input bits are labeled with the name of the flip-flop and a .D extension. It is common to refer to an entire flip-flop by the name of its Q output, and this notation provides a way to designate a particular flip-flop and one of its ports.

Think of Table 7-7 as a set of truth tables (one for each FF) specifying the necessary D inputs for each PRESENT state of the Q outputs. The final step using D flip-flops would be to write an equation for each flip-flop's D input. These equations are presented below in unsimplified SOP form.

$$Q_2 \cdot D = \overline{Q_2} Q_1 Q_0$$

$$Q_1 \cdot D = \overline{Q_2} \overline{Q_1} Q_0 + \overline{Q_2} Q_1 \overline{Q_0}$$

$$Q_2 \cdot D = \overline{Q_2} \overline{Q_1} \overline{Q_0} + \overline{Q_2} Q_1 \overline{Q_0}$$

We can then simplify each of the D input equations to reduce the amount of gating needed. In HDL, this is accomplished by the compiler software. Techniques similar to those we used in Chapter 5 can be used to implement these logic equations using HDL in the structural level of abstraction.

TABLE 7-7
D flip-flop PRESENT state–NEXT state table.

Clock Pulse	PRESENT State			NEXT State			D Inputs		
	Q_2	Q_1	Q_0	Q_2	Q_1	Q_0	Q_2.D	Q_1.D	Q_0.D
0	0	0	0	0	0	1	0	0	1
1	0	0	1	0	1	0	0	1	0
2	0	1	0	0	1	1	0	1	1
3	0	1	1	1	0	0	1	0	0
4	1	0	0	0	0	0	0	0	0
Unused state	1	0	1	0	0	0	0	0	0
Unused state	1	1	0	0	0	0	0	0	0
Unused state	1	1	1	0	0	0	0	0	0

State Transition Description Methods

The next method of describing circuits that we need to examine uses tables. This method is not concerned with connecting ports of components but rather with assigning values to objects like ports, signals, and variables. In other words, it describes how the output data relates to the input data throughout the circuit. We have already used this method in several of the introductory circuits in Chapters 3 and 4, in the form of truth tables. With sequential counter circuits, the equivalent of the truth table is the PRESENT state/NEXT state table, as we saw in the last section. We can use the HDL essentially to describe the PRESENT state/NEXT state table in a way like that used to describe a truth table and thus avoid the tedious details of generating the Boolean equations.

STATE DESCRIPTIONS IN AHDL

As an example of a simple counter circuit, we will implement the MOD-5 counter of Figure 7-36 in AHDL. The inputs and outputs are defined in the SUBDESIGN section of Figure 7-40, as always. Note that the variable *count[ ]* on line 7 is defined as a three-bit array of DFF primitives. Because this is a synchronous counter, we need to tie all the DFF *clk* inputs to the SUBDE-SIGN's *clock* input. This is accomplished in AHDL by the following statement:

```
count[].clk = clock;
```

Count is a three-bit array of D flip-flops, each of which has a *clk* input. This notation assigns the same *clock* input from the SUBDESIGN section to each of these *clk* inputs. To implement the PRESENT state/NEXT state table, a CASE construct is used. For each of the possible values of the register *count[ ]*, we determine the value that should be placed on the *D* inputs of the flip-flops, which will determine the NEXT state of the counter. The statement on line 21 assigns the value on *count[ ]* to the output pins.

FIGURE 7-40 AHDL
MOD-5 counter.

```
1    SUBDESIGN fig7_40
2    (
3         clock          :INPUT;
4         q[2..0]        :OUTPUT;
5    )
6    VARIABLE
7         count[2..0]  :DFF;          --create a 3-bit register
8    BEGIN
9         count[].clk = clock;        --connect all clocks in parallel
10
11             CASE count[] IS
12   --              Present            Next
13   ----------------------------------------------------------
14             WHEN   0      =>     count[].d = 1;
15             WHEN   1      =>     count[].d = 2;
16             WHEN   2      =>     count[].d = 3;
17             WHEN   3      =>     count[].d = 4;
18             WHEN   4      =>     count[].d = 0;
19             WHEN OTHERS   =>     count[].d = 0;
20           END CASE;
21         q[] = count[];              -- assign register to output pins
22   END;
```

A unique aspect of AHDL is its ability to imply association of similarly named objects. For example, the description in Figure 7-40 has been rewritten in Figure 7-41, with the register of D flip-flops on line 7 named *q* instead of *count*. The *q* outputs of this register have been associated automatically with the *q*[] outputs, making line 21 of Figure 7-40 unnecessary. In Figure 7-41, we have also rewritten the code to demonstrate the use of the AHDL TABLE to describe the PRESENT state/NEXT state transitions.

FIGURE 7-41 Another version of the MOD-5 counter in Figure 7-40.

```
1   SUBDESIGN fig7_41
2   (
3       clock      :INPUT;
4       q[2..0]    :OUTPUT;
5   )
6   VARIABLE
7       q[2..0]     :DFF;    -- create a 3-bit register
8   BEGIN
9       q[].clk = clock;    -- connect all clocks in parallel
10      TABLE
11         q[].q =>    q[].d;
12          0      =>    1;
13          1      =>    2;
14          2      =>    3;
15          3      =>    4;
16          4      =>    0;
17          5      =>    0;
18          6      =>    0;
19          7      =>    0;
20      END TABLE;
21  END;
```

STATE DESCRIPTIONS IN VHDL

As an example of a simple counter circuit, we will implement the MOD-5 counter of Figure 7-36 in VHDL. Our purpose in this example is to demonstrate a counter using a control structure similar to a PRESENT state/NEXT state table. Two key tasks must be accomplished in VHDL: detecting the desired clock edge, and assigning the proper NEXT state to the counter. Recall from our study of flip-flops that a PROCESS can be used to respond to a transition of an input signal. Also, we have learned that a CASE construct can evaluate an expression and, for any valid input value, assign a corresponding value to another signal. The code in Figure 7-42 uses a PROCESS and a CASE construct to implement this counter. The inputs and outputs are defined in the ENTITY declaration, as in the past.

When VHDL is used to describe a counter, we must find a way to "store" the state of the counter between clock pulses (i.e., the action of a flip-flop). This is done in one of two ways: using SIGNALs, or using VARIABLEs. We have used SIGNALs extensively in previous examples that operated concurrently. A SIGNAL in VHDL holds the last value that was assigned to it, very much like a flip-flop. Consequently, we can use a SIGNAL as the data object

```
1    ENTITY fig7_42 IS
2    PORT  (
3            clk    :IN BIT;
4            q      :OUT BIT_VECTOR[2 DOWNTO 0]
5          );
6    END fig7_42 ;
7
8    ARCHITECTURE a OF fig7_42    IS
9    BEGIN
10      PROCESS (clk)                              -- respond to clk input
11      VARIABLE count: BIT_VECTOR(2 DOWNTO 0);  -- create a 3-bit register
12      BEGIN
13        IF (clk = '1' AND clk'EVENT) THEN     -- rising edge trigger
14          CASE count IS
15   --          Present              Next
16   -------------------------------------------------------------------
17            WHEN "000"    =>     count := "001";
18            WHEN "001"    =>     count := "010";
19            WHEN "010"    =>     count := "011";
20            WHEN "011"    =>     count := "100";
21            WHEN "100"    =>     count := "000";
22            WHEN OTHERS   =>     count := "000";
23          END CASE;
24        END IF;
25        q <= count;           -- assign register to output pins
26      END PROCESS;
27   END a;
```

FIGURE 7-42   VHDL MOD-5 counter.

representing the counter value. This SIGNAL can then be used to connect the counter value to any other elements in the architecture description.

In this design, we have chosen to use a **VARIABLE** instead of a SIGNAL as the data object that stores the counter value. VARIABLEs are not exactly like SIGNALs because they are not used to connect various parts of the design. Instead, they are used as a local place to "store" a value. Variables are considered to be local data objects because they are recognized only within the PROCESS in which they are declared. On line 11 of Figure 7-42, the variable named *count* is declared within the PROCESS before BEGIN. Its type is the same as the output port *q.* The keyword PROCESS on line 10 is followed by the sensitivity list containing the input signal *clk.* Whenever *clk* changes state, the PROCESS is invoked, and the circuit performs the sequence of tasks described between BEGIN and END. The qualifier 'EVENT evaluates TRUE if the signal preceding it has just changed states. Line 13 states that if *clk* has just changed states and right now it is '1', then we know it was a rising edge. To implement the PRESENT state/NEXT state table, a CASE construct is used. For each of the possible values of the variable *count,* we determine the NEXT state of the counter. Notice that the := operator is used to assign a value to a variable. Line 25 assigns the value stored in *count* to the output pins. Because *count* is a local variable, this assignment must be done before END PROCESS on line 26.

Behavioral Description

The **behavioral level of abstraction** is a way to describe a circuit by describing its behavior in terms very similar to the way you might describe its operation in English. Think about the way a counter circuit's operation might be described by someone who knows nothing about flip-flops or logic gates. Perhaps that person's description would sound something like, "When the counter input changes from LOW to HIGH, the number on the output counts up by 1." This level of description deals more with cause-and-effect relationships than with the path of data flow or wiring details. However, we cannot really use just any description in English to describe the circuit's behavior. The proper syntax must be used within the constraints of the HDL.

AHDL

In AHDL, the first important step in this description method is to declare the counter output pins properly. They should be declared as a bit array, with indices decreasing left to right and with 0 as the least significant index in the array, as opposed to individual bits named a, b, c, d, and so on. In this way, the numeric value associated with the bit array's name is interpreted as a binary number upon which certain arithmetic operations can be performed. For example, the bit array *count* shown in Figure 7-43 might contain the bits 1001, as shown. The AHDL compiler interprets this bit pattern as having the value of 9 in decimal.

In order to make this four-bit array named *count* in AHDL, a register of flip-flop primitives must be declared and associated with the name of the counter, as shown in Figure 7-44, line 7. Notice that the variable *count* is the same size as the output declaration and is declared to be made up of D flip-flops. Recall from Figure 7-41 that in AHDL, the name of the set of flip-flops (named *count* in Figure 7-44) can be the same as the name of the output pins (*q*). In other words, the statement on line 7 can be changed to

```
q[2..0] :DFF; .
```

If this were done, all references to *count* thereafter would be changed to *q*. This can make the code shorter, but it does not demonstrate universal HDL concepts as clearly. In AHDL, all the clocks can be specified as being tied together and connected to a common clock source using the statement on line 10, *count[ ].clk = clock*. In this example, *count[ ].clk* refers to the clock input of each flip-flop in the array called *count*.

The behavioral description of this counter is very simple. The current state of the counter is evaluated (*count[ ].q*) on line 11, and if it is less than the highest desired count value, it uses the description *count[ ].d = count.q + 1* (line 12). This means that the current state of the D inputs must be equal to a value one count greater than the current state of the Q outputs. When the current state of the counter has reached the highest desired state (or higher),

FIGURE 7-43 The elements of a D register storing the number 9.

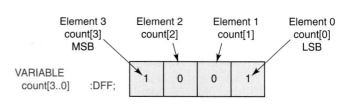

```
1    SUBDESIGN fig7_44
2    (
3       clock    :INPUT;
4       q[2..0]  :OUTPUT;   -- declare 3-bit array of output bits
5    )
6    VARIABLE
7       count[2..0]  :DFF; -- declare a register of D flip flops.
8
9    BEGIN
10      count[].clk = clock;  -- connect all clocks to synchronous source
11      IF count[].q < 4 THEN    -- note; count[] is the same as count[].q
12         count[].d = count[].q + 1; -- increment current value by one
13      ELSE count[].d = 0;         -- recycle to zero: force unused states to 0
14      END IF;
15      q[] = count[];             -- transfer register contents to outputs
16   END;
```

FIGURE 7-44   Behavioral description of a counter in AHDL.

the *IF* statement test will be false, resulting in the NEXT state being reset to zero (line 13). The last statement on line 15 simply connects the counter value to the output pins of the device.

VHDL

In VHDL, the first important step in this description method is to declare properly the counter output port, as shown in Figure 7-45. The data type of

```
1    ENTITY fig7_45 IS
2    PORT( clk   :IN BIT;
3          q   :OUT INTEGER RANGE 0 TO 7  );
4    END  fig7_45;
5
6    ARCHITECTURE a OF fig7_45 IS
7    BEGIN
8       PROCESS (clk)
9       VARIABLE count: INTEGER RANGE 0 to 7;   -- define a numeric VARIABLE
10         BEGIN
11            IF (clk = '1' AND clk'EVENT) THEN    -- rising edge?
12               IF count < 4 THEN            -- less than max?
13                  count := count + 1;       -- increment value
14               ELSE                         -- must be at max or bigger
15                  count := 0;               -- recycle to zero
16               END IF;
17            END IF;
18       q <= count;                 -- transfer register contents to outputs
19       END PROCESS;
20   END a;
```

FIGURE 7-45   Behavioral description of a counter in VHDL.

the output port (line 3) must match the type of the counter variable (line 9), and it must be a type that allows arithmetic operations. Recall that VHDL treats BIT_VECTORS as just a string of bits, not as a binary numeric quantity. In order to recognize the signal as a numeric quantity, the data object must be typed as an INTEGER. The compiler looks at the RANGE 0 TO 7 clause on line 3 and knows that the counter needs three bits. A similar declaration is needed for the register variable on line 9 that will actually be counting up. This is called *count*. The first statement after BEGIN in the PROCESS responds to the rising edge of the clock as in the previous examples. It then uses behavioral description methods to define the counter's response to the clock edge. If the counter has not reached its maximum (line 12), then it should be incremented (line 13). Otherwise (line 14), it should recycle the counter to zero (line 15). The last statement on line 18 simply connects the counter value to the output pins of the device.

1. What type of table is used to describe a counter's operation?
2. When designing a counter with D flip-flops, what is applied to the D inputs in order to drive it to the NEXT state on the next active clock edge?
3. How would you refer to a D input of a flip-flop named *data1* using HDL notation?
4. Which method describes the circuit's operation using cause-and-effect relationships?

7-16   FULL-FEATURED COUNTERS IN HDL

The examples we have chosen so far have been very basic counters. All they do is count up to four and then roll over to zero. The traditional IC counters that we have examined have many other features that make them very useful. For example, think about the 74193 that was covered in Section 7-9. It can count up or down, or it can be loaded (preset to any count) or cleared. It can be cascaded synchronously with other 74193 ICs. In this section, we will explore the techniques that allow us to include these features in an HDL counter.

The counter that we will create is not exactly like a 74193. It is actually more similar to a 74191, but the purpose of this example is to demonstrate the methods of making a counter with features that specifically suit our needs. When we use HDL, we are not limited to features that happen to be included with a certain IC. The focus will be on several issues:

How do I make the counter count up and down?
How can I clear it asynchronously?
How can I load it synchronously?
How can I include synchronous cascade controls?

Before starting the discussion, let's review some concepts. Counting up and down is pretty straightforward, but like anything else, there are different ways to get this job done. The 74193 used two separate clock inputs, one for up and one for down. This design will take a different approach and use a single clock input. The count will change state only on the rising edge of the clock input. When the clock edge is detected, however, this counter uses a direction control input that makes it count up when it is LOW and count down when it is HIGH. The next issue is the asynchronous clear operation. Asynchronous means *not synchronized* with the clock. In other words, it must clear immediately when this control input is activated. On the other hand, the synchronous load feature is intended to take the number on the data inputs and load it into the *q* outputs of the counter whenever the *load* input is activated *during a positive clock transition.* This event is synchronized with the clock. As we will see, the operation of these features is determined by the way we write the HDL code.

The last issue is the set of control inputs and outputs that allow these counter modules to be combined into a larger counter. We will see the benefit of this in Chapter 10, when we discuss modular, hierarchical projects. Most of the flip-flop counters we have looked at so far were asynchronous or ripple counters. The *q* outputs did not all update (change states) at the same time because a clock edge had to ripple or propagate through all the stages. Synchronous counters, you will recall, have a common clock signal for every FF stage, so all the outputs update simultaneously. In order to cascade counter stages synchronously, there must be a way for the higher-order stages to ignore the clock pulses until the lower-order stage is *rolling over.* This concept of rolling over simply means that the count sequence has reached its limit and must start over at the beginning of the sequence. If the counter is counting up, it rolls over from its maximum value to zero. If it is counting down, it rolls over from zero to its maximum value. Counters that can be cascaded synchronously have a *count enable* input that causes them to ignore clock edges unless they are enabled. They also have a *terminal count* output that indicates that the sequence is at its limit and will roll over on the next clock. To cascade these stages, the terminal count output is connected to the next higher-order stage's enable input. The other control inputs of each counter stage are connected in parallel to a common source.

AHDL FULL-FEATURED COUNTER

The code in Figure 7-46 implements all of the features we have discussed. This is a four-bit counter, but it can easily be expanded in size. Read through the inputs and outputs on lines 3 and 4 to make sure you understand what each one is supposed to do. If you do not, reread the previous paragraphs of this section. Line 7 defines a four-bit register of D flip-flops that will serve as the counter. It should be noted again here that this register could have been named the same as the output variable (*q*). The code is written with different names to distinguish between ports (inputs and outputs) of the circuit and the devices that are operating within the circuit. The clock input is connected to all the *clk* inputs of all the D flip-flops on line 10. All the active-LOW clear inputs (*clrn*) to the DFF primitive are connected to the complement of the *clear* input signal. This clears the flip-flops immediately when the *clear* input goes HIGH because the *prn* and *clrn* inputs to the DFF primitive are not dependent on the clock (i.e., they are asynchronous).

In order to make the load function synchronously, the *D* inputs to the flip-flops must be controlled so that the input data (*din*) is present on the *D*

AHDL

```
1    SUBDESIGN fig7_46
2    (
3       clock, clear, load, cntenabl, down, din[3..0]      :INPUT;
4       q[3..0], term_ct :OUTPUT;    -- declare 4-bit array of output bits
5    )
6    VARIABLE
7       count[3..0]    :DFF;        -- declare a register of D flip flops
8
9    BEGIN
10      count[].clk = clock;         -- connect all clocks to synch source
11      count[].clrn= !clear;        -- connect for asynch active HIGH clear
12      IF load THEN count[].d = din[]; -- synchronous load
13         ELSIF !cntenabl THEN count[].d = count[].q; -- hold count
14         ELSIF !down THEN count[].d = count[].q + 1; -- increment
15         ELSIF count[].d = count[].q - 1;            -- decrement
16      END IF;
17      IF ((count[] == 0) & down # (count[] == 15) & !down)& cntenabl
18      THEN      term_ct = VCC;     -- synchronous cascade output signal
19      ELSE term_ct = GND;
20      END IF;
21      q[] = count[];               -- transfer register contents to outputs
22   END;
```

FIGURE 7-46 Full-featured counter in AHDL.

inputs when the load line is HIGH. This way, when the next active clock edge comes along, the data will be loaded into the counter. This action must happen regardless of whether the counter is enabled or not. Consequently, the first conditional decision (IF) on line 12 evaluates the load input. Recall from Chapter 4 that the IF/ELSE decision structure gives precedence to the first condition that is found to be true because, once it finds a condition that is true, it does not go on to evaluate the conditions in subsequent ELSE clauses. In this case, it means that if the load line is activated, it does not matter whether the count is enabled, or it is trying to count up or down. It will do a parallel load on the next clock edge.

Assuming that the load line is not active, the ELSIF clause on line 13 is evaluated to see if the count is disabled. In AHDL, it is very important to realize that the Q output must be fed back to the D input so that, on the next clock edge, the register will hold its previous value. Forgetting to insert this clause results in the D inputs defaulting to zero, thus resetting the counter. If the counter is enabled, the ELSIF clause on line 14 is evaluated and either increments *count* (line 14) or decrements *count* (line 15). To summarize these decisions, first decide if it is time to load, next decide if the count should hold or change, then decide whether to count up or down.

The next issue is detecting the terminal count. Lines 17–20 decide whether the terminal count has been reached going up or down and drives the output to the appropriate level. After all these decisions are made, *count* should have the right value in the register, and line 21 effectively connects the register to the output pins.

VHDL FULL-FEATURED COUNTER

The code in Figure 7-47 implements all the features we have discussed. This is a four-bit counter, but it can easily be expanded in size. Read through the inputs and outputs on lines 2–5 to make sure you understand what each one is supposed to do. If you do not, reread the previous paragraphs of this section. The PROCESS statement on line 10 is the key to all clocked circuits described in VHDL, but it also plays an important role in determining whether the circuit responds synchronously or asynchronously to its inputs. We want this circuit to respond immediately to transitions on the *clock, clear,* and *down* inputs. With these signals in the sensitivity list, we assure that the code inside the PROCESS will be evaluated as soon as any of these inputs change states. The variable *count* is defined on line 11 as an INTEGER so it can be incremented and decremented easily. Variables are declared within the PROCESS and can be used within the PROCESS only.

```
1   ENTITY fig7_47 IS
2   PORT( clock, clear, load, cntenabl, down   :IN BIT;
3         din          :IN INTEGER RANGE 0 TO 15;
4         q            :OUT INTEGER RANGE 0 TO 15;
5         term_ct      :OUT BIT);
6   END fig7_47;
7
8   ARCHITECTURE a OF fig7_47 IS
9      BEGIN
10        PROCESS ( clock, clear, down)
11        VARIABLE count :INTEGER RANGE 0 to 15;    -- define a numeric signal
12           BEGIN
13              IF clear = '1' THEN count := 0;    -- asynch clear
14              ELSIF (clock = '1' AND clock'EVENT)  THEN  -- rising edge?
15                 IF load = '1' THEN count := din;    -- parallel load
16                 ELSIF cntenabl = '1' THEN         -- enabled?
17                    IF down = '0' THEN count := count + 1;  -- increment
18                    ELSE         count := count - 1;  -- decrement
19                    END IF;
20                 END IF;
21              END IF;
22              IF (((count = 0) AND (down = '1')) OR
23                 ((count = 15) AND (down = '0'))) AND cntenabl = '1'
24                 THEN  term_ct <= '1';
25              ELSE     term_ct <= '0';
26              END IF;
27              q <= count;    -- transfer register contents to outputs
28           END PROCESS;
29      END a;
```

FIGURE 7-47   Full-featured counter in VHDL.

VHDL

The *clear* input is given precedence by evaluating it with the first IF statement on line 13. Recall from Chapter 4 that the IF/ELSE decision structure gives precedence to the first condition that is found to be true because it does not go on to evaluate the conditions in subsequent ELSE clauses. In this case, if the *clear* is active, the other conditions will not matter. The output will be zero. In order to make the *load* function operate synchronously, it must be evaluated after detecting the clock edge. The clock edge is detected on line 14, and the circuit checks immediately to see if *load* is active. If *load* is active, the *count* is loaded from *din,* regardless of whether or not the counter is enabled. Consequently, the conditional decision (IF) on line 15 evaluates the *load* input; only if it is inactive does it evaluate line 16 to see if the counter is enabled. If the counter is enabled, the *count* will be incremented or decremented (lines 17 and 18, respectively).

The next issue is detecting the terminal count. Lines 22–25 decide whether the maximum or minimum terminal count has been reached and drive the output to the appropriate level. The decision-making structure here is very important because we want to evaluate this situation, regardless of whether the decision-making process was invoked by *clock, clear,* or *down.* Notice that this decision is not another ELSE branch of the previous IF decisions but is evaluated for each signal in the sensitivity list *after* the clearing or counting has occurred. After all these decisions are made, *count* should have the right value in the register, and line 27 effectively connects the register to the output pins.

REVIEW QUESTIONS

1. What is the difference between asynchronous clear and synchronous load?
2. What is the difference between asynchronous cascading and synchronous cascading?
3. How are the clocks connected for synchronous cascading?
4. What are the key control signals needed to allow synchronous cascading of counters?

7-17 LPM COUNTERS

In Chapter 6, we demonstrated the use of an LPM adder/subtractor. It used only one parameter (the width). This section demonstrates the use of an LPM to create a counter of any width and any modulus. For this example, the width is seven bits and the modulus is 100. The LPM counter in the Altera library has controls that allow it to count up or down and to perform synchronous parallel load, asynchronous clear, and synchronous cascading.

AHDL

Figure 7-48 shows the AHDL hardware description for the MOD-100 universal counter circuit. Notice that the definition of the LPM_COUNTER must be included as shown on line 1. The inputs and outputs that we are using to connect to the counter are named and sized to fit the application in lines 5–7. On line 11, the two parameters are specified to establish this as a seven-bit, MOD-100 counter. Lines 13–20 simply connect the defined ports of the LPM_COUNTER named *counter* to the various signals defined in this block description.

FIGURE 7-48   MOD-100 LPM counter in AHDL.

```
1    INCLUDE "1pm_counter.inc";
2
3    SUBDESIGN fig7_48
4    (
5       clk, load, enable, clear, updn    :INPUT;
6       d[6..0]]                          :INPUT;
7       q[6..0], tc                       :OUTPUT;
8    )
9    VARIABLE
10      counter           :LPM_COUNTER WITH
11                        (LPM_WIDTH=7, LPM_MODULUS="100");
12   BEGIN
13      counter.clock = clk;
14      counter.cnt_en = enable;
15      counter.sload = load;
16      counter.aclr = !clear;
17      counter.updown = updn;
18      counter.data[] = d[];
19      q[] = counter.q[];
20      tc = counter.cout & enable;
21   END;
```

VHDL

Figure 7-49 shows the VHDL hardware description for the MOD-100 universal counter circuit. Notice that the libraries must be used as shown on lines 1–4. The inputs and outputs that we are using to connect to the counter are named and sized to fit the application in lines 8–11. On line 19, the two parameters are specified to establish this as a seven-bit, MOD-100 counter. Lines 20–22 simply connect the defined ports of the LPM_COUNTER named *counter* to the various signals defined in this block description. On lines 23 and 24, when logic functions are needed to condition the inputs and outputs, it is necessary to declare and use SIGNALS to connect the various ports.

```
1     LIBRARY ieee;
2     USE ieee.std_logic_1164.all;
3     LIBRARY 1pm;
4     USE 1pm.1pm_components.all;
5
6     ENTITY fig7_49 IS
7     PORT (
8        clk, load, enable, clear, updn        :IN STD_LOGIC;
9        d                          :IN STD_LOGIC_VECTOR (6 DOWNTO 0);
10       q                          :OUT STD_LOGIC_VECTOR (6 DOWNTO 0);
11       tc                         :OUT STD_LOGIC
12       );
13    END fig7_49;
14
15    ARCHITECTURE 1pm OF fig7_49 IS
16    SIGNAL terminal, notclear        :STD_LOGIC;
17    BEGIN
18       counter       :LPM_COUNTER
19              GENERIC MAP (LPM_WIDTH => 7, LPM_MODULUS => 100)
20              PORT MAP (clock => clk, cnt_en => enable, sload => load,
21                     aclr => notclear, updown => updn,
22                     data => d, q => q, cout => terminal);
23    notclear <= NOT clear;
24    tc <= '1' WHEN (terminal = '1' AND enable = '1') ELSE '0';
25    END 1pm;
```

FIGURE 7-49 MOD-100 LPM counter in VHDL.

REVIEW QUESTIONS

1. What does LPM stand for? (Refer back to Section 6-22.)
2. What is the name Altera uses for its generic counter megafunction?
3. What are the two parameters passed to the generic counter megafunction?

7-18   STATE MACHINES

The idea of a counter that does not follow a numeric counting sequence (e.g., 1, 2, 3, etc.) was presented in the section on synchronous counter design (Section 7-14). In that section, the objective was to drive a stepper motor that required a specific sequence of states. There are many instances when predefined sequences are necessary, like they are in stepper drivers. There are also many instances when the binary value of each state is of little interest because decoding logic will be used to identify specific states and to generate the desired signals. Counter circuits used for such purposes are often referred to as **state machines**. For these situations, HDL techniques can make state machines easy and intuitive to describe. As an oversimplified example that everyone can relate to, the following hardware description deals with four states through which a typical washing machine might progress. Although even a real washing machine is more complex than this example, it will serve to demonstrate the techniques. This washing machine is idle until the start button is pressed, then it fills with water until the tub is full, then it runs the agitator until a timer expires, and finally it spins the tub until the water is spun out, at which time it goes back to idle. The point of this example focuses on the use of a set of named states for which no binary values are defined. The name of the counter variable is *wash*, which can be in any of the named states: *idle, fill, agitate,* or *spin.*

AHDL STATE MACHINES

The AHDL code in Figure 7-50 shows the syntax for declaring a counter with named states on lines 6 and 7. The name of this counter is *cycle.* The keyword **MACHINE** is used in AHDL to define *cycle* as a state machine. The number of bits needed for this counter to produce the named states will be

FIGURE 7-50 State machine example using AHDL.

```
1   SUBDESIGN fig7_50
2   (  clock, start, full, timesup, dry    :INPUT;
3      water_valve, ag_mode, sp_mode       :OUTPUT;
4   )
5   VARIABLE
6   cycle:  MACHINE
7           WITH STATES (idle, fill, agitate, spin);
8   BEGIN
9   cycle.clk = clock;
10
11     CASE cycle IS
12        WHEN idle =>IF start THEN cycle = fill;
13                    ELSE     cycle = idle;
14                    END IF;
15        WHEN fill =>IF full THEN cycle = agitate;
16                    ELSE     cycle = fill;
17                    END IF;
18        WHEN agitate=> IF timesup THEN cycle = spin;
19                    ELSE     cycle = agitate;
20                    END IF;
21        WHEN spin => IF dry THEN cycle = idle;
22                    ELSE     cycle = spin;
23                    END IF;
24        WHEN OTHERS => cycle = idle;
25     END CASE;
26
27     TABLE
28        cycle      => water_valve,    ag_mode, sp_mode;
29        idle    => gnd,             gnd,     gnd;
30        fill    => vcc,             gnd,     gnd;
31        agitate => gnd,             vcc,     gnd;
32        spin    => gnd,             gnd,     vcc;
33     END TABLE;
34   END;
```

determined by the compiler. Notice that in line 7, the states are named, but the binary value for each state is also left for the compiler to determine. The designer does not need to worry about this level of detail. The CASE structure on lines 11–25 and the decoding logic that drives the outputs (lines 27–33) refer to the states by name. This makes the description easy to read and allows the compiler more freedom to minimize the circuitry. If the design requires the state machine also to be connected to an output port, then line 6 can be changed to:

```
cycle: MACHINE OF BITS (st[1..0])
```

and the output port *st[1..0]* can be added to the SUBDESIGN section. A second state machine option that is available is the ability for the designer to define a binary value for each state. This can be accomplished in this example by changing line 7 to:

```
WITH STATES (idle = B"00", fill = B"01", agitate = B"11", spin =
B"10");
```

VHDL STATE MACHINES

The VHDL code in Figure 7-51 shows the syntax for declaring a counter with named states. On line 7, a data object is declared named *state_machine.* Notice the keyword TYPE. This is called an **enumerated type** in VHDL, which means that the designer decides all the possible values that can be assigned to the data. Notice that on line 7, the states are named, but the binary value for each state is left for the compiler to determine. The designer does not need to worry about this level of detail. The CASE structure on lines 13–30 and the decoding logic that drives the outputs (lines 32–38) refer to the states by name. This makes the description easy to read and allows the compiler more freedom to minimize the circuitry.

By this time, you may be wondering why there are so many ways to describe logic circuits. If one way is easier than the others, why not just study that one? The answer, of course, is that each level of abstraction offers advantages over the others in certain cases. The structural method provides the most complete control over interconnections and allows the use of multiple standard parts, like TTL equivalent ICs in the same design file. The use of Boolean equations, truth tables, and PRESENT state/NEXT state tables allows us to describe the way data flows through the circuit using HDL. Finally, the behavioral method allows a more abstract description of the circuit's operation in terms of cause and effect. In practice, each source file may have portions that can be categorized under each level of abstraction. Choosing the right level when writing code is not an issue of right and wrong as much as it is an issue of style and preference.

There are also several ways to approach any task from a standpoint of choosing control structures. Should we use selected signal assignments or Boolean equations, IF/ELSE or CASE, sequential processes or concurrent statements, macrofunctions or megafunctions? Or should we write our own code? The answers to these questions ultimately define your personal strategy in solving the problem. Your preferences and the advantages you find in using one method over another will be established with practice and experience.

```
1    ENTITY fig7_51 IS
2    PORT (   clock, start, full, timesup, dry : IN BIT;
3          drivers :OUT bit_vector (2 DOWNTO 0)); --water, ag_mode, sp_mode
4    END fig7_51;
5
6    ARCHITECTURE vhdl OF fig7_51 IS
7    TYPE state_machine IS (idle, fill, agitate, spin);
8    BEGIN
9       PROCESS (clock, start)
10      VARIABLE cycle : state_machine;
11      BEGIN
12         IF (clock'EVENT AND clock = '1') THEN
13            CASE cycle IS
14               WHEN idle=>
15                  IF (start = '1') THEN   cycle := fill;
16                  ELSE                    cycle := idle;
17                  END IF;
18               WHEN fill=>
19                  IF (full = '1') THEN    cycle := agitate;
20                  ELSE                    cycle := fill;
21                  END IF;
22               WHEN agitate=>
23                  IF (timesup = '1') THEN cycle := spin;
24                  ELSE                    cycle := agitate;
25                  END IF;
26               WHEN spin =>
27                  IF (dry = '1') THEN     cycle := idle;
28                  ELSE                    cycle := spin;
29                  END IF;
30            END CASE;
31         END IF;
32         CASE cycle IS              -- ("water, ag_mode, sp_mode")
33            WHEN idle         => drivers <= "000";
34            WHEN spin         => drivers <= "001";
35            WHEN agitate      => drivers <= "010";
36            WHEN fill         => drivers <= "100";
37            WHEN OTHERS       => drivers <= "000";
38         END CASE;
39      END PROCESS;
40   END vhdl;
```

FIGURE 7-51 State machine example using VHDL.

REVIEW QUESTIONS

1. What is the difference between describing a counter and describing a state machine?
2. How are the actual binary states of a state machine assigned?
3. What is the advantage of using state machine description?

PART 1 SUMMARY

1. In asynchronous (ripple) counters, the clock signal is applied to the LSB FF, and all other FFs are clocked by the output of the preceding FF.

2. A counter's MOD number is the number of stable states in its counting cycle; it is also the maximum frequency-division ratio.

3. The normal (maximum) MOD number of a counter is 2^N. One way to modify a counter's MOD number is to add circuitry that will cause it to recycle before it reaches its normal last count.

4. Counters can be cascaded (chained together) to produce greater counting ranges and frequency-division ratios.

5. In a synchronous (parallel) counter, all of the FFs are simultaneously clocked from the input clock signal.

6. The maximum clock frequency for an asynchronous counter, f_{max}, decreases as the number of bits increases. For a synchronous counter, f_{max} remains the same, regardless of the number of bits.

7. A decade counter is any MOD-10 counter. A BCD counter is a decade counter that sequences through the 10 BCD codes (0–9).

8. A presettable counter can be loaded with any desired starting count.

9. An up/down counter can be commanded to count up or count down.

10. Logic gates can be used to decode (detect) any or all states of a counter.

11. Asynchronous counters can produce glitches in decoding gates due to the counter propagation delays. Synchronous counters are less likely to cause decoding glitches. Strobing is a technique for eliminating the effects of decoding glitches.

12. Synchronous counters with arbitrary counting sequences can be implemented by following a standard design procedure.

13. Counters can be described in many different ways using HDL, including structural wiring descriptions, PRESENT state/NEXT state tables, and behavioral descriptions.

14. All the features available on the various standard ICs, such as asynchronous or synchronous loading, clearing, and cascading, can be described using HDL.

15. Generic models of various common digital circuits, called libraries of parameterized modules (LPMs), are available. For counters, the number of bits and the modulus can be determined by simply assigning a number to the specified parameter for the LPM function.

16. State machines can be represented in HDL using descriptive names for each state rather than specifying a numeric sequence of states.

PART 1 IMPORTANT TERMS

asynchronous (ripple) counter	up/down counters	J-K excitation table
MOD number	presettable counters	circuit excitation table
decade counter	parallel load	VARIABLE
BCD counter	multistage counters	behavioral level of abstraction
up counter	decoding	state machines
down counter	decoding glitches	MACHINE
synchronous (parallel) counters	strobing	enumerated type
	cascading	
	sequential circuit design	

PART 2

7-19 INTEGRATED-CIRCUIT REGISTERS

The various types of registers can be classified according to the manner in which data can be entered into the register for storage and the manner in which data are outputted from the register. The various classifications are listed below.

1. Parallel in/parallel out (PIPO)
2. Serial in/serial out (SISO)
3. Parallel in/serial out (PISO)
4. Serial in/parallel out (SIPO)

Each of these types and several variations are available in IC form so that a logic designer can usually find exactly what is required for a given application. In the following sections, we will examine a representative IC from each of the above categories.

7-20 PARALLEL IN/PARALLEL OUT—THE 74ALS174/74HC174

A group of flip-flops used to store a binary value is referred to as a **parallel in/parallel out** register. Figure 7–52(a) shows the logic diagram for the 74ALS174 (also the 74HC174), a six-bit register that has parallel inputs D_5

(a)

(b)

FIGURE 7-52 (a) Circuit diagram of the 74ALS174; (b) logic symbol. (Courtesy of Fairchild, a Schlumberger company) through D_0 and parallel outputs Q_5 through Q_0. Parallel data are loaded into the register on the PGT of the clock input *CP*. A master reset input $\overline{MR}$ can be used to reset asynchronously all of the register FFs to 0. The logic symbol for the 74ALS174 is shown in Figure 7–52(b). This symbol is used in circuit diagrams to represent the circuitry of Figure 7–52(a).

The 74ALS174 is normally used for synchronous parallel data transfer whereby the logic levels present at the *D* inputs are transferred to the corresponding *Q* outputs when a PGT occurs at the clock *CP*. This IC, however, can be wired for serial data transfer, as the following examples will show.

EXAMPLE 7-18

Show how to connect the 74ALS174 so that it operates as a serial shift register with data shifting on each PGT of *CP* as follows: $D_5 \rightarrow D_4 \rightarrow D_3 \rightarrow D_2 \rightarrow D_1 \rightarrow D_0$. In other words, serial data will enter at D_5 and will output at Q_0.

Solution

Looking at Figure 7-52(a), we can see that to connect the six FFs as a serial shift register, we have to connect the *Q* output of one to the *D* input of the next so that data is transferred in the required manner. Figure 7-53 shows how this is accomplished. Note that data shifts left to right, with input data applied at D_5 and output data appearing at Q_0.

FIGURE 7-53 Example 7-18: The 74ALS174 wired as a shift register.

EXAMPLE 7-19

How would you connect two 74ALS174s to operate as a 12-bit shift register?

Solution

Connect a second 74ALS174 IC as a shift register, and connect Q_0 from the first IC to D_5 of the second IC. Also connect the *CP* inputs of both ICs so that they will be clocked from the same signal.

7-21 SERIAL IN/SERIAL OUT—THE 4731B

The 4731B is a CMOS *quad* 64-bit **serial in/serial out** shift register. It contains four identical 64-bit shift registers on one chip. Figure 7-54 shows the logic diagram for one of the 64-bit registers. It has a serial input, D_S, a clock input $\overline{CP}$ that responds to NGTs, and a serial output from the last FF, Q_{63}. This is the only output that is externally accessible. Note that this output goes through a *buffer* circuit (triangle symbol with no inversion bubble). A buffer does not change the signal's logic level; it is used to provide a greater output-current capability than normal. Also note that there is no means for parallel data entry into the register FFs.

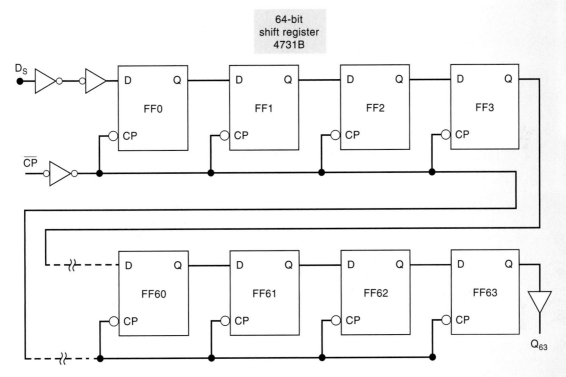

FIGURE 7-54 Logic diagram for one of four 64-bit shift registers on a 4731B. (Courtesy of Fairchild, a Schlumberger company)

EXAMPLE 7-20

A shift register is often used as a way to delay a digital signal by an integral number of clock cycles. The digital signal is applied to the shift register's serial input and is shifted through the shift register by successive clock pulses until it reaches the end of the shift register, where it appears as the output signal. This is illustrated in Figure 7-55 using one of the 64-bit registers on the 4731B chip.

Let's assume that the serial input D_S has been LOW for a long time as clock pulses have been applied, so that the register is filled with all 0s, and Q_{63} starts out LOW, as shown. Then D_S goes HIGH just prior to t_0. The NGTs at $\overline{CP}$ will cause this HIGH to shift into and through the shift register, making each FF go HIGH in succession until finally, at t_{63}, the Q_{63} output goes HIGH. The net effect, then, is that the change in the D_S signal is not felt at the Q_{63} output until approximately 64 clock cycles later.

This method for delaying the effect of a digital signal is common in the digital communications field. For instance, the digital signal might be the *digitized* version of an audio signal that is to be delayed before it is transmitted.

FIGURE 7-55   Waveforms showing how one of the 4731B shift registers can be used to delay a digital signal.

7-22   PARALLEL IN/SERIAL OUT—THE 74ALS165/74HC165

The logic symbol for the 74HC165 is shown in Figure 7-56(a). This IC is an eight-bit **parallel in/serial out** register. It actually has serial data entry via D_S and asynchronous parallel data entry via P_0 through P_7. The register contains eight FFs—Q_0 through Q_7—internally connected as a shift register, but the

FIGURE 7-56   (a) Logic symbol for the 74HC165 parallel in/serial out register; (b) function table.

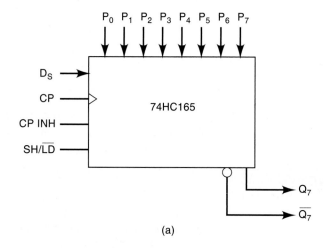

(a)

Function Table

Inputs				
SH/$\overline{LD}$	CP	CP INH	Operation	
L	X	X	Parallel load	
H	H	X	No change	
H	X	H	No change	
H	⌁	L	Shifting	
H	L	⌁	Shifting	

H = high level
L = low level
X = immaterial
⌁ = PGT

(b)

only accessible FF outputs are Q_7 and $\overline{Q}_7$. *CP* is the clock input used for the shifting operation. The clock inhibit input, *CP INH* is used to inhibit the effect of the *CP* input. The shift/load input, *SH/$\overline{LD}$*, controls which operation is taking place—shifting or parallel loading. The function table in Figure 7-56(b) shows how the various input combinations determine what operation, if any, is being performed.

EXAMPLE 7-21

Examine the 74HC165 function table and determine (a) the conditions necessary to load the register with parallel data; (b) the conditions necessary for the shifting operation.

Solution

(a) The first entry in the table shows that the *SH/$\overline{LD}$* input has to be LOW for the parallel load operation. When this input is LOW, the data present at the *P* inputs are *asynchronously* loaded into the register FFs, independent of the *CP* and the *CP INH* inputs. Of course, only the outputs from the last FF are externally available.

(b) The shifting operation cannot take place unless the *SH/$\overline{LD}$* input is HIGH and a PGT occurs at *CP* while *CP INH* is LOW [see the fourth table entry in Figure 7-56(b)]. A HIGH at *CP INH* will inhibit the effect of any clock pulses. Note that the roles of the *CP* and *CP INH* inputs can be reversed, as indicated by the last table entry, because these two signals are ORed together inside the IC.

EXAMPLE 7-22

If we connect a 74HC165 so that $\overline{Q}_7$ is connected to D_S, and we parallel load it with all 0s, what signal will appear at Q_7 when a 200-kHz signal is applied to *CP* while *CP INH* is LOW?

Solution

By connecting $\overline{Q}_7$ to D_S, we have constructed an eight-bit Johnson counter. From our earlier work, we know that an eight-bit Johnson counter will produce a square wave at a frequency that is $\frac{1}{16}$ of the clock frequency, in this case, 12.5 kHz.

7-23   SERIAL IN/PARALLEL OUT—THE 74ALS164/74HC164

The logic diagram for the 74ALS164 is shown in Figure 7-57(a). It is an eight-bit **serial in/parallel out** shift register with each FF output externally accessible. Instead of a single serial input, an AND gate combines inputs *A* and *B* to produce the serial input to flip-flop Q_0.

The shift operation occurs on the PGTs of the clock input *CP*. The $\overline{MR}$ input provides asynchronous resetting of all FFs on a LOW level.

The logic symbol for the 74ALS164 is shown in Figure 7-57(b). Note that the & symbol is used inside the block to indicate that the *A* and *B* inputs are ANDed inside the IC and the result is applied to the *D* input of Q_0.

(a)

(b)

FIGURE 7-57   (a) Logic diagram for the 74ALS164; (b) logic symbol.
(Courtesy of Fairchild, a Schlumberger company)

EXAMPLE 7-23

Assume that the initial contents of the 74ALS164 register in Figure 7-58(a) are 00000000. Determine the sequence of states as clock pulses are applied.

(a)

FIGURE 7-58   Example 7-24.

Input pulse number	Q_0	Q_1	Q_2	Q_3	Q_4	Q_5	Q_6	Q_7	
0	0	0	0	0	0	0	0	0	
1	1	0	0	0	0	0	0	0	
2	1	1	0	0	0	0	0	0	
3	1	1	1	0	0	0	0	0	
4	1	1	1	1	0	0	0	0	Recycles
5	1	1	1	1	1	0	0	0	
6	1	1	1	1	1	1	0	0	
7	1	1	1	1	1	1	1	0	
8	1	1	1	1	1	1	1	1	

Temporary state

(b)

Solution

The correct sequence is given in Figure 7-58(b). With $A = B = 1$, the serial input is 1, so that 1s will shift into the register on each PGT of CP. Because Q_7 is initially at 0, the $\overline{MR}$ input is inactive.

On the eighth pulse, the register tries to go to the 11111111 state as the 1 from Q_6 shifts into Q_7. This state occurs only momentarily because $Q_7 = 1$ produces a LOW at $\overline{MR}$ that immediately resets the register back to 00000000. The sequence is then repeated on the next eight clock pulses.

The following is a list of some other register ICs that are variations on those already presented:

- 74194/ALS194/HC194. This is a four-bit *bidirectional universal shift-register* IC that can perform shift-left, shift-right, parallel in, and parallel out operations. These operations are selected by a two-bit mode-select code applied as inputs to the device. (Problem 7-60 will provide you with a chance to find out more about this versatile chip.)

- 74373/ALS373/HC373/HCT373. This is an eight-bit (octal) parallel in/ parallel out register containing eight D latches with *tristate* outputs. A tristate output is a special type of logic circuit output that allows device outputs to be tied together safely. We will cover the characteristics of tristate devices such as the 74373 in the next chapter.

- 74374/ALS374/HC374/HCT374. This is an eight-bit (octal) parallel in/parallel out register containing eight edge-triggered D flip-flops with tristate outputs.

The IC registers that have been presented here are representative of the various types that are commercially available. Although there are many variations on these basic registers, most of them should now be relatively easy to understand from the manufacturers' data sheets.

We will present several register applications in the end-of-chapter problems and in the material covered in subsequent chapters.

REVIEW QUESTIONS

1. What kind of register can have a complete binary number loaded into it in one operation, and then have it shifted out one bit at a time?
2. *True or false:* A serial in/parallel out register can have all of its bits displayed at one time.
3. What type of register can have data entered into it only one bit at a time, but has all data bits available as outputs?
4. In what type of register do we have access to only one FF output?
5. How does the parallel data entry differ for the 74165 and the 74174?
6. How does the *CP INH* input of the 74ALS165 work?

7-24 IEEE/ANSI REGISTER SYMBOLS

We will present two examples of the IEEE/ANSI symbols for register ICs. First, let's consider the 74174 parallel in/parallel out IC whose internal logic and traditional logic symbol was shown in Figure 7-52. The IEEE/ANSI symbol for the

FIGURE 7-59 IEEE/ANSI symbols for (a) the 74174 parallel in/parallel out register and (b) the 74164 serial in/parallel out shift register.

74174 is given in Figure 7-59(a). Its outline consists of the notched common-control block and the six narrow rectangles representing the six FFs.

The common-control block has the inputs that are common to all of the elements in the IC; in this case, the $\overline{MR}$ and CP inputs are common to the six FFs Q_0 through Q_5 that make up the register. The internal label for the $\overline{MR}$ input is shown as an R to indicate that its function is to *reset* each FF. The internal label for the CP input is C1, which tells us that this input controls the entry of data into any storage element that has a prefix of 1 in its input label. Each FF's D input has an internal label of 1D (shown only for Q_0, but assumed the same for each FF). The "1" in C1 and 1D establishes the dependency of the flip-flop D inputs on the common clock input CP.

The IEEE/ANSI symbol for the 74164 serial in/parallel out shift register is presented in Figure 7-59(b). Its outline consists of the common-control block and the eight FFs that make up the register. The notation SRG 8 identifies this IC as being an eight-bit shift register.

The CP input has the internal label C1/→. The slash (/) is used to separate the two functions C1 and → performed by this input. The C1 indicates that CP controls the data entry into flip-flop Q_0 because Q_0 has the input label 1D. Note that the data bit entered into Q_0 is indicated as the AND combination of inputs A and B. Also note that, because there are no external data inputs to Q_1 through Q_7, CP does not control data entry into these FFs. The → denotes that the active transition of CP will produce the shift-right operation (from Q_0 toward Q_7).

REVIEW QUESTIONS

1. What does the slash (/) mean when it appears in an input label?
2. What notation would be used to describe the function performed by one of the registers on the 4731B IC of Figure 7-51?

7-25 SHIFT-REGISTER COUNTERS

In Section 5-18, we saw how to connect FFs in a shift-register arrangement to transfer data left to right, or vice versa, one bit at a time (serially). Shift-register counters use *feedback,* which means that the output of the last FF in the register is connected back to the first FF in some way.

Ring Counter

The simplest shift-register counter is essentially a **circulating shift register** connected so that the last FF shifts its value into the first FF. This arrangement is shown in Figure 7-60 using D-type FFs (J-K flip-flops can also be

(a)

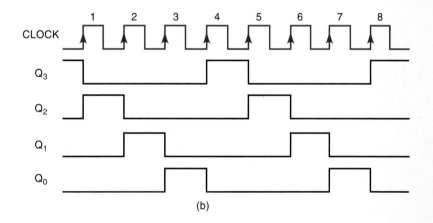

(b)

Q_3	Q_2	Q_1	Q_0	CLOCK pulse
1	0	0	0	0
0	1	0	0	1
0	0	1	0	2
0	0	0	1	3
1	0	0	0	4
0	1	0	0	5
0	0	1	0	6
0	0	0	1	7
.	.	.	.	.
.	.	.	.	.

(c)

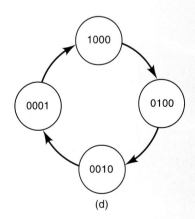

(d)

FIGURE 7-60   (a) Four-bit ring counter; (b) waveforms; (c) sequence table; (d) state diagram.

used). The FFs are connected so that information shifts from left to right and back around from Q_0 to Q_3. In most instances, only a single 1 is in the register, and it is made to circulate around the register as long as clock pulses are applied. For this reason, it is called a **ring counter**.

The waveforms, sequence table, and state diagram in Figure 7-60 show the various states of the FFs as pulses are applied, assuming a starting state of $Q_3 = 1$ and $Q_2 = Q_1 = Q_0 = 0$. After the first pulse, the 1 has shifted from Q_3 to Q_2 so that the counter is in the 0100 state. The second pulse produces the 0010 state, and the third pulse produces the 0001 state. On the *fourth* clock pulse, the 1 from Q_0 is transferred to Q_3, resulting in the 1000 state, which is, of course, the initial state. Subsequent pulses cause the sequence to repeat.

This counter functions as a MOD-4 counter because it has *four* distinct states before the sequence repeats. Although this circuit does not progress through the normal binary counting sequence, it is still a counter because each count corresponds to a unique set of FF states. Note that each FF output waveform has a frequency equal to one-fourth of the clock frequency because this is a MOD-4 ring counter.

Ring counters can be constructed for any desired MOD number; a MOD-N ring counter uses N flip-flops connected in the arrangement of Figure 7-60. In general, a ring counter requires more FFs than a binary counter for the same MOD number; for example, a MOD-8 ring counter requires eight FFs, while a MOD-8 binary counter requires only three.

Despite the fact that it is less efficient in the use of FFs, a ring counter is still useful because it can be decoded without the use of decoding gates. The decoding signal for each state is obtained at the output of its corresponding FF. Compare the FF waveforms of the ring counter with the decoding waveforms in Figure 7-27. In some cases, a ring counter might be a better choice than a binary counter with its associated decoding gates. This is especially true in applications where the counter is being used to control the sequencing of operations in a system.

Starting a Ring Counter

To operate properly, a ring counter must start off with only one FF in the 1 state and all the others in the 0 state. Because the starting states of the FFs will be unpredictable on power-up, the counter must be preset to the required starting state before clock pulses are applied. One way to do this is to apply a momentary pulse to the asynchronous $\overline{PRE}$ input of one of the FFs (e.g., Q_3 in Figure 7-60) and to the $\overline{CLR}$ input of all other FFs. Another method is shown in Figure 7-61. On power-up, the capacitor will charge up relatively slowly toward $+V_{CC}$. The output of Schmitt-trigger INVERTER 1 will stay

FIGURE 7-61 Circuit for ensuring that the ring counter of Figure 7-60 starts in the 1000 state on power-up.

HIGH, and the output of INVERTER 2 will remain LOW until the capacitor voltage exceeds the positive-going threshold voltage (V_{T+}) of the INVERTER 1 input (about 1.7 V). This will hold the $\overline{PRE}$ input of Q_3 and the $\overline{CLR}$ inputs of Q_2, Q_1, and Q_0 in the LOW state long enough during power-up to ensure that the counter starts at 1000.

Johnson Counter

The basic ring counter can be modified slightly to produce another type of shift-register counter, which will have somewhat different properties. The **Johnson** or **twisted-ring counter** is constructed exactly like a normal ring counter except that the *inverted* output of the last FF is connected to the input of the first FF. A three-bit Johnson counter is shown in Figure 7-62. Note that the $\overline{Q}_0$ output is connected back to the D input of Q_2, which means that the *inverse* of the level stored in Q_0 will be transferred to Q_2 on the clock pulse.

FIGURE 7-62 (a) MOD-6 Johnson counter; (b) waveform; (c) sequence table; (d) state diagram.

(a)

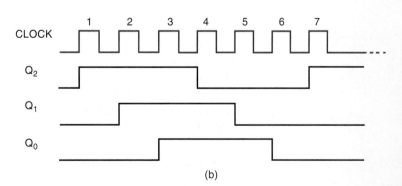

(b)

Q_2	Q_1	Q_0	CLOCK pulse
0	0	0	0
1	0	0	1
1	1	0	2
1	1	1	3
0	1	1	4
0	0	1	5
0	0	0	6
1	0	0	7
1	1	0	8
.	.	.	.
.	.	.	.
.	.	.	.

(c)

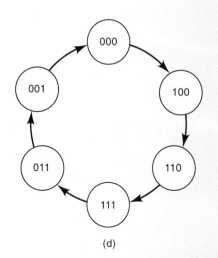

(d)

The Johnson-counter operation is easy to analyze if we realize that on each positive clock-pulse transition, the level at Q_2 shifts into Q_1, the level at Q_1 shifts into Q_0, and the *inverse* of the level at Q_0 shifts into Q_2. Using these ideas and assuming that all FFs are initially 0, the waveforms, sequence table, and state diagram of Figure 7-62 can be generated.

Examination of the waveforms and sequence table reveals the following important points:

1. This counter has six distinct states—000, 100, 110, 111, 011, and 001—before it repeats the sequence. Thus, it is a MOD-6 Johnson counter. Note that it does not count in a normal binary sequence.

2. The waveform of each FF is a square wave (50 percent duty cycle) at one-sixth the frequency of the clock. In addition, the FF waveforms are shifted by one clock period with respect to each other.

The MOD number of a Johnson counter will always be equal to *twice* the number of FFs. For example, if we connect five FFs in the arrangement of Figure 7-62, the result is a MOD-10 Johnson counter, where each FF output waveform is a square wave at one-tenth the clock frequency. Thus, it is possible to construct a MOD-N counter (where N is an even number) by connecting $N/2$ flip-flops in a Johnson-counter arrangement.

Decoding a Johnson Counter

For a given MOD number, a Johnson counter requires only half the number of FFs that a ring counter requires. However, a Johnson counter requires decoding gates, whereas a ring counter does not. As in the binary counter, the Johnson counter uses one logic gate to decode for each count, but each gate requires only two inputs, regardless of the number of FFs in the counter. Figure 7-63 shows the decoding gates for the six states of the Johnson counter of Figure 7-62.

Notice that each decoding gate has only two inputs, even though there are three FFs in the counter, because for each count, two of the three FFs are in a unique combination of states. For example, the combination $Q_2 = Q_0 = 0$ occurs only once in the counting sequence, at the count of 0. Thus, AND gate 0, with inputs $\overline{Q_2}$ and $\overline{Q_0}$, can be used to decode for this count. This same characteristic is shared by all of the other states in the sequence, as the reader can verify. In fact, for *any size* Johnson counter, the decoding gates will have only two inputs.

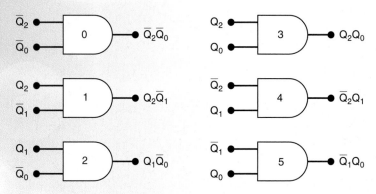

FIGURE 7-63   Decoding logic for a MOD-6 Johnson counter.

Johnson counters represent a middle ground between ring counters and binary counters. A Johnson counter requires fewer FFs than a ring counter but generally more than a binary counter; it has more decoding circuitry than a ring counter but less than a binary counter. Thus, it sometimes represents a logical choice for certain applications.

IC Shift-Register Counters

Very few ring counters or Johnson counters are available as ICs because it is relatively simple to take a shift-register IC and to wire it as either a ring counter or a Johnson counter. Some of the CMOS Johnson-counter ICs (74HC4017, 74HC4022) include the complete decoding circuitry on the same chip as the counter.

REVIEW QUESTIONS

1. Which shift-register counter requires the most FFs for a given MOD number?
2. Which shift-register counter requires the most decoding circuitry?
3. How can a ring counter be converted to a Johnson counter?
4. *True or false:*
   (a) The outputs of a ring counter are always square waves.
   (b) The decoding circuitry for a Johnson counter is simpler than for a binary counter.
   (c) Ring and Johnson counters are synchronous counters.
5. How many FFs are needed in a MOD-16 ring counter? How many are needed in a MOD-16 Johnson counter?

7-26 TROUBLESHOOTING

Flip-flops, counters, and registers are the major components in **sequential logic systems**. A sequential logic system, because of its storage devices, has the characteristic that its outputs and sequence of operations depend on both the present inputs and the inputs that occurred earlier. Even though sequential logic systems are generally more complex than combinational logic systems, the essential procedures for troubleshooting apply equally well to both types of systems. Sequential systems suffer from the same types of failures (open circuits, shorts, internal IC faults, and the like) as do combinational systems.

Many of the same steps used to isolate faults in a combinational system can be applied to sequential systems. One of the most effective troubleshooting techniques begins with the troubleshooter observing the system operation and, by analytical reasoning, determining the possible causes of the system malfunction. Then he or she uses available test instruments to isolate the exact fault. The following examples will show the kinds of analytical reasoning that should be the initial step in troubleshooting sequential systems. After studying these examples, you should be ready to tackle the troubleshooting/problems at the end of the chapter.

EXAMPLE 7-24

Figure 7-64(a) shows a 74LS293 wired as a MOD-10 counter. A technician tests the counter operation by applying a 1-kHz clock signal and observing the Q outputs with an oscilloscope. The displayed waveforms are shown in Figure 7-64(b). Determine the possible causes for the incorrect circuit behavior.

(a)

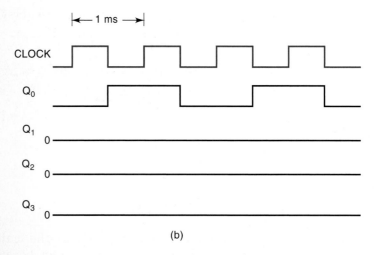

(b)

FIGURE 7-64   Example 7-24.

Solution

The waveforms show that Q_0 is toggling in response to the NGTs of the clock, but all other FFs are stuck in the LOW state. Several possible faults could produce this operation.

1. The Q_1 output is internally or externally shorted to ground. Referring to the 74LS293 diagram in Figure 7-8(a), we can see that this would prevent Q_2 and Q_3 from toggling because Q_1 is the clock signal for Q_2 and Q_2 is the clock signal for Q_3.
2. The connection from Q_0 to $\overline{CP}_1$ is open, so that Q_1 receives no clock signal.
3. There is an internal fault in the IC that prevents Q_1 from toggling.
4. The MR_1 input is internally shorted to ground, forcing $Q_1 = 0$.

EXAMPLE 7-25

After analyzing the situation described in Example 7-24, the technician proceeds to isolate the fault. He performs ohmmeter checks and verifies that Q_1 is not shorted to ground and that Q_0 is connected to $\overline{CP_1}$. This verification eliminates the first two possible faults. Concluding that the IC is bad, he replaces it. To his surprise, the circuit operation exhibits the same symptoms. Scratching his head, he decides to take a closer look at the FF waveforms by displaying them using a 10-ns/cm time scale. On this scale, he can see a very narrow glitch occurring on the Q_1 signal at the time when Q_0 makes a NGT (see Figure 7-65). What is the probable fault?

FIGURE 7-65 Example 7-25.

Solution

When the counter is operating correctly, there should be a glitch at Q_1 when the counter goes to the 1010 (10) state, at which point the HIGHs at Q_3 and Q_1 cause the *MR* inputs to clear the count back to 0000. The waveforms in Figure 7-65, however, do not show Q_3 HIGH when the glitch at Q_1 occurs. The most probable fault is an open connection at MR_2 because this would be interpreted as a constant logic HIGH by the TTL integrated circuit. Thus, as soon as Q_1 goes HIGH, the *MR* inputs are both HIGH and the counter resets to 0000.

7-27 HDL REGISTERS

The various options of serial and parallel data transfer within registers were described thoroughly in Sections 7-19 through 7-23, and some examples of ICs that perform these operations have also been described. The beauty of using HDL to describe a register is in the fact that a circuit can be given any of these options and as many bits as are needed by simply changing a few words. One easy way to accomplish this is by using LPMs, as we demonstrated for arithmetic circuits and also counters. There are LPMs available for a generic shift register. Examples of a four-bit PISO register in both AHDL

FIGURE 7-66 AHDL
LPM for a parallel
in/serial out register.

```
1    INCLUDE "1pm_shiftreg.inc";
2
3    SUBDESIGN fig7_66
4    (
5       clock        :INPUT;
6       din[3..0]    :INPUT;
7       ld/sh        :INPUT;
8       ser_out      :OUTPUT;
9    )
10   VARIABLE
11      shiftreg  :LPM_SHIFTREG WITH
12                      (LPM_WIDTH = 4);
13   BEGIN
14      shiftreg.clock    = clock;
15      shiftreg.data[]   = din[];
16      shiftreg.load     = ld/sh;
17      shiftreg.shiftin  = GND;
18      ser_out           = shiftreg.shiftout;
19   END;
```

FIGURE 7-67 VHDL
LPM to make a parallel
in/serial out register.

```
1    LIBRARY ieee;
2    USE ieee.std_logic_1164.all;
3    LIBRARY 1pm;
4    USE 1pm.1pm_components.all;
5
6    ENTITY piso_v IS
7    PORT  (
8            clock        :IN STD_LOGIC;
9            din          :IN STD_LOGIC_VECTOR (3 DOWNTO 0);
10           ld_sh        :IN STD_LOGIC;
11           ser_out      :OUT STD_LOGIC;
12         );
13   END piso_v;
14
15   ARCHITECTURE vhdl OF piso_v IS
16   SIGNAL low        :STD_LOGIC;
17   BEGIN
18      low <= '0';
19      shiftreg :LPM_SHIFTREG GENERIC MAP (LPM_WIDTH => 4)
20           PORT MAP (clock => clock, data => din, load => ld_sh,
21                     shiftin => low, shiftout => ser_out);
22   END vhdl;
```

(Figure 7-66) and VHDL (Figure 7-67) are presented here. Other configurations are left for you to practice.

The focus of the rest of this section is on the HDL techniques that use bit arrays to describe a register's data and the methods used to transfer that data in a parallel or serial format. To understand how data is shifted in HDL, con-

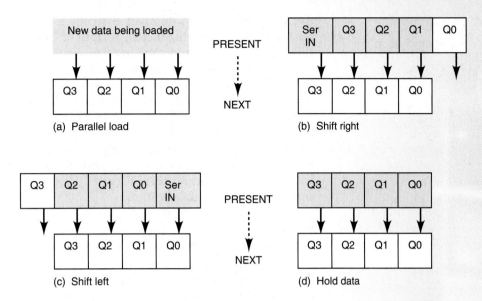

FIGURE 7-68   Data transfers made in shift registers: (a) parallel load; (b) shift right; (c) shift left; (d) hold data.

sider the diagrams in Figure 7-68, which shows four flip-flops performing transfer operations of parallel load, shift right, shift left, and hold data. For all of these diagrams, the bits are transferred synchronously, which means that they all move simultaneously on a single clock edge. In Figure 7-68(a), the data that is to be parallel loaded into the register is presented to the D inputs, and on the next clock pulse, it will be transferred to the q outputs. Shifting data right means that each bit is transferred to the bit location to its immediate right, while a new bit is transferred in on the left end and the last bit on the right end is lost. This situation is depicted in Figure 7-68(b). Notice that the data set that we want in the NEXT state is made up of the new serial input and three of the four bits in the PRESENT state array. This data simply needs to move over and overwrite the four data bits of the register. The same operation occurs in Figure 7-68(c), but it is moving data to the left. The last possibility is called the hold data mode and is shown in Figure 7-68(d). It may seem unnecessary because registers (flip-flops) hold data by their very nature. We must consider, however, what must be done to a register in order to hold its value as it is clocked. The Q outputs must be tied back to the D inputs for each flip-flop so that the old data is reloaded on each clock. This is another example of the use of concatenation to make up a desired set of data. The three PRESENT state data bits are grouped with the Serial input bit, and these four bits are loaded into the four data bits of the register. These four options provide the setting for the following design example (Example 7-26).

EXAMPLE 7-26

Suppose we want to design a universal four-bit shift register, using HDL, that has four synchronous modes of operation: Hold Data, Shift Left, Shift Right, and Parallel Load. Two input bits will select the operation that is to be performed on each rising edge of the clock. To implement a shift register, we can use structural code to describe a string of flip-flops. Making the shift register versatile by allowing it to shift right or left or to parallel load would make this file quite long and thus hard to read and understand using structural methods. A much better approach is to use the more abstract and intuitive methods available in HDL to describe the circuit concisely. To do this, we must develop a strategy that will create the shifting action. The concept is very similar to the one presented in Example 7-18, where a D flip-flop register chip (74174) was wired to form a shift register. Rather than thinking of the shift register as a serial string of flip-flops, we consider it as a parallel register whose contents are being transferred in parallel to a set of bits that is offset by one bit position. Figure 7-68 demonstrates the concept of each transfer needed in this design.

Solution

A very reasonable first step is to define a two-bit input named *mode* with which we can specify mode 0, 1, 2, or 3. The next challenge is deciding how to choose among the four operations using HDL. Several methods can work here. The CASE structure was chosen because it allows us to choose a different set of HDL statements for each and every possible mode value. There is no priority associated with checking for the existing mode settings or overlapping ranges of mode numbers, so we do not need the advantages of the IF/ELSE construct. The HDL solutions are given in Figures 7-69 and 7-70. The same inputs and outputs are defined in each approach: a clock, four bits of parallel load data, a single bit for the serial input to the register, two bits for the mode selection, and four output bits.

```
1    SUBDESIGN fig7_69
2    (
3        clock       :INPUT;
4        din[3..0]   :INPUT;   -- parallel data in
5        ser_in      :INPUT;   -- serial data in from Left or Right
6        model[1..0] :INPUT;   -- MODE Select: 0=hold, 1=right, 2=left, 3=load
7        q[3..0]     :OUTPUT;
8    )
9    VARIABLE
10       ff[3..0] :DFF;        -- define register set
11   BEGIN
12       ff[].clk = clock;     -- synchronous clock
13       CASE mode[] IS
14          WHEN 0  => ff[].d     = ff[].q;         -- hold shift
15          WHEN 1  => ff[2..0].d = ff[3..1].q);    -- shift right
16                     ff[3].d    = ser_in;         -- new data from left
17          WHEN 2  => ff[3..1].d = ff[2..0].q;     -- shift left
18                     ff[0].d    = ser_in;         -- new data bit from right
19          WHEN 3  => ff[].d     = din[];          -- parallel load
20       END CASE;
21       q[] = ff[];                    -- update outputs
22   END;
```

FIGURE 7-69   AHDL bidirectional shift register.

```
1    ENTITY fig7_70 IS
2    PORT(
3       clock      :IN BIT;
4       din        :IN BIT_VECTOR (3 DOWNTO 0);   -- parallel data in
5       ser_in     :IN BIT;                        -- serial data in L or R
6       mode       :IN INTEGER RANGE 0 TO 3;       -- 0=hold 1=rt 2=left 3=load
7       q          :OUT BIT_VECTOR (3 DOWNTO 0)
8       );
9    END fig7_70;
10
11   ARCHITECTURE a OF fig7_70 IS
12   SIGNAL ff   : BIT_VECTOR (3 DOWNTO 0);
13
14   BEGIN
15      PROCESS (clock)      -- respond to clock
16      BEGIN
17         IF (clock = '1' AND clock'EVENT) THEN
18            CASE mode IS
19            WHEN 0  => ff                <= ff;              -- hold shift
20            WHEN 1  => ff(2 DOWNTO 0) <= ff (3 DOWNTO 1);  -- shift right
21                       ff(3)             <= ser_in;
22            WHEN 2  => ff(3 DOWNTO 1) <= ff(2 DOWNTO 0);   -- shift left
23                       ff(0)             <= ser_in;
24            WHEN OTHERS =>   ff <= din;                     -- parallel load
25            END CASE;
26         END IF;
27      END PROCESS;
28      q <= ff;                -- update outputs
29   END a;
```

FIGURE 7-70   VHDL bidirectional shift register.

AHDL SOLUTION

The AHDL solution of Figure 7-69 uses a register of D flip-flops declared by the name *ff* on line 10, representing the current state of the register. Because the flip-flops all need to be clocked at the same time (synchronously), all the clock inputs are assigned to *clock* on line 12. The CASE construct selects a different transfer configuration for each value of the *mode* inputs. Mode 0 (hold data) uses a direct parallel transfer from the current state to the same bit positions on the *D* inputs to produce the identical NEXT state. Mode 1 (shift right), which is described on lines 15 and 16, transfers bits 3, 2, and 1 to bit positions 2, 1, and 0, respectively, and loads bit 3 from the serial input. Mode 2 (shift left) performs a similar operation in the opposite direction (see lines 17 and 18). Mode 3 (parallel load) transfers the value on the parallel data inputs to become the NEXT state of the register. The code creates the circuitry that chooses one of these logical operations on the actual register, and the proper data is transferred to the output pins on the next clock. This code can be shortened by combining lines 15 and 16 into a single statement that con-

AHDL

catenates the ser_in with the three data bits and groups them as a set of four bits. The statement that can replace lines 15 and 16 is:

```
WHEN 1 => ff[].d = (ser_in, ff[3..1].q);
```

Lines 17 and 18 can also be replaced by:

```
WHEN 2 => ff[].d = (ff[2..0].q, ser_in);
```

VHDL SOLUTION

The VHDL solution of Figure 7-70 defines an intermediate signal by the name *ff* on line 12, representing the current state of the register. Because all the transfer operations need to take place in response to a rising clock edge, a PROCESS is used, with *clock* specified in the sensitivity list. The CASE construct selects a different transfer configuration for each value of the *mode* inputs. Mode 0 (hold data) uses a direct parallel transfer from the current state to the same bit positions to produce the identical NEXT state. Mode 1 (shift right) transfers bits 3, 2, and 1 to bit positions 2, 1, and 0, respectively (line 20), and loads bit 3 from the serial input (line 21). Mode 2 (shift left) performs a similar operation in the opposite direction. Mode 3 (parallel load) transfers the value on the parallel data inputs to the NEXT state of the register. After choosing one of these operations on the actual register, the data is transferred to the output pins on line 28. This code can be shortened by combining lines 20 and 21 into a single statement that concatenates the ser_in with the three data bits and groups them as a set of four bits. The statement that can replace lines 20 and 21 is:

```
WHEN 1 => ff <= ser_in & ff(3 DOWNTO 1);
```

Lines 22 and 23 can also be replaced by:

```
WHEN 2 => ff <= ff(2 DOWNTO 0) & ser_in;
```

REVIEW QUESTIONS

1. What is the name of the Altera LPM used to make shift registers?
2. Write an AHDL equation that can implement a shift left of an eight-bit array *reg[ ]* with serial input *dat*.
3. Write a VHDL equation that can implement a shift left of an eight-bit array *reg[ ]* with serial input *dat*.
4. Why is it necessary to reload the current data during the hold data mode on a shift register?

7-28   HDL RING COUNTERS

In Section 7-25, we used a shift register to make a counter that circulates a single active logic level through all of its flip-flops. This was referred to as a ring counter. One characteristic of ring counters is that the modulus is equal to the number of flip-flops in the register and thus there are always many unused and invalid states. We have already discussed ways of describing counters using the CASE construct to specify PRESENT state and NEXT state transitions. In those examples, we took care of invalid states by including them under "others." This method also works for ring counters. In this section, however, we look at a more intuitive way to describe shift counters.

These methods use the same techniques described in Section 7-27 in order to make the register shift one position on each clock. The main feature of this code is the method of completing the "ring" by driving the *ser_in* line of the shift register. With a little planning, we should also be able to ensure that the counter eventually reaches the desired sequence, no matter what state it is in initially. For this example, we re-create the operation of the ring counter whose state diagram is shown in Figure 7-60(d). In order to make this counter self-start without using asynchronous inputs, we control the *ser_in* line of the shift register using an IF/ELSE construct. Any time we detect that the upper three bits are all LOW, we assume the lowest order bit is HIGH, and on the next clock, we want to shift in a HIGH to *ser_in*. For all other states (valid and invalid), we shift in a LOW. Regardless of the state to which the counter is initialized, it eventually fills with zeros; at which time, our logic shifts in a HIGH to start the ring sequence.

AHDL RING COUNTER

The AHDL code shown in Figure 7-71 should look familiar by now. Lines 11 and 12 control the serial input using the strategy we just described. Notice the use of the double equals (==) operator on line 11. This operator evaluates whether the expressions on each side are equal or not. Remember, the single equals (=) operator assigns (i.e., connects) one object to another. Line 14 implements the shift right action that we described in the previous section.

FIGURE 7-71 AHDL four-bit ring counter.

```
1    SUBDESIGN fig7_71
2    (
3       clk          :INPUT;
4       q[3..0]      :OUTPUT;
5    )
6    VARIABLE
7       ff[3..0]     :DFF;
8       ser_in       :NODE;
9    BEGIN
10      ff[].clk = clk;
11      IF ff[3..1] == B"000" THEN ser_in = VCC;   -- self start
12      ELSE ser_in = GND;
13      END IF;
14      ff[3..0].d = (ser_in, ff[3..1].q);          -- shift right
15      q[] = ff[];
16   END;
```

VHDL RING COUNTER

The VHDL code shown in Figure 7-72 should look familiar by now. Lines 13 and 14 control the serial input using the strategy we just described. Line 16 implements the shift right action that we described in the previous section.

```
1    ENTITY fig7_72 IS
2    PORT ( clk        :IN BIT;
3           q          :OUT BIT_VECTOR (3 DOWNTO 0));
4    END fig7_72;
5
6    ARCHITECTURE vhdl OF fig7_72 IS
7    BEGIN
8       PROCESS (clk)
9       VARIABLE ff         :BIT_VECTOR (3 DOWNTO 0);
10      VARIABLE ser_in     :BIT;
11      BEGIN
12         IF (clk'EVENT AND clk = '1') THEN
13            IF (ff (3 DOWNTO 1) = "000") THEN ser_in := '1'; -- self start
14            ELSE ser_in :='0';
15            END IF;
16            ff(3 DOWNTO 0)  := (ser_in & ff(3 DOWNTO 1));   -- shift right
17         END IF;
18         q <= ff;
19      END PROCESS;
20   END vhdl;
```

FIGURE 7-72   VHDL four-bit ring counter.

REVIEW QUESTIONS

1. What does it mean for a ring counter to self-start?
2. Which lines of Figure 7-71 ensure that the ring counter self-starts?
3. Which lines of Figure 7-72 ensure that the ring counter self-starts?

7-29   HDL ONE-SHOTS

Another important circuit that we have studied is the one-shot. We can apply the concept of a counter to implement a **digital one-shot** using HDL. Recall from Chapter 5 that one-shots were devices that produce a pulse of a predefined width every time the trigger input is activated. A *nonretriggerable* one-shot ignores the trigger input as long as the pulse output is still active. A *retriggerable* one-shot starts a pulse in response to a trigger and restarts the internal pulse timer every time a subsequent trigger edge occurs before the pulse is complete. The first example we investigate is a nonretriggerable, HIGH-level-triggered digital one-shot. The one-shots that we studied in Chapter 5 used a resistor and capacitor as the internal pulse timing mechanism. In order to create a one-shot using HDL techniques, we use a four-bit counter to determine the width of the pulse. The inputs are a clock signal, trigger, clear, and pulse width value. The only output is the pulse out, *Q*. The idea is quite simple. Whenever a trigger is detected, make the pulse go HIGH and load a down-counter with a number from the pulse width inputs. The larger this number, the longer it will take to count down to zero. The advantage of this one-shot is that the pulse width can be adjusted easily by changing the value loaded into the counter. As you read the sections below, consider the following question: "What makes this circuit nonretriggerable and what makes it level-triggered?"

SIMPLE AHDL ONE-SHOTS

A nonretriggerable, level-sensitive, one-shot description in AHDL is shown in Figure 7-73. A register of four flip-flops is created on line 8, and it serves as the counter that counts down during the pulse. The *clock* is connected in parallel to all the flip-flops on line 10. The reset function is implemented by connecting the *reset* control line directly to the asynchronous clear input of each flip-flop on line 11. After these assignments, the first condition that is tested is the trigger. If it is activated (HIGH) at any time while the count value is 0 (i.e., the previous pulse is done), then the delay value is loaded into the counter. On line 14, it tests to see if the pulse is done by checking to see if the counter is down to zero. If it is, then the counter should not roll over but rather stay at zero. If the count is not at zero, then it must be counting, so line 15 sets up the flip-flops to decrement on the next clock. Finally, line 17 generates the output pulse. This Boolean expression can be thought of as follows: "Make the pulse (*Q*) HIGH when the *count* is anything other than zero."

FIGURE 7-73 AHDL nonretriggerable one-shot.

```
1   SUBDESIGN fig7_73
2   (
3       clock, trigger, reset   : INPUT;
4       delay[3..0]             : INPUT;
5       q                       : OUTPUT;
6   )
7   VARIABLE
8       count[3..0]     : DFF;
9   BEGIN
10      count[].clk = clock;
11      count[].clrn = reset;
12      IF trigger & count[].q == b"0000" THEN
13              count[].d = delay[];
14      ELSIF count[].q == B"0000" THEN count[].d = B"0000";
15      ELSE count[].d = count[].q - 1;
16      END IF;
17      q = count[].q != B"0000";   -- make output pulse
18  END;
```

SIMPLE VHDL ONE-SHOTS

A nonretriggerable, level-sensitive, one-shot description in VHDL is shown in Figure 7-74. The inputs and outputs are shown on lines 3–5, as previously described. In the architecture description, a PROCESS is used (line 11) to respond to either of two inputs: the clock, or the reset. Within this PROCESS, a variable is used to represent the value on the counter. The input that should have overriding precedence is the *reset* signal. This is tested first (line 14) and if it is active, the *count* is cleared immediately. If the *reset* is not active, line 15 is evaluated and looks for a rising edge on the *clock*. Line 16 checks for the trigger. If it is activated at any time while the count value is 0 (i.e., the previous pulse is done), then the width value is loaded into the counter. On line 18, it tests to see if the pulse is done by checking to see if the counter is down to zero. If it is, then the counter should not roll over but rather stay at zero. If the *count* is not at zero, then it must be counting, so line 19 sets up the flip-flops to decrement on the next clock. Finally, lines 22 and 23 generate the output pulse. This Boolean expression can be thought of as follows: "Make the pulse (q) HIGH when the count is anything other than zero."

```
1    ENTITY fig7_74 IS
2    PORT (
3         clock, trigger, reset    :IN BIT;
4         delay                    :IN INTEGER RANGE 0 TO 15;
5         q                        :OUT BIT
6         );
7    END fig 7_74;
8
9    ARCHITECTURE vhdl OF fig7_74 IS
10   BEGIN
11      PROCESS (clock, reset)
12      VARIABLE count       : INTEGER RANGE 0 TO 15;
13      BEGIN
14         IF reset = '0' THEN count := 0;
15         ELSIF (clock'EVENT AND clock = '1' ) THEN
16            IF trigger = '1' AND count = 0 THEN
17               count := delay;                        -- load counter
18            ELSIF count = 0 THEN count := 0;
19            ELSE count := count - 1;
20            END IF;
21         END IF;
22         IF count /= 0 THEN q <= '1';
23         ELSE q <= '0';
24         END IF;
25      END PROCESS;
26   END vhdl;
```

FIGURE 7-74   VHDL nonretriggerable one-shot.

Now that we have reviewed the code that describes this one-shot, let's evaluate its performance. Converting a traditionally analog circuit to digital usually offers some advantages and some disadvantages. On a standard one-shot chip, the output pulse starts immediately after the trigger. For the digi-

FIGURE 7-75 Simulation of the nonretriggerable one-shots.

tal one-shot described here, the output pulse starts on the next clock edge and lasts as long as the counter is greater than zero. This situation is shown in Figure 7-75 within the first ms of the simulation. Notice that the trigger goes high almost 0.5 ms before the q out responds. If another trigger event occurs while it is counting down (like the one just before 3 ms), it is ignored. This is the nonretriggerable characteristic.

Another point to make for this digital one-shot is that the trigger pulse must be long enough to be seen as a HIGH on the rising clock edge. At about the 4.5-ms mark, a pulse occurs on the trigger input but goes LOW before the rising edge of the clock. This circuit does *not* respond to this input event. At just past 5 ms, the trigger goes HIGH and stays there. The pulse lasts exactly 6 ms, but because the trigger input remains HIGH, it responds with another output pulse one clock later. The reason for this situation is that this circuit is level-triggered rather than edge-triggered, like most of the conventional one-shot ICs.

Retriggerable, Edge-Triggered One-Shots in HDL

Many applications of one-shots require the circuit to respond to an edge rather than a level. How can HDL code be used to make the circuit respond once to each positive transition on its trigger input? The technique described here is called *edge-trapping* and has been a very useful tool in programming microcontrollers for years. As we will see, it is equally useful for describing edge-triggering for a digital circuit using HDL. This section illustrates an example of a retriggerable one-shot while also demonstrating edge-trapping, which can be useful in many other situations.

The general operation of this retriggerable one-shot requires that it responds to a rising edge of the trigger input. As soon as the edge is detected, it should start timing the pulse. In the digital one-shot, this means that it loads the counter as soon as possible after the trigger edge and starts counting down toward zero. If another trigger event (rising edge) occurs before the pulse is terminated, the counter is immediately reloaded, and the pulse timing starts again from the beginning, thus sustaining the pulse. Activating the clear at any point should force the counter to zero and terminate the pulse. The minimum output pulse width is simply the number applied to the width input multiplied by the clock period.

The strategy behind edge-trapping for a one-shot is demonstrated in Figure 7-76. On each active clock edge are two important pieces of information that are needed. The first is the state of the *trigger* input *now* and the second is the state of the *trigger* input when the last active clock edge occurred. Start with point *a* on the diagram of Figure 7-76 and determine these two values, then move to point *b*, and so on. By completing this task, you should have

FIGURE 7-76   Detecting edges.

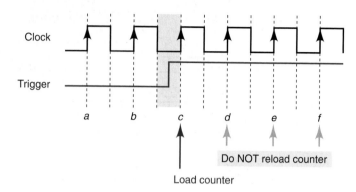

concluded that, at point *c*, a unique result has been obtained. The *trigger* is HIGH now but it was LOW on the last active clock edge. This is the point where we have detected the *trigger* edge event.

In order to know what the *trigger* was on the last active clock edge, the system must remember the last value that *the trigger had at that point*. This is done by storing the value of the trigger bit in a flip-flop. Recall that we discussed a similar concept in Chapter 5 when we talked about using a flip-flop to detect a sequence. The code for a one-shot is written so that the counter is loaded only after a rising edge is detected on the *trigger* input.

AHDL RETRIGGERABLE, EDGE-TRIGGERED ONE-SHOT

The first five lines of Figure 7-77 are identical to the previous nonretriggerable example. In AHDL, the only way to remember a value obtained in the past is to store the value on a flip-flop. This section uses a flip-flop named *trig_was* (line 9) to store the value that was on the trigger on the last active

FIGURE 7-77   AHDL retriggerable one-shot with edge trigger.

```
1    SUBDESIGN fig7_77
2    (
3        clock, trigger, reset   : INPUT;
4        delay[3..0]             : INPUT;
5        q                       : OUTPUT;
6    )
7    VARIABLE
8            count[3..0]   : DFF;
9            trig_was      : DFF;
10   BEGIN
11       count[].clk = clock;
12       count[].clrn = reset;
13       trig_was.clk = clock;
14       trig_was.d = trigger;
15
16       IF trigger & !trig_was.q THEN
17            count[].d = delay[];
18       ELSIF count[].q == B"0000" THEN count[].d = B"0000";
19       ELSE count[].d = count[].q - 1;
20       END IF;
21       q = count[].q != B"0000";
22   END;
```

clock edge. This flip-flop is simply connected so that the trigger is on its *D* input (line 13) and the clock is connected to its *clk* input (line 13). The *Q* output of *trig_was* remembers the value of the *trigger* right up to the next clock edge. At this point, we use line 16 to evaluate if a triggering edge has occurred. If *trigger* is HIGH (now), but *trigger* was LOW (last clock), it is time to load the counter (line 17). Line 18 ensures that, once the *count* reaches zero, it will remain at zero until a new trigger comes along. If the decisions allow line 19 to be evaluated, it means that there is a value loaded into the counter and it is not zero, so it needs to be decremented. Finally, the output pulse is made HIGH any time a value other than 0000 is still on the counter, like we saw previously.

VHDL RETRIGGERABLE, EDGE-TRIGGERED ONE-SHOT

The ENTITY description in Figure 7-78 is exactly like the previous nonretriggerable example. In fact, the only differences between this example and the one shown in Figure 7-74 have to do with the logic of the decision process. When we want to remember a value in VHDL, it must be stored in a VARIABLE. Recall that we can think of a PROCESS as a piece of code that describes what happens each time a signal in the sensitivity list changes state.

```
1    ENTITY fig7_78 IS
2    PORT (   clock, trigger, reset   : IN BIT;
3          delay                     : IN INTEGER RANGE 0 TO 15;
4          q                         : OUT BIT);
5    END fig7_78;
6
7    ARCHITECTURE vhdl OF fig7_78 IS
8    BEGIN
9       PROCESS (clock, reset)
10      VARIABLE count      : INTEGER RANGE 0 TO 15;
11      VARIABLE trig_was   : BIT;
12      BEGIN
13         IF reset = '0' THEN count := 0;
14         ELSIF (clock'EVENT AND clock = '1' ) THEN
15            IF trigger = '1' AND trig_was = '0' THEN
16               count := delay;              -- load counter
17               trig_was := '1';             -- "remember" edge detected
18            ELSIF count = 0 THEN count := 0; -- hold @ 0
19            ELSE count := count - 1;         -- decrement
20            END IF;
21            IF trigger = '0' THEN trig_was := '0';
22            END IF;
23         END IF;
24         IF count /= 0  THEN q <= '1';
25         ELSE q <= '0';
26         END IF;
27      END PROCESS;
28   END vhdl;
```

FIGURE 7-78   VHDL retriggerable one-shot with edge trigger.

A VARIABLE retains the last value assigned to it between the times the process is invoked. In this sense, it acts like a flip-flop. For the one-shot, we need to store a value that tells us what the trigger was on the last active clock edge. Line 11 declares a variable bit to serve this purpose. The first decision (line 13) is the overriding decision that checks and responds to the *reset* input. Notice that this is an asynchronous control because it is evaluated before the clock edge is detected on line 14. Line 14 determines that a rising clock edge has occurred, and then the main logic of this process is evaluated between lines 15 and 20.

When a clock edge occurs, one of three conditions exists:

1. A trigger edge has occurred and we must load the counter.
2. The counter is zero and we need to keep it at zero.
3. The counter is not zero and we need to count down by one.

Recall that it is very important to consider the order in which questions are asked and assignments are made in VHDL PROCESS statements because the *sequence* affects the operation of the circuit we are describing. The code that updates the *trig_was* variable must occur after the evaluation of its previous condition. For this reason, the conditions necessary to detect a rising edge on *trigger* are evaluated on line 15. If an edge occurred, then the counter is loaded (line 16) and the variable is updated (line 17) to remember this for the next time. If a trigger edge has not occurred, the code either holds at zero (line 18) or counts down (line 19). Line 21 makes sure that, as soon as the trigger input goes LOW, the variable *trig_was* remembers this by resetting. Finally, lines 24–25 are used to create the output pulse during the time the counter is not zero.

The two improvements that were made in this one-shot over the last example are the edge-triggering and the retriggerable feature. Figure 7-79 evaluates the new performance features. Notice in the first ms of the timing diagram that a trigger edge is detected, but the response is not immediate. The output pulse goes high on the next clock edge. This is a drawback to the digital one-shot. The retriggerable feature is demonstrated at about the 2-ms mark. Notice that *trigger* goes high and on the next clock edge, the *count* starts again at 5, sustaining the output pulse. Also notice that even after the *q* output pulse is complete and the *trigger* is still HIGH, the one-shot does not fire another pulse because it is not level-triggered but rather rising edge-triggered. At the 6-ms mark, a short trigger pulse occurs but is ignored because it does not stay HIGH until the next clock. On the other hand, an

FIGURE 7-79   Simulation of the edge-triggered retriggerable one-shot.

even shorter trigger pulse occurring just after the 7-ms mark does fire the one-shot because it is present during the rising clock edge. The resulting output pulse lasts exactly 5 clock cycles because no other triggers occur during this period.

To minimize the effects of delayed response to trigger edges and the possibility of missing trigger edges that are too short, this circuit can be improved quite simply. The clock frequency and the number of bits used to load the delay value can both be increased to provide the same range of pulse widths (with more precise control) while reducing the minimum trigger pulse width. In order to cure this problem completely, the one-shot must respond asynchronously to the trigger input. This is possible in both AHDL and VHDL, but it will always result in a pulse that fluctuates in width by up to one clock period.

REVIEW QUESTIONS

1. Which control input signal holds the highest priority for each of the one-shot descriptions?
2. Name two factors that determine how long a pulse from a digital one-shot will last.
3. For the one-shots shown in this section, are the counters loaded synchronously or asynchronously?
4. What is the advantage of loading a counter synchronously?
5. What is the advantage of loading the counter asynchronously?
6. What two pieces of information are necessary to detect an edge?

PART 2 SUMMARY

1. Numerous IC registers are available and can be classified according to whether their inputs are parallel (all bits entered simultaneously), serial (one bit at a time), or both. Likewise, registers can have outputs that are parallel (all bits available simultaneously) or serial (one bit available at a time).

2. A sequential logic system uses FFs, counters, and registers, along with logic gates. Its outputs and sequence of operations depend on present and past inputs.

3. Troubleshooting a sequential logic system begins with observation of the system operation, followed by analytical reasoning to determine the possible causes of any malfunction, and finally test measurements to isolate the actual fault.

4. A ring counter is actually an N-bit shift register that recirculates a single 1 continuously, thereby acting as a MOD-N counter. A Johnson counter is a modified ring counter that operates as MOD-$2N$ counter.

5. Shift registers can be implemented with HDL using the built-in LPM megafunctions or by writing custom descriptions of their operation.

6. An understanding of bit arrays/bit vectors and their notation is very important in describing shift register operations.

7. Shift register counters such as Johnson and ring counters can be implemented easily in HDL. Decoding and self-starting features are easy to write into the description.

8. Digital one-shots are implemented with a counter loaded with a delay value when the trigger input is detected and counts down to zero. During the countdown time, the output pulse is held HIGH.

9. With strategic placement of the hardware description statements, these circuits can be made edge- or level-triggered and retriggerable or nonretriggerable. They produce an output pulse that responds synchronously or asynchronously to the trigger.

PART 2 IMPORTANT TERMS

parallel in/parallel out	circulating shift register	sequential logic system
serial in/serial out	ring counter	digital one-shot
parallel in/serial out	Johnson counter (twisted	
serial in/parallel out	ring counter)	

PROBLEMS

PART 1

SECTIONS 7-1 AND 7-2

B 7-1. Add another flip-flop, E, to the counter of Figure 7-1. The clock signal is an 8-MHz square wave.

   (a) What will be the frequency at the E output? What will be the duty cycle of this signal?

   (b) Repeat (a) if the clock signal has a 20 percent duty cycle.

   (c) What will be the frequency at the C output?

   (d) What is the MOD number of this counter?

B 7-2. Construct a binary counter that will convert a 64-kHz pulse signal into a 2-kHz square wave.

B 7-3. Assume that a five-bit binary counter starts in the 00000 state. What will be the count after 144 input pulses?

D 7-4. Use J-K flip-flops and any other necessary logic to construct a MOD-24 asynchronous counter.

7-5. Draw the waveforms for all the FFs in the decade counter of Figure 7-6(b) in response to a 1-kHz clock frequency. Show any glitches that might appear on any of the FF outputs. Determine the frequency at the D output.

7-6. Repeat Problem 7-5 for the counter of Figure 7-6(a).

7-7. Change the inputs to the NAND gate of Figure 7-7 so that the counter divides the frequency by 50. Repeat for a frequency division of 100.

N 7-8. A counter or a group of counters is often used to divide a high-frequency clock signal down to a lower-frequency output. When these counters are binary counters (i.e., they count in the binary sequence), the output will not be a symmetrical square wave if the binary sequence has been shortened in order to produce the desired MOD number. For example, refer to the C waveform of the MOD-6 counter in Figure 7-4.

FIGURE 7-80   Problem 7-8.

When a counter is being used only for frequency division, it is not necessary that it count in a binary sequence as long as it has the desired MOD number. A symmetrical square-wave output can be obtained for any *even* MOD number by breaking the MOD number into the product of two MOD numbers, one of which is a power of 2. For example, a MOD-6 counter can be formed from a MOD-3 counter and MOD-2 counter as shown in Figure 7-80.

Here flip-flops A and B and the NAND gate make up the MOD-3 counter, whose B output has one-third the frequency of the input pulses. This B output is connected to the input of flip-flop C, which is acting as a MOD-2 counter to divide the frequency down to one-sixth the frequency of the input pulses.

(a) Assume that all FFs are initially LOW, and sketch the waveforms at each FF output for 12 cycles of the input.

(b) Construct the state transition diagram, and show that it is not a normal binary sequence.

SECTION 7-3

B   7-9. In Figure 7-8, connect Q_0 to $\overline{CP_1}$ and MR_1, and connect Q_3 to MR_2. If 180-kHz pulses are appled to $\overline{CP_0}$, determine the following: (a) the count sequence; (b) MOD number; (c) frequency at Q_3.

D   7-10. Show how a 74LS293 counter can be used to produce a 1.2-kpps output from an 18-kpps input.

D   7-11. Show how two 74LS293s can be connected to divide an input frequency by 60 while producing a symmetrical square-wave output.

C   7-12. Determine the frequency at output X in Figure 7-81.

FIGURE 7-81   Problems 7-12 and 7-69.

D   7-13. (a) Add the necessary logic to a 74HC4024 so that it operates as a MOD-100 counter.

        (b) Use 74HC4024s and any necessary logic to convert a 10-kpps signal to 1 pps.

SECTION 7-4

B   7-14. (a) Draw the diagram for a MOD-16 *down* counter.

        (b) Construct the state transition diagram.

        (c) If the counter is initially in the 0110 state, what state will it be in after 37 clock pulses?

7-15. Refer to the counter shown in Figure 7-82. How can you tell that it is a down counter? It has been modified so that it does *not* count through the entire binary sequence 111 to 000. Determine the actual sequence it counts through.

All J, K inputs are HIGH

FIGURE 7-82   Problem 7-15.

SECTION 7-5

7-16. A four-bit ripple counter is driven by a 20-MHz clock signal. Draw the waveforms at the output of each FF if each FF has $t_{pd} = 20$ ns. Determine which counter states, if any, will not occur because of the propagation delays.

7-17. (a) What is the maximum clock frequency that can be used with the counter of Problem 7-16?

    (b) What would f_{max} be if the counter were expanded to six bits?

SECTIONS 7-6 AND 7-7

B   7-18. (a) Draw the circuit diagram for a MOD-64 parallel counter.

    (b) Determine f_{max} for this counter if each FF has $t_{pd} = 20$ ns and each gate has $t_{pd} = 10$ ns.

C   7-19. Figure 7-83 shows a four-bit parallel counter designed so that it does not sequence through the entire 16 binary states. Analyze its operation by determining its counting sequence, and then draw the waveforms at each FF output. (See Section 5-23 to review the analysis procedure.) Assume that all FFs are initially in the 0 state.

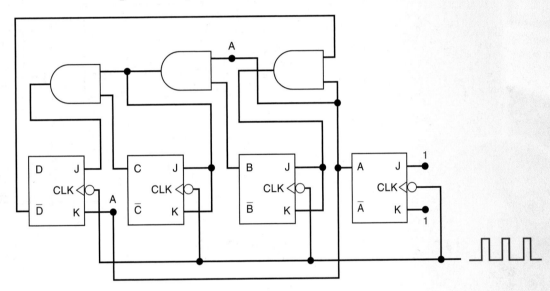

FIGURE 7-83   Problem 7-19.

7-20. Simplify the counter of Figure 7-18(a) so that it becomes a MOD-8 synchronous *down* counter.

C, T   7-21. Describe how the up/down counter of Figure 7-18 would operate if the INVERTER output were stuck HIGH.

SECTIONS 7-8 AND 7-9

B   7-22. Modify the circuit of Figure 7-22 so that the counter is preset to 0101 and counts *down* to 0000. Draw the waveforms at each FF output and at the $\overline{TC}_D$ output for 10 clock cycles.

N, C   7-23. Figure 7-84 shows how a presettable down counter can be used in a *programmable timer* circuit. The input clock frequency is an accurate 1 Hz derived from the 60-Hz line frequency after division by 60. Switches S1 to S4 are used to preset the counter to a desired starting count when a momentary pulse is applied to $\overline{PL}$. The timer operation is initiated by depressing the START push-button switch. Flip-flop Z is used to eliminate the effects of bounce in the START switch. The OS is used to provide a very narrow pulse to the $\overline{PL}$ input. The output of flip-flop X will be a waveform that goes HIGH for a number of seconds equal to the number set on the switches.

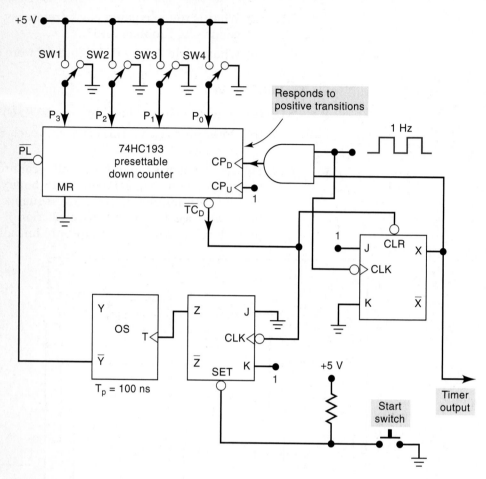

FIGURE 7-84   Problems 7-23, 7-68, and 7-79.

(a) Assume that all FFs and the counter are in the 0 state, and analyze and explain the circuit operation, showing waveforms when necessary, for the case where S1 and S4 are LOW and S2 and S3 are HIGH. Be sure to explain the function of flip-flop X.

(b) Why can't the timer output be taken at the $\overline{TC}_D$ output?

(c) Why can't the START switch be used to trigger the OS directly?

(d) What will happen if the START switch is held down too long? Add the necessary logic needed to ensure that holding the START switch down will not affect the timer operation.

7-24. Modify the circuit of Figure 7-24 so that it functions as a MOD-10 counter. The frequency at the Q_3 output should be one-tenth the frequency of the CP_D input. Draw the waveforms at Q_3, Q_2, Q_1, Q_0, and $\overline{TC}_D$.

7-25. Change the parallel data inputs in Figure 7-24 to 1001. Draw the waveforms at Q_3, Q_2, Q_1, Q_0, and $\overline{TC}_D$. What is the MOD number?

7-26. Draw the waveforms for the input signals required to perform the following sequence of operations in the circuit of Figure 7-25: (1) clear the count to 0; (2) count up to 24_{10}; (3) preset the count to 76_{10}; (4) count down to 0.

FIGURE 7-85   Problem 7-27.

SECTION 7-10

7-27. Figure 7-85 shows the IEEE/ANSI symbol for a 7490 or 74290 counter IC. Examine the symbol and determine the following.

(a) The overall MOD number

(b) The function performed by the *MR* inputs

(c) The function performed by the *MS* inputs

(d) Is this an up counter or a down counter?

(e) How would you connect it to function as a BCD counter? (Refer to the data sheet on the CD ROM.)

(f) How would you connect it to divide the clock frequency by 10 and produce a symmetrical square wave?

7-28. The 74192 counter IC operates exactly like the 74193 except for the following differences:

- The 74192 is a BCD counter that either counts up from 0 to 9 or counts down from 9 to 0.

- The $\overline{TC}_U$ output is activated when the count is 9 and the CP_U input is LOW.

Modify the IEEE/ANSI symbol of Figure 7-26 so that it represents the 74192.

SECTIONS 7-11 AND 7-12

B   7-29. Draw the gates necessary to decode all of the states of a MOD-16 counter using active-LOW outputs.

B   7-30. Draw the AND gates necessary to decode the 10 states of the BCD counter of Figure 7-6(b).

7-31. Figure 7-86 shows a ripple counter being used to help generate control waveforms. Control waveforms 1 and 2 can be used for many purposes, including control of motors, solenoids, valves, and heaters. Determine the control waveforms, assuming that all FFs are initially LOW. Ignore decoding glitches. Assume that clock frequency = 1 kpps.

FIGURE 7-86   Problem 7-31.

7-32. Draw the complete waveforms at the output of the decoding gates of a MOD-16 *ripple* counter, including any glitches or spikes that can occur due to the FF delays. Why are the gates that are decoding for *even* numbers the only ones that have glitches?

7-33. The circuit of Figure 7-86 might malfunction because of glitches at the outputs of the decoding NAND gates.

   (a) Determine at what point(s) the glitches can cause erroneous operation.

   (b) What are two ways that can be used to eliminate the possibility of erroneous operation?

SECTION 7-13

B   7-34. How many FFs are used in Figure 7-32? Indicate the states of each of these FFs after 795 pulses have occurred once the counters have been cleared.

7-35. How many cascaded BCD counters are needed to be able to count up to 8000? How many FFs does this operation require? Compare this with the number of FFs required for a normal binary counter to count up to 8000. Even though it uses more FFs, why is the cascaded BCD method used?

SECTION 7-14

D   7-36. (a) Design a synchronous counter that has the following sequence: 000, 010, 101, 110, and repeat. The undesired (unused) states 001, 011, 100, and 111 must always go to 000 on the NEXT clock pulse.

   (b) Redesign the counter of part (a) without any requirement on the unused states; that is, their NEXT states can be don't-cares. Compare with the design from (a).

C, D   7-37. Use the synchronous counter design procedure to design a four-bit synchronous down counter that counts through all states from 1111 down to 0000. Compare your result with the synchronous down counter described in Section 7-7.

C, D   7-38. Using a procedure similar to the one followed in the design of the counter to drive the stepper motor (Figure 7-39), design a three-bit synchronous counter that will count up or count down under the control of Direction input, D. It should count up when $D = 1$ and count down when $D = 0$. (*Hint:* This is a *four*-variable problem.)

         Compare your final circuit to the synchronous up/down counter of Figure 7-18.

SECTION 7-15

B   7-39. Draw the PRESENT state/NEXT state table for a MOD-8 binary up-counter using D flip-flops.

B   7-40. Write the Boolean equations for the input of each D flip-flop in Problem 7-39.

H   7-41. Write an HDL description in the language of your choice to describe the counter in Problem 7-39. Use PRESENT state/NEXT state description techniques.

H   7-42. Write an HDL description in the language of your choice to describe the counter in Problem 7-39. Use behavioral description techniques.

SECTION 7-16

H, D   7-43. Starting with the HDL counter description in Problem 7-42, add a *synchronous* reset input feature that allows the counter to be cleared (to zero) on the next clock after the *reset* input is pulled LOW.

H, D   7-44. Starting with the HDL counter description in Problem 7-42, add an *asynchronous* load feature that allows the value on three data inputs to be loaded into the counter immediately when the *load* input is pulled LOW.

H, D   7-45. Starting with the HDL counter description in Problem 7-42, add the controls necessary to allow two of these modules to be cascaded synchronously. Draw the block diagram to describe a 12-bit synchronous counter made up of four of these HDL descriptions cascaded together.

SECTION 7-18

H, D   7-46. Modify the HDL description in Figure 7-50 or Figure 7-51 to add a rinse sequence after the clothes are washed. The new sequence should be *idle* => *wash_fill* => *wash_agitate* => *wash_spin* => *rinse_fill* => *rinse_agitate* => *rinse_spin* => *idle.* Use hot water to wash, and cold water to rinse (add output bits to control two water valves).

GENERAL COUNTER ANALYSIS AND DESIGN PROBLEMS

The following problems involve applying the concepts covered in Part 1 of Chapter 7 to circuits using various standard IC counters. You must obtain data sheets from the Web or CD ROM in order to complete these problems.

B   7-47. Refer to the diagrams of the two counters shown in Figure 7-87.

FIGURE 7-87   Problem 7-47.

(a) Draw the state transition diagram for each counter.

(b) Determine the modulus for each counter.

(c) Which output is the MSB for each counter?

(d) How would the MSB output signal frequency compare to the frequency of the clock input for each counter?

(e) Sometimes it is necessary to design a counter with an unusual count sequence for a specific application. Which of these counter circuits seem(s) to have an unusual count sequence?

(f) How are the MSB signals different for the two counters?

B   7-48. Refer to the diagram in Figure 7-88.

    (a) Draw the timing diagram for this counter.

    (b) What is the counter's modulus?

    (c) What is the count sequence?

    (d) Is this an up-counter or a down-counter?

    (e) Can we produce the same modulus with a 74190 counter chip?

    (f) Can we produce the same count sequence with a 74190 counter chip?

FIGURE 7-88 Problem 7-48.

D   7-49. Design recycling, MOD-6 counters using the standard IC counter and the specifications given in each of the following:

    (a) 74160 using the CLR control

    (b) 74162 using the CLR control

    (c) 74162 with the count sequence 1, 2, 3, 4, 5, 6

    (d) 74190 with the down count sequence 5, 4, 3, 2, 1, 0

    (e) 74190 with the down count sequence 6, 5, 4, 3, 2, 1

D   7-50. Design a MOD-100 BCD counter using two 74162 chips. Use the same clock signal to clock synchronously the two counter chips and the RCO output from one of the counters to cascade them together. Add an active-HIGH count ENABLE control to your design. Note that the two enable controls (ENT and ENP) on this counter chip are slightly different from each other. Both enable pins must be high to enable the count, but ENT also controls the RCO output. Label the ONES digit and TENS digit outputs on your schematic.

7-51. Determine the signal frequency at each of the nine labeled outputs in Figure 7-89 if all three clocks (CLK1, CLK2, CLK3) are driven by the same 6-MHz signal. What are the frequency division factors for each output?

FIGURE 7-89 Problem 7-51.

7-52. Cascade the 74163 and 74162 counters together, as shown in Figure 7-90, and determine the resulting output frequencies. A 6-MHz signal is applied to CLK1. What signal frequencies are produced at the nine labeled outputs?

D, C   7-53. Design a frequency divider circuit that will produce the following three signal frequencies: 1.5 MHz, 150 kHz, and 100 kHz. Use 74162 and 74163 counter chips. A 12-MHz signal is available.

FIGURE 7-90   Problem 7-52.

PART 2

SECTIONS 7-20 AND 7-21

7-54. In Example 7-20 we saw how a 74ALS174 can be wired as a shift register. Show how to connect the 74ALS174 (and any necessary logic) so that it operates as a Johnson counter. What is its MOD number?

7-55. Suppose a 74ALS174 is connected as follows:

$$\overline{MR} = \text{HIGH}; Q_5 \rightarrow D_2; Q_4 \rightarrow D_1; Q_3 \rightarrow D_0$$
$$D_5 = D_3 = \text{HIGH}; D_4 = \text{LOW}$$

Assume all FFs have a zero hold-time and are initially LOW.

(a) Determine the states of each FF after a single pulse is applied to *CP*.

(b) Repeat for a second clock pulse.

7-56. Consider the situation depicted by the waveforms in Figure 7-55. If D_S goes LOW just prior to t_{247}, when will Q_{63} go LOW?

7-57. Show how the 4731B chip can be connected as a 256-bit shift register.

SECTIONS 7-22 AND 7-23

C, D   7-58. Modify the circuit of Figure 7-58 so that the INVERTER output is connected to the A input instead of to $\overline{MR}$.

(a) Draw the waveforms at each FF output in response to the input waveforms shown in Figure 7-91.

(b) Add the necessary logic to produce a timing-signal output that goes HIGH only during time intervals t_1 to t_2 and t_8 to t_9.

(c) Add the necessary logic to produce a timing signal that goes LOW only during the interval t_4 to t_7.

FIGURE 7-91   Problem 7-58.

7-59. A 74HC165 is connected as shown in Figure 7-92. Assume that prior to t_0, pulses have been applied to CP, and the $SH/\overline{LD}$ input has been held HIGH for a long time. Draw the Q_7 waveform in response to the CP and $SH/\overline{LD}$ waveforms beginning at t_0.

FIGURE 7-92   Problem 7-59.

N   7-60. While examining the schematic for a piece of equipment, a technician or an engineer will often come across an IC that is unfamiliar. In such cases, it is often necessary to consult the manufacturer's IC data book for specifications on the device. The information on the IC data sheets is always complete, but it is sometimes difficult to understand, especially by someone with very little experience. This problem will give you practice in obtaining information about a fairly complex IC—the 74194 bidirectional universal shift register. Consult the CD ROM to answer the following questions. Support your answers.

(a) Is the CLR input asynchronous or synchronous?

(b) *True or false:* When CLK is LOW, the S_0 and S_1 levels have no effect on the register.

(c) Assume the following conditions:

$$Q_A Q_B Q_C Q_D = 1\ 0\ 1\ 1$$
$$ABCD = 0\ 1\ 1\ 0$$
$$\overline{CLR} = 1$$
$$SR\ SER = 0$$
$$SL\ SER = 1$$

If $S_0 = 0$ and $S_1 = 1$, what will the register outputs be after one CLK pulse? After two CLK pulses? After three? After four?

(d) Use the same conditions except $S_0 = 1$, $S_1 = 0$, and repeat part (c).

(e) Repeat part (c) with $S_0 = S_1 = 1$.

(f) Repeat part (c) with $S_0 = S_1 = 0$.

(g) Use the same conditions as in part (c), except assume that output Q_A is connected to SL SER. What will be the register outputs after four CLK pulses?

(h) Show how to connect this IC to function as a ring counter that counts through the following $Q_A\ Q_B\ Q_C\ Q_D$ sequence: 0001, 0010, 0100, 1000, and repeat.

SECTION 7-25

7-61. Draw the diagram for a five-bit ring counter using J-K flip-flops.

7-62. Combine the ring counter of Problem 7-61 with a *single* J-K flip-flop to produce a MOD-10 counter. Determine the sequence of states for this counter. This is an example of a decade counter that is not a BCD counter.

7-63. Draw the diagram for a MOD-10 Johnson counter using J-K flip-flops, and determine its counting sequence. Draw the decoding circuit needed to decode each of the 10 states. This is another example of a decade counter that is not a BCD counter.

7-64. Determine the frequency of the pulses at points w, x, y, and z in the circuit of Figure 7-93.

FIGURE 7-93   Problem 7-64.

7-65. (a) A group of eight display lights on a pinball machine are controlled by the FFs of an eight-bit ring counter that is being clocked by a 2-pps signal. Describe the visual effect that is produced.

(b) Repeat for an eight-bit Johnson counter.

SECTION 7-26

T 7-66. A technician tests the counter of Figure 7-64(a) by applying a low-frequency clock signal and monitoring the FF outputs on indicator LEDs. He observes a repetitive sequence indicated by the LEDs (Table 7-8). What are the possible reasons why the counter is not counting properly?

TABLE 7-8
Problem 7-66.

Q_3	Q_2	Q_1	Q_0	
0	0	0	0	←
0	0	0	1	
0	0	1	0	
0	0	1	1	recycle and
0	1	0	0	repeat
0	1	0	1	
0	1	1	0	
0	1	1	1	

T 7-67. Refer to the up/down counter of Figure 7-18. Describe how each of the following circuit faults will affect the count-up and count-down operations.

(a) The AND gate 4 output is internally shorted to V_{CC}.

(b) A solder bridge is shorting the AND gate 1 output to the AND gate 3 output.

T 7-68. A technician runs a test on the timer circuit of Figure 7-84 and records the results shown in Table 7-9. Examine the recorded data and determine the possible causes of the faulty operation.

TABLE 7-9
Problem 7-68

S1	S2	S3	S4	Timer Output(s)
5 V	0 V	0 V	0 V	10
0 V	0 V	5 V	5 V	3
5 V	0 V	5 V	5 V	11
5 V	5 V	5 V	0 V	14
0 V	5 V	0 V	5 V	7
5 V	5 V	0 V	0 V	14

T 7-69. A technician wires up the counter circuit of Figure 7-81. He applies an accurate 8.64-kpps signal to the input and measures a frequency of 54 pps at X instead of the expected 60 pps. What is the most probable wiring error that he made?

SECTION 7-27

H, B 7-70. Use the LPM_SHIFTREG to implement an eight-bit PIPO register in the HDL of your choice.

H, B 7-71. Use the LPM_SHIFTREG to implement an eight-bit SIPO register in the HDL of your choice.

H, B 7-72. Use the LPM_SHIFTREG to implement an eight-bit SISO register in the HDL of your choice.

C, D 7-73. Sketch the block symbol for the bidirectional shift register HDL code of Figure 7-69 (AHDL) or Figure 7-70 (VHDL). In order to cascade (graphically) two of these four-bit registers to create an eight-bit bidirectional shift register, it is necessary to add some external logic to drive the serial inputs with the correct signal, depending on the mode (shift left/shift right). The Q0 and Q3 outputs must also be individually accessible. Draw the logic circuit necessary to cascade them.

H, D 7-74. Rewrite the HDL code of Figure 7-69 or Figure 7-70 to make an eight-bit bidirectional shift register.

SECTION 7-28

H, D 7-75. Write the HDL source code in the language of your choice to create an eight-bit self-starting Johnson counter.

SECTION 7-29

H, D, C 7-76. Write the AHDL source code for a retriggerable one-shot that is truly edge-triggered because the output pulse goes HIGH in *immediate* response to the trigger edge (asynchronous relative to the clock) and goes LOW when the counter is back down to zero.

H 7-77. The clock and trigger inputs in Figure 7-94 are applied to a digital one-shot. The clock frequency is 1 KHz. The delay value is set to "0011." Draw the output pulse waveform and measure the width of the pulse for each of the following AHDL circuits:

(a) The nonretriggerable circuit (Figure 7-73)

(b) The retriggerable circuit (Figure 7-77)

(c) The asynchronous edge-triggered circuit of Problem 7-76

FIGURE 7-94 Problem 7-77.

DRILL QUESTION

B 7-78. For each of the following statements, indicate the type(s) of counter being described.

(a) Each FF is clocked at the same time.

(b) Each FF divides the frequency at its *CLK* input by 2.

(c) The counting sequence is 111, 110, 101, 100, 011, 010, 001, 000.

(d) The counter has 10 distinct states.

(e) The total switching delay is the sum of the individual FFs' delays.

(f) This counter requires no decoding logic.

(g) The MOD number is always twice the number of FFs.

(h) This counter divides the input frequency by its MOD number.

(i) This counter can begin its counting sequence at any desired starting state.

(j) This counter can count in either direction.

(k) This counter can suffer from decoding glitches.

(l) This counter counts from 0 to 99.

(m) This counter can be designed to count through arbitrary sequences by determining logic needed at each flip-flop's J and K input.

MICROCOMPUTER APPLICATION

C, D   7-79. A microprocessor that is used in a control application must often control the timing of external events, like the turning on and off of devices such as solenoids, motors, and relays. Such actions can be timed using software by repetitively executing a program loop a specific number of times. This, however, places a heavy burden on the MPU because it can't do anything else while it is executing the loop over and over. For this reason, most timed intervals are usually generated by hardware that is *under control of the MPU.* In other words, the MPU will send data to the hardware to tell it how long a time interval to generate.

In Problem 7-23, we saw how the 74HC193 IC can be used in a timer circuit (Figure 7-84) to generate accurate time intervals corresponding to the binary data from four switches. This circuit can be modified so that the binary data come from an MPU instead of the switches. In Section 5-20, we saw how an MPU can transfer data to an external device using its address, data, and clock outputs (Figure 5-48).

Show how to combine these two circuits so that the timer output X generates a HIGH level for a time interval (in seconds) equal to the binary number that the MPU presets into the 74HC193 counter. You can eliminate any circuitry that is not needed. Assume that the MPU's $\overline{CP}$ signal is a 1-MHz square wave. Remember that $\overline{PL}$ is an asynchronous input.

ANSWERS TO SECTION REVIEW QUESTIONS

PART 1

SECTION 7-1

1. False   2. 0000   3. 128

SECTION 7-2

1. D, C, and A 2. True, because a BCD counter has 10 distinct states   3. 5 kHz

SECTION 7-3

1. 250 Hz   2. $f_{in}/60$ 3. 4096   4. The counter is MOD-64 and divides the frequency by 64.   5. Q_6, Q_5, Q_4, Q_3

SECTION 7-4

1. In an up counter, the count is increased by 1 with each clock pulse; in a down counter, the count is decreased by 1 with each pulse.   2. The inverted output of each FF is connected to the *CLK* input of the following FF.

SECTION 7-5

1. Each FF adds its propagation delay to the total counter delay in response to a clock pulse.   2. MOD-256

SECTION 7-6

1. Can operate at higher clock frequencies and has more complex circuitry   2. Six FFs and four AND gates   3. *ABCDE*

SECTION 7-8

1. It can be preset to any desired starting count.   2. Asynchronous presetting is independent of the clock input, while synchronous presetting occurs on the active edge of the clock signal.

SECTION 7-9

1. When $\overline{PL}$ is pulsed LOW, the counter is preset to the binary number present at inputs P_0 to P_3.   2. A HIGH at *MR* overrides all other inputs to reset the counter to 0000.   3. True   4. 1, 1, 0, respectively   5. 0 to 65,535

SECTION 7-10

1. See appropriate text.   2. (a) Up-counter operation. (b) This input is ANDed with any other input or output that has a "4" in its label. (c) This input controls the effect of any inputs that have "5" in their label. (d) Data input that is controlled by input labeled C5.

SECTION 7-11

1. Sixty-four   2. A six-input NAND gate with inputs A, B, C, $\overline{D}$, E, and $\overline{F}$

SECTION 7-12

1. The glitches are caused by the FFs' changing states one at a time during counter state transitions.   2. The strobe signal inhibits the decoding gates until all FFs have completed their transitions.

SECTION 7-14

1. See text.   2. It associates every possible PRESENT state with its desired NEXT state.   3. It shows the necessary levels at each flip-flop's J and K input to produce the counter's state transitions.   4. True

SECTION 7-15

1. PRESENT state/NEXT state tables   2. The desired NEXT state   3. *data1.d*
4. Behavioral description

SECTION 7-16

1. Asynchronous clear causes the counter to clear immediately. Synchronous load occurs on the next clock edge.   2. Asynchronous cascading uses one stage to clock the next and produces a ripple effect. Synchronous cascading results in all bits updating simultaneously on the clock edge.   3. All clocks are tied together.
4. A terminal count output and a count enable input

SECTION 7-17

1. Library of parameterized modules.   2. lpm_counter   3. lpm_width and lpm_modulus

SECTION 7-18

1. A state machine can be described using symbols to describe its states rather than actual binary states.   2. The compiler assigns the optimal values to minimize the circuitry.   3. The description is much easier to write and understand.

PART 2

SECTION 7-23

1. Parallel in/serial out   2. True   3. Serial in/parallel out   4. Serial in/serial out   5. The 74165 uses asynchronous parallel data transfer; the 74174 uses synchronous.   6. A HIGH prevents shifting on *CP*.

SECTION 7-24

1. It separates the two indicated functions performed by that input.   2. SRG 64

SECTION 7-25

1. Ring counter   2. Johnson counter   3. The inverted output of the last FF is connected to the input of the first FF.   4. (a) False   (b) True   (c) True
5. Sixteen; eight

SECTION 7-27

1. LPM_SHIFTREG   2. reg[].D = (reg[6..0], dat)   3. reg <= reg(6 DOWNTO 0) & dat;   4. Because the register may continue to receive clock edges during hold

SECTION 7-28

1. It can start in any state, but it will eventually reach the desired ring sequence.
2. Lines 11 and 12   3. Lines 13 and 14

SECTION 7-29

1. The reset input   2. The clock frequency and the delay value loaded into the counter   3. Synchronously   4. The output pulse width is very consistent.
5. The output pulse responds to the trigger edge immediately.   6. The state of the trigger on the current clock edge and its state on the previous clock edge.

INTEGRATED-CIRCUIT LOGIC FAMILIES

■ OUTLINE

■ OBJECTIVES

Upon completion of this chapter, you will be able to:

- Read and understand digital IC terminology as specified in manufacturers' data sheets.
- Compare the characteristics of standard TTL and the various TTL series.
- Determine the fan-out for a particular logic device.
- Use logic devices with open-collector outputs.
- Analyze circuits containing tristate devices.
- Compare the characteristics of the various CMOS series.
- Analyze circuits that use a CMOS bilateral switch to allow a digital system to control analog signals.
- Describe the major characteristics of and differences among TTL, ECL, MOS, and CMOS logic families.
- Cite and implement the various considerations that are required when interfacing digital circuits from different logic families.
- Use voltage comparators to allow a digital system to be controlled by analog signals.
- Use a logic pulser and a logic probe as digital circuit troubleshooting tools.

■ INTRODUCTION

As we described in Chapter 4, digital IC technology has advanced rapidly from small-scale integration (SSI), with fewer than 12 gates per chip; through medium-scale integration (MSI), with 12 to 99 equivalent gates per chip; on to large-scale and very large-scale integration (LSI and VLSI, respectively), which can have tens of thousands of gates per chip; and, most recently, to ultra-large-scale integration (ULSI), with over 100,000 gates per chip, and giga-scale integration (GSI), with 1 million or more gates.

Most of the reasons that modern digital systems use integrated circuits are obvious. ICs pack a lot more circuitry in a small package, so that the overall size of almost any digital system is reduced. The cost is dramatically reduced because of the economies of mass-producing large volumes of similar devices. Some of the other advantages are not so apparent.

ICs have made digital systems more reliable by reducing the number of external interconnections from one device to another. Before we had ICs, every circuit connection was from one discrete component (transistor, diode, resistor, etc.) to another. Now most of the connections are internal to the ICs, where they are protected from poor soldering, breaks or shorts in connecting paths on a circuit board, and other physical problems. ICs have also drastically reduced the amount of electrical power needed to perform a given function because their miniature circuitry typically requires less power than their discrete counterparts. In addition to the savings in power-supply costs, this reduction in power has also meant that a system does not require as much cooling.

There are some things that ICs cannot do. They cannot handle very large currents or voltages because the heat generated in such small spaces would cause temperatures to rise beyond acceptable limits. In addition, ICs cannot easily implement certain electrical devices such as inductors, transformers, and large capacitors. For these reasons, ICs are principally used to perform low-power circuit operations that are commonly called *information processing*. The operations that require high power levels or devices that cannot be integrated are still handled by discrete components.

With the widespread use of ICs comes the necessity to know and understand the electrical characteristics of the most common IC logic families. Remember that the various logic families differ in the major components that they use in their circuitry. TTL and ECL use *bipolar* transistors as their major circuit element; PMOS, NMOS, and CMOS use unipolar *MOSFET* transistors as their principal component. In this chapter, we will present the important characteristics of each of these IC families and their subfamilies. The most important point is understanding the nature of the input circuitry and output circuitry for each logic family. Once these are understood, you will be much better prepared to do analysis, troubleshooting, and some design of digital circuits that contain any combination of IC families. We will study the inner workings of devices in each family with the simplest circuitry that conveys the critical characteristics of all members of the family.

8-1   DIGITAL IC TERMINOLOGY

Although there are many digital IC manufacturers, much of the nomenclature and terminology is fairly standardized. The most useful terms are defined and discussed below.

Current and Voltage Parameters (See Figure 8-1)

- V_{IH}(min)—**High-Level Input Voltage.** The minimum voltage level required for a logical 1 at an *input*. Any voltage below this level will not be accepted as a HIGH by the logic circuit.

- V_{IL}(max)—**Low-Level Input Voltage.** The maximum voltage level required for a logic 0 at an *input*. Any voltage above this level will not be accepted as a LOW by the logic circuit.

- V_{OH}(min)—**High-Level Output Voltage.** The minimum voltage level at a logic circuit *output* in the logical 1 state under defined load conditions.

- V_{OL}(max)—**Low-Level Output Voltage.** The maximum voltage level at a logic circuit *output* in the logical 0 state under defined load conditions.

- I_{IH}—**High-Level Input Current.** The current that flows into an input when a specified high-level voltage is applied to that input.

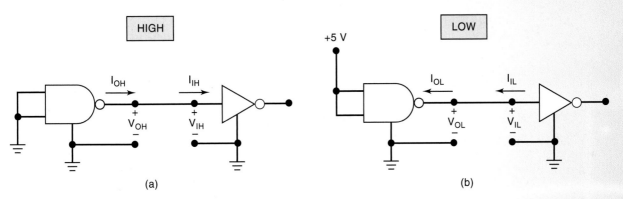

FIGURE 8-1   Currents and voltages in the two logic states.

- I_{IL}—**Low-Level Input Current.** The current that flows into an input when a specified low-level voltage is applied to that input.
- I_{OH}—**High-Level Output Current.** The current that flows from an output in the logical 1 state under specified load conditions.
- I_{OL}—**Low-Level Output Current.** The current that flows from an output in the logical 0 state under specified load conditions.

Note: The actual current directions may be opposite to those shown in Figure 8-1, depending on the logic family. All descriptions of current flow in this text refer to conventional current flow (from higher potential to lower potential). In keeping with the conventions of most data books, current flowing into a node or device is considered positive, and current flowing out of a node or device is considered negative.

Fan-Out

In general, a logic-circuit output is required to drive several logic inputs. Sometimes all ICs in the digital system are from the same logic family, but many systems have a mix of various logic families. The **fan-out** (also called *loading factor*) is defined as the *maximum* number of logic inputs that an output can drive reliably. For example, a logic gate that is specified to have a fan-out of 10 can drive 10 logic inputs. If this number is exceeded, the output logic-level voltages cannot be guaranteed. Obviously, fan-out depends on the nature of the input devices that are connected to an output. Unless a different logic family is specified as the load device, fan-out is assumed to refer to load devices of the same family as the driving output.

Propagation Delays

A logic signal always experiences a delay in going through a circuit. The two propagation delay times are defined as follows:

- t_{PLH}. Delay time in going from logical 0 to logical 1 state (LOW to HIGH)
- t_{PHL}. Delay time in going from logical 1 to logical 0 state (HIGH to LOW)

Figure 8-2 illustrates these propagation delays for an INVERTER. Note that t_{PHL} is the delay in the output's response as it goes from HIGH to LOW. It is measured between the 50 percent points on the input and output transitions. The t_{PLH} value is the delay in the output's response as it goes from LOW to HIGH.

FIGURE 8-2 Propagation delays.

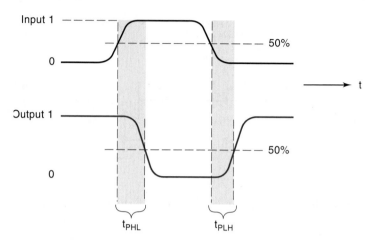

In some logic circuits, t_{PHL} and t_{PLH} are not the same value, and both will vary depending on capacitive loading conditions. The values of propagation times are used as a measure of the relative speed of logic circuits. For example, a logic circuit with values of 10 ns is a faster logic circuit than one with values of 20 ns under specified load conditions.

Power Requirements

Every IC requires a certain amount of electrical power to operate. This power is supplied by one or more power-supply voltages connected to the power pin(s) on the chip labeled V_{CC} (for TTL) or V_{DD} (for MOS devices).

The amount of power that an IC requires is determined by the current, I_{CC} (or I_{DD}), that it draws from the V_{CC} (or V_{DD}) supply, and the actual power is the product $I_{CC} \times V_{CC}$. For many ICs, the current drawn from the supply varies depending on the logic states of the circuits on the chip. For example, Figure 8-3(a) shows a NAND chip where *all* of the gate *outputs* are HIGH. The current drain on the V_{CC} supply for this case is called I_{CCH}. Likewise, Figure 8-3(b) shows the current when *all* of the gate *outputs* are LOW. This current is called I_{CCL}. The values are always measured with the outputs open circuit (no load) because the size of the load will also have an effect on I_{CCH}.

FIGURE 8-3 I_{CCH} and I_{CCL}.

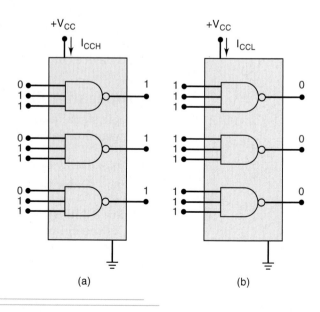

(a)   (b)

In some logic circuits, I_{CCH} and I_{CCL} will be different values. For these devices, the average current is computed based on the assumption that gate outputs are LOW half the time and HIGH half the time.

$$I_{CC}(\text{avg}) = \frac{I_{CCH} + I_{CCL}}{2}$$

This equation can be rewritten to calculate average power dissipated:

$$P_D(\text{avg}) = I_{CC}(\text{avg}) \times V_{CC}$$

Speed–Power Product

Digital IC families have historically been characterized for both speed and power. It is generally more desirable to have shorter gate propagation delays (higher speed) and lower values of power dissipation. As we shall soon see, the various logic families and subfamilies provide a wide spectrum of speed and power ratings. A common means for measuring and comparing the overall performance of an IC family is the **speed–power product**, which is obtained by multiplying the gate propagation delay by the gate power dissipation. For example, suppose that an IC family has an average propagation delay of 10 ns and an average power dissipation of 5 mW. The speed–power product is

$$10 \text{ ns} \times 5 \text{ mW} = 50 \times 10^{-12} \text{ watt-second}$$
$$= 50 \text{ picojoules (pJ)}$$

Note that when propagation delay is in nanoseconds and power is in milliwatts, the speed–power product is in picojoules.

Clearly, a low value of speed–power product is desirable. IC designers are continually striving to reduce the speed–power product by increasing the speed of an IC (i.e., reducing propagation delay) or by decreasing its power dissipation. Because of the nature of transistor switching circuits, it is difficult to do both.

Noise Immunity

Stray electric and magnetic fields can induce voltages on the connecting wires between logic circuits. These unwanted, spurious signals are called *noise* and can sometimes cause the voltage at the input to a logic circuit to drop below $V_{IH}(\text{min})$ or rise above $V_{IL}(\text{max})$, which could produce unpredictable operation. The noise immunity of a logic circuit refers to the circuit's ability to tolerate noise without causing spurious changes in the output voltage. A quantitative measure of noise immunity is called **noise margin** and is illustrated in Figure 8-4.

Figure 8-4(a) is a diagram showing the range of voltages that can occur at a logic-circuit output. Any voltages greater than $V_{OH}(\text{min})$ are considered a logic 1, and any voltages lower than $V_{OL}(\text{max})$ are considered a logic 0. Voltages in the indeterminate range should not appear at a logic circuit output under normal conditions. Figure 8-4(b) shows the voltage requirements at a logic circuit input. The logic circuit responds to any input greater than $V_{IH}(\text{min})$ as a logic 1, and it responds to voltages lower than $V_{IL}(\text{max})$ as a logic 0. Voltages in the indeterminate range produce an unpredictable response and should not be used.

The *high-state noise margin* V_{NH} is defined as

$$V_{NH} = V_{OH}(\text{min}) - V_{IH}(\text{min}) \tag{8-1}$$

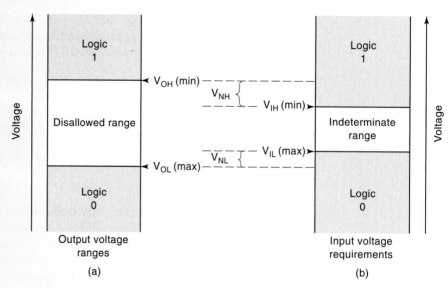

FIGURE 8-4 dc noise margins.

and is illustrated in Figure 8-4. V_{NH} is the difference between the lowest possible HIGH output and the minimum input voltage required for a HIGH. When a HIGH logic output is driving a logic-circuit input, any negative noise spikes greater than V_{NH} appearing on the signal line can cause the voltage to drop into the indeterminate range, where unpredictable operation can occur.

The *low-state noise margin* V_{NL} is defined as

$$V_{NL} = V_{IL}(\text{max}) - V_{OL}(\text{max}) \tag{8-2}$$

and it is the difference between the largest possible LOW output and the maximum input voltage required for a LOW. When a LOW logic output is driving a logic input, any positive noise spikes greater than V_{NL} can cause the voltage to rise into the indeterminate range.

EXAMPLE 8-1

The input/output voltage specifications for the standard TTL family are listed in Table 8-1. Use these values to determine the following.

(a) The maximum-amplitude noise spike that can be tolerated when a HIGH output is driving an input.

(b) The maximum-amplitude noise spike that can be tolerated when a LOW output is driving an input.

TABLE 8-1

Parameter	Min (V)	Typical (V)	Max (V)
V_{OH}	2.4	3.4	
V_{OL}		0.2	0.4
V_{IH}	2.0*		
V_{IL}			0.8*

*Normally only the minimum V_{IH} and maximum V_{IL} values are given.

Solution

(a) When an output is HIGH, it may be as low as $V_{OH}(min) = 2.4$ V. The minimum voltage that an input responds to as a HIGH is $V_{IH}(min) = 2.0$ V. A negative noise spike can drive the actual voltage below 2.0 V if its amplitude is greater than

$$V_{NH} = V_{OH}(min) - V_{IH}(min)$$
$$= 2.4 \text{ V} - 2.0 \text{ V} = 0.4 \text{ V}$$

(b) When an output is LOW, it may be as high as $V_{OL}(max) = 0.4$ V. The maximum voltage that an input responds to as a LOW is $V_{IL}(max) = 0.8$ V. A positive noise spike can drive the actual voltage above the 0.8-V level if its amplitude is greater than

$$V_{NL} = V_{IL}(max) - V_{OL}(max)$$
$$= 0.8 \text{ V} - 0.4 \text{ V} = 0.4 \text{ V}$$

Invalid Voltage Levels

For proper operation the input voltage levels to a logic circuit must be kept outside the indeterminate range shown in Figure 8-4(b); that is, they must be either lower than $V_{IL}(max)$ or higher than $V_{IH}(min)$. For the standard TTL specifications given in Example 8-1, this means that the input voltage must be less than 0.8 V or greater than 2.0 V. An input voltage between 0.8 and 2.0 V is considered an *invalid* voltage that will produce an unpredictable output response, and so must be avoided. In normal operation, a logic input voltage will not fall into the invalid region because it comes from a logic output that is within the stated specifications. However, when this logic output is malfunctioning or is being overloaded (i.e., its fan-out is being exceeded), then its voltage may be in the invalid region. Invalid voltage levels in a digital circuit can also be caused by power-supply voltages that are outside the acceptable range. It is important to know the valid voltage ranges for the logic family being used so that invalid conditions can be recognized when testing or troubleshooting.

Current-Sourcing and Current-Sinking Action

Logic families can be described according to how current flows between the output of one logic circuit and the input of another. Figure 8-5(a) illustrates **current-sourcing** action. When the output of gate 1 is in the HIGH state, it supplies a current I_{IH} to the input of gate 2, which acts essentially as a resistance to ground. Thus, the output of gate 1 is acting as a *source* of current for the gate 2 input. We can think of it as being like a faucet that acts as a *source* of water.

   Current-sinking action is illustrated in Figure 8-5(b). Here the input circuitry of gate 2 is represented as a resistance tied to $+V_{CC}$, the positive terminal of a power supply. When the gate 1 output goes to its LOW state, current will flow in the direction shown from the input circuit of gate 2 back through the output resistance of gate 1 to ground. In other words, in the LOW state, the circuit output that drives the input of gate 2 must be able to *sink* a current, I_{IL}, coming from that input. We can think of this as acting like a *sink* into which water is flowing.

FIGURE 8-5   Comparison of current-sourcing and current-sinking actions.

The distinction between current sourcing and current sinking is an important one, which will become more apparent as we examine the various logic families.

IC Packages

Developments and advancements in integrated circuits continue at a rapid pace. The same is true of IC packaging. There are various types of packages, which differ in physical size, the environmental and power-consumption conditions under which the device can be operated reliably, and the way in which the IC package is mounted to the circuit board. Figure 8-6 shows five representative IC packages.

The package in Figure 8-6(a) is the **DIP** (dual-in-line package), which has been around for a long time. Its pins (or leads) run down the two long sides of the rectangular package. The device shown is a 24-pin DIP. Note the presence of the notch on one end, which is used to locate pin 1. Some DIPs use a small dot on the top surface of the package to locate pin 1. The leads extend straight out of the DIP package so that the IC can be plugged into an IC socket or inserted into holes drilled through a printed circuit board. The spacing between pins **(lead pitch)** is typically 100 mils (a mil is a thousandth of an inch). DIP packages are still the most popular package for prototyping, breadboarding, and educational experimentation.

Nearly all new circuit boards that are produced using automated manufacturing equipment have moved away from using DIP packages whose leads are inserted through holes in the board. New manufacturing methods use **surface-mount technology**, which places an IC onto conductive pads on the surface of the board. They are held in place by a solder paste, and the entire board is heated to create a soldered connection. The precision of the placement machine allows for very tight lead spacing. The leads on these surface-mount packages are bent out from the plastic case, providing adequate surface area for the solder joint. The shape of these leads has resulted in the

FIGURE 8-6 Common IC packages. (Courtesy of Texas Instruments)

nickname of "gull-wing" package. Many different packages are available for surface-mount devices. Some of the most common packages used for logic ICs are shown in Figure 8-6. Table 8-2 gives the definition of each abbreviation along with its dimensions.

The need for more and more connections to a complex IC has resulted in another very popular package that has pins on all four sides of the chip. The PLCC has J-shaped leads that curl under the IC, as shown in Figure 8-6(c). These devices can be surface-mounted to a circuit board but can also be placed in a special PLCC socket. This is commonly used for components that are likely to need to be replaced for repair or upgrade, such as programmable logic devices or central processing units in computers. The QFP and TQFP packages have pins on all four sides in a gull-wing surface-mount package, as shown in Figure 8-6(d). The ball grid array (BGA) shown in Figure 8-6(e) is a surface-mount package that offers even more density. The pin grid array

TABLE 8-2 IC packages.

Abbreviation	Package Name	Height	Lead Pitch
DIP	Dual-in-line package	200 mils (5.1 mm)	100 mils (2.54 mm)
SOIC	Small outline integrated circuit	2.65 mm	50 mils (1.27 mm)
SSOP	Shrink small outline package	2.0 mm	0.65 mm
TSSOP	Thin shrink small outline package	1.1 mm	0.65 mm
TVSOP	Thin very small outline package	1.2 mm	0.4 mm
PLCC	Plastic leaded chip carrier	4.5 mm	1.27 mm
QFP	Quad flat pack	4.5 mm	0.635 mm
TQFP	Thin quad flat pack	1.6 mm	0.5 mm
LFBGA	Low-profile fine-pitch ball grid array	1.5 mm	0.8 mm

(PGA) is a similar package that is used when components must be in a socket to allow easy removal. The PGA has a long pin instead of a contact ball (BGA) at each position in the grid.

The proliferation of small, handheld consumer equipment such as digital video cameras, cellular phones, computers (PDAs), portable audio systems, and other devices has created a need for logic circuits in very small packages. Logic gates are now available in individual surface-mount packages containing one, two, or three gates (1G, 2G, 3G, respectively). These devices may have as few as five or six pins (power, ground, two to three inputs, and an output) and take up less space than an individual letter on this page.

REVIEW QUESTIONS

1. Define each of the following: V_{OH}, V_{IL}, I_{OL}, I_{IH}, t_{PLH}, t_{PHL}, I_{CCL}, I_{CCH}.
2. *True or false:* If a logic circuit has a fan-out of 5, the circuit has five outputs.
3. *True or false:* The HIGH-stage noise margin is the difference between $V_{IH}(\text{min})$ and V_{CC}.
4. *True or false:* A logic family with $t_{pd}(\text{avg}) = 12$ ns and $P_D(\text{avg}) = 15$ mW has a greater speed–power product than one with 8 ns and 30 mW.
5. Describe the difference between current sinking and current sourcing.
6. Which IC package can be plugged into sockets?
7. Which package has leads bent under the IC?
8. How do surface-mount packages differ from DIPs?
9. Will a standard TTL device work with an input level of 1.7 V?

8-2 THE TTL LOGIC FAMILY

At this writing, many small- to medium-scale ICs (SSI and MSI) can still be obtained in the standard **TTL** technology series that has been available for over 30 years. This original series of devices and their descendants in the TTL family have had a tremendous influence on the characteristics of all logic devices today. TTL devices are still used as "glue" logic that connects the more complex devices in digital systems. They are also used as interface circuits to devices that require high current drive. Even though the bipolar TTL family as a whole is on the decline, we will begin our discussion of logic ICs with the devices that shaped digital technology.

The basic TTL logic circuit is the NAND gate, shown in Figure 8-7(a). Even though the standard TTL family is nearly obsolete, we can learn a great deal about the more current family members by studying the original circuitry in its simplest form. The characteristics of TTL inputs come from the multiple-emitter (diode junction) configuration of transistor Q_1. Forward biasing either (or both) of these diode junctions will turn on Q_1. Only when all junctions are reverse biased will the transistor be off. This *multiple-emitter* input transistor can have up to eight emitters for an eight-input NAND gate.

Also note that on the output side of the circuit, transistors Q_3 and Q_4 are in a **totem-pole** arrangement. The totem pole is made up of two transistor switches, Q_3 and Q_4. The job of Q_3 is to connect V_{CC} to the output, making a logic HIGH. The job of Q_4 is to connect the output to ground, making a logic LOW. As we will see shortly, in normal operation, either Q_3 or Q_4 will be conducting, depending on the logic state of the output.

(a)   (b)

FIGURE 8-7   (a) Basic TTL NAND gate; (b) diode equivalent for Q_1.

Circuit Operation—LOW State

Although this circuit looks extremely complex, we can simplify its analysis somewhat by using the diode equivalent of the multiple-emitter transistor Q_1, as shown in Figure 8-7(b). Diodes D_2 and D_3 represent the two E–B junctions of Q_1, and D_4 is the collector-base (C–B) junction. In the following analysis, we will use this representation for Q_1.

First, let's consider the case where the output is LOW. Figure 8-8(a) shows this situation with inputs A and B both at +5 V. The +5 V at the cathodes of D_2 and D_3 will turn these diodes off, and they will conduct almost no current. The +5 V supply will push current through R_1 and D_4 into the base of Q_2, which turns on. Current from Q_2's emitter will flow into the base of Q_4 and turn Q_4 on. At the same time, the flow of Q_2 collector current produces a voltage drop across R_2 that reduces Q_2's collector voltage to a low value that is insufficient to turn Q_3 on.

The voltage at Q_2's collector is shown as approximately 0.8 V. This is because Q_2's emitter is at 0.7 V relative to ground due to Q_4's E–B forward voltage, and Q_2's collector is at 0.1 V relative to its emitter due to V_{CE}(sat). This 0.8 V at Q_3's base is not enough to forward-bias both Q_3's E–B junction and diode D_1. In fact, D_1 is needed to keep Q_3 off in this situation.

With Q_4 on, the output terminal, X, will be at a very low voltage because Q_4's ON-state resistance will be low (1 to 25 Ω). Actually, the output voltage,

Input conditions		Output conditions
A and B are both HIGH ($\geqslant$ 2 V)		Q_3 OFF
Input currents are very low I_{IH} = 10 μA		Q_4 ON so that V_X is LOW ($\leqslant$ 0.4 V)

(a)   LOW output

Input conditions		Output conditions
A or B or both are LOW ($\leqslant$ 0.8 V)		Q_4 OFF
Current flows back to ground through LOW input terminal. I_{IL} = 1.1 mA		Q_3 acts as emitter-follower and $V_{OH} \geqslant$ 2.4 V, typically 3.6 V

(b)   HIGH output

FIGURE 8-8   TTL NAND gate in its two output states.

V_{OL}, will depend on how much collector current Q_4 conducts. With Q_3 off, there is no current coming from the +5 V terminal through R_4. As we shall see, Q_4's collector current will come from the TTL inputs that terminal X is connected to.

It is important to note that the HIGH inputs at A and B will have to supply only a very small diode leakage current. Typically, this current I_{IH} is only around 10 μA at room temperature.

Circuit Operation—HIGH State

Figure 8-8(b) shows the situation where the circuit output is HIGH. This situation can be produced by connecting either or both inputs LOW. Here, input B is connected to ground. This will forward-bias D_3 so that current will flow from the +5 V source terminal, through R_1 and D_3, and through terminal B to ground. The forward voltage across D_3 will hold point Y at approximately 0.7 V. This voltage is not enough to forward-bias D_4 and the E–B junction of Q_2 sufficiently for conduction.

With Q_2 off, there is no base current for Q_4, and it turns off. Because there is no Q_2 collector current, the voltage at Q_3's base will be large enough to forward-bias Q_3 and D_1, so that Q_3 will conduct. Actually, Q_3 acts as an emitter follower because output terminal X is essentially at its emitter. With no load connected from point X to ground, V_{OH} will be around 3.4 to 3.8 V because two 0.7-V diode drops (E–B of Q_3, and D_1) subtract from the 5 V applied to Q_3's base. This voltage will decrease under load because the load will draw emitter current from Q_3, which draws base current through R_2, thereby increasing the voltage drop across R_2.

It's important to note that there is a substantial current flowing back through input terminal B to ground when B is held LOW. This current, I_{IL}, is determined by the value of resistor R_1, which will vary from series to series. For standard TTL, it is about 1.1 mA. The LOW B input acts as a *sink* to ground for this current.

Current-Sinking Action

A TTL output acts as a current sink in the LOW state because it *receives* current from the input of the gate that it is driving. Figure 8-9 shows one TTL gate driving the input of another gate (the load) for both output voltage states. In the output LOW state situation depicted in Figure 8-9(a), transistor Q_4 of the driving gate is on and essentially "shorts" point X to ground. This LOW voltage at X forward-biases the emitter–base junction of Q_1, and current flows, as shown, back through Q_4. Thus, Q_4 is performing a current-sinking action that derives its current from the input current (I_{IL}) of the load gate. We will often refer to Q_4 as the **current-sinking transistor** or as the **pull-down transistor** because it brings the output voltage down to its LOW state.

Current-Sourcing Action

A TTL output acts as a current source in the HIGH state. This is shown in Figure 8-9(b), where transistor Q_3 is supplying the input current, I_{IH}, required by the Q_1 transistor of the load gate. As stated above, this current is a small reverse-bias leakage current (typically 10 μA). We will often refer to Q_3 as the **current-sourcing transistor** or **pull-up transistor**. In some of the more modern TTL series, the pull-up circuit is made up of two transistors, rather than a transistor and diode.

FIGURE 8-9   (a) When the TTL output is in the LOW state, Q_4 acts as a current sink, deriving its current from the load. (b) In the output HIGH state, Q_3 acts as a current source, providing current to the load gate.

Totem-Pole Output Circuit

Several points should be mentioned concerning the totem-pole arrangement of the TTL output circuit, as shown in Figure 8-9, because it is not readily apparent why it is used. The same logic can be accomplished by eliminating Q_3 and D_1 and connecting the bottom of R_4 to the collector of Q_4. But this arrangement would mean that Q_4 would conduct a fairly heavy current in its saturation state (5 V/130 $\Omega \approx$ 40 mA). With Q_3 in the circuit, there will be no current through R_4 in the output LOW state. This is important because it keeps the circuit power dissipation down.

Another advantage of this arrangement occurs in the output HIGH state. Here Q_3 is acting as an emitter follower with its associated low output impedance (typically 10 Ω). This low output impedance provides a short time constant for charging up any capacitive load on the output. This action (commonly called *active pull-up*) provides very fast rise-time waveforms at TTL outputs.

A disadvantage of the totem-pole output arrangement occurs during the transition from LOW to HIGH. Unfortunately, Q_4 turns off more slowly than Q_3 turns on, and so there is a period of a few nanoseconds during which both transistors are conducting and a relatively large current (30 to 40 mA) will be drawn from the 5-V supply. This can present a problem that will be examined later.

TTL NOR Gate

Figure 8-10 shows the internal circuit for a TTL NOR gate. We will not go through a detailed analysis of this circuit, but it is important to note how it

FIGURE 8-10   TTL NOR gate
circuit.

compares to the NAND circuit of Figure 8-8. On the input side, we can see that the NOR circuit *does not use a multiple-emitter* transistor; instead, each input is applied to the emitter of a separate transistor. On the output side, the NOR circuit uses the same totem-pole arrangement as the NAND circuit.

Summary

All TTL circuits have a similar structure. NAND and AND gates use multiple-emitter transistor or multiple diode junction inputs; NOR and OR gates use separate input transistors. In either case, the input will be the cathode (N-region) of a P–N junction, so that a HIGH input voltage will turn off the junction and only a small leakage current (I_{IH}) will flow. Conversely, a LOW input voltage turns on the junction, and a relatively large current (I_{IL}) will flow back through the signal source. Most, but not all, TTL circuits will have some type of totem-pole output configuration. There are some exceptions that will be discussed later.

REVIEW QUESTIONS

1. *True or false:* A TTL output acts as a current sink in the LOW state.
2. In which TTL input state does the largest amount of input current flow?
3. State the advantages and disadvantages of a totem-pole output.
4. Which TTL transistor is the pull-up transistor in the NAND circuit?
5. Which TTL transistor is the pull-down transistor in the NOR circuit?
6. How does the TTL NOR circuit differ from the NAND circuit?

8-3   TTL DATA SHEETS

In 1964, Texas Instruments Corporation introduced the first line of standard TTL ICs. The 54/74 series, as it is called, has been one of the most widely used IC logic families. We will simply refer to it as the 74 series because the major difference between the 54 and 74 versions is that devices in the 54 series can operate over a wider range of temperatures and power-supply voltages. Many semiconductor manufacturers still produce TTL ICs. Fortunately, they all use the same numbering system, so that the basic IC number is the same from one

EXAMPLE 8-4

Which TTL series can drive the most device inputs of the same series?

Solution

The 74AS series has the highest fan-out (40), which means that a 74AS00 NAND gate can drive 40 inputs of other 74AS devices. If we want to determine the number of inputs of a *different* TTL series that an output can drive, we will need to know the input and output currents of the two series. This will be dealt with in the next section.

REVIEW QUESTIONS

1. (a) Which TTL series is the best at high frequencies?
   (b) Which TTL series has the largest HIGH-state noise margin?
   (c) Which series has essentially become obsolete in new designs?
   (d) Which series uses a special diode to reduce switching time?
   (e) Which series would be best for a battery-powered circuit operating at 10 MHz?

2. Assuming the same cost for each, why should you choose to use a 74ALS193 counter over a 74LS193 or a 74AS193 in a circuit operating from a 40-MHz clock?

3. Identify the pull-up and pull-down transistors for the 74S circuit in Figure 8-12.

8-5 TTL LOADING AND FAN-OUT

It is important to understand what determines the fan-out or load drive capability of an IC output. Figure 8-13(a) shows a standard TTL output in the LOW state connected to drive several standard TTL inputs. Transistor Q_4 is on and is acting as a current sink for an amount of current I_{OL} that is the sum of the I_{IL} currents from each input. In its ON state, Q_4's collector–emitter resistance is very small, but it is not zero, and so the current I_{OL} will produce a voltage drop V_{OL}. This voltage must not exceed the $V_{OL}(\text{max})$ limit of the IC, which limits the maximum value of I_{OL} and thus the number of loads that can be driven.

To illustrate, suppose that the ICs are in the 74 series and each I_{IL} is 1.6 mA. From Table 8-6, we see that the 74 series has $V_{OL}(\text{max}) = 0.4$ V and $V_{IL}(\text{max}) = 0.8$ V. Let's suppose further that Q_4 can sink up to 16 mA before its output voltage reaches $V_{OL}(\text{max}) = 0.4$ V. This means that it can sink the current from up to 16 mA/1.6 mA = 10 loads. If it is connected to more than 10 loads, its I_{OL} will increase and cause V_{OL} to increase above 0.4 V. This is usually undesirable because it reduces the noise margin at the IC inputs [remember, $V_{NL} = V_{IL}(\text{max}) - V_{OL}(\text{max})$]. In fact, if V_{OL} rises above $V_{IL}(\text{max}) = 0.8$ V, it will be in the indeterminate range.

A similar situation occurs in the HIGH state depicted in Figure 8-13(b). Here, Q_3 is acting as an emitter follower that is sourcing (supplying) a total current I_{OH} that is the sum of the I_{IH} currents of the different TTL inputs. If too many loads are being driven, this current I_{OH} will become large enough to cause the voltage drops across R_2, Q_3's emitter–base junction, and D_1 to bring

FIGURE 8-13 Currents when a TTL output is driving several inputs.

V_{OH} below V_{OH}(min). This too is undesirable because it reduces the HIGH-state noise margin and could even cause V_{OH} to go into the indeterminate range.

What this all means is that a TTL output has a limit, I_{OL}(max), on how much current it can sink in the LOW state. It also has a limit, I_{OH}(max), on how much current it can source in the HIGH state. These output current limits must not be exceeded if the output voltage levels are to be maintained within their specified ranges.

Determining the Fan-Out

To determine how many different inputs an IC output can drive, you need to know the current drive capability of the output [i.e., I_{OL}(max) and I_{OH}(max)] and the current requirements of each input (i.e., I_{IL} and I_{IH}). This information is always presented in some form on the manufacturer's IC data sheet. The following examples will illustrate one type of situation.

EXAMPLE 8-5

How many 74ALS00 NAND gate inputs can be driven by a 74ALS00 NAND gate output?

Solution

We will consider the LOW state first as depicted in Figure 8-14. Refer to the 74ALS00 data sheet in Figure 8-11 and find

$$I_{OL}(\text{max}) = 8 \text{ mA}$$
$$I_{IL}(\text{max}) = 0.1 \text{ mA}$$

This says that a 74ALS00 output can sink a maximum of 8 mA and that each 74ALS00 input will source a maximum of 0.1 mA back through the driving

FIGURE 8-14   Example 8-5.

*All gates are 74ALS00 NAND gates.

gate's output. Thus, the number of inputs that can be driven in the LOW state is obtained as

$$\text{fan-out (LOW)} = \frac{I_{OL}(\text{max})}{I_{IL}(\text{max})}$$
$$= \frac{8 \text{ mA}}{0.1 \text{ mA}}$$
$$= 80$$

(*Note:* The entry for I_{IL} is actually −0.1 mA. The negative sign is used to indicate that this current flows *out of* the input terminal; we can ignore this sign for our purposes here.) The HIGH state is analyzed in the same manner. Refer to the data sheet to find values for I_{OH} and I_{IH}, ignoring any negative signs.

$$I_{OH}(\text{max}) = 0.4 \text{ mA} = 400 \text{ } \mu\text{A}$$
$$I_{IH}(\text{max}) = 20 \text{ } \mu\text{A}$$

Thus, the number of inputs that can be driven in the HIGH state is

$$\text{fan-out (HIGH)} = \frac{I_{OH}(\text{max})}{I_{IH}(\text{max})}$$
$$= \frac{400 \text{ } \mu\text{A}}{20 \text{ } \mu\text{A}}$$
$$= 20$$

If fan-out (LOW) and fan-out (HIGH) are not the same, as will sometimes occur, the fan-out is chosen as the smaller of the two. Thus, the 74ALS00 NAND gate can drive up to 20 other 74ALS00 NAND gates.

EXAMPLE 8-6

Refer to the data sheet on the TI CD ROM (or Table 8-7) and determine how many 74AS20 NAND gates can be driven by the output of another 74AS20.

Solution

The 74AS20 data sheet gives the following values:

$$I_{OH}(\text{max}) = 2 \text{ mA}$$
$$I_{OL}(\text{max}) = 20 \text{ mA}$$
$$I_{IH}(\text{max}) = 20 \text{ } \mu\text{A}$$
$$I_{IL}(\text{max}) = 0.5 \text{ mA}$$

Considering the HIGH state first, we have

$$\text{fan-out (HIGH)} = \frac{2 \text{ mA}}{20 \text{ } \mu\text{A}} = 100$$

For the LOW state, we have

$$\text{fan-out (LOW)} = \frac{20 \text{ mA}}{0.5 \text{ mA}} = 40$$

In this case, the overall fan-out is chosen to be 40 because it is the lower of the two values. Thus, one 74AS20 can drive 40 other 74AS20 inputs.

In older equipment, you will notice that most of the logic ICs were often chosen from the same logic family. In today's digital systems, there is much more likely to be a combination of various logic families. Consequently, loading and fan-out calculations are not as straightforward as they once were. A good method for determining the loading of any digital output is as follows:

Step 1. Add the I_{IH} for all inputs connected to an output. This sum must be less than the output's I_{OH} specification.

Step 2. Add the I_{IL} for all inputs connected to an output. This sum must be less than the output's I_{OL} specification.

Table 8-7 shows the limiting specifications for input and output currents in simple logic gates of the various TTL families. Notice that some of the current values are given as negative numbers. This convention is used to show the direction of current flow. Positive values indicate current flowing into the specified node, whether it is an input or an output. Negative values indicate current flowing out of the specified node. Consequently, all I_{OH} values are negative as current flows out of the output (sourcing current), and all I_{OL} values are positive as load current flows into the output pin on its way to ground (sinking current). Likewise, I_{IH} is positive, while I_{IL} is negative. When calculating loading and fan-out as described above, you should ignore these signs.

TABLE 8-7 Current ratings of TTL series logic gates.[*]

	Outputs		Inputs	
TTL Series	I_{OH}	I_{OL}	I_{IH}	I_{IL}
74	−0.4 mA	16 mA	40 μA	−1.6 mA
74S	−1 mA	20 mA	50 μA	−2 mA
74LS	−0.4 mA	8 mA	20 μA	−0.4 mA
74AS	−2 mA	20 mA	20 μA	−0.5 mA
74ALS	−0.4 mA	8 mA	20 μA	−0.1 mA
74F	−1 mA	20 mA	20 μA	−0.6 mA

[*] Some devices may have different input or output current ratings. Always consult the data sheet.

EXAMPLE 8-7

A 74ALS00 NAND gate output is driving three 74S gate inputs and one 7406 input. Determine if there is a loading problem.

Solution

1. Add all of the I_{IH} values:

$$3 \cdot (I_{IH} \text{ for 74S}) + 1 \cdot (I_{IH} \text{ for 74})$$
$$\text{Total} = 3 \cdot (50 \ \mu A) + 1 \cdot (40 \ \mu A) = 190 \ \mu A$$

The I_{OH} for the 74ALS output is 400 μA (max), which is greater than the sum of the loads (190 μA). This poses no problem when the output is HIGH.

2. Add all of the I_{IL} values:

$$3 \cdot (I_{IL} \text{ for 74S}) + 1 \cdot (I_{IL} \text{ for 74})$$
$$\text{Total} = 3 \cdot (2 \text{ mA}) + 1 \cdot (1.6 \text{ mA}) = 7.6 \text{ mA}$$

The I_{OH} for the 74ALS output is 8 mA (max), which is greater than the sum of the loads (7.6 mA). This poses no problem when the output is LOW.

EXAMPLE 8-8

The 74ALS00 NAND gate output in Example 8-7 needs to be used to drive some 74ALS inputs in addition to the load inputs described in Example 8-7. How many additional 74ALS inputs could the output drive without being overloaded?

Solution

From the calculations of Example 8-7, only in the LOW state are we close to being overloaded. A 74ALS input has an I_{IL} of 0.1 mA. The maximum sink current (I_{OL}) is 8 mA, and the load current is 7.6 mA (as calculated in Example 8-7). The additional current that the output can sink is found by

$$\text{additional current} = I_{OLmax} - \text{sum of loads} (I_{IL})$$
$$= 8 \text{ mA} - 7.6 \text{ mA} = 0.4 \text{ mA}$$

This output can drive up to four more 74ALS inputs that have an I_{IL} of 0.1 mA.

EXAMPLE 8-9

The output of a 74AS04 inverter is providing the CLEAR signal to a parallel register made up of 74AS74 D flip-flops. What is the maximum number of FF *CLR* inputs that this gate can drive?

Solution

The input specifications for flip-flop inputs are not always the same as those for a logic gate input in the same family. Refer to the 74AS74 data sheet on the TI CD ROM. The clock and *D* inputs are similar to the gate inputs in Table 8-7. However, the *PRE* and *CLR* inputs have specifications of $I_{IH} = 40 \ \mu A$ and $I_{IL} = 1.8$ mA. The 74AS04 has specifications of $I_{OH} = 2$ mA and $I_{OL} = 20$ mA.

$$\text{Maximum number of inputs (HIGH)} = 2 \text{ mA}/40 \ \mu A = 50$$
$$\text{Maximum number of inputs (LOW)} = 20 \text{ mA}/1.8 \text{ mA} = 11.11$$

We must limit the fan-out to 11 *CLR* inputs.

REVIEW QUESTIONS

1. What factors determine the I_{OL}(max) rating of a device?
2. How many 7407 inputs can a 74AS chip drive?
3. What can happen if a TTL output is connected to more gate inputs than it is rated to handle?
4. How many 74S112 $\overline{CP}$ inputs can be driven by a 74LS04 output? By a 74F00 output?

8-6 OTHER TTL CHARACTERISTICS

Several other characteristics of TTL logic must be understood if one is to use TTL intelligently in a digital-system application.

Unconnected Inputs (Floating)

Any input to a TTL circuit that is left disconnected (open) acts exactly like a logical 1 applied to that input because in either case the emitter–base junction or diode at the input will not be forward-biased. This means that on *any* TTL IC, *all* of the inputs are 1s if they are not connected to some logic signal or to ground. When an input is left unconnected, it is said to be **floating**.

Unused Inputs

Frequently, not all of the inputs on a TTL IC are being used in a particular application. A common example is when not all the inputs to a logic gate are needed for the required logic function. For example, suppose that we needed the logic operation $\overline{AB}$ and we were using a chip that had a three-input NAND gate. The possible ways of accomplishing this are shown in Figure 8-15.

In Figure 8-15(a), the unused input is left disconnected, which means that it acts as a logical 1. The NAND gate output is therefore $x = \overline{A \cdot B \cdot 1} = \overline{A \cdot B}$, which is the desired result. Although the logic is correct, it is highly undesirable to leave an input disconnected because it will act like an antenna, which is liable to pick up stray radiated signals that could cause the gate to operate improperly. A better technique is shown in Figure 8-15(b). Here, the unused input is connected to +5 V through a 1-kΩ resistor, so that the logic level is a 1. The 1-kΩ resistor is simply for current protection of the emitter–base junctions of the gate inputs in case of spikes on the power-supply line. This same technique can be used for AND gates because a 1 on an unused input will not affect the output. As many as 30 unused inputs can share the same 1-kΩ resistor tied to V_{CC}.

FIGURE 8-15   Three ways to handle unused logic inputs.

A third possibility is shown in Figure 8-15(c), where the unused input is tied to a used input. This is satisfactory provided that the circuit driving input B is not going to have its fan-out exceeded. This technique can be used for *any* type of gate. For OR gates and NOR gates, the unused inputs cannot be left disconnected or tied to +5 V because this would produce a constant-output logic level (1 for OR, 0 for NOR) regardless of the other inputs. Instead, for these gates, the unused inputs must either be connected to ground (0 V) for a logic 0 or be tied to a used input, as in Figure 8-15(c).

Tied-Together Inputs

When two (or more) TTL inputs on the same gate are connected together to form a common input, as in Figure 8-15(c), the common input will generally represent a load that is the sum of the load current rating of each individual input. The only exception is for NAND and AND gates. For these gates, the LOW-state input load *will be the same as a single input* no matter how many inputs are tied together.

To illustrate, assume that each input of the three-input NAND gate in Figure 8-15(c) is rated at 0.5 mA for I_{IL} and 20 μA for I_{IH}. The common input B will therefore represent an input load of 40 μA in the HIGH state but only 0.5 mA in the LOW state. The same would be true if this were an AND gate. If it were an OR or a NOR gate, the common B input would present an input load 40 μA in the HIGH state and 1 mA in the LOW state.

The reason for this characteristic can be found by looking back at the circuit diagram of the TTL NAND gate in Figure 8-8(b). The current I_{IL} is limited by the resistance R_1. Even if inputs A and B were tied together and grounded, this current would not change; it would merely divide and flow through the parallel paths provided by diodes D_2 and D_3. The situation is different for OR and NOR gates because they do not use multiple-emitter transistors but rather have a separate input transistor for each input, as we saw in Figure 8-10.

EXAMPLE 8-10

Determine the load that the X output is driving in Figure 8-16. Assume that each gate is a 74LS series device with $I_{IH} = 20$ μA and $I_{IL} = 0.4$ mA.

Loading on gate 1 output				
HIGH			LOW	
Load Current	Gate		Load Current	Gate
40 μA	2		0.4 mA	2
20 μA	3		0.4 mA	3
60 μA	4		1.2 mA	4
120 μA	Total		2.0 mA	Total

FIGURE 8-16   Example 8-10.

Solution

The loading on the output of gate 1 is equivalent to six 74LS input loads in the HIGH state but only five 74LS input loads in the LOW state because the NAND gate represents only a single input load in the LOW state.

FIGURE 8-17

Biasing TTL Inputs Low

Occasionally, the situation arises where a TTL input must be held normally LOW and then caused to go HIGH by the actuation of a mechanical switch. This situation is illustrated in Figure 8-17 for the input to a one-shot. This OS triggers on a positive transition that occurs when the switch is momentarily closed. The resistor R serves to keep the T input LOW while the switch is open. Care must be taken to keep the value of R low enough so that the voltage developed across it by the current I_{IL} that flows out of the OS input to ground will not exceed $V_{IL}(max)$. Thus, the largest value of R is given by

$$I_{IL} \times R_{max} = V_{IL}(max)$$
$$R_{max} = \frac{V_{IL}(max)}{I_{IL}} \qquad (8\text{-}3)$$

R must be kept below this value to ensure that the OS input will be at an acceptable LOW level while the switch is open. The minimum value of R is determined by the current drain on the 5-V supply when the switch is closed. In practice, this current drain should be minimized by keeping R just slightly below R_{max}.

EXAMPLE 8-11

Determine an acceptable value for R if the OS is a 74LS TTL IC with an I_{IL} input rating of 0.4 mA.

Solution

The value of I_{IL} will be a maximum of 0.4 mA. This maximum value should be used to calculate R_{max}. From Table 8-6, $V_{IL}(max) = 0.8$ V for the 74LS series. Thus, we have

$$R_{max} = \frac{0.8 \text{ V}}{0.4 \text{ mA}} = 2000 \ \Omega$$

A good choice here would be $R = 1.8$ kΩ, a standard resistor value.

Current Transients

TTL logic circuits suffer from internally generated current transients or spikes because of the totem-pole output structure. When the output is switching from the LOW state to the HIGH state (see Figure 8-18), the two output transistors are changing states: Q_3 OFF to ON, and Q_4 ON to OFF. Because Q_4 is changing from the saturated condition, it will take longer than

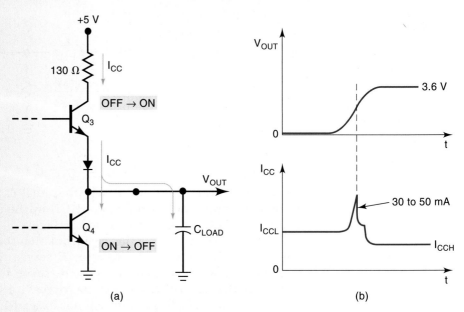

FIGURE 8-18   A large current spike is drawn from V_{CC} when a totem-pole output switches from LOW to HIGH.

Q_3 to switch states. Thus, there is a short interval of time (about 2 ns) during the switching transition when both transistors are conducting and a relatively large surge of current (30 to 50 mA) is drawn from the +5 V supply. The duration of this current transient is extended by the effects of any load capacitance on the circuit output. This capacitance consists of stray wiring capacitance and the input capacitance of any load circuits and must be charged up to the HIGH-state output voltage. This overall effect can be summarized as follows:

> **Whenever a totem-pole TTL output goes from LOW to HIGH, a high-amplitude current spike is drawn from the V_{CC} supply.**

In a complex digital circuit or system, there may be many TTL outputs switching states at the same time, each one drawing a narrow spike of current from the power supply. The accumulative effect of all of these current spikes will be to produce a voltage spike on the common V_{CC} line, mostly due to the distributed inductance on the supply line [remember: $V = L(di/dt)$ for inductance, and di/dt is very large for a 2-ns current spike]. This voltage spike can cause serious malfunctions during switching transitions unless some type of filtering is used. The most common technique uses small radio-frequency capacitors connected from V_{CC} to GROUND essentially to "short out" these high-frequency spikes. This is called **power-supply decoupling**.

It is standard practice to connect a 0.01-μF or 0.1-μF low-inductance, ceramic disk capacitor between V_{CC} and ground near each TTL IC on a circuit board. The capacitor leads are kept very short to minimize series inductance.

In addition, it is standard practice to connect a single large capacitor (2 to 20 μF) between V_{CC} and ground on each board to filter out relatively low-frequency variations in V_{CC} caused by the large changes in I_{CC} levels as outputs switch states.

1. What will be the logic output of a TTL NAND gate that has all of its inputs unconnected?
2. What are two acceptable ways to handle unused inputs to an AND gate?
3. Repeat question 2 for a NOR gate.
4. *True or false:* When NAND gate inputs are tied together, they are always treated as a single load on the signal source.
5. What is power-supply decoupling? Why is it used?

8-7 MOS TECHNOLOGY

MOS (metal-oxide-semiconductor) technology derives its name from the basic MOS structure of a metal electrode over an oxide insulator over a semiconductor substrate. The transistors of MOS technology are field-effect transistors called **MOSFETs**. This means that the electric *field* on the *metal* electrode side of the *oxide* insulator has an *effect* on the resistance of the substrate. Most of the MOS digital ICs are constructed entirely of MOSFETs and no other components.

The chief advantages of the MOSFET are that it is relatively simple and inexpensive to fabricate, it is small, and it consumes very little power. The fabrication of MOS ICs is approximately one-third as complex as the fabrication of bipolar ICs (TTL, ECL, etc.). In addition, MOS devices occupy much less space on a chip than do bipolar transistors. More important, MOS digital ICs normally do not use the IC resistor elements that take up so much of the chip area of bipolar ICs.

All of this means that MOS ICs can accommodate a much larger number of circuit elements on a single chip than bipolar ICs. This advantage is illustrated by the fact that MOS ICs have dominated bipolar ICs in the area of large-scale integration (LSI, VLSI). The high packing density of MOS ICs makes them especially well suited for complex ICs such as microprocessor and memory chips. Improvements in MOS IC technology have led to devices that are faster than 74, 74LS, and 74ALS TTL with comparable current drive characteristics. Consequently, MOS devices (specifically CMOS) have also become dominant in the SSI and MSI market. The 74AS TTL family is still as fast as the best CMOS devices, but at the price of much greater power dissipation.

The principal disadvantage of MOS devices is their susceptibility to static-electricity damage. Although this can be minimized by proper handling procedures, TTL is still more durable for laboratory experimentation. Consequently, you are likely to see TTL devices used in education as long as they are available.

The MOSFET

There are presently two general types of MOSFETs: *depletion* and *enhancement*. MOS digital ICs use enhancement MOSFETs exclusively, and so only this type will be considered in the following discussion. Furthermore, we will concern ourselves only with the operation of these MOSFETs as on/off switches.

Figure 8-19 shows the schematic symbols for the N-channel and P-channel enhancement MOSFETs, where the direction of the arrow indicates either

FIGURE 8-19   Schematic symbols for enhancement MOSFETs.

P- or N-channel. The symbols show a broken line between the *source* and the *drain* to indicate that there is *normally* no conducting channel between these electrodes. The symbol also shows a separation between the *gate* and the other terminals to indicate the very high resistance (typically around $10^{12}\ \Omega$) of the oxide layer between the gate and the channel, which is formed in the substrate.

Basic MOSFET Switch

Figure 8-20 shows the switching operation of an N-channel *MOSFET*. For the N-channel device, the drain is always biased positive relative to the source. The gate-to-source voltage V_{GS} is the input voltage, which is used to control the resistance between drain and source (i.e., the channel resistance) and therefore determines whether the device is on or off.

When $V_{GS} = 0$ V, there is no conductive channel between source and drain, and the device is off, as shown in Figure 8-20(b). Typically the channel resistance in this OFF state is $10^{10}\ \Omega$, which for most purposes is an *open circuit*. The MOSFET will remain off as long as V_{GS} is zero or negative. As V_{GS} is made positive (gate positive relative to source), a threshold voltage (V_T) is reached, at which point a conductive channel begins to form between source and drain. Typically $V_T = +1.5$ V for N-MOSFET, and so any $V_{GS} \geq 1.5$ V will cause the MOSFET to conduct. Generally, a value of V_{GS} much larger than V_T is used to turn on the MOSFET more completely. As shown in Figure 8-20(c), when $V_{GS} = +5$ V, the channel resistance between source and drain has dropped to a value of $R_{ON} = 1000\ \Omega$.

In essence, then, the N-MOSFET will switch from a very high resistance to a low resistance as the gate voltage switches from a LOW voltage to a HIGH voltage. It is helpful simply to think of the MOSFET as a switch that is either opened or closed between source and drain.

FIGURE 8-20   N-channel MOSFET switching states.

The P-channel MOSFET operates in exactly the same manner as the N-channel except that it uses voltages of the opposite polarity. For P-MOSFETs, the drain is connected to $-V_{DD}$ so that it is biased negative relative to the source. To turn the P-MOSFET on, a negative voltage that exceeds V_T must be applied to the gate.

Table 8-8 summarizes the P- and N-channel switching characteristics.

TABLE 8-8

	Drain-to-Source Bias	Gate-to-Source Voltage (V_{GS}) Needed for Conduction	R_{ON} (Ω)	R_{OFF} (Ω)
P-channel	Negative	Typically more negative than -1.5 V	1000 (typical)	10^{10}
N-channel	Positive	Typically more positive than $+1.5$ V	1000 (typical)	10^{10}

8-8 DIGITAL MOSFET CIRCUITS

Digital circuits employing MOSFETs are broken down into three categories: (1) **P-MOS**, which uses *only* P-channel enhancement MOSFETs; (2) **N-MOS**, which uses *only* N-channel enhancement MOSFETs; and (3) **CMOS** (complementary MOS), which uses both P- and N-channel devices.

N-MOS circuits use an N-channel MOSFET as a switch and also implement all resistors using the channel resistance of a MOSFET with the gate connected to the drain (always turned ON). Figure 8-21(a) shows an N-MOS INVERTER circuit. Q_1 serves as a pull-up resistor of typically 100 kΩ, and Q_2 switches on and off depending on its gate voltage (V_{IN}). Figure 8-21(b) summarizes the two states. P-MOS circuits are similar but use P-channel MOSFETs. The simplicity of these circuits and their fabrication processes caused them to dominate the early LSI and VLSI markets. The speed and power advantages that current CMOS fabrication technology offers have made CMOS the leader across all levels of integration.

FIGURE 8-21 N-MOS INVERTER.

V_{IN}	Q_1	Q_2	$V_{OUT} = \overline{V_{IN}}$
0 V (logic 0)	$R_{ON} =$ 100 kΩ	$R_{OFF} =$ 10^{10} Ω	+5 V (logic 1)
+5 V (logic 1)	$R_{ON} =$ 100 kΩ	$R_{ON} =$ 1 kΩ	+0.05 V (logic 0)

(a)   (b)

FIGURE 8-24 CMOS NOR gate.

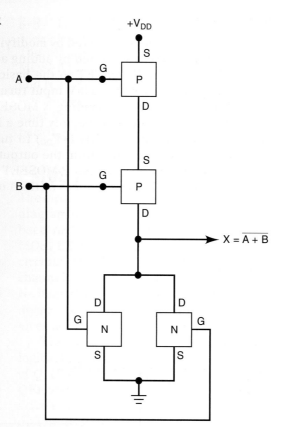

A	B	X
LOW	LOW	HIGH
LOW	HIGH	LOW
HIGH	LOW	LOW
HIGH	HIGH	LOW

1. How does CMOS internal circuitry differ from N-MOS?
2. How many P-channel MOSFETs are in a CMOS INVERTER?
3. How many MOSFETs are in a three-input CMOS NAND gate?

8-10 CMOS SERIES CHARACTERISTICS

The CMOS family of integrated circuits competes directly with TTL in the small- and medium-scale integration (SSI, MSI) areas. As CMOS technology has produced better and better performance characteristics, CMOS has gradually taken over the field that has been dominated by TTL for so long. TTL devices will be around for a long time, but more and more new equipment is using primarily CMOS logic circuits.

CMOS ICs provide not only all of the same logic functions that are available in TTL but also several special-purpose functions not provided by TTL. Several different CMOS series have been developed over time because manufacturers have sought to improve performance characteristics. Before we look at the various CMOS series, it will be helpful to define a few terms that are used when ICs from different families or series are to be used together or as replacements for one another.

- **Pin-compatible.** Two ICs are pin-compatible when their pin configurations are the same. For example, pin 7 on both ICs is GROUND, pin 1 on both is an input to the first INVERTER, and so on.

- **Functionally equivalent.** Two ICs are functionally equivalent when the logic functions they perform are exactly the same. For example, both contain four two-input NAND gates, or both contain six D flip-flops with positive-edge clock triggering.

■ **Electrically compatible.** Two ICs are electrically compatible when they can be connected directly to each other without taking any special measures to ensure proper operation.

4000/14000 Series

The oldest CMOS series is the 4000 series first introduced by RCA, and its functionally equivalent 14000 series from Motorola. Devices in the 4000/14000 series have very low power dissipation and can operate over a wide range of power-supply voltages (3 to 15 V). They are very slow compared to TTL and other CMOS series and have very low output current capabilities. They are not pin-compatible or electrically compatible with any TTL series. The 4000/14000 series devices are rarely used in new designs except when a special-purpose IC is available that is not available in other series.

74C Series

This CMOS series is pin-compatible with and functionally equivalent to TTL devices having the same number. For example, a 74C74 is a dual edge-triggered D flip-flop that has the same pin configuration as the TTL 7474 dual edge-triggered D flip-flop IC. Many but not all functions that are available in TTL are also available in this CMOS series. This makes it possible to replace some TTL circuits with an equivalent CMOS design. The performance characteristics of the 74C series are about the same as those of the 4000 series.

74HC/HCT (High-Speed CMOS)

This is an improved version of the 74C series, which has a tenfold increase in switching speed, comparable to that of the 74LS devices, and a much higher output current capability than that of the 74C. 74HC/HCT ICs are pin-compatible with and functionally equivalent to TTL ICs with the same device number. 74HCT devices are electrically compatible with TTL, but 74HC devices are not. This means, for example, that a 74HCT04 hex-INVERTER chip can replace a 74LS04 chip, and vice versa. It also means that a 74HCT IC can be connected directly to any TTL IC. The 74HC/HCT series has made the 74C series obsolete.

74AC/ACT (Advanced CMOS)

This series is often referred to as ACL for advanced CMOS logic. The series is functionally equivalent to the various TTL series but is *not* pin-compatible with TTL because the pin placements on 74AC or 74ACT chips have been chosen to improve noise immunity so that the device inputs are less sensitive to signal changes occurring on other IC pins. 74AC devices are not electrically compatible with TTL; 74ACT devices can be connected directly to TTL. ACL offers advantages over the HC series in noise immunity, propagation delay, and maximum clock speed.

Device numbering for this series differs slightly from TTL, 74C, and 74HC/HCT numbering. It uses a five-digit device number beginning with the digits 11. The following examples illustrate:

$$74AC11 \; \boxed{004} \; \equiv 74HC \; \boxed{04}$$
$$74ACT11 \; \boxed{293} \; \equiv 74HCT \; \boxed{293}$$

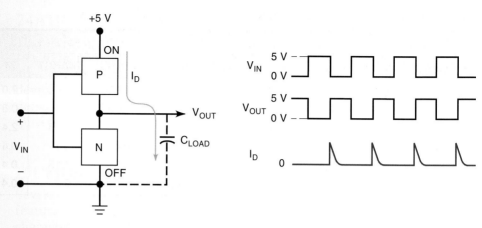

FIGURE 8-25   Current spikes are drawn from the V_{DD} supply each time the output switches from LOW to HIGH. This is due mainly to the charging current of the load capacitance.

tance consists of the combined input capacitances of any loads being driven and the device's own output capacitance. These narrow spikes of current are supplied by V_{DD} and can have a typical amplitude of 5 mA and a duration of 20 to 30 ns. Clearly, as the switching frequency increases, there will be more of these current spikes occurring per second, and the average current drawn from V_{DD} will increase. Even with very low capacitive loads, there is a brief point in the transition from LOW to HIGH or HIGH to LOW when the two output transistors are partially turned on. This effectively lowers the resistance from the supply to ground, causing a current spike as well.

Thus, at higher frequencies, CMOS begins to lose some of its advantage over other logic families. As a general rule, a CMOS gate will have the same average P_D as a 74LS gate at frequencies near 2 to 3 MHz. Above these frequencies, TTL power also increases with frequency because of the current required to reverse the charge on the load capacitance. For MSI chips, the situation is somewhat more complex than stated here, and a logic designer must do a detailed analysis to determine whether or not CMOS has a power-dissipation advantage at a particular frequency of operation.

Fan-Out

Like N-MOS and P-MOS, CMOS inputs have an extremely large resistance ($10^{12}\ \Omega$) that draws essentially no current from the signal source. Each CMOS input, however, typically presents a 5-pF load to ground. This input capacitance limits the number of CMOS inputs that one CMOS output can drive (see Figure 8-26). The CMOS output must charge and discharge the parallel combination of all of the input capacitances, so that the output switching time will be increased in proportion to the number of loads being driven. Typically, each CMOS load increases the driving circuit's propagation delay by 3 ns. For example, NAND gate 1 in Figure 8-26 might have a t_{PLH} of 25 ns if it were driving no loads; this would increase to 25 ns + 20(3 ns) = 85 ns if it were driving *twenty* loads.

Thus, CMOS fan-out depends on the permissible maximum propagation delay. Typically, CMOS outputs are limited to a fan-out of 50 for low-frequency operation (≤ 1 MHz). Of course, for higher-frequency operation, the fan-out would have to be less.

FIGURE 8-26 Each CMOS input adds to the total load capacitance seen by the driving gate's output.

Gate 1 output drives a total C_{LOAD} of N X 5 pF

To other loads

Switching Speed

Although CMOS, like N-MOS and P-MOS, must drive relatively large load capacitances, its switching speed is somewhat faster because of its low output resistance in each state. An N-MOS output must charge the load capacitance through a relatively large (100-kΩ) resistance. In the CMOS circuit, the output resistance in the HIGH state is the R_{ON} of the P-MOSFET, which is typically 1 kΩ or less. This allows more rapid charging of load capacitance.

A 4000 series NAND gate will typically have an average t_{pd} of 50 ns at $V_{DD} = 5$ V, and 25 ns at $V_{DD} = 10$ V. The reason for the improvement in t_{pd} as V_{DD} is increased is that the R_{ON} on the MOSFETs decreases significantly at higher supply voltages. Thus, it appears that V_{DD} should be made as large as possible for operation at higher frequencies. However, the larger V_{DD} will result in increased power dissipation.

A typical NAND gate in the 74HC or 74HCT series has an average t_{pd} of around 8 ns when operated at $V_{DD} = 5$ V. A 74AC/ACT NAND gate has an average t_{pd} of around 4.7 ns. A 74AHC NAND gate has an average t_{pd} of around 4.3 ns.

Unused Inputs

CMOS inputs should never be left disconnected. All CMOS inputs must be tied either to a fixed voltage level (0 V or V_{DD}) or to another input.

This rule applies even to the inputs of extra unused logic gates on a chip. An unconnected CMOS input is susceptible to noise and static charges that could easily bias both the P-channel and the N-channel MOSFETs in the conductive state, resulting in increased power dissipation and possible overheating.

Static Sensitivity

All electronic devices, to varying degrees, are sensitive to damage by static electricity. The human body is a great storehouse of electrostatic charges. For example, when you walk across a carpet, a static charge of over 30,000 V can be built up on your body. If you then touch an electronic device, some of this sizable charge can be transferred to the device. The

MOS logic families (and all MOSFETs) are especially susceptible to static-charge damage. All of this potential difference (static charge) applied across the thin oxide film overcomes the film's dielectric insulation capability. When it breaks down, the resulting flow of current (discharge) is like a lightning strike, blowing a hole in the oxide layer and permanently damaging the device.

Electrostatic discharge (ESD) is responsible for billions of dollars of damage to electronic equipment annually, and equipment manufacturers have devoted considerable attention to developing special handling procedures for all electronic devices and circuits. Even though most modern ICs have on-chip resistor–diode networks to protect inputs and outputs from the effects of ESD, the following precautions are used by most engineering labs, production facilities, and field service departments:

1. Connect the chassis of all test instruments, soldering-iron tips, and your workbench (if metal) to earth ground (i.e., the round prong in the 120 VAC plug). This prevents the buildup of static charge on these devices that could be transferred to any circuit board or IC that they come in contact with.

2. Connect yourself to earth ground with a special wrist strap. This will allow potentially dangerous charges from your body to be discharged to ground. The wrist strap contains a 1-MΩ resistor that limits current to a nonlethal value should you accidentally touch a "live" voltage while working with the equipment.

3. Keep ICs (especially MOS) in conductive foam or aluminum foil. This will keep all IC pins shorted together so that no dangerous voltages can be developed between any two pins.

4. Avoid touching IC pins, and insert the IC into the circuit immediately after removing it from the protective carrier.

5. Place shorting straps across the edge connectors of PC boards when the boards are being carried or transported. Avoid touching the edge connectors. Store PC boards in conductive plastic or metallic envelopes.

6. Do not leave any unused IC inputs unconnected because open inputs tend to pick up stray static charges.

Latch-Up

Because of the unavoidable existence of *parasitic* (unwanted) PNP and NPN transistors embedded in the substrate of CMOS ICs, a condition known as **latch-up** can occur under certain circumstances. If these parasitic transistors on a CMOS chip are triggered into conduction, they will latch-up (stay ON permanently), and a large current may flow and destroy the IC. Most modern CMOS ICs are designed with protection circuitry that helps prevent latch-up, but it can still occur when the device's maximum voltage ratings are exceeded. Latch-up can be triggered by high-voltage spikes or ringing at the device inputs and outputs. Clamping diodes can be connected externally to protect against such transients, especially when the ICs are used in industrial environments where high-voltage and/or high-current load switching takes place (motor controllers, relays, etc.). A well-regulated power supply will minimize spikes on the V_{DD} line; if the supply also has current limiting, it will limit current should latch-up occur. Modern CMOS fabrication techniques have greatly reduced ICs' susceptibility to latch-up.

REVIEW QUESTIONS

1. Which CMOS series is pin-compatible with TTL?
2. Which CMOS series is electrically compatible with TTL?
3. Which CMOS series is functionally equivalent to TTL?
4. What logic family combines the best features of CMOS and bipolar logic?
5. What factors determine CMOS fan-out?
6. What precautions should be taken when handling CMOS ICs?
7. Which IC family (CMOS, TTL) is best suited for battery-powered applications?
8. *True or false:*
    (a) CMOS power drain increases with operating frequency.
    (b) Unused CMOS inputs can be left unconnected.
    (c) TTL is better suited than CMOS for operation in high-noise environments.
    (d) CMOS switching speed increases with operating frequency.
    (e) CMOS switching speed increases with supply voltage.
    (f) The latch-up condition is an advantage of CMOS over TTL.

8-11 LOW-VOLTAGE TECHNOLOGY

IC manufacturers are continually looking for ways to put semiconductor devices (diodes, resistors, transistors, etc.) closer together on a chip, that is, to increase the chip density. This higher density has at least two major benefits. First, it allows more circuits to be packed onto the chip; second, with the circuits closer together, the time for signals to propagate from one circuit to another will decrease, thereby improving overall circuit operating speed. There are also drawbacks to higher chip density. When circuits are placed closer together, the insulating material that isolates one circuit from another is narrower. This decreases the amount of voltage that the device can withstand before dielectric breakdown occurs. Increasing the chip density increases the overall chip power dissipation, which can raise the chip temperature above the maximum level allowed for reliable operation.

These drawbacks can be neutralized by operating the chip at lower voltage levels, thereby reducing power dissipation. Several series of logic on the market operate on 3.3 volts. The newer series are optimized to run on 2.5 volts. This low-voltage technology may very well signal the beginning of a gradual transition in the digital equipment field that will eventually find all digital ICs operating from a new low-voltage standard.

Low-voltage devices are currently designed for applications ranging from electronic games to engineering workstations. The newer CPUs are 2.5-V devices, and 3.3-V dynamic RAM chips are used in memory modules for personal computers.

Several low-voltage logic series are currently available. It is not possible to cover all of the families and series from all manufacturers, so we will describe those currently offered by Texas Instruments.

CMOS Family

■ The *74LVC (Low-Voltage CMOS)* series contains the widest assortment of the familiar SSI gates and MSI functions of the 5-V families, along with many bus-interface devices such as buffers, latches, drivers, and so on. This series can handle 5-V logic levels on its inputs, so it can convert from 5-V systems to 3-V systems. As long as the current drive is kept low enough to keep the output voltage within acceptable limits, the 74LVC can also drive 5-V TTL inputs. The V_{IH} input requirements of 5-V CMOS parts such as the 74HC/AHC do not allow LVC devices to drive them.

- The *74ALVC (Advanced Low-Voltage CMOS)* series currently offers the highest performance. The devices in this series are intended primarily for bus-interface applications that use 3.3-V logic only.

- The *74LV (Low-Voltage)* series offers CMOS technology and many of the common SSI gates and MSI logic functions, along with some popular octal buffers, latches, and flip-flops. It is intended to operate only with other 3.3-V devices.

- The *74AVC (Advanced Very-Low-Voltage CMOS)* series has been introduced with tomorrow's systems in mind. It is optimized for 2.5-V systems, but it can operate on supplies as low as 1.2 V or as high as 3.3 V. This broad range of supply voltage makes it useful in mixed-voltage systems. It has propagation delays of less than 2 ns, which rivals 74AS bipolar devices. It has many of the bus interface features of the BiCMOS series that will make it useful in future generations of low-voltage workstations, PCs, networks, and telecommunications equipment.

- The *74AUC (Advanced Ultra-Low-Voltage CMOS)* family is optimized to operate at 1.8-V logic levels.

- The *74CBT (Cross Bar Technology)* series offers high-speed bus interface circuits that can switch quickly when enabled and not load the bus when they are disabled.

- The *74CBTLV (Cross Bar Technology Low Voltage)* is the 3.3-V complement to the 74CBT series.

- The *74GTLP1394 (Gunning Transceivers Logic Plus)* is made specifically as a transceiver on a standard high-speed serial bus (1394).

- The *74SSTV (Stub Series Terminated Logic)* is useful in the high-speed advanced memory systems of today's computers.

- The *74TVC (Translation Voltage Clamp)* series is used to protect the inputs and outputs of sensitive devices from voltage overshoot on the bus lines.

BiCMOS Family

- The *74LVT (Low-Voltage BiCMOS Technology)* contains BiCMOS parts that are intended for 8- and 16-bit bus-interface applications. As with the LVC series, the inputs can handle 5-V logic levels and serve as a 5-V to 3-V translator. Because the output levels [V_{OH}(min) and V_{OL}(max)] are equivalent to TTL levels, they are fully electrically compatible with TTL. Table 8-10 compares the various features.

- The *74ALVT (Advanced Low-Voltage BiCMOS Technology)* series is an improvement over the LVT series. It offers 3.3-V or 2.5-V operation at 3 ns and is pin-compatible with existing ABT and LVT series. It is also intended for bus interface applications.

- The *74ALB (Advanced Low-Voltage BiCMOS)* series is designed for 3.3-V bus interface applications. It provides 25 mA output drive and propagation delays of only 2.2 ns.

Digital technicians and engineers can no longer assume that every IC in a digital circuit, system, or piece of equipment is operating at 5 V, and they

TABLE 8-10 Low-voltage series characteristics.

	LV	ALVC	AVC	ALVT	ALB
V_{CC} (recommended)	2.7–3.6	2.3–3.6	1.65–3.6	2.3–2.7	3–3.6
T_{PD} (ns)	18	3	1.9	3.5	2
V_{IH} (V)	2 to V_{CC} + 0.5	2.0 to 4.6	1.2 to 4.6	2 to 7	2.2 to 4.6
V_{IL} (V)	0.8	0.8	0.7	0.8	0.6
I_{OH} (mA)	6	12	8	32	25
I_{OL} (mA)	6	12	8	32	25

must be prepared to deal with the necessary interfacing considerations in mixed-voltage systems. The interfacing skills you learn in this chapter will allow you to accomplish this, regardless of what develops as low-voltage systems become more common.

The continued development of low-voltage technology promises to bring about a complete revolution from the original 5-V system, to mixed-voltage systems, and finally to pure 3.3-V, 2.5-V, or even lower voltage digital systems. To put all of this in perspective, Figure 8-27 shows Texas Instruments' perception of the life cycle of the various logic families.

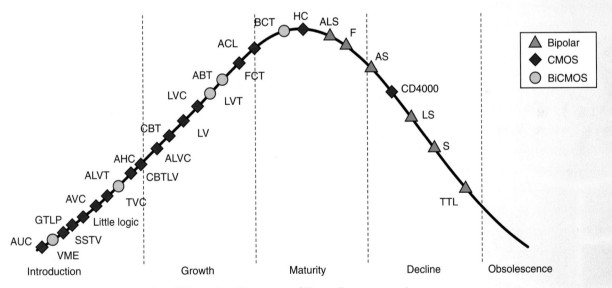

FIGURE 8-27 Logic product life cycle. (Courtesy of Texas Instruments)

REVIEW QUESTIONS

1. What are the two advantages of higher-density ICs?
2. What are the drawbacks?
3. What is the minimum HIGH voltage at a 74LVT input?
4. Which low-voltage series can work only with other low-voltage series ICs?
5. Which low-voltage series is fully electrically compatible with TTL?

8-12 OPEN-COLLECTOR/OPEN-DRAIN OUTPUTS

Several digital devices must sometimes share the use of a single wire in order to transmit a signal to some destination device, very much like several neighbors sharing the same street. This means that several devices must have their outputs connected to the same wire, which essentially connects them all to each other. For all of the logic devices we have considered so far, this presents a problem. Each output has two states, HIGH and LOW. When one output is HIGH while the other is LOW and when they are connected together, we have a HIGH/LOW conflict. Which one will win? Just like arm wrestling, the stronger of the two wins. In this case, the transistor circuit whose output transistor has the lowest "ON" resistance will pull the output voltage in its direction.

Figure 8-28 shows a generic block diagram of two logic devices with their outputs connected to a common wire. If the two logic devices were CMOS, then the ON resistance of the pull-up circuit that outputs the HIGH would be approximately the same as the ON resistance of the pull-down circuit that outputs the LOW. The voltage on the common wire will be about half the supply voltage. This voltage is in the indeterminate range for most CMOS series and is unacceptable for driving a CMOS input. Furthermore, the current through the two conducting MOSFETs will be much greater than normal, especially at higher values of V_{DD}, and it can damage the ICs.

Conventional CMOS outputs should never be connected together.

If the two devices were TTL totem-pole outputs, as shown in Figure 8-29, a similar situation would occur but with different results because of the difference in output circuitry. Suppose that gate A output is in the HIGH state (Q_{3A} ON, Q_{4A} OFF) and the gate B output is in the LOW state (Q_{3B} OFF, Q_{4B} ON). In this situation, Q_{4B} is a very low resistance load on Q_{3A} and will draw a current that is far greater than it is rated to handle. This current might not damage Q_{3A} or Q_{4B} immediately, but over a period of time it can cause overheating and deterioration in performance and eventual device failure.

Another problem caused by this relatively high current flowing through Q_{4B} is that it will produce a larger voltage drop across the transistor collector

FIGURE 8-28   Two outputs contending for control of a wire.

FIGURE 8-29 Totem-pole outputs tied together can produce harmful current through Q_4.

emitter, making V_{OL} of between 0.5 and 1 V. This is greater than the allowable V_{OL} (max). For these reasons:

TTL totem pole outputs should never be tied together.

Open-Collector/Open-Drain Outputs

One solution to the problem of sharing a common wire among gates is to remove the active pull-up transistor from each gate's output circuit. In this way, none of the gates will ever try to assert a logic HIGH. TTL outputs that have been modified in this way are called **open-collector outputs**. CMOS output circuits that have been modified in this way are called open-drain outputs. The output is taken at the drain of the N-channel pull-down MOSFET, which is an open circuit (i.e., not connected to any other circuitry).

The TTL equivalent is called an open-collector output because the collector of the bottom transistor in the totem pole is connected directly to the output pin and nowhere else, as shown in Figure 8-30(a). The open-collector structure eliminates the pull-up transistors Q_3, D_1, and R_4. In the output LOW state, Q_4 is ON (has base current and is essentially a short between collector and emitter); in the output HIGH state, Q_4 is OFF (has no base current and is essentially an open between collector and emitter). Because this circuit has no internal way to pull the output HIGH, the circuit designer must connect an external pull-up resistor R_P to the output, as shown in Figure 8-30(b).

When Q_4 is ON, it pulls the output voltage down to a LOW. When Q_4 is OFF, R_P pulls the output of the gate HIGH. Note that without the pull-up resistor, the output voltage would be indeterminate (floating). The value of the resistor R_P is usually chosen to be 10 kΩ. This value is small enough so that, in the HIGH state, the voltage dropped across it due to load current will not lower the output voltage below the minimum V_{OH}. It is large enough so that, in the LOW state, it will limit the current through Q_4 to a value below I_{OL}(max).

(a)

(b)

FIGURE 8-30   (a) Open-collector TTL circuit; (b) with external pull-up resistor.

When several open-collector or open-drain gates share a common connection, as shown in Figure 8-31, the common wire is HIGH by default due to the pull-up resistor. When any one (or more) of the gate outputs pulls it LOW, the 5 V are dropped across R_P and the common connection is in the LOW state. Because the common output is HIGH only when all the outputs are in the HIGH state, connecting the outputs in this way essentially implements the logic AND function. This is called a **wired-AND** connection. This is shown symbolically by the dotted AND gate symbol. There is no actual AND gate

FIGURE 8-31   Wired-AND operation using open-collector gates.

there. *A wired AND can be implemented only with open-collector TTL and open-drain CMOS logic devices.*

To summarize, the open-collector/open-drain circuits cannot actively make their outputs HIGH; they can only pull them LOW. This feature can be used to allow several devices to share the same wire for transmitting a logic level to another device or to combine the outputs of the devices effectively in a logic AND function. As we mentioned before, the purpose of the active pull-up transistor in the output circuit of conventional gates is to charge up the load capacitance rapidly and allow for fast switching. Open-collector and open-drain devices have a much slower switching speed from LOW to HIGH and consequently are not used in high-speed applications.

Open-Collector/Open-Drain Buffer/Drivers

The applications of open-collector/drain outputs that we have described were more prevalent in the early days of logic circuits than they are today. A more common use of these circuits now is as a **buffer/driver**. A buffer or a driver is a logic circuit that is designed to have a greater output current and/or voltage capability than an ordinary logic circuit. They allow a weaker output circuit to drive a heavy load. Open-collector/drain circuits offer some unique flexibility as buffer/drivers.

Due to their high I_{OL} and V_{OH} specifications, the 7406 and 7407 are the only standard TTL devices that are still being recommended for new designs. The 7406 is an open-collector buffer/driver IC that contains six IN-VERTERs with open-collector outputs that can sink up to 40 mA in the LOW state. In addition, the 7406 can handle output voltages up to 30 V in the HIGH state. This means that the output can be connected to a load that operates on a voltage greater than 5 V. This is illustrated in Figure 8-32, where a 7406 is used as a buffer between a 74LS112 flip-flop and an incandescent indicator lamp that is rated at 24 V, 25 mA. The 7406 controls the lamp's ON/OFF status to indicate the state of FF output Q. Note that the lamp is powered from +24 V, and it acts as the pull-up resistor for the open-collector output.

When $Q = 1$, the 7406 output goes LOW, its output transistor sinks the 25 mA of lamp current supplied by the 24-V source, and the lamp is on. When $Q = 0$, the 7406 output transistor turns off; there is no path for current, and the lamp turns off. In this state, the full 24 V will appear across

FIGURE 8-32 An open-collector buffer/driver drives a high-current, high-voltage load.

the OFF output transistor so that $V_{OH} = 24$ V, which is lower than the 7406 maximum V_{OH} rating.

Open-collector outputs are often used to drive indicator LEDs, as shown in Figure 8-33(a). The resistor is used to limit the current to a safe value. When the INVERTER output is LOW, its output transistor provides a low-resistance path to ground for the LED current, so that the LED is on. When the INVERTER output is HIGH, its output transistor is off, and there is no path for LED current; in this state, the LED is off.

The 7407 is an open-collector, noninverting buffer with the same voltage and current ratings as a 7406.

The 74HC05 is an open-drain hex inverter with 25 mA current sink capability. Figure 8-33(b) shows a way to interface a 74AHC74 D-FF to a control relay. A control relay is an electromagnetic switch. The contacts close magnetically when the rated current flows through the coil. The 74HC05 can handle the relay's relatively high voltage and current so that the 74AHC74 output can turn the relay on and off.

FIGURE 8-33 (a) An open-collector output can be used to drive an LED indicator; (b) an open-drain CMOS output.

IEEE/ANSI Symbol for Open-Collector/Drain Outputs

The new IEEE/ANSI symbology uses a distinctive notation to identify open-collector/drain outputs. Figure 8-34 shows the standard IEEE/ANSI designation for an open-collector/drain output. It is an underlined diamond. Although we will not normally use the complete IEEE/ANSI symbology in this book, we will use this underlined diamond to indicate open-collector and open-drain outputs.

FIGURE 8-34 IEEE/ANSI notation for open-collector and open-drain outputs.

1. When does a HIGH/LOW conflict occur?
2. Why shouldn't totem-pole outputs be tied together?
3. How do open-collector outputs differ from totem-pole outputs?
4. Why do open-collector outputs need a pull-up resistor?
5. What is the logic expression for the wired-AND connection of six 7406 outputs?
6. Why are open-collector outputs generally slower than totem-pole?
7. What is the IEEE/ANSI symbol for open-collector outputs?

8-13 TRISTATE (THREE-STATE) LOGIC OUTPUTS

The **tristate** configuration is a third type of output circuitry used in TTL and CMOS families. It takes advantage of the high-speed operation of the pull-up/pull-down output arrangement, while allowing outputs to be connected together to share a common wire. It is called tristate because it allows three possible output states: HIGH, LOW, and high-impedance (Hi-Z). The Hi-Z state is a condition in which both the pull-up and the pull-down transistors are turned OFF so that the output terminal is a high impedance to both ground and the power supply $+V$. Figure 8-35 illustrates these three states for a simple inverter circuit.

Devices with tristate outputs have an *enable* input. It is often labeled E for enable or OE for output enable. When $OE = 1$, as shown in Figures 8-35(a) and (b), the circuit operates as a normal INVERTER because the HIGH logic level at OE enables the output. The output will be either HIGH or LOW, depending on the input level. When $OE = 0$, as shown in Figure 8-35(c), the circuit's output is *disabled*. It goes into its Hi-Z state with both transistors in the nonconducting state. In this state, the output terminal is essentially an open circuit (not connected to anything).

Advantage of Tristate

The outputs of tristate ICs can be connected together (share the use of a common wire) without sacrificing switching speed because a tristate output, when enabled, operates as a totem pole for TTL or an active pull-up/pull-down

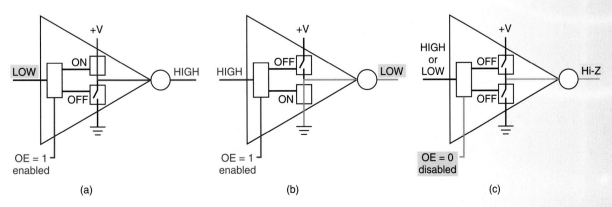

FIGURE 8-35 Three output conditions of tristate.

CMOS output with its associated low-impedance, high-speed characteristics. It is important to realize, however, that when tristate outputs are connected together, only one of them should be enabled at one time. Otherwise, two active outputs could fight for control of the common wire, as we discussed earlier, causing damaging currents to flow and producing invalid logic levels.

In our discussion of open-collector/open-drain and tristate circuits, we have referred to cases when the outputs of several devices must share a single wire to transmit information to another device. The shared wire is referred to as a bus wire. An entire bus is made up of several wires that are used to carry digital information between two or more devices that share the use of the bus.

Tristate Buffers

A *tristate buffer* is a circuit used to control the passage of a logic signal from input to output. Some tristate buffers also invert the signal as it goes through. The circuits in Figure 8-35 can be called *inverting tristate buffers.*

Two commonly used tristate buffer ICs are the 74LS125 and the 74LS126. Both contain four *noninverting* tristate buffers like those shown in Figure 8-36. The 74LS125 and 74LS126 differ only in the active state of their ENABLE inputs. The 74LS125 allows the input signal A to reach the output when $\bar{E} = 0$, while the 74LS126 passes the input when $E = 1$.

Tristate buffers have many applications in circuits where several signals are connected to common lines (buses). We will examine some of these applications in Chapter 9, but we can get the basic idea from Figure 8-37(a). Here, we have three logic signals A, B, and C connected to a common bus line through 74AHC126 tristate buffers. This arrangement permits us to transmit any one of these signals over the bus line to other circuits by enabling the appropriate buffer.

For example, consider the situation in Figure 8-37(b), where $E_B = 1$ and $E_A = E_C = 0$. This disables the upper and lower buffers so that their outputs are in the Hi-Z state and are essentially disconnected from the bus. This is symbolized by the X's on the diagram. The middle buffer is enabled so that its input, B, is passed through to its output and onto the bus, from which it is routed to other circuits connected to the bus. When tristate outputs are connected together as in Figure 8-37, it is important to remember that no more than one output should be enabled at one time. Otherwise, two or more active totem-pole outputs would be connected, which could produce damaging currents. Even if damage did not occur, this situation would produce a signal

FIGURE 8-36 Tristate noninverting buffers.

(a)   (b)

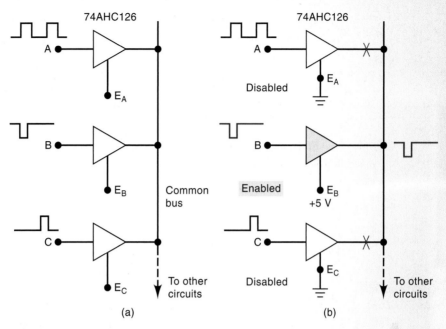

FIGURE 8-37 (a) Tristate buffers used to connect several signals to a common bus; (b) conditions for transmitting *B* to the bus.

FIGURE 8-38 If two enabled CMOS outputs are connected together, the bus will be at approximately $V_{DD}/2$ when the outputs are trying to be different.

on the bus that is a combination of more than one signal. This is commonly referred to as **bus contention**. Figure 8-38 shows the effect of enabling outputs *A* and *B* simultaneously. In Figure 8-37, when inputs *A* and *B* are in opposite states, they contend for control of the bus. The resulting voltage on the bus is an invalid logic state. In tristate bus systems, the designer must make sure that the enable signals do not allow bus contention to occur.

Tristate ICs

In addition to tristate buffers, many ICs are designed with tristate outputs. For example, the 74LS374 is an octal D-type FF register IC with tristate outputs. This means that it is an eight-bit register made up of D-type FFs whose outputs are connected to tristate buffers. This type of register can be connected to common bus lines along with the outputs from other, similar devices to allow efficient transfer of data over the bus. We examine this *tristate data bus* arrangement in Chapter 9. Other types of logic devices that are available with tristate outputs include decoders, multiplexers, analog-to-digital converters, memory chips, and microprocessors.

IEEE/ANSI Symbol for Tristate Outputs

The traditional logic symbology has no special notation for tristate outputs. Figure 8-39 shows the notation used in the IEEE/ANSI symbology to indicate a tristate output. It is a triangle that points downward. Although it is not part of the traditional symbology, we will use this triangle to designate tristate outputs throughout the remainder of the book.

FIGURE 8-39   IEEE/ANSI notation for tristate outputs.

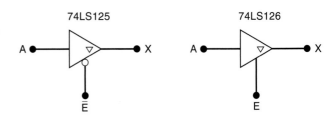

1. What are the three possible states of a tristate output?
2. What is the state of a tristate output when it is disabled?
3. What is bus contention?
4. What conditions are necessary to transmit signal *C* onto the bus in Figure 8-37?
5. What is the IEEE/ANSI designation for tristate outputs?

8-14   HIGH-SPEED BUS INTERFACE LOGIC

Many digital systems use a shared bus to transfer digital signals and data between the various components of the system. As you can see from our discussion of CMOS technology development, systems are getting faster and faster. Many of the newer high-speed logic series are designed specifically to interface to a tristate bus system. The components in these series are primarily tristate buffers, bidirectional transceivers, latches, and high-current line drivers.

A significant distance often physically separates the components in these systems. If this distance is more than about 4 inches, the bus wires between them need to be viewed as a transmission line. Although transmission line theory could fill up a whole book and is beyond the scope of this text, the general idea is simple enough. Wires have inductance, capacitance, and resistance, which means that for changing signals (ac), they have a characteristic impedance that can affect a signal placed on one end and distort it by the time it reaches the other end. At the high speeds we are discussing, the travel time down the wire and the effects of reflected waves (like echoes) and ringing become real concerns. There are several ways to combat the problems associated with transmission lines. In order to prevent reflected pulse waves, the end of the bus must be terminated with a resistance that is equal to the line impedance (about 50 Ω), as shown in Figure 8-40(a). This method is not feasible because too much current is required to maintain logic level voltages across such a low resistance. Another technique uses a capacitor to block the dc current when the line is not changing, but effectively appears as just a resistor to the rising or falling pulse. This method is shown in Figure 8-40(b).

Using a voltage divider, as in Figure 8-40(c), with resistances larger than the line, impedance helps reduce reflections, but with hundreds of individ-

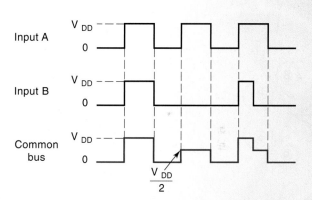

FIGURE 8-40 Bus termination techniques.

ual bus lines, it obviously makes a heavy load on the system power supply. The diode termination shown in Figure 8-40(d) simply clips off or clamps the overshoot/undershoot of the ringing caused by the reactive LC nature of the line. Series termination at the source, as shown in Figure 8-40(e), slows down the switching speed, which reduces the frequency limits of the bus but substantially improves the reliability of the bus signals.

As you can see, none of these methods are ideal. IC manufacturers are designing new series of logic circuits that overcome many of these problems. Texas Instruments' bus interface logic series offers new output circuits that dynamically lower the output impedance during signal transition to provide fast transition times, then raises the impedance during the steady state (like a series termination) to damp any ringing and reduce reflections on the bus line. The BTL (backplane transceiver logic) series of bus interface devices is specially designed to drive the relatively long buses that connect modules of a large digital system. The backplane refers to the interconnections between modules in the back of an industry standard, 19-inch rack mounting system. GTL (gunning transceiver logic) is more suitable for high-speed buses within a single circuit board or between boards in small enclosures, like a personal computer case.

REVIEW QUESTIONS

1. How close together do components need to be to ignore "transmission line" effects?
2. What three characteristics of real wires add up to distort signals that move through them?
3. What is the purpose of bus terminations?

8-15   THE ECL DIGITAL IC FAMILY

The TTL family uses transistors operating in the saturated mode. As a result, their switching speed is limited by the storage delay time associated with a transistor that is driven into saturation. Another *bipolar* logic family has been developed that prevents transistor saturation, thereby increasing overall switching speed. This logic family is called **emitter-coupled logic (ECL)**, and it operates on the principle of current switching whereby a fixed bias current less than $I_C(\text{sat})$ is switched from one transistor's collector to another. Because of this current-mode operation, this logic form is also referred to as *current-mode logic* (CML).

Basic ECL Circuit

The basic circuit for emitter-coupled logic is essentially the differential amplifier configuration of Figure 8-41(a). The V_{EE} supply produces an essentially fixed current I_E, which remains around 3 mA during normal operation. This current is allowed to flow through either Q_1 or Q_2, depending on the voltage level at V_{IN}. In other words, this current switches between Q_1's collector and Q_2's collector as V_{IN} switches between its two logic levels of -1.7 V (logical 0 for ECL) and -0.8 V (logical 1 for ECL). The table in Figure 8-41(a) shows the resulting output voltages for these two conditions at V_{IN}. Two important points should be noted: (1) V_{C1} and V_{C2} are the *complements* of each other, and (2) the output voltage levels are not the same as the input logic levels.

The second point noted above is easily taken care of by connecting V_{C1} and V_{C2} to emitter-follower stages (Q_3 and Q_4), as shown in Figure 8-41(b). The emitter followers perform two functions: (1) they subtract approximately 0.8 V from V_{C1} and V_{C2} to shift the output levels to the correct ECL logic levels; and (2) they provide a very low output impedance (typically 7 Ω), which provides for large fan-out and fast charging of load capacitance. This circuit produces two complementary outputs: V_{OUT1}, which is equal to $\overline{V_{IN}}$, and V_{OUT2}, which is equal to V_{IN}.

ECL OR/NOR Gate

The basic ECL circuit of Figure 8-41(b) can be used as an INVERTER if the output is taken at V_{OUT1}. This basic circuit can be expanded to more than one input by paralleling transistor Q_1 with other transistors for the other inputs, as in Figure 8-42(a). Here, either Q_1 or Q_3 can cause the current to be switched out of Q_2, resulting in the two outputs V_{OUT1} and V_{OUT2} being the logical NOR and OR operations, respectively. This OR/NOR gate is symbolized in Figure 8-42(b) and is the fundamental ECL gate.

ECL Characteristics

The latest ECL series by Motorola is called ECLin PS. This stands for ECL in pico seconds. This logic series boasts a maximum gate propagation delay of 500 ps (that's half a nanosecond!) and FF toggle rates of 1.4 GHz. Some devices in this series have gate delays of only 100 ps at an average power of 5 mW, for a speed power rating of 0.5 pJ. The following are the most important characteristics of the ECLin PS series of Motorola's MECL family of logic circuits:

1. The transistors never saturate, and so switching speed is very high. Typical propagation delay time is 360 ps, which makes ECL faster than any TTL or CMOS family members.

Operating States		
V_{IN}	Outputs	
−1.7 V (logic 0)	$V_{C1} = 0$ V $V_{C2} = -0.9$ V	Q_2 conducts
−0.8 V (logic 1)	$V_{C1} = -0.9$ V $V_{C2} = 0$ V	Q_1 conducts

(a)

(b)

FIGURE 8-41 (a) Basic ECL circuit; (b) with addition of emitter followers.

2. The logic levels are nominally −0.8 V and −1.7 V for the logical 1 and 0, respectively. ECLin PS is fully voltage-compatible with former series of ECL.

3. Worst-case ECL noise margins are approximately 150 mV. These low noise margins make ECL somewhat unreliable for use in heavy industrial environments.

4. An ECL logic block usually produces an output and its complement. This eliminates the need for inverters. High current complementary drive

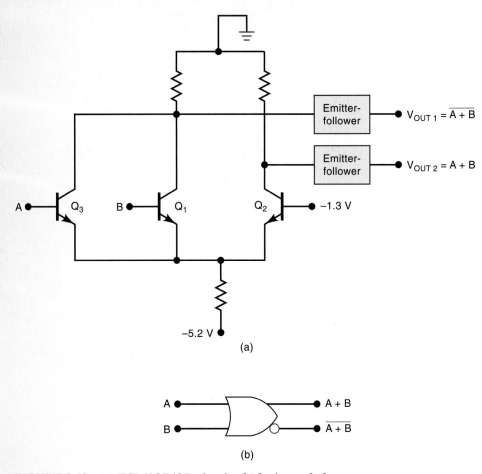

$$V_{OUT\ 1} = \overline{A + B}$$

$$V_{OUT\ 2} = A + B$$

(a)

(b)

FIGURE 8-42   (a) ECL NOR/OR circuit; (b) logic symbol.

also makes ECL an excellent line driver for twisted pair cable (like tele-phone wires).

5. Fan-out is typically around 25, owing to the low-impedance emitter–follower outputs.

6. Typical power dissipation is 25 mW, somewhat higher than the 74AS series.

7. The total current flow in an ECL circuit remains relatively constant, re-gardless of its logic state. This helps to maintain an unvarying current drain on the power supply even during switching transitions. Thus, no noise spikes will be generated internally, like those produced by TTL and CMOS switching.

Table 8-11 shows how ECL compares with the important TTL logic fam-ilies. The ECL family of ICs does not include a wide range of general-purpose logic devices, as do the TTL and CMOS families. ECL does include complex, special-purpose ICs used in applications such as high-speed data transmission, high-speed memories, and high-speed arithmetic units. The relatively low noise margins and high power drain of ECL are disadvan-tages compared with TTL and CMOS. Another drawback is its negative power-supply voltage and logic levels, which are not compatible with those of the other logic families. This makes it difficult to use ECL devices in con-junction with TTL and/or CMOS ICs; special level-shifting circuits must be connected between ECL devices and the TTL (or CMOS) devices on both input and output.

TABLE 8-11   High-speed logic comparison.

Logic Family	t_{pd}(ns)	P_D(mW) <100 kHz	Worst-Case Noise Margin (mV)	Maximum Clock Rate (MHz)	Speed–Power Product (pJ)
74AS	1.7	8	300	200	13.6
74F	3.8	6	300	100	22.8
74AHC	3.7	.006	550	130	.02
74AVC	2	.006	250	*	.012
74ALVT	2.4	0.33	400	*	.79
74ALB	2.2	1	400	*	2.2
ECL	0.3	25	150	1400	7.5

* Flip-flops not available in this series.

REVIEW QUESTION

1. *True or false:*
   (a) ECL obtains high-speed operation by preventing transistor saturation.
   (b) ECL circuits usually have complementary outputs.
   (c) The noise margins for ECL circuits are larger than TTL noise margins.
   (d) ECL circuits do not generate noise spikes during state transitions.
   (e) ECL devices require less power than standard TTL.
   (f) ECL can easily be used with TTL.

8-16   CMOS TRANSMISSION GATE (BILATERAL SWITCH)

A special CMOS circuit that has no TTL or ECL counterpart is the **transmission gate** or **bilateral switch**, which acts essentially as a single-pole, single-throw switch controlled by an input logic level. This transmission gate passes signals in both directions and is useful for digital and analog applications.

Figure 8-43(a) is the basic arrangement for the bilateral switch. It consists of a P-MOSFET and an N-MOSFET in parallel so that both polarities of input voltage can be switched. The CONTROL input and its inverse are used to turn the switch on (closed) and off (open). When the CONTROL is HIGH, both MOSFETs are turned on and the switch is closed. When CONTROL is LOW, both MOSFETs are turned off and the switch is open. Ideally, this circuit operates like an electromechanical relay. In practice, however, it is not a perfect short circuit when the switch is closed; the switch resistance R_{ON} is typically 200 Ω. In the open state, the switch resistance is very large, typically 10^{12} Ω, which for most purposes is an open circuit. The symbol in Figure 8-43(b) is used to represent the bilateral switch.

This circuit is called a *bilateral* switch because the input and output terminals can be interchanged. The signals applied to the switch input can be either digital or analog signals, provided that they stay within the limits of 0 to V_{DD} volts.

Figure 8-44(a) shows the traditional logic diagram for a 4016 quad bilateral switch IC, which is also available in the 74HC series as a 74HC4016. The IC contains four bilateral switches that operate as described above. Each

Control input	Switch
0 V	Open (OFF)
+V_{DD}	Closed (ON)

(a)   (b)

FIGURE 8-43   CMOS bilateral switch (transmission gate).

FIGURE 8-44   The 4016/74HC4016 quad bilateral switch.

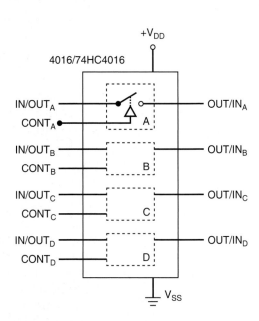

switch is independently controlled by its own control input. For example, the ON/OFF status of the top switch is controlled by input CONT$_A$. Because the switches are bidirectional, either switch terminal can serve as input or output, as the labeling indicates.

EXAMPLE 8-12

Describe the operation of the circuit in Figure 8-45.

Solution

Here, two of the bilateral switches are connected so that a common analog input signal can be switched to either output X or output Y, depending on the logic state of the OUTPUT SELECT input. When OUTPUT SELECT is LOW, the upper switch is closed and the lower one is open so that V_{IN} is connected to output X. When OUTPUT SELECT is HIGH, the upper switch is open and the lower one is closed so that V_{IN} is connected to output Y. Figure 8-45(b)

FIGURE 8-45   Example 8-12: 74HC4016 bilateral switches used to switch an analog signal to two different outputs.

shows some typical waveforms. Note that for proper operation, V_{IN} must be within the range 0 V to $+V_{DD}$.

The 4016/74HC4016 bilateral switch can switch only input voltages that lie between 0 V and V_{DD}, and so it could not be used for signals that were both positive and negative relative to ground. The 4316 and 74HC4316 ICs are quad bilateral switches that can switch *bipolar* analog signals. These devices have a second power-supply terminal called V_{EE}, which can be made negative with respect to ground. This permits input signals that can range from V_{EE} to V_{DD}. For example, with $V_{EE} = -5$ V and $V_{DD} = +5$ V, the analog input signal can be anywhere from -5 V to $+5$ V.

REVIEW QUESTIONS

1. Describe the operation of a CMOS bilateral switch.
2. *True or false:* There is no TTL bilateral switch.

8-17   IC INTERFACING

Interfacing means connecting the output(s) of one circuit or system to the input(s) of another circuit or system that has different electrical characteristics. Often a direct connection cannot be made because of the difference in the electrical characteristics of the *driver* circuit that is providing the output signal and the *load* circuit that is receiving the signal.

An *interface circuit* is a circuit connected between the driver and the load; its function is to take the driver output signal and condition it so that it is compatible with requirements of the load.

In the following sections we will address the problems involved in interfacing devices from one logic family to those of a different logic family. This type of interfacing occurs quite often in the more complex digital systems, where designers utilize different logic families for different parts of the system in order to take advantage of the strong points of each family. For example, high-speed TTL (74AS, 74S) might be used in those parts of the system that are operating at the highest frequency, 74HCT in the slower parts of the system, and N-MOS for the LSI and VLSI parts of the system.

ICs from the same logic series are designed to be connected together without any special considerations, provided that the fan-out limitation of each output is not exceeded. When you connect the output of an IC to the input of an IC from a different logic family or a different series within the same logic family, you generally must be concerned with the voltage and current parameters of the two devices. This usually involves checking the device data sheets for values of input and output current/voltage parameters. Table 8-9, which we have been using, has the input/output voltage parameters for the various IC series. These values will generally be valid for most devices in the series indicated. Table 8-12 lists the input/output current values for *standard* devices in the various CMOS and TTL series, that is, for ICs with no special input or output circuitry. These values are not valid for devices such as buffers, which have greater output current capability, or for ICs whose external inputs are internally connected to more than one gate on the chip. We use these two tables in the following sections to demonstrate the ideas and procedures for IC interfacing, but it should be understood that, in practice, it is best to consult individual IC data sheets to get the actual voltage and current values.

TABLE 8-12 Input/output currents for standard devices with a supply voltage of 5 V.

Parameter	CMOS				TTL				
	4000B	74HC/HCT	74AC/ACT	74AHC/AHCT	74	74LS	74AS	74ALS	74F
I_{IH}(max)	1 μA	1 μA	1 μA	1 μA	40 A	20 μA	20 μA	20 μA	20 μA
I_{IL}(max)	1 μA	1 μA	1 μA	1 μA	1.6 mA	0.4 mA	0.5 mA	100 μA	0.6 mA
I_{OH}(max)	0.4 mA	4 mA	24 mA	8 mA	0.4 mA	0.4 mA	2 mA	400 mA	1.0 mA
I_{OL}(max)	0.4 mA	4 mA	24 mA	8 mA	16 mA	8 mA	20 mA	8 mA	20 mA

8-18 TTL DRIVING CMOS

When interfacing different types of ICs, we must check that the driving device can meet the current and voltage requirements of the load device. Examination of Table 8-12 indicates that the input current values for CMOS are extremely low compared with the output current capabilities of any TTL series. Thus, TTL has no problem meeting the CMOS input current requirements.

There is a problem, however, when we compare the TTL output voltages with the CMOS input voltage requirements. Table 8-9 shows that V_{OH}(min) of every TTL series is too low when compared with the V_{IH}(min) requirement of the 4000B, 74HC, and 74AC series. For these situations, something must be done to raise the TTL output voltage to an acceptable level for CMOS.

The most common solution to this interface problem is shown in Figure 8-46, where the TTL output is connected to +5 V with a pull-up resistor. The presence of the pull-up resistor causes the TTL output to rise to approxi-

FIGURE 8-46 External pull-up resistor is used when TTL drives CMOS.

mately 5 V in the HIGH state, thereby providing an adequate CMOS input voltage level. This pull-up resistor is not required if the CMOS device is a 74HCT or a 74ACT because these series are designed to accept TTL outputs directly, as Table 8-9 shows.

TTL Driving High-Voltage CMOS

If the CMOS IC is operating with V_{DD} greater than 5 V, the situation becomes somewhat more difficult. For example, with $V_{DD} = 10$ V, the CMOS input requires $V_{IH}(min) = 7$ V. The outputs of many TTL devices cannot be operated at more than 5 V, and so a pull-up resistor connected to +10 V is prohibited. The LS-TTL series from certain manufacturers (e.g., Fairchild) can operate with an output pull-up to 10 V. In general, the device's data sheet should be checked before using a pull-up to more than 5 V.

When the TTL output cannot be pulled up to V_{DD}, there are some alternatives. One common solution is shown in Figure 8-47, where a 7407 open-collector buffer is used as the interface between a TTL totem-pole output and CMOS operating at $V_{DD} > 5$ V. The 7407 is the noninverting counterpart of the 7406 and has an output voltage rating of 30 V.

Another solution is to utilize a **voltage level-translator** circuit such as the 4504B. This is a CMOS chip designed to take a low-voltage input (e.g., from TTL) and translate it to a high-voltage output for CMOS.

FIGURE 8-47 A 7407 open-collector buffer can be used to interface TTL to high-voltage CMOS.

REVIEW QUESTIONS

1. What must be done to interface a standard TTL output to a 74AC or a 74HC input? Assume that $V_{DD} = +5$ V.
2. What are the various ways to interface TTL to high-voltage CMOS?

8-19   CMOS DRIVING TTL

Before we consider the problem of interfacing CMOS outputs to TTL inputs, it will be helpful to review the CMOS output characteristics for the two logic states. Figure 8-48(a) shows the equivalent output circuit in the HIGH state. The R_{ON} of the P-MOSFET connects the output terminal to V_{DD} (remember, the N-MOSFET is off). Thus, the CMOS output circuit acts like a V_{DD} source with a source resistance of R_{ON}. The value of R_{ON} typically ranges from 100 to 1000 Ω.

Figure 8-48(b) shows the equivalent output circuit in the LOW state. The R_{ON} of the N-MOSFET connects the output terminal to ground (remember, the P-MOSFET is off). Thus, the CMOS output acts as a low resistance to ground; that is, it acts as a current sink.

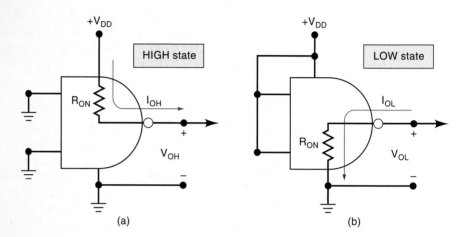

FIGURE 8-48   Equivalent CMOS output circuits for both logic states.

CMOS Driving TTL in the HIGH State

Table 8-9 shows that CMOS outputs can easily supply enough voltage (V_{OH}) to satisfy the TTL input requirement in the HIGH state (V_{IH}). Table 8-12 shows that CMOS outputs can supply more than enough current (I_{OH}) to meet the TTL input current requirements (I_{IH}). Thus, no special consideration is needed for the HIGH state.

CMOS Driving TTL in the LOW State

Table 8-12 shows that TTL inputs have a relatively high input current in the LOW state, ranging from 100 μA to 2 mA. The 74HC and 74HCT series can sink up to 4 mA, and so they would have no trouble driving a *single* TTL load of any series. The 4000B series, however, is much more limited. Its low I_{OL} capability is not sufficient to drive even one input of the 74 or 74AS series. The 74AHC series has output drive comparable to that of the 74LS series.

EXAMPLE 8-13

How many 74LS inputs can be driven by a 74HC output? Repeat for a 4000B output.

Solution

The 74LS series has $I_{IL}(\text{max}) = 0.4$ mA. The 74HC can sink up to $I_{OL}(\text{max}) = 4$ mA. Thus, the 74HC can drive *ten* 74LS loads (4 mA/0.4 mA = 10).

The 4000B can sink only 0.4 mA, and so it can drive only *one* 74LS input.

EXAMPLE 8-14

How many 74ALS inputs can be driven by a 74HC output? Repeat for 74AS inputs.

Solution

The 74ALS series has $I_{IL}(\text{max}) = 100 \ \mu\text{A}$. Thus, the 74HC can drive *forty* 74ALS inputs (4 mA/100 μA = 40). A 74HC can drive only eight 74AS inputs (4 mA/0.5 mA = 8).

EXAMPLE 8-15

What's wrong with the circuit in Figure 8-49(a)?

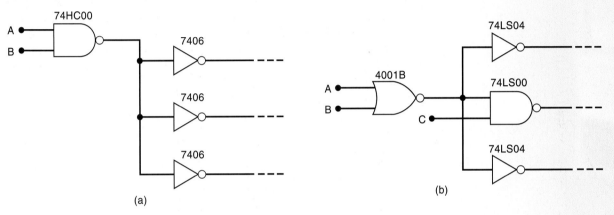

FIGURE 8-49   Examples 8-15 and 8-16.

Solution

The 74HC00 can sink 4 mA, but the three 7406 inputs require 3 × 1.6 mA = 4.8 mA.

EXAMPLE 8-16

What's wrong with the circuit in Figure 8-49(b)?

Solution

The 4001B NOR gate can sink 0.4 mA, but the three 74LS inputs require 3 × 0.4 mA = 1.2 mA.

Situations like those in Figure 8-49 can occur when older equipment is being interfaced to newer equipment or when several digital modules are being combined to implement a system. The simplest solution to the case of Figure 8-49(a) is to replace the 74HC00 gate with a 74AHC00. If the device is already soldered to a board or if a higher drive device is not available in a pin-compatible package, as in Figure 8-49(b), then some type of interface circuit is necessary. The interface circuit should have a low input current requirement and a sufficiently high output current rating to drive the loads. Figure 8-50 shows a possible interface for the case of Figure 8-49(b). In Figure 8-50, the 74LS125 is a permanently enabled, noninverting, tristate buffer that can be driven by the 4001B. Its output can easily drive the 74LS loads. In this circuit, the interface buffer simply passes the 4001B output signal to the 74LS loads with the assurance that a logic "0" will pull LS inputs to their LOW state.

FIGURE 8-50   A buffer used to interface low-current CMOS to 74LS inputs.

High-Voltage CMOS Driving TTL

Some IC manufacturers have produced several 74LS TTL devices that can withstand input voltages as high as 15 V. These devices can be driven directly from CMOS outputs operating at $V_{DD} = 15$ V. Most TTL inputs cannot handle more than 7 V before becoming damaged, and so an interface is necessary if they are to be driven from high-voltage CMOS. The interface functions as a voltage-level translator that converts the high-voltage input to a 5-V output that can be connected to TTL. Figure 8-51 shows how the 4050B performs this level translation between 15 and 5 V.

FIGURE 8-51   A 4050B buffer can also serve as a level translator between high-voltage CMOS and TTL.

REVIEW QUESTIONS

1. What is the function of an *interface* circuit?
2. *True or false:* All CMOS outputs can drive TTL in the HIGH state.
3. *True or false:* Any CMOS output can drive any single TTL input.
4. Which CMOS series can TTL drive without a pull-up resistor?
5. How many 7400 inputs can be driven from a 74HCT00 output?
6. How is high-voltage CMOS interfaced to TTL?

8-20   ANALOG VOLTAGE COMPARATORS

Another very useful device for interfacing to digital systems is the **analog voltage comparator**. It is especially useful in systems that contain both analog voltages as well as digital components. An analog voltage comparator compares two voltages. If the voltage on the (+) input is greater than the voltage on the (−) input, the output is HIGH. If the voltage on the (−) input is greater than the voltage on the (+) input, the output is LOW. The inputs to a comparator can be thought of as analog inputs, but the output is digital because it will always be either HIGH or LOW. For this reason, the comparator is often referred to as a one-bit analog-to-digital (A/D) converter. We will examine A/D converters in detail in Chapter 10.

An LM339 is an analog linear IC that contains four voltage comparators. The output of each comparator is an open-collector transistor just like an open-collector TTL output. V_{CC} can range from 2 to 36 V but is usually set slightly greater than the analog input voltages that are being compared. A pull-up resistor must be connected from the output to the same supply that the digital circuits use (normally 5 V).

EXAMPLE 8-17

Suppose that an incubator must have an emergency alarm to warn if the temperature exceeds a dangerous level. The temperature-measuring device is an LM34 that puts out a voltage directly proportional to the temperature. The output voltage goes up 10 mV per degree F. The digital system alarm must sound when the temperature exceeds 100°F. Design a circuit to interface the temperature sensor to the digital circuit.

Solution

We need to compare the voltage from the sensor with a fixed threshold voltage. First, we must calculate the proper threshold voltage. We want the comparator output to go HIGH when the temperature exceeds 100°F. The voltage out of the LM34 at 100°F will be

$$100°F \cdot 10 \text{ mV/°F} = 1.0 \text{ V}$$

This means that we must set the (−) input pin of the comparator to 1.0 V and connect the LM34 to the (+) input. In order to create a 1.0-V reference voltage, we can use a voltage-divider circuit and choose a bias current of 100 μA. The LM339 input current will be relatively negligible because it will draw less than 1 μA. This means that $R_1 + R_2$ must total 10 kΩ. In this example, we can operate everything from a +5 V power supply. Figure 8-52 shows the com-

FIGURE 8-52 A temperature-limit detector using an LM339 analog voltage comparator.

plete circuit. The calculations are as follows:

$$V_{R2} = V_{CC} \cdot \frac{R_2}{R_1 + R_2}$$
$$R_2 = V_{R2} \cdot (R_1 + R_2)/V_{CC}$$
$$= 1.0 \text{ V}(10 \text{ k}\Omega)/5 \text{ V} = 2 \text{ k}\Omega$$
$$R_1 = 10 \text{ k}\Omega - R_2 = 10 \text{ k}\Omega - 2 \text{ k}\Omega = 8 \text{ k}\Omega$$

REVIEW QUESTIONS

1. What causes the output of a comparator to go to the HIGH logic state?
2. What causes the output of a comparator to go to the LOW logic state?
3. Is an LM339 output more similar to a TTL totem-pole or an open-collector output?

8-21   TROUBLESHOOTING

A **logic pulser** is a testing and troubleshooting tool that generates a short-duration pulse when manually actuated, usually by pressing a push button. The logic pulser shown in Figure 8-53 has a needle-shaped tip that is touched to the circuit node that is to be pulsed. The logic pulser is designed so that it senses the existing voltage level at the node and produces a voltage pulse in the opposite direction. In other words, if the node is LOW, the logic pulser produces a narrow positive-going pulse; if the node is HIGH, it produces a narrow negative-going pulse.

The logic pulser is used to change the logic level at a circuit node momentarily, even though the output of another device may be connected to that same node. In Figure 8-53, the logic pulser is contacting node X, which is also connected to the output of the NAND gate. The logic pulser has a very low output impedance (typically 2 Ω or less), so that it can overcome the NAND gate's output and can change the voltage at the node. The logic pulser, however, cannot produce a voltage pulse at a node that is shorted directly to ground or V_{CC} (e.g., through a solder bridge).

FIGURE 8-53   A logic pulser can inject a pulse at any node that is not shorted directly to ground or V_{CC}.

Using a Logic Pulser and Probe to Test a Circuit

A logic pulser can be used to inject a pulse or a series of pulses manually into a circuit in order to test the circuit's response. A logic probe is almost always used to monitor the circuit's response to the logic pulser. In Figure 8-53, the J-K flip-flop's toggle operation is being tested by applying pulses from the logic pulser and monitoring Q with the logic probe. This logic pulser/logic probe combination is very useful for checking the operation of a logic device while it is wired into a circuit. Note that the logic pulser is applied to the circuit node *without* disconnecting the output of the NAND gate that is driving that node. When the probe is placed on the same node as the pulser, the logic level indicators appear to remain unchanged (LOW in this example), but the yellow pulse indicator flashes once each time the pulser button is pressed. When the probe is placed on the Q output, the pulse LED flashes once (indicating a transition), and the logic level indicators change state each time the pulser button is pushed.

Finding Shorted Nodes

The logic pulser and logic probe can be used to check for nodes that are shorted directly to ground or V_{CC}, as shown in Figure 8-54. When you touch a logic pulser and a logic probe to the same node and press the logic pulser button, the logic probe should indicate the occurrence of a pulse at the node. If the probe indicates a constant LOW and the pulse LED does not flash, the node is shorted to ground, as shown in Figure 8-54(a). If the probe indicates a constant HIGH and the pulse LED does not flash, the node is shorted to V_{CC} as shown in Figure 8-54(b).

(a)

(b)

FIGURE 8-54   A logic pulser and a logic probe can be used to trace shorted nodes.

1. What is the function of a logic pulser?
2. *True or false:* A logic pulser will produce a voltage pulse at any node.
3. *True or false:* A logic pulser can force a node LOW or HIGH for extended periods of time.
4. How does a logic probe respond to the logic pulser?

SUMMARY

1. All digital logic devices are similar in nature but very much different regarding the details of their characteristics. An understanding of the terms used to describe these characteristics is important and allows us to compare and contrast the performance of devices. By understanding the capabilities and limitations of each type of device, we can intelligently combine devices to take advantage of each device's strengths in building reliable digital systems.

2. The TTL family of logic devices has been in use for over 30 years. The circuitry uses bipolar transistors. This family offers many SSI logic gates, and MSI devices. Numerous series of similarly numbered devices have been developed because advances in technology have offered improved characteristics.

3. When you are connecting devices together, it is vital to know how many inputs a given output can drive without compromising reliability. This is referred to as *fan-out*.

4. Open-collector and open-drain outputs can be wired together to implement a wired-AND function. Tristate outputs can be wired together to allow numerous devices to share a common data path known as a *bus.* In such a case, only one device is allowed to assert a logic level on the bus (i.e., drive the bus) at any one time.

5. The *fastest* logic devices are from a family that uses emitter-coupled logic (ECL). This technology also uses bipolar transistors but is not as widely used as TTL due to inconvenient input/output characteristics.

6. MOSFET transistors can also be used to implement logic functions. The main advantage of MOS logic is lower power and greater packing density.

7. The use of complementary MOSFETs has produced CMOS logic families. CMOS technology has captured the market due to its very low power and competitive speed.

8. The ongoing need to reduce power and size has led to several new series of devices that operate on 3.3 V and 2.5 V.

9. Logic devices that use various technologies cannot always be directly connected together and operate reliably. The voltage and current characteristics of inputs and outputs must be considered and precautions taken to ensure proper operation.

10. CMOS technology allows a digital system to control analog switches called *transmission gates.* These devices can pass or block an analog signal, depending on the digital logic level that controls it.

11. Analog voltage comparators offer another bridge between analog signals and digital systems. These devices compare analog voltages and output a digital logic level based on which voltage is greater. They allow an analog system to control a digital system.

IMPORTANT TERMS

fan-out
speed–power product
noise immunity
noise margin
current sourcing
current sinking
DIP
lead pitch
surface mount
TTL
totem pole
current-sinking transistor
  (pull-down transistor)

current-sourcing transistor
  (pull-up transistor)
floating inputs
power-supply decoupling
MOSFETs
P-MOS
N-MOS
CMOS
electrostatic discharge
  (ESD)
latch-up
open-collector output
wired-AND

buffer/driver
tristate
bus contention
emitter-coupled logic
  (ECL)
transmission gate
  (bilateral switch)
interfacing
voltage-level translator
analog voltage
  comparator
logic pulser

PROBLEMS

SECTIONS 8-1 TO 8-3

B   8-1. Two different logic circuits have the characteristics shown in Table 8-13.

    (a) Which circuit has the best LOW-state dc noise immunity? The best HIGH-state dc noise immunity?

    (b) Which circuit can operate at higher frequencies?

    (c) Which circuit draws the most supply current?

TABLE 8-13

	Circuit A	Circuit B
V_{supply} (V)	6	5
$V_{IH}(min)$ (V)	1.6	1.8
$V_{IL}(max)$ (V)	0.9	0.7
$V_{OH}(min)$ (V)	2.2	2.5
$V_{OL}(max)$ (V)	0.4	0.3
t_{PLH} (ns)	10	18
t_{PHL} (ns)	8	14
P_D (mW)	16	10

B   8-2. Look up the IC data sheets, and use *maximum* values to determine $P_D(avg)$, $t_{pd}(avg)$, and the speed–power product for one gate on each of the following TTL ICs. (See Example 8-2 in Section 8-3.)

    (a) 7432

    (b) 74S32

    (c) 74LS20

    (d) 74ALS20

    (e) 74AS20

8-3. A certain logic family has the following voltage parameters:

$$V_{IH}(min) = 3.5 \text{ V} \qquad V_{IL}(max) = 1.0 \text{ V}$$
$$V_{OH}(min) = 4.9 \text{ V} \qquad V_{OL}(max) = 0.1 \text{ V}$$

(a) What is the largest positive-going noise spike that can be tolerated?

(b) What is the largest negative-going noise spike that can be tolerated?

DRILL QUESTION

B   8-4. For each statement, indicate the term or parameter being described.

(a) Current at an input when a logic 1 is applied to that input

(b) Current drawn from the V_{CC} source when all outputs are LOW

(c) Time required for an output to switch from the 1 to the 0 state

(d) A common measure used to compare overall performance of different IC families

(e) The size of the voltage spike that can be tolerated on a HIGH input without causing indeterminate operation

(f) An IC package that does not require holes to be drilled in the printed circuit board

(g) When a LOW output receives current from the input of the circuit it is driving

(h) Number of different inputs that an output can safely drive

(i) Arrangement of output transistors in a standard TTL circuit

(j) Another term that describes pull-down transistor Q_4

(k) Range of V_{CC} values allowed for TTL

(l) $V_{OH}(min)$ and $V_{IH}(min)$ for the 74ALS series

(m) $V_{IL}(max)$ and $V_{OL}(max)$ for the 74ALS series

(n) When a HIGH output supplies current to a load

SECTION 8-4

8-5. (a) From Table 8-6, determine the noise margins when a 74LS device is driving a 74ALS input.

(b) Repeat part (a) for a 74ALS driving a 74LS.

(c) What will be the overall noise margin of a logic circuit that uses 74LS and 74ALS circuits in combination?

(d) A certain logic circuit has $V_{IL}(max) = 450$ mV. Which TTL series can be used with this circuit?

SECTIONS 8-5 AND 8-6

B   8-6. DRILL QUESTION

(a) Define *fan-out*.

(b) In which type of gates do tied-together inputs always count as a single input load in the LOW state?

(c) Define "floating" inputs.

(d) What causes current spikes in TTL? What undesirable effect can they produce? What can be done to reduce this effect?

(e) When a TTL output drives a TTL input, where does I_{OL} come from? Where does I_{OH} go?

8-7. Use Table 8-12 to find the fan-out for interfacing the first logic family to drive the second.

(a) 74AS to 74AS

(b) 74F to 74F

(c) 74AHC to 74AS

(d) 74HC to 74ALS

B   8-8. Refer to the data sheet for the 74LS112 J-K flip-flop.

(a) Determine the HIGH and LOW load current at the J and K inputs.

(b) Determine the HIGH and LOW load current at the clock and clear inputs.

(c) How many 74LS112 clock inputs can the output of one 74LS112 drive?

B   8-9. Figure 8-55(a) shows a 74LS112 J-K flip-flop whose output is required to drive a total of eight standard TTL inputs. Because this exceeds the fan-out of the 74LS112, a buffer of some type is needed. Figure 8-55(b) shows one possibility using one of the NAND gates from the 74LS37 quad NAND buffer, which has a much higher fan-out than the 74LS112. Note that $\overline{Q}$ is used because the NAND is acting as an IN-VERTER. Refer to the data sheet for the 74LS37.

(a) Determine its maximum fan-out to standard TTL.

(b) Determine its maximum sink current in the LOW state.

(a)   (b)

FIGURE 8-55   Problems 8-9 and 8-10.

D   8-10. Buffer gates are generally more expensive than ordinary gates, and sometimes there are unused ordinary gates available that can be used to solve a loading problem such as that in Figure 8-55(a). Show how 74LS00 NAND gates can be used to solve this problem.

B 8-11. Refer to the logic diagram of Figure 8-56, where the 74LS86 exclusive-OR output is driving several 74LS20 inputs. Determine whether the fan-out of the 74LS86 is being exceeded, and explain. Repeat using all 74AS devices. Use Table 8-7.

FIGURE 8-56 Problems 8-11 and 8-13.

B 8-12. How long does it take for the output of a typical 74LS04 to change states in response to a positive-going transition at its input?

C 8-13. For the circuit of Figure 8-56, determine the longest time it will take for a change in the *A* input to be felt at output *W*. Use all worst-case conditions and maximum values of gate propagation delays. (*Hint:* Remember that NAND gates are inverting gates.) Repeat using all 74ALS devices. Use the TI CD ROM.

8-14. (a) Figure 8-57 shows a 74LS193 counter with its active-HIGH master reset input activated by a push-button switch. Resistor *R* is used to hold *MR* LOW while the switch is open. What is the maximum value that can be used for *R*? Use the TI CD ROM.

(b) Repeat part (a) for the 74ALS193.

FIGURE 8-57 Problem 8-14.

C, T 8-15. Figure 8-58(a) shows a circuit used to convert a 60-Hz sine wave to a 60-pps signal that can reliably trigger FFs and counters. This type of circuit might be used in a digital clock.

(a) Explain the circuit operation.

(b) A technician is testing this circuit and observes that the 74LS14 output stays LOW. He checks the waveform at the INVERTER input, and it appears as shown in Figure 8-58(b). Thinking that the INVERTER is faulty, he replaces the chip and observes the same results. What do you think is causing the problem, and how can it be fixed? (*Hint:* Examine the v_x waveform carefully.)

(a)

(b)

FIGURE 8-58 Problem 8-15.

T 8-16. For each waveform in Figure 8-59, determine *why* it will *not* reliably trigger a 74LS112 flip-flop at its *CLK* input.

(a)

(b)

(c)

FIGURE 8-59 Problem 8-16.

T 8-17. A technician breadboards a logic circuit for testing. As she tests the circuit's operation, she finds that many of the FFs and counters are triggering erratically. Like any good technician, she checks the V_{CC} line with a dc meter and reads 4.97 V, which is acceptable for TTL. She then checks all circuit wiring and replaces each IC one by one, but the problem persists. Finally she decides to observe V_{CC} on the scope and sees the waveform shown in Figure 8-60. What is the probable cause of the noise on V_{CC}? What did the technician forget to include when she breadboarded the circuit?

FIGURE 8-60 Problem 8-17. V_{CC} 4.97 V

1.3 V

SECTIONS 8-7 TO 8-11

B   8-18. Which type of MOSFET is turned on by placing
  (a) 5 V on the gate and 0 V on the source?
  (b) 0 V on the gate and 5 V on the source?

B   8-19. Which of the following are advantages that CMOS generally has over TTL?
  (a) Greater packing density
  (b) Higher speed
  (c) Greater fan-out
  (d) Lower output impedance
  (e) Simpler fabrication process
  (f) More suited for LSI
  (g) Lower P_D (below 1 MHz)
  (h) Transistors as only circuit element
  (i) Lower input capacitance
  (j) Less susceptible to ESD

   8-20. Which of the following operating conditions will probably result in the lowest average P_D for a CMOS logic system? Explain.
  (a) $V_{DD} = 5$ V, switching frequency $f_{max} = 1$ MHz
  (b) $V_{DD} = 5$ V, $f_{max} = 10$ kHz
  (c) $V_{DD} = 10$ V, $f_{max} = 10$ kHz

C   8-21. The output of each INVERTER on a 74LS04 IC is driving two 74HCT08 inputs. The input to each INVERTER is LOW over 99 percent of the time. What is the maximum power that the 74LS04 chip is dissipating?

   8-22. Use the values from Table 8-9 to calculate the HIGH-state noise margin when a 74HC gate drives a standard 74LS input.

   8-23. What will cause latch-up in a CMOS IC? What might happen in this condition? What precautions should be taken to prevent latch-up?

   8-24. Refer to the data sheet for the 74HC20 NAND gate IC. Use maximum values to calculate P_D(avg), t_{pd}(avg), and the speed–power product. Compare with the values calculated in Problem 8-2 for TTL.

SECTIONS 8-12 AND 8-13

B   8-25. **DRILL QUESTION**
  (a) Define wired-AND.
  (b) What is a pull-up resistor? Why is it used?
  (c) What types of TTL outputs can safely be tied together?
  (d) What is bus contention?

D   8-26. The 74LS09 TTL IC is a quad two-input AND with open-collector outputs. Show how 74LS09s can be used to implement the operation $x = A \cdot B \cdot C \cdot D \cdot E \cdot F \cdot G \cdot H \cdot I \cdot J \cdot K \cdot M$.

B   8-27. Determine the logic expression for output X in Figure 8-61.

FIGURE 8-61   Problem 8-27.

C   8-28. Which of the following would be most likely to destroy a TTL totem-pole output while it is trying to switch from HIGH to LOW?

(a) Tying the output to +5 V

(b) Tying the output to ground

(c) Applying an input of 7 V

(d) Tying the output to another TTL totem-pole output

D   8-29. Figure 8-62(a) shows a 7406 open-collector inverting buffer used to control the ON/OFF status of an LED to indicate the state of FF output Q. The LED's nominal specification is $V_F = 2.4$ V at $I_F = 20$ mA, and $I_F(\text{max}) = 30$ mA.

(a) What voltage will appear at the 7406 output when $Q = 0$?

(b) Choose an appropriate value for the series resistor for proper operation.

(a)   (b)

FIGURE 8-62   Problems 8-29 and 8-30.

8-30. In Figure 8-62(b), the 7406 output is used to switch current to a relay.

(a) What voltage will be at the 7406 output when $Q = 0$?

(b) What is the largest current relay that can be used?

(c) How can we modify this circuit to use a 7407?

N   8-31. Figure 8-63 shows how two tristate buffers can be used to construct a *bidirectional transceiver* that allows digital data to be transmitted in either direction (*A* to *B*, or *B* to *A*). Describe the circuit operation for the two states of the DIRECTION input.

FIGURE 8-63   Problem 8-31.

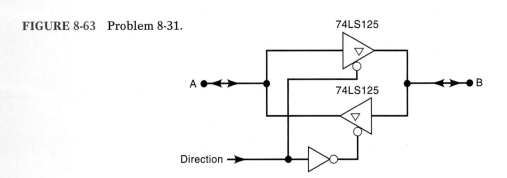

8-32. The circuit of Figure 8-64 is used to provide the enable inputs for the circuit of Figure 8-37.

    (a) Determine which of the data inputs (*A*, *B*, or *C*) will appear on the bus for each combination of inputs x and y.

    (b) Explain why the circuit will not work if the NOR is changed to an XNOR.

FIGURE 8-64   Problem 8-32.

8-33. What type of counter circuit from Chapter 7 could control the enables in Figure 8-37 so that only one buffer is on at any time, and the buffers are enabled sequentially?

SECTION 8-15

B   8-34. **DRILL QUESTIONS**

    (a) Which logic family must be used if maximum speed is of utmost importance?

    (b) Which logic family uses the most power?

    (c) Which TTL series is the fastest?

    (d) Which pure CMOS series is the fastest?

    (e) Which family has the best speed–power product?

8-35. Name two radical differences between ECL outputs and either TTL or CMOS outputs.

SECTION 8-16

8-36. Determine the approximate values of V_{OUT} for both states of the CONTROL input in Figure 8-65.

FIGURE 8-65   Problem 8-36.

8-37. Determine the waveform at output X in Figure 8-66 for the given input waveforms. Assume that $R_{ON} \approx 200 \ \Omega$ for the bilateral switch.

FIGURE 8-66   Problem 8-37.

N, D, C   8-38. Determine the gain of the op-amp circuit of Figure 8-67 for the two states of the GAIN SELECT input. This circuit shows the basic principle of digitally controlled signal amplification.

FIGURE 8-67   Problem 8-38.

SECTIONS 8-17 TO 8-19

B   8-39. **DRILL QUESTION**

(a) Which CMOS series can have its inputs driven directly from a TTL output?

(b) What is the function of a level translator? When is it used?

(c) Why is a buffer required between some CMOS outputs and TTL inputs?

(d) *True or false:* Most CMOS outputs have trouble supplying the TTL HIGH-state input current.

T   8-40. Refer to Figure 8-68(a), where a 74LS TTL output, Q, is driving a CMOS INVERTER operating at $V_{DD} = 10$ V. The waveforms at Q and X appear as shown in Figure 8-68(b). Which of the following is a possible reason why X stays HIGH?

(a) The 10-V supply is faulty.

(b) The pull-up resistor is too large.

(c) The 74LS112 output breaks down at well below 10 V and maintains a 5.5-V level in the HIGH state. This is in the indeterminate range for the CMOS input.

(d) The CMOS input is loading down the TTL output.

FIGURE 8-68   Problem 8-40.

8-41. (a) Use Table 8-12 to determine how many 74AS inputs can typically be driven by a 4000B output.

(b) Repeat part (a) for a 74HC output.

T   8-42. Figure 8-69 is a logic circuit that was poorly designed. It contains at least eight instances where the characteristics of the ICs have not been properly taken into account. Find as many of these as you can.

FIGURE 8-69   Problems 8-42 and 8-43.

T   8-43. Repeat Problem 8-42 with the following changes in the circuit:
- Each TTL IC is replaced with its 74LS equivalent.
- The 4001B is replaced with a 74HCT02.

8-44. Use Table 8-12 to explain why the circuit of Figure 8-70 will not work as it is. How can the problem be corrected?

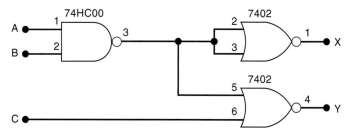

FIGURE 8-70   Problem 8-44.

SECTION 8-20

D   8-45. The gas tank on your car has a fuel-level sending unit that works like a potentiometer. A float moves up and down with the gasoline level, changing the variable resistor setting and producing a voltage proportional to the gas level. A full tank produces 12 V, and an empty tank produces 0 V. Design a circuit using an LM339 that turns on the "Fuel Low" indicator lamp when the voltage level from the sending unit gets below 0.5 V.

D   8-46. The over-temperature comparator circuit in Figure 8-52 is modified by replacing the LM34 temperature sensor with an LM35 that outputs 10 mV per degree Celsius. The alarm must still be activated (HIGH) when the temperature is over 100°F, which is equal to approximately 38°C. Recalculate the values of R_1 and R_2 to complete the modification.

SECTION 8-21

T   8-47. The circuit in Figure 8-71 uses a 74HC05 IC that contains six open-drain INVERTERs. The INVERTERs are connected in a wired-AND arrangement. The output of the NAND gate is always HIGH, regardless of the inputs A–H. Describe a procedure that uses a logic probe and pulser to isolate this fault.

FIGURE 8-71   Problem 8-47.

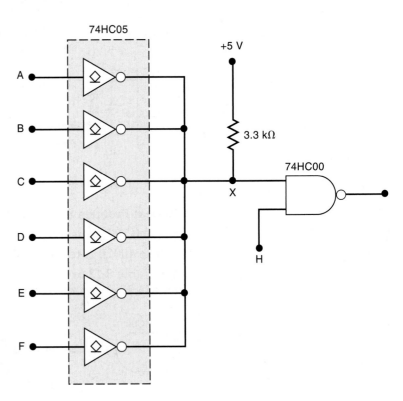

T   8-48. The circuit of Figure 8-53 has a solder bridge to ground somewhere between the output of the NAND gate and the input of the FF. Describe a procedure for a test that could be performed to indicate that the fault is on the circuit board and probably not in either the NAND or the FF ICs.

T   8-49. In Figure 8-46, a logic probe indicates that the lower end of the pull-up resistor is stuck in the LOW state. Which of the following is the possible fault?

(a) The TTL gate's current-sourcing transistor is open.

(b) The TTL gate's current-sinking transistor has a collector–emitter short.

(c) There is a break in the connection from R_P to the CMOS gate.

MICROCOMPUTER APPLICATION

C, N   8-50. In Chapter 5, we saw how a microprocessor (MPU), under software control, transfers data to an external register. The circuit diagram is repeated in Figure 8-72. Once the data are in the register, they are stored there and used for whatever purpose they are needed. Sometimes, each individual bit in the register has a unique function. For example, in automobile computers, each bit could represent the status of a different physical variable being monitored by the MPU. One bit might indicate when the engine temperature is too high. Another bit might signal when oil pressure is too low. In other applications, the bits in the register are used to produce an analog output that can be used to drive devices requiring analog inputs that have many different voltage levels.

FIGURE 8-72 Problem 8-50.

Figure 8-73 shows how we can use the MPU to generate an analog voltage by taking the register data from Figure 8-72 and using them to control the inputs to a summing amplifier. Assume that the MPU is executing a program that is transferring a new set of data to the register every 10 μs according to Table 8-14. Sketch the resulting waveform at V_{OUT}.

FIGURE 8-73   Problem 8-50.

TABLE 8-14   Problem 8-50.

Time (μs)	MPU Data
0	0000
10	0010
20	0100
30	0111
40	1010
50	1110
60	1111
70	1111
80	1110
90	1100
100	1000

ANSWERS TO SECTION REVIEW QUESTIONS

SECTION 8-1

1. See text.   2. False   3. False; V_{NH} is the difference between $V_{OH}(\text{min})$ and $V_{IH}(\text{min})$.   4. False   5. Current sinking: an output actually receives (sinks) current from the input of the circuit it is driving. Current sourcing: an output supplies (sources) current to the circuit it is driving.   6. DIP   7. J-lead   8. Its leads are bent.   9. No

SECTION 8-2

1. True   2. LOW   3. Fast switching times, low power dissipation; large current spike during switching from LOW to HIGH   4. Q_3 5. Q_6 6. No multiple-emitter transistor

SECTION 8-4

1. (a) 74AS (b) 74S, 74LS (c) standard 74 (d) 74S, 74LS, 74AS, 74ALS (e) 74ALS   2. All three can operate at 40 MHz, but the 74ALS193 will use less power.   3. Q_4, Q_5, respectively

SECTION 8-5

1. Q_4's ON-state resistance and V_{OL}(max)   2. 12   3. Its output voltages may not remain in the allowed logic 0 and 1 ranges.   4. Two; five

SECTION 8-6

1. LOW   2. Connect to $+V_{CC}$ through a 1-kΩ resistor; connect to another used input 3. Connect to ground; connect to another used input   4. False; only in the LOW state   5. Connecting small RF capacitors between V_{CC} and ground near each TTL IC to filter out voltage spikes caused by rapid current changes during output transitions from LOW to HIGH

SECTION 8-9

1. CMOS uses both P- and N-channel MOSFETs.   2. One   3. Six

SECTION 8-10

1. 74C, HC, HCT, AHC, AHCT   2. 74ACT, HCT, AHCT   3. 74C, HC/HCT, AC/ACT, AHC/AHCT   4. BiCMOS   5. Maximum permissible propagation delay; input capacitance of each load   6. See text.   7. CMOS   8. (a) True (b) False (c) False (d) False (e) True (f) False

SECTION 8-11

1. More circuits on chip; higher operating speed   2. Can't handle higher voltages: increased power dissipation can overheat the chip.   3. Same as standard TTL: 2.0 V 4. 74ALVC, 74LV   5. 74LVT

SECTION 8-12

1. When two or more circuit outputs are connected together   2. Damaging current can flow; V_{OL} exceeds V_{OL}(max).   3. Current-sinking transistor Q_4's collector is unconnected (there is no Q_3).   4. To produce a V_{OH} level   5. $\overline{A}\,\overline{B}\,\overline{C}\,\overline{D}\,\overline{E}\,\overline{F}$ 6. No active pull-up transistor   7. See Figure 8-34.

SECTION 8-13

1. HIGH, LOW, Hi-Z   2. Hi-Z   3. When two or more tristate outputs tied to a common bus are enabled at the same time   4. $E_A = E_B = 0, E_C = 1$ 5. See Figure 8-39.

SECTION 8-14

1. Less than 4 inches   2. Resistance, capacitance, inductance   3. To reduce reflections and reactive ringing on the line.

SECTION 8-15

1. (a) True (b) True (c) False (d) True (e) False (f) False

SECTION 8-16

1. The logical level at the control input controls the open/closed status of a bidirectional switch that can pass analog signals in either direction.   2. True

SECTION 8-17

1. The common output voltage will be in the indeterminate range.   2. Output stage consists only of an N-channel MOSFET whose drain is unconnected.

SECTION 8-18

1. A pull-up resistor must be connected to +5 V at the TTL output.   2. A high-voltage buffer (7407) or a voltage-level translator (4504 B) should be connected between TTL output and CMOS input.

SECTION 8-19

1. It takes the output from a driver circuit and conditions it so that it is compatible with the input requirements of the load.   2. True   3. False; for example, the 4000B series cannot sink I_{IL} of a 74 or a 74AS device.   4. 74HCT and ACT 5. Two   6. Through a voltage-level translator such as the 4050B

SECTION 8-20

1. $V^{(+)} > V^{(-)}$ 2. $V^{(-)} > V^{(+)}$ 3. Open-collector

SECTION 8-21

1. It injects a voltage pulse of selected polarity at a node that is not shorted to V_{CC} or ground.   2. False   3. False   4. The pulse LED flashes once each time the pulser is activated.

CHAPTER 9

MSI LOGIC CIRCUITS

■ OUTLINE

■ OBJECTIVES

Upon completion of this chapter, you will be able to:

- Analyze and use decoders and encoders in various types of circuit applications.
- Compare the advantages and disadvantages of LEDs and LCDs.
- Utilize the observation/analysis technique for troubleshooting digital circuits.
- Understand the operation of multiplexers and demultiplexers by analyzing several circuit applications.
- Compare two binary numbers by using the magnitude comparator circuit.
- Understand the function and operation of code converters.
- Cite the precautions that must be considered when connecting digital circuits using the data bus concept.
- Use HDL to implement the equivalent of MSI logic circuits.

■ INTRODUCTION

Digital systems obtain binary-coded data and information that are continuously being operated on in some manner. Some of the operations include: (1) *decoding and encoding,* (2) *multiplexing,* (3) *demultiplexing,* (4) *comparison,* (5) *code conversion,* and (6) *data busing.* All of these operations and others have been facilitated by the availability of numerous ICs in the MSI (medium-scale-integration) category.

In this chapter, we will study many of the common types of MSI devices. For each type, we will start with a brief discussion of its basic operating principle and then introduce specific ICs. We then show how they can be used alone or in combination with other ICs in various applications.

9-1 DECODERS

A **decoder** is a logic circuit that accepts a set of inputs that represents a binary number and activates only the output that corresponds to that input number. In other words, a decoder circuit looks at its inputs, determines which binary number is present there, and activates the one output that corresponds to that number; all other outputs remain inactive. The diagram for a general decoder is shown in Figure 9-1 with *N* inputs and *M* outputs. Because each of the *N* inputs can be 0 or 1, there are 2^N possible input combinations or codes. For each of these input combinations, only one of the *M* outputs will be active (HIGH); all the other outputs are LOW. Many decoders are designed to produce active-LOW outputs, where only the selected output is LOW while all others are HIGH. This situation is indicated by the presence of small circles on the output lines in the decoder diagram.

Some decoders do not utilize all of the 2^N possible input codes but only certain ones. For example, a BCD-to-decimal decoder has a four-bit input code and *ten* output lines that correspond to the *ten* BCD code groups 0000 through 1001. Decoders of this type are often designed so that if any of the unused codes are applied to the input, *none* of the outputs will be activated.

In Chapter 7, we saw how decoders are used in conjunction with counters to detect the various states of the counter. In that application, the FFs in the counter provided the binary code inputs for the decoder. The same basic decoder circuitry is used no matter where the inputs come from. Figure 9-2 shows the circuitry for a decoder with three inputs and $2^3 = 8$ outputs. It uses all AND gates, and so the outputs are active-HIGH. Note that for a given input code, the only output that is active (HIGH) is the one corresponding to the decimal equivalent of the binary input code (e.g., output O_6 goes HIGH only when $CBA = 110_2 = 6_{10}$).

This decoder can be referred to in several ways. It can be called a *3-line-to-8-line decoder* because it has three input lines and eight output lines. It can also be called a *binary-to-octal decoder* or *converter* because it takes a three-bit binary input code and activates one of the eight (octal) outputs corresponding to that code. It is also referred to as a *1-of-8 decoder* because only 1 of the 8 outputs is activated at one time.

FIGURE 9-1   General decoder diagram.

ENABLE Inputs

Some decoders have one or more ENABLE inputs that are used to control the operation of the decoder. For example, refer to the decoder in Figure 9-2 and visualize having a common ENABLE line connected to a fourth input of each gate. With this ENABLE line held HIGH, the decoder will function normally, and the *A, B, C* input code will determine which output is HIGH. With ENABLE held LOW, however, *all* of the outputs will be forced to the LOW state regardless of the levels at the *A, B, C* inputs. Thus, the decoder is enabled only if ENABLE is HIGH.

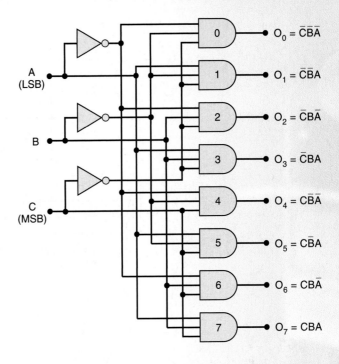

FIGURE 9-2 Three-line-to-8-line (or 1-of-8) decoder.

$O_0 = \overline{C}\,\overline{B}\,\overline{A}$

$O_1 = \overline{C}\,\overline{B}\,A$

$O_2 = \overline{C}\,B\,\overline{A}$

$O_3 = \overline{C}\,B\,A$

$O_4 = C\,\overline{B}\,\overline{A}$

$O_5 = C\,\overline{B}\,A$

$O_6 = C\,B\,\overline{A}$

$O_7 = C\,B\,A$

A (LSB)

B

C (MSB)

C	B	A	O_7	O_6	O_5	O_4	O_3	O_2	O_1	O_0
0	0	0	0	0	0	0	0	0	0	1
0	0	1	0	0	0	0	0	0	1	0
0	1	0	0	0	0	0	0	1	0	0
0	1	1	0	0	0	0	1	0	0	0
1	0	0	0	0	0	1	0	0	0	0
1	0	1	0	0	1	0	0	0	0	0
1	1	0	0	1	0	0	0	0	0	0
1	1	1	1	0	0	0	0	0	0	0

Figure 9-3(a) shows the logic diagram for the 74ALS138 decoder as it appears in the Fairchild *TTL Data Book*. By examining this diagram carefully, we can determine exactly how this decoder functions. First, notice that it has NAND gate outputs, so its outputs are active-LOW. Another indication is the labeling of the outputs as $\overline{O}_7$, $\overline{O}_6$, $\overline{O}_5$, and so on; the overbar indicates active-LOW outputs.

The input code is applied at A_2, A_1, and A_0, where A_2 is the MSB. With three inputs and eight outputs, this is a 3-to-8 decoder or, equivalently, a 1-of-8 decoder.

Inputs $\overline{E}_1$, $\overline{E}_2$, and E_3 are separate enable inputs that are combined in the AND gate. In order to enable the output NAND gates to respond to the input code at $A_2A_1A_0$, this AND gate output must be HIGH. This will occur only when $\overline{E}_1 = \overline{E}_2 = 0$ and $E_3 = 1$. In other words, $\overline{E}_1$ and $\overline{E}_2$ are active-LOW, E_3 is active-HIGH, and all three must be in their active states to activate the decoder outputs. If one or more of the enable inputs is in its inactive state, the AND output will be LOW, which will force all NAND outputs to their inactive HIGH state regardless of the input code. This operation is summarized in the truth table in Figure 9-3(b). Recall that *x* represents the don't-care condition.

$\bar{E}_1$	$\bar{E}_2$	E_3	Outputs
0	0	1	Respond to input code $A_2A_1A_0$
1	X	X	Disabled – all HIGH
X	1	X	Disabled – all HIGH
X	X	0	Disabled – all HIGH

(b)

(c)

FIGURE 9-3 (a) Logic diagram for the 74ALS138 decoder; (b) truth table; (c) logic symbol. (Courtesy of Fairchild, a Schlumberger company)

The logic symbol for the 74ALS138 is shown in Figure 9-3(c). Note how the active-LOW outputs are represented and how the enable inputs are represented. Even though the enable AND gate is shown as external to the decoder block, it is part of the IC's internal circuitry. The 74HC138 is the high-speed CMOS version of this decoder.

EXAMPLE 9-1

Indicate the states of the 74ALS138 outputs for each of the following sets of inputs.

(a) $E_3 = \bar{E}_2 = 1, \bar{E}_1 = 0, A_2 = A_1 = 1, A_0 = 0$
(b) $E_3 = 1, \bar{E}_2 = \bar{E}_1 = 0, A_2 = 0, A_1 = A_0 = 1$

Solution

(a) With $\overline{E}_2 = 1$, the decoder is disabled and all of its outputs will be in their inactive HIGH state. This can be determined from the truth table or by following the input levels through the circuit logic.

(b) All of the enable inputs are activated, so the decoding portion is enabled. It will decode the input code $011_2 = 3_{10}$ to activate output $\overline{O}_3$. Thus, $\overline{O}_3$ will be LOW and all other outputs will be HIGH.

EXAMPLE 9-2

Figure 9-4 shows how four 74ALS138s and an INVERTER can be arranged to function as a 1-of-32 decoder. The decoders are labeled Z_1 to Z_4 for easy reference, and the eight outputs from each one are combined into 32 outputs. Z_1's outputs are $\overline{O}_0$ to $\overline{O}_7$; Z_2's outputs $\overline{O}_0$ to $\overline{O}_7$ are renamed $\overline{O}_8$ to $\overline{O}_{15}$, respectively; Z_3's outputs are renamed $\overline{O}_{16}$ to $\overline{O}_{23}$; and Z_4's are renamed $\overline{O}_{24}$ to $\overline{O}_{31}$. A five-bit input code $A_4A_3A_2A_1A_0$ will activate only one of these 32 outputs for each of the 32 possible input codes.

(a) Which output will be activated for $A_4A_3A_2A_1A_0 = 01101$?

(b) What range of input codes will activate the Z_4 chip?

FIGURE 9-4 Four 74ALS138s forming a 1-of-32 decoder.

Solution

(a) The five-bit code has two distinct portions. The A_4 and A_3 bits determine which one of the decoder chips Z_1 to Z_4 will be enabled, while $A_2A_1A_0$ determine which output of the enabled chip will be activated. With $A_4A_3 = 01$, only Z_2 has all of its enable inputs activated. Thus, Z_2 responds to the $A_2A_1A_0 = 101$ code and activates its $\overline{O}_5$ output, which has been renamed $\overline{O}_{13}$. Thus, the input code 01101, which is the binary equivalent of decimal 13, will cause output $\overline{O}_{13}$ to go LOW, while all others stay HIGH.

(b) To enable Z_4, both A_4 and A_3 must be HIGH. Thus, all input codes ranging from 11000 (24_{10}) to 11111 (31_{10}) will activate Z_4. This corresponds to outputs $\overline{O}_{24}$ to $\overline{O}_{31}$.

BCD-to-Decimal Decoders

Figure 9-5(a) shows the logic diagram for a 7442 **BCD-to-decimal decoder**. It is also available as a 74LS42 and a 74HC42. Each output goes LOW only when its corresponding BCD input is applied. For example, $\overline{O}_5$ will go LOW only when inputs $DCBA = 0101$; $\overline{O}_8$ will go LOW only when $DCBA = 1000$. For input combinations that are invalid for BCD, none of the outputs will be activated. This decoder can also be referred to as a *4-to-10 decoder* or a *1-of-10 decoder*. The logic symbol and the truth table for the 7442 are also shown in the figure. Note that this decoder does not have an enable input. In Problem 9-7, we will see how the 7442 can be used as a 3-to-8 decoder, with the D input used as an enable input.

Inputs				Active Output
D	C	B	A	
L	L	L	L	$\overline{O}_0$
L	L	L	H	$\overline{O}_1$
L	L	H	L	$\overline{O}_2$
L	L	H	H	$\overline{O}_3$
L	H	L	L	$\overline{O}_4$
L	H	L	H	$\overline{O}_5$
L	H	H	L	$\overline{O}_6$
L	H	H	H	$\overline{O}_7$
H	L	L	L	$\overline{O}_8$
H	L	L	H	$\overline{O}_9$
H	L	H	L	None
H	L	H	H	None
H	H	L	L	None
H	H	L	H	None
H	H	H	L	None
H	H	H	H	None

H = HIGH Voltage Level
L = LOW Voltage Level

(c)

(a)

FIGURE 9-5   (a) Logic diagram for the 7442 BCD-to-decimal decoder; (b) logic symbol; (c) truth table. (Courtesy of Fairchild, a Schlumberger company)

BCD-to-Decimal Decoder/Driver

The TTL 7445 is a BCD-to-decimal decoder/**driver**. The term *driver* is added to its description because this IC has open-collector outputs that can operate at higher current and voltage limits than a normal TTL output. The 7445's outputs can sink up to 80 mA in the LOW state, and they can be pulled up to 30 V in the HIGH state. This makes them suitable for directly driving loads such as indicator LEDs or lamps, relays, or dc motors.

Decoder Applications

Decoders are used whenever an output or a group of outputs is to be activated only on the occurrence of a specific combination of input levels. These input levels are often provided by the outputs of a counter or a register. When the decoder inputs come from a counter that is being continually pulsed, the decoder outputs will be activated sequentially, and they can be used as timing or sequencing signals to turn devices on or off at specific times. An example of this operation is shown in Figure 9-6 using the 74LS293 counter and the 7445 decoder/driver described above.

(a)

(b)

FIGURE 9-6 Example 9-3: counter/decoder combination used to provide timing and sequencing operations.

EXAMPLE 9-3

Describe the operation of the circuit in Figure 9-6(a).

Solution

The counter is being pulsed by a 1-pps signal so that it will sequence through the binary counts at the rate of 1 count/s. The counter FF outputs are connected as the inputs to the decoder. The 7445 open-collector outputs $\overline{O}_3$ and $\overline{O}_6$ are used to switch relays K_1 and K_2 on and off. For instance, when $\overline{O}_3$ is in its inactive HIGH state, its output transistor will be off (nonconducting) so that no current can flow through relay K_1 and it will be deenergized. When $\overline{O}_3$ is in its active-LOW state, its output transistor is on and acts as a current sink for current through K_1 so that K_1 is energized. Note that the relays operate from +24 V. Also note the presence of the diodes across the relay coils; these protect the decoder's output transistors from the large "inductive kick" voltage that would be produced when coil current is stopped abruptly.

The timing diagram in Figure 9-6(b) shows the sequence of events. If we assume that the counter is in the 0000 state at time 0, then both outputs $\overline{O}_3$ and $\overline{O}_6$ are initially in the inactive HIGH state, where their output transistors are off and both relays are deenergized. As clock pulses are applied, the counter will be incremented once per second. On the NGT of the third pulse (time 3), the counter will go to the 0011 (3) state. This will activate decoder output $\overline{O}_3$ and thereby energize K_1. On the NGT of the fourth pulse, the counter goes to the 0100 (4) state. This will deactivate $\overline{O}_3$ and deenergize relay K_1.

Similarly, at time 6, the counter will go to the 0110 (6) state; this will make $\overline{O}_6 = 0$ and energize K_2. At time 7, the counter goes to 0111 (7) and deactivates $\overline{O}_6$ to deenergize K_2.

The counter will continue counting as pulses are applied. After 16 pulses, the sequence just described will start over.

Decoders are widely used in the memory system of a computer where they respond to the address code generated by the central processor to activate a particular memory location. Each memory IC contains many registers that can store binary numbers (data). Each register needs to have its own unique address to distinguish it from all the other registers. A decoder is built into the memory IC's circuitry and allows a particular storage register to be activated when a unique combination of inputs (i.e., its address) is applied. In a system, there are usually several memory ICs combined to make up the entire storage capacity. A decoder is used to select a memory chip in response to a range of addresses by decoding the most significant bits of the system address and enabling (selecting) a particular chip. We will examine this application in Problem 9-63, and we will study it in much more depth when we read about memories in Chapter 12.

In more complicated memory systems, the memory chips are arranged in multiple banks that must be selected individually or simultaneously, depending on whether the microprocessor wants one or more bytes at a time. This means that under certain circumstances, more than one output of the decoder must be activated. For systems such as this, a programmable logic device is often used to implement the decoder because a simple 1-of-8 decoder alone is not sufficient. Programmable logic devices can be used easily for custom decoding applications.

REVIEW QUESTIONS

1. Can more than one decoder output be activated at one time?
2. What is the function of a decoder's enable input(s)?
3. How does the 7445 differ from the 7442?
4. The 74154 is a 4-to-16 decoder with two active-LOW enable inputs. How many pins (including power and ground) does this IC have?

9-2   BCD-TO-7-SEGMENT DECODER/DRIVERS

Most digital equipment has some means for displaying information in a form that can be understood readily by the user or operator. This information is often numerical data but can also be alphanumeric (numbers and letters). One of the simplest and most popular methods for displaying numerical digits uses a 7-segment configuration [Figure 9-7(a)] to form the decimal characters 0 through 9 and sometimes the hex characters A through F. One common arrangement uses light-emitting diodes (LEDs) for each segment. By controlling the current through each LED, some segments will be light and others will be dark so that the desired character pattern will be generated. Figure 9-7(b) shows the segment patterns that are used to display the various digits. For example, to display a "6," the segments *a, c, d, e, f,* and *g* are made bright while segment *b* is dark.

FIGURE 9-7   (a) 7-segment arrangement; (b) active segments for each digit.

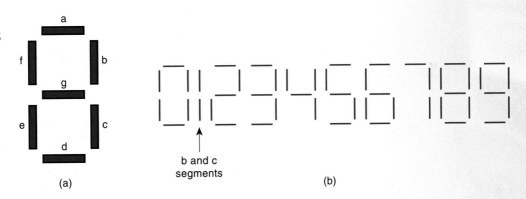

A **BCD-to-7-segment decoder/driver** is used to take a four-bit BCD input and provide the outputs that will pass current through the appropriate segments to display the decimal digit. The logic for this decoder is more complicated than the logic of decoders that we have looked at previously because each output is activated for more than one combination of inputs. For example, the *e* segment must be activated for any of the digits 0, 2, 6, and 8, which means whenever any of the codes 0000, 0010, 0110, or 1000 occurs.

Figure 9-8(a) shows a BCD-to-7-segment decoder/driver (TTL 7446 or 7447) being used to drive a 7-segment LED readout. Each segment consists of one or two LEDs. The anodes of the LEDs are all tied to V_{CC} (+5 V). The cathodes of the LEDs are connected through current-limiting resistors to the appropriate outputs of the decoder/driver. The decoder/driver has active-LOW outputs that are open-collector driver transistors and can sink a fairly large current because LED readouts may require 10 to 40 mA per segment, depending on their type and size.

To illustrate the operation of this circuit, let us suppose that the BCD input is $D = 0, C = 1, B = 0, A = 1$, which is BCD for 5. With these inputs, the decoder/driver outputs $\bar{a}, \bar{f}, \bar{g}, \bar{c}$, and $\bar{d}$ will be driven LOW (connected to ground), allowing current to flow through the *a, f, g, c,* and *d* LED segments and thereby displaying the numeral 5. The $\bar{b}$ and $\bar{e}$ outputs will be HIGH (open), so that LED segments *b* and *e* cannot conduct.

The 7446/47 decoder/drivers are designed to activate specific segments even for non-BCD input codes (greater than 1001). Figure 9-8(b) shows the activated segment patterns for all possible input codes from 0000 to 1111. Note that an input code of 1111 (15) will blank out all the segments.

FIGURE 9-8 (a) BCD-to-7-segment decoder/driver driving a common-anode 7-segment LED display; (b) segment patterns for all possible input codes.

Seven-segment decoder/drivers such as the 7446/47 are exceptions to the rule that decoder circuits activate only one output for each combination of inputs. Rather, they activate a unique pattern of outputs for each combination of inputs.

Common-Anode Versus Common-Cathode LED Displays

The LED display used in Figure 9-8 is a **common-anode** type because the anodes of all of the segments are tied together to V_{CC}. Another type of 7-segment LED display uses a **common-cathode** arrangement where the cathodes of all of the segments are tied together and connected to ground. This type of display must be driven by a BCD-to-7-segment decoder/driver with active-HIGH outputs that apply a HIGH voltage to the anodes of those segments that are to be activated. Because each segment requires 10 to 20 mA of current to light it, TTL and CMOS devices are normally not used to drive the common-cathode display directly. Recall from Chapter 8 that TTL and CMOS outputs are not able to source large amounts of current. A transistor interface circuit is often used between decoder chips and the common-cathode display.

EXAMPLE 9-4

Each segment of a typical 7-segment LED display is rated to operate at 10 mA at 2.7 V for normal brightness. Calculate the value of the current-limiting resistor needed to produce approximately 10 mA per segment.

Solution

Referring to Figure 9-8(a), we can see that the series resistor must have a voltage drop equal to the difference between $V_{CC} = 5$ V and the segment voltage of 2.7 V. This 2.3 V across the resistor must produce a current of about 10 mA. Thus, we have

$$R_S = \frac{2.3 \text{ V}}{10 \text{ mA}} = 230 \text{ }\Omega$$

A standard resistor value close to this can be used. A 220-Ω resistor would be a good choice.

REVIEW QUESTIONS

1. Which LED segments will be on for a decoder/driver input of 1001?
2. *True or false:* More than one output of a BCD-to-7-segment decoder/driver can be active at one time.

9-3 LIQUID-CRYSTAL DISPLAYS

An LED display generates or emits light energy as current is passed through the individual segments. A liquid-crystal display **(LCD)** controls the reflection of available light. The available light may simply be ambient (surrounding) light such as sunlight or normal room lighting; *reflective* LCDs use ambient light. Or the available light might be provided by a small light source that is part of the display unit; *backlit* LCDs use this method. In any case, LCDs have gained wide acceptance because of their very low power consumption compared to LEDs, especially in battery-operated equipment such as calculators, digital watches, and portable electronic measuring instruments. LEDs have the advantage of a much brighter display that, unlike reflective LCDs, is easily visible in dark or poorly lit areas.

Basically, LCDs operate from a low-voltage (typically 3 to 15 V rms), low-frequency (25 to 60 Hz) ac signal and draw very little current. They are often arranged as 7-segment displays for numerical readouts as shown in Figure 9-9(a). The ac voltage needed to turn on a segment is applied between the segment and the **backplane**, which is common to all segments. The segment and the backplane form a capacitor that draws very little current as long as the ac frequency is kept low. It is generally not lower than 25 Hz because this would produce visible flicker.

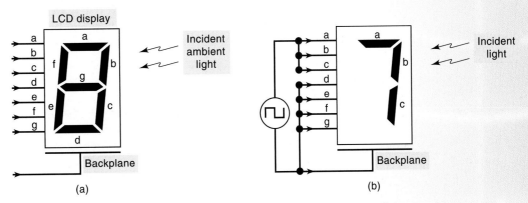

FIGURE 9-9 Liquid-crystal display: (a) basic arrangement; (b) applying a voltage between the segment and the backplane turns ON the segment. Zero voltage turns the segment OFF.

An admittedly simplified explanation of how an LCD operates goes something like this. When there is no difference in voltage between a segment and the backplane, the segment is said to be *nonactivated* (OFF). Segments *d, e, f,* and *g* in Figure 9-9(b) are OFF and will reflect incident light so that they appear invisible against their background. When an appropriate ac voltage is applied between a segment and the backplane, the segment is activated (ON). Segments *a, b,* and *c* in Figure 9-9(b) are ON and will not reflect the incident light, and thus they appear dark against their background.

Driving an LCD

An LCD segment will turn ON when an ac voltage is applied between the segment and the backplane, and will turn OFF when there is no voltage between the two. Rather than generating an ac signal, it is common practice to produce the required ac voltage by applying out-of-phase square waves to the segment and the backplane. This is illustrated in Figure 9-10(a) for one segment. A 40-Hz square wave is applied to the backplane and also to the input of a CMOS 74HC86 XOR. The other input to the XOR is a CONTROL input that will control whether the segment is ON or OFF.

When the CONTROL input is LOW, the XOR output will be exactly the same as the 40-Hz square wave, so that the signals applied to the segment and the backplane are equal. Because there is no difference in voltage, the segment will be OFF. When the CONTROL input is HIGH, the XOR output will be the INVERSE of the 40-Hz square wave, so that the signal applied to the segment is out of phase with the signal applied to the backplane. As a result, the segment voltage will alternately be at +5 V and at −5 V relative to the backplane. This ac voltage will turn ON the segment.

This same idea can be extended to a complete 7-segment LCD display, as shown in Figure 9-10(b). Here, the CMOS 74HC4511 BCD-to-7-segment decoder/driver supplies the CONTROL signals to each of seven XOR for the seven segments. The 74HC4511 has active-HIGH outputs because a HIGH is required to turn on a segment. The decoder/driver and XOR gates of Figure 9-10(b) are available on a single chip. The CMOS 74HC4543 is one such device. It takes the BCD input code and provides the outputs to drive the LCD segments directly.

In general, CMOS devices are used to drive LCDs for two reasons: (1) they require much less power than TTL and are more suited to the battery-operated applications where LCDs are used; (2) the TTL LOW-state voltage is not exactly 0 V and can be as much as 0.4 V. This will produce a dc component of voltage between the segment and the backplane that considerably shortens the life of an LCD.

Types of LCDs

Liquid crystals are available as multidigit 7-segment decimal numeric displays. They come in many sizes and with many special characters such as colons (:) for clock displays, + and − indicators for digital voltmeters, decimal points for calculators, and battery-low indicators because many LCD devices are battery-powered. These displays must be driven by a decoder/driver chip such as the 74HC4543.

A more complicated but readily available LCD display is the alphanumeric LCD module. These modules are available from many companies in numerous formats such as 1-line-by-16-characters up to 4-lines-by-40-characters. The in-

FIGURE 9-10 (a) Method for driving an LCD segment; (b) driving a 7-segment display.

terface to these modules has been standardized so that an LCD module from any manufacturer will use the same signals and data format. The module includes some VLSI chips that make this device simple to use. Eight data lines are used to send the ASCII code for whatever you wish to display. These data lines also carry special control codes to the LCD command register. Three other inputs (Register Select, Read/Write, and Enable) are used to control the location, direction, and timing of the data transfer. As characters are sent to the module, it stores them in its own memory and types them across the display screen.

Other LCD modules allow the user to create a graphical display by controlling individual dots on the screen called **pixels**. Larger LCD panels can be scanned at a high rate, producing high-quality video motion pictures. In these displays, the control lines are arranged in a grid of rows and columns. At the intersection of each row and column is a pixel that acts like a "window" or "shutter" that can be electronically opened and closed to control the amount of light that is transmitted through the cell. The voltage from a row to a column determines the brightness of each pixel. In a laptop computer, a binary number for each pixel is stored in the "video" memory. These numbers are converted to voltages that are applied to the display.

Each pixel on a color display is actually made up of three subpixels. These subpixels control the light that passes through a red, green, or blue filter to produce the color of each pixel. On a 640-by-480 LCD screen there would be 640×3 connections for columns and 480 connections for rows, for

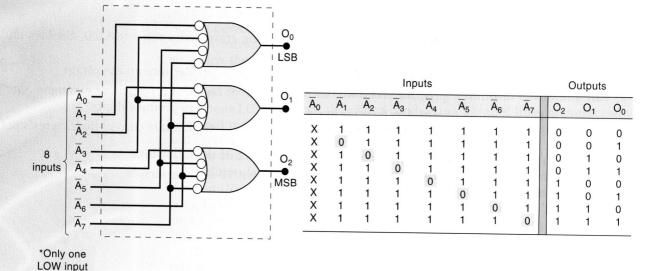

Inputs								Outputs		
$\overline{A}_0$	$\overline{A}_1$	$\overline{A}_2$	$\overline{A}_3$	$\overline{A}_4$	$\overline{A}_5$	$\overline{A}_6$	$\overline{A}_7$	O_2	O_1	O_0
X	1	1	1	1	1	1	1	0	0	0
X	0	1	1	1	1	1	1	0	0	1
X	1	0	1	1	1	1	1	0	1	0
X	1	1	0	1	1	1	1	0	1	1
X	1	1	1	0	1	1	1	1	0	0
X	1	1	1	1	0	1	1	1	0	1
X	1	1	1	1	1	0	1	1	1	0
X	1	1	1	1	1	1	0	1	1	1

8 inputs: $\overline{A}_0$, $\overline{A}_1$, $\overline{A}_2$, $\overline{A}_3$, $\overline{A}_4$, $\overline{A}_5$, $\overline{A}_6$, $\overline{A}_7$

O_0 LSB O_1 O_2 MSB

*Only one LOW input at a time

FIGURE 9-13   Logic circuit for an octal-to-binary (8-line-to-3-line) encoder. For proper operation, only one input should be active at one time.

EXAMPLE 9-5

Determine the outputs of the encoder in Figure 9-13 when $\overline{A}_3$ and $\overline{A}_5$ are simultaneously LOW.

Solution

Following through the logic gates, we see that the LOWs at these two inputs will produce HIGHs at each output, in other words, the binary code 111. Clearly, this is not the code for either activated input.

Priority Encoders

This last example identifies a drawback of the simple encoder circuit of Figure 9-13 when more than one input is activated at one time. A modified version of this circuit, called a **priority encoder**, includes the necessary logic to ensure that when two or more inputs are activated, the output code will correspond to the highest-numbered input. For example, when both $\overline{A}_3$ and $\overline{A}_5$ are LOW, the output code will be 101 (5). Similarly, when $\overline{A}_6$, $\overline{A}_2$, and $\overline{A}_0$ are all LOW, the output code is 110 (6). The 74148, 74LS148, and 74HC148 are all octal-to-binary priority encoders.

74147 Decimal-to-BCD Priority Encoder

Figure 9-14 shows the logic symbol and the truth table for the 74147 (74LS147, 74HC147), which functions as a decimal-to-BCD priority encoder. It has nine active-LOW inputs representing the decimal digits 1 through 9, and it produces the *inverted* BCD code corresponding to the highest-numbered activated input.

Let's examine the truth table to see how this IC works. The first line in the table shows all inputs in their inactive HIGH state. For this condition, the

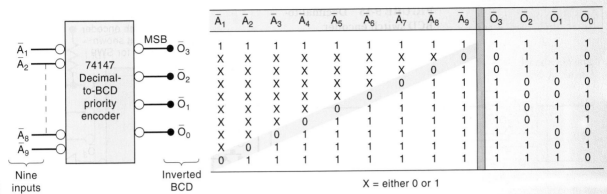

$\overline{A}_1$	$\overline{A}_2$	$\overline{A}_3$	$\overline{A}_4$	$\overline{A}_5$	$\overline{A}_6$	$\overline{A}_7$	$\overline{A}_8$	$\overline{A}_9$	$\overline{O}_3$	$\overline{O}_2$	$\overline{O}_1$	$\overline{O}_0$
1	1	1	1	1	1	1	1	1	1	1	1	1
X	X	X	X	X	X	X	X	0	0	1	1	0
X	X	X	X	X	X	X	0	1	0	1	1	1
X	X	X	X	X	X	0	1	1	1	0	0	0
X	X	X	X	X	0	1	1	1	1	0	0	1
X	X	X	X	0	1	1	1	1	1	0	1	0
X	X	X	0	1	1	1	1	1	1	0	1	1
X	X	0	1	1	1	1	1	1	1	1	0	0
X	0	1	1	1	1	1	1	1	1	1	0	1
0	1	1	1	1	1	1	1	1	1	1	1	0

X = either 0 or 1

FIGURE 9-14 74147 decimal-to-BCD priority encoder.

outputs are 1111, which is the inverse of 0000, the BCD code for 0. The second line in the table indicates that a LOW at $\overline{A}_9$, regardless of the states of the other inputs, will produce an output code of 0110, which is the inverse of 1001, the BCD code for 9. The third line shows that a LOW at $\overline{A}_8$, provided that $\overline{A}_9$ is HIGH, will produce an output code of 0111, the inverse of 1000, the BCD code for 8. In a similar manner, the remaining lines in the table show that a LOW at any input, provided that all higher-numbered inputs are HIGH, will produce the inverse of the BCD code for that input.

The 74147 outputs will normally be HIGH when none of the inputs are activated. This corresponds to the decimal 0 input condition. There is no $\overline{A}_0$ input because the encoder assumes the decimal 0 input state when all other inputs are HIGH. The 74147 inverted BCD outputs can be converted to normal BCD by putting each one through an INVERTER.

EXAMPLE 9-6

Determine the states of the outputs in Figure 9-14 when $\overline{A}_5$, $\overline{A}_7$, and $\overline{A}_3$ are LOW and all other inputs are HIGH.

Solution

The truth table shows that when $\overline{A}_7$ is LOW, the levels at $\overline{A}_5$ and $\overline{A}_3$ do not matter. Thus, the outputs will each be 1000, the inverse of 0111 (7).

Switch Encoder

Figure 9-15 shows how a 74147 can be used as a *switch encoder*. The 10 switches might be the keyboard switches on a calculator representing digits 0 through 9. The switches are of the normally open type, so that the encoder inputs are all normally HIGH and the BCD output is 0000 (note the INVERTERs). When a digit key is depressed, the circuit will produce the BCD code for that digit. Because the 74LS147 is a priority encoder, simultaneous key depressions will produce the BCD code for the higher-numbered key.

The switch encoder of Figure 9-15 can be used whenever BCD data must be entered manually into a digital system. A prime example would be in an electronic calculator, where the operator depresses several keyboard switches in succession to enter a decimal number. In a simple, basic calculator, the BCD code for each decimal digit is entered into a four-bit storage register. In other words, when the first key is depressed, the BCD code for that

4. The "3" key is released and the OR gate output returns LOW. The "0" key is then depressed. This produces the BCD code of 0000, which is fed to the inputs of the three registers.

5. The OR output goes HIGH in response to the "0" key (note the INVERTER) and triggers the OS for 20 ms. After 20 ms, the ring counter shifts to the 010 state (Y goes HIGH). The positive transition at Y is fed to the CLK inputs of Q_4 to Q_7 and transfers the 0000 to these FFs. Note that flip-flops Q_0 to Q_3 and Q_8 to Q_{11} are not affected by the Y transition.

6. The "0" key is released and the OR output returns LOW. The "9" key is depressed, producing BCD outputs 1001, which are fed to the storage registers.

7. The OR output goes HIGH again, triggering the OS, which in turn clocks the ring counter to the 001 state (Z goes HIGH). The positive transition at Z is fed to the CLK inputs of Q_0 to Q_3 and transfers the 1001 into these FFs. The other storage FFs are unaffected.

8. At this point, the storage register contains 001100001001, beginning with Q_{11}. This is the BCD code of 309. These register outputs feed decoder/drivers that drive appropriate displays for indicating the decimal digits 309.

9. The storage FF outputs are also fed to other circuits in the system. In a calculator, for example, these outputs would be sent to the arithmetic section to be processed.

Several problems at the end of the chapter will deal with some other aspects of this circuit, including troubleshooting exercises.

REVIEW QUESTIONS

1. How does an encoder differ from a decoder?
2. How does a priority encoder differ from an ordinary encoder?
3. What will the outputs be in Figure 9-15 when SW6, SW5, and SW2 are all closed?
4. Describe the functions of each of the following parts of the keyboard entry circuit of Figure 9-16.

   (a) OR gate   (d) Flip-flops X, Y, Z

   (b) 74147 encoder   (e) Flip-flops Q_0 to Q_{11}

   (c) One-shot

9-5 TROUBLESHOOTING

As circuits and systems become more complex, the number of possible causes of failure obviously increases. Whereas the procedure for fault isolation and correction remains essentially the same, the application of the **observation/analysis** process is more important for complex circuits because it helps the troubleshooter narrow the location of the fault to a small area of the circuit. This reduces to a reasonable amount the testing steps and resulting data that must be analyzed. By understanding the circuit operation, observing the symptoms of the failure, and reasoning through the operation, the troubleshooter can often predict the possible faults before ever picking up a logic probe or an oscilloscope. This observation/analysis process is one that inexperienced troubleshooters are hesitant to apply, probably because of the great variety and capabilities of modern test equipment available to them. It is easy to become overly reliant on these tools while not adequately utilizing the human brain's reasoning and analytical skills.

The following examples illustrate how the observation/analysis process can be applied. Many of the end-of-chapter troubleshooting problems will provide you with the opportunity to develop your skill at applying this process.

Another vital strategy in troubleshooting is known as **divide-and-conquer**. It is used to identify the location of the problem after observation/analysis has generated several possibilities. A less efficient method would be to investigate each possible cause, one by one. The divide-and-conquer method finds a point in the circuit that can be tested, thereby dividing the total possible number of causes in half. In simple systems, this may seem unnecessary, but as complexity increases, the total number of possible causes also increases. If there are eight possible causes, then a test should be performed that eliminates four of them. The next test should eliminate two more, and the third test should identify the problem.

EXAMPLE 9-7

A technician tests the circuit of Figure 9-4 by using a set of switches to apply the input code at A_4 through A_0. She runs through each possible input code and checks the corresponding decoder output to see if it is activated. She observes that all of the odd-numbered outputs respond correctly, but all of the even-numbered outputs fail to respond when their code is applied. What are the most probable faults?

Solution

In a situation where so many outputs are failing, it is unreasonable to expect that each of these outputs has a fault. It is much more likely that some faulty input condition is causing the output failures. What do all of the even-numbered outputs have in common? The input codes for several of them are listed in Table 9-1.

Clearly, each even-numbered output requires an input code with an $A_0 = 0$ in order to be activated. Thus, the most probable faults would be those that prevent A_0 from going LOW. These include:

1. A faulty switch connected to the A_0 input
2. A break in the path between the switch and the A_0 line
3. An external short from the A_0 line to V_{CC}
4. An internal short to V_{CC} at the A_0 inputs of any one of the decoder chips

Through observation and analysis, the technician has identified several possible causes. Potential causes 1 and 2 are in the switches generating the address. Causes 3 and 4 are in the decoder circuit itself. The circuit can be divided by opening the connection between the least significant switch and the A_0 input, as shown in Figure 9-17. A logic probe can be used to see if the switch can generate a LOW as well as a HIGH. Regardless of the outcome, two of the four possible causes have been eliminated.

Thus, the fault is narrowed to a specific area of the circuit. The exact fault can be traced with the testing and measurement techniques that we are already familiar with.

TABLE 9-1

Output	Input Code
$\overline{O}_0$	00000
$\overline{O}_4$	00100
$\overline{O}_{14}$	01110
$\overline{O}_{18}$	10010

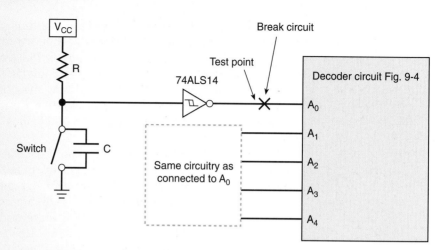

FIGURE 9-17 Troubleshooting circuitry in Example 9-7.

EXAMPLE 9-8

A technician wires the outputs from a BCD counter to the inputs of the decoder/driver of Figure 9-8. He applies pulses to the counter at a very slow rate and observes the LED display, which is shown below, as the counter counts up from 0000 to 1001. Examine this observed sequence carefully and try to predict the most probable fault.

COUNT	0	1	2	3	4	5	6	7	8	9
Observed display	0	1	⊐	3	4	5	6	7	8	⊓
Expected display	0	1	2	3	4	5	6	7	8	9

Solution

Comparing the observed display with the expected display for each count, we see several important points:

- For those counts where the observed display is incorrect, the observed display is not one of the segment patterns that correspond to counts greater than 1001.

- This rules out a faulty counter or faulty wiring from the counter to the decoder/driver.

- The correct segment patterns (0, 1, 3, 6, 7, and 8) have the common property that segments e and f are either both on or both off.

- The incorrect segment patterns have the common property that segments e and f are in opposite states, and if we interchange the states of these two segments, the correct pattern is obtained.

Giving some thought to these points should lead us to conclude that the technician has probably "crossed" the connections to the e and f segments.

9-6 MULTIPLEXERS (DATA SELECTORS)

A modern home stereo system may have a switch that selects music from one of four sources: a cassette tape, a compact disc (CD), a radio tuner, or an auxilliary input such as audio from a VCR or DVD. The switch selects one of the electronic signals from one of these four sources and sends it to the power amplifier and speakers. In simple terms, this is what a **multiplexer (MUX)** does: it selects one of several input signals and passes it on to the output.

A *digital multiplexer* or *data selector* is a logic circuit that accepts several digital data inputs and selects one of them at any given time to pass on to the output. The routing of the desired data input to the output is controlled by SELECT inputs (often referred to as ADDRESS inputs). Figure 9-18 shows the functional diagram of a general digital multiplexer. The inputs and outputs are drawn as wide arrows rather than lines; this indicates that they may actually be more than one signal line.

The multiplexer acts like a digitally controlled multiposition switch where the digital code applied to the SELECT inputs controls which data inputs will be switched to the output. For example, output Z will equal data input I_0 for some particular SELECT input code, Z will equal I_1 for another particular SELECT input code, and so on. Stated another way, a multiplexer selects 1 out of N input data sources and transmits the selected data to a single output channel. This is called **multiplexing**.

FIGURE 9-18 Functional diagram of a digital multiplexer (MUX).

SELECT input code determines which input is transmitted to output Z.

Basic Two-Input Multiplexer

Figure 9-19 shows the logic circuitry for a two-input multiplexer with data inputs I_0 and I_1 and SELECT input S. The logic level applied to the S input determines which AND gate is enabled so that its data input passes through the OR gate to output Z. Looking at it another way, the Boolean expression for the output is

$$Z = I_0\overline{S} + I_1S$$

With $S = 0$, this expression becomes

$$Z = I_0 \cdot 1 + I_1 \cdot 0 \qquad \text{[gate 2 enabled]}$$
$$= I_0$$

which indicates that Z will be identical to input signal I_0, which in turn can be a fixed logic level or a time-varying logic signal. With $S = 1$, the expression becomes

$$Z = I_0 \cdot 0 + I_1 \cdot 1 = I_1 \qquad \text{[gate 1 enabled]}$$

showing that output Z will be identical to input signal I_1.

FIGURE 9-19   Two-input multiplexer.

An example of where a two-input MUX could be used is in a digital system that uses two different MASTER CLOCK signals: a high-speed clock (say, 10 MHz) in one mode and a slow-speed clock (say, 4.77 MHz) for the other. Using the circuit of Figure 9-19, the 10-MHz clock would be tied to I_0, and the 4.77-MHz clock would be tied to I_1. A signal from the system's control logic section would drive the SELECT input to control which clock signal appears at output Z for routing to the other parts of the circuit.

Four-Input Multiplexer

The same basic idea can be used to form the four-input multiplexer shown in Figure 9-20. Here, four inputs are selectively transmitted to the output according to the four possible combinations of the S_1S_0 select inputs. Each data input is gated with a different combination of select input levels. I_0 is gated with $\overline{S_1}\,\overline{S_0}$ so that I_0 will pass through its AND gate to output Z only when $S_1 = 0$ and $S_0 = 0$. The table in the figure gives the outputs for the other three input-select codes.

FIGURE 9-20   Four-input multiplexer.

Two-, four-, eight-, and 16-input multiplexers are readily available in the TTL and CMOS logic families. These basic ICs can be combined for multiplexing a larger number of inputs.

Eight-Input Multiplexer

Figure 9-21(a) shows the logic diagram for the 74ALS151 (74HC151) eight-input multiplexer. This multiplexer has an enable input $\bar{E}$ and provides both the normal and the inverted outputs. When $\bar{E} = 0$, the select inputs $S_2S_1S_0$ will select one data input (from I_0 through I_7) for passage to output Z. When $\bar{E} = 1$, the multiplexer is disabled so that $Z = 0$ regardless of the select input code. This operation is summarized in Figure 9-21(b), and the 74151 logic symbol is shown in Figure 9-21(c).

(a)

Inputs				Outputs	
$\bar{E}$	S_2	S_1	S_0	$\bar{Z}$	Z
H	X	X	X	H	L
L	L	L	L	$\bar{I_0}$	I_0
L	L	L	H	$\bar{I_1}$	I_1
L	L	H	L	$\bar{I_2}$	I_2
L	L	H	H	$\bar{I_3}$	I_3
L	H	L	L	$\bar{I_4}$	I_4
L	H	L	H	$\bar{I_5}$	I_5
L	H	H	L	$\bar{I_6}$	I_6
L	H	H	H	$\bar{I_7}$	I_7

(b)

(c)

FIGURE 9-21   (a) Logic diagram for the 74ALS151 multiplexer; (b) truth table; (c) logic symbol. (Courtesy of Fairchild, a Schlumberger company)

EXAMPLE 9-9

The circuit in Figure 9-22 uses two 74HC151s, an INVERTER, and an OR gate. Describe this circuit's operation.

FIGURE 9-22   Example 9-9: two 74HC151s combined to form a 16-input multiplexer.

Solution

This circuit has a total of 16 data inputs, eight applied to each multiplexer. The two multiplexer outputs are combined in the OR gate to produce a single output X. The circuit functions as a 16-input multiplexer. The four select inputs $S_3 S_2 S_1 S_0$ will select one of the 16 inputs to pass through to X.

The S_3 input determines which multiplexer is enabled. When $S_3 = 0$, the top multiplexer is enabled, and the $S_2 S_1 S_0$ inputs determine which of its data inputs will appear at its output and pass through the OR gate to X. When $S_3 = 1$, the bottom multiplexer is enabled, and the $S_2 S_1 S_0$ inputs select one of its data inputs for passage to output X.

Quad Two-Input MUX (74ALS157/HC157)

The 74ALS157 is a very useful multiplexer IC that contains four two-input multiplexers like the one in Figure 9-19. The logic diagram for the 74ALS157 is shown in Figure 9-23(a). Note the manner in which the data inputs and outputs are labeled.

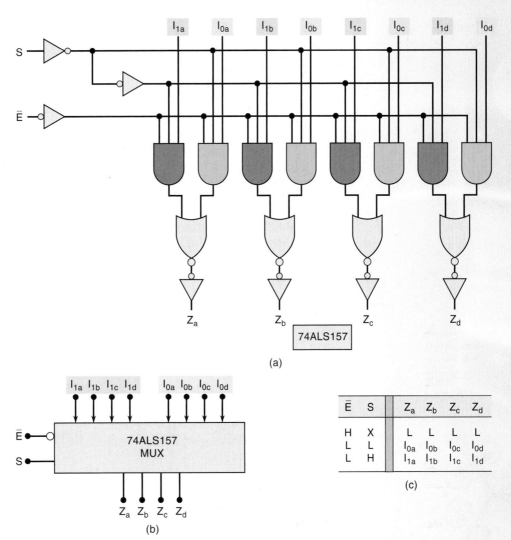

FIGURE 9-23   (a) Logic diagram for the 74ALS157 multiplexer; (b) logic symbol; (c) truth table. (Courtesy of Fairchild, a Schlumberger company)

EXAMPLE 9-10

Determine the input conditions required for each Z output to take on the logic level of its corresponding I_0 input. Repeat for I_1.

Solution

First of all, the enable input must be active; that is, $\overline{E} = 0$. In order for Z_a to equal I_{0a}, the select input must be LOW. These same conditions will produce $Z_b = I_{0b}$, $Z_c = I_{0c}$, and $Z_d = I_{0d}$.

With $\overline{E} = 0$ and $S = 1$, the Z outputs will follow the set of I_1 inputs; that is, $Z_a = I_{1a}$, $Z_b = I_{1b}$, $Z_c = I_{1c}$, and $Z_d = I_{1d}$.

All of the outputs will be disabled (LOW) when $\overline{E} = 1$.

It is helpful to think of this multiplexer as being a simple two-input multiplexer, but one in which each input is four lines and the output is four lines. The four output lines switch back and forth between the two sets of four input lines under the control of the select input. This operation is represented by the 74ALS157's logic symbol in Figure 9-23(b).

1. What is the function of a multiplexer's select inputs?
2. A certain multiplexer can switch one of 32 data inputs to its output. How many different inputs does this MUX have?

9-7 MULTIPLEXER APPLICATIONS

Multiplexer circuits find numerous and varied applications in digital systems of all types. These applications include data selection, data routing, operation sequencing, parallel-to-serial conversion, waveform generation, and logic-function generation. We shall look at some of these applications here and several more in the problems at the end of the chapter.

Data Routing

Multiplexers can route data from one of several sources to one destination. One typical application uses 74ALS157 multiplexers to select and display the contents of either of two BCD counters using a *single* set of decoder/drivers and LED displays. The circuit arrangement is shown in Figure 9-24.

FIGURE 9-24   System for displaying two multidigit BCD counters one at a time.

Each counter consists of two cascaded BCD stages, and each one is driven by its own clock signal. When the COUNTER SELECT line is HIGH, the outputs of counter 1 will be allowed to pass through the multiplexers to the decoder/drivers to be displayed on the LED readouts. When COUNTER SELECT = 0, the outputs of counter 2 will pass through the multiplexers to the displays. In this way, the decimal contents of one counter or the other will be displayed under the control of the COUNTER SELECT input. A common situation where this might be used is in a digital watch. The digital watch circuitry contains many counters and registers that keep track of seconds, minutes, hours, days, months, alarm settings, and so on. A multiplexing scheme such as this one allows different data to be displayed on the limited number of decimal readouts.

The purpose of the multiplexing technique, as it is used here, is to *time-share* the decoder/drivers and display circuits between the two counters rather than have a separate set of decoder/drivers and displays for each counter. This results in a significant saving in the number of wiring connections, especially when more BCD stages are added to each counter. Even more important, it represents a significant decrease in power consumption because decoder/drivers and LED readouts typically draw relatively large amounts of current from the V_{CC} supply. Of course, this technique has the limitation that only one counter's contents can be displayed at a time. However, in many applications, this limitation is not a drawback. A mechanical switching arrangement could have been used to perform the function of switching first one counter and then the other to the decoder/drivers and displays, but the number of required switch contacts, the complexity of wiring, and the physical size could all be disadvantages over the completely logic method of Figure 9-24.

Parallel-to-Serial Conversion

Many digital systems process binary data in parallel form (all bits simultaneously) because it is faster. When data are to be transmitted over relatively long distances, however, the parallel arrangement is undesirable because it requires a large number of transmission lines. For this reason, binary data or information in parallel form is often converted to serial form before being transmitted to a remote destination. One method for performing this **parallel-to-serial conversion** uses a multiplexer, as illustrated in Figure 9-25 on page 612.

The data are present in parallel form at the outputs of the X register and are fed to the eight-input multiplexer. A three-bit (MOD-8) counter is used to provide the select code bits $S_2S_1S_0$ so that they cycle through from 000 to 111 as clock pulses are applied. In this way, the output of the multiplexer will be X_0 during the first clock period, X_1 during the second clock period, and so on. The output Z is a waveform that is a serial representation of the parallel input data. The waveforms in the figure are for the case where $X_7X_6X_5X_4X_3X_2X_1X_0 = 10110101$. This conversion process takes a total of eight clock cycles. Note that X_0 (the LSB) is transmitted first and the X_7 (MSB) is transmitted last.

Operation Sequencing

The circuit of Figure 9-26 (found on page 613) uses an eight-input multiplexer as part of a control sequencer that steps through seven steps, each of which actuates some portion of the physical process being controlled. This could be, for example, a process that mixes two liquid ingredients and then

FIGURE 9-25 (a) Parallel-to-serial converter; (b) waveforms for $X_7X_6X_5X_4X_3X_2X_1X_0$ = 10110101.

cooks the mixture. The circuit also uses a 3-line-to-8-line decoder and a MOD-8 binary counter. The operation is described as follows:

1. Initially the counter is reset to the 000 state. The counter outputs are fed to the select inputs of the multiplexer and to the inputs of the decoder. Thus, the decoder output $\overline{O}_0 = 0$ and the others are all 1, so that all the ACTUATOR inputs of the process are LOW. The SENSOR outputs of the process all start out LOW. The multiplexer output $\overline{Z} = \overline{I}_0 = 1$ because the S inputs are 000.

2. The START pulse initiates the sequencing operation by setting flip-flop Q_0 HIGH, bringing the counter to the 001 state. This causes decoder output $\overline{O}_1$ to go LOW, thereby activating actuator 1, which is the first step in the process (opening fill valve 1).

3. Some time later, SENSOR output 1 goes HIGH, indicating the completion of the first step (the float switch indicates that the tank is full). This HIGH is now present at the I_1 input of the multiplexer. It is inverted and reaches the $\overline{Z}$ output because the select code from the counter is 001.

4. The LOW transition at $\overline{Z}$ is fed to the CLK of flip-flop Q_0. This negative transition advances the counter to the 010 state.

5. Decoder output $\overline{O}_2$ now goes LOW, activating actuator 2, which is the second step in the process (opening fill valve 2). $\overline{Z}$ now equals $\overline{I}_2$ (the select code is 010). Because SENSOR output 2 is still LOW, $\overline{Z}$ will go HIGH.

FIGURE 9-26 Seven-step control sequencer.

6. When the second process step is complete, SENSOR output 2 goes HIGH, producing a LOW at $\overline{Z}$ and advancing the counter to 011.

7. This same action is repeated for each of the other steps. When the seventh step is completed, SENSOR output 7 goes HIGH, causing the counter to go from 111 to 000, where it will remain until another START pulse reinitiates the sequence.

Logic Function Generation

Multiplexers can be used to implement logic functions directly from a truth table without the need for simplification. When a multiplexer is used for this purpose, the select inputs are used as the logic variables, and each data input is connected permanently HIGH or LOW as necessary to satisfy the truth table.

Figure 9-27 illustrates how an eight-input multiplexer can be used to implement the logic circuit that satisfies the given truth table. The input variables A, B, C are connected to S_0, S_1, S_2, respectively, so that the levels on these inputs determine which data input appears at output Z. According to the truth table, Z is supposed to be LOW when $CBA = 000$. Thus, multiplexer input I_0 should be connected LOW. Likewise, Z is supposed to be LOW for $CBA = 011$, 100, 101, and 110, so that inputs I_3, I_4, I_5, and I_6 should also be connected LOW. The other sets of CBA conditions must produce $Z = 1$, and so multiplexer inputs I_1, I_2, and I_7 are connected permanently HIGH.

It is easy to see that any three-variable truth table can be implemented with this eight-input multiplexer. This method of implementation is often more efficient than using separate logic gates. For example, if we can write the sum-of-products expression for the truth table in Figure 9-27, we have

$$Z = A\overline{B}\,\overline{C} + \overline{A}B\overline{C} + ABC$$

This *cannot* be simplified either algebraically or by K mapping, and so its gate implementation would require three INVERTERs and four NAND gates, for a total of two ICs.

There is an even more efficient method for using multiplexers to implement logic functions. This method will allow the logic designer to use a multiplexer with three select inputs (e.g., a 74HC151) to implement a *four-variable* logic function. We will present this method in Problem 9-37.

(a)

$$Z = A\overline{B}\,\overline{C} + \overline{A}B\overline{C} + ABC$$

C	B	A	Z
0	0	0	0
0	0	1	1
0	1	0	1
0	1	1	0
1	0	0	0
1	0	1	0
1	1	0	0
1	1	1	1

(b)

FIGURE 9-27   Multiplexer used to implement a logic function described by the truth table.

1. What are some of the major applications of multiplexers?
2. *True or false:* When a multiplexer is used to implement a logic function, the logic variables are applied to the multiplexer's data inputs.
3. What type of circuit provides the select inputs when a MUX is used as a parallel-to-serial converter?

9-8 DEMULTIPLEXERS (DATA DISTRIBUTORS)

A multiplexer takes several inputs and transmits *one* of them to the output. A **demultiplexer (DEMUX)** performs the reverse operation: it takes a single input and distributes it over several outputs. Figure 9-28 shows the functional diagram for a digital demultiplexer. The large arrows for inputs and outputs can represent one or more lines. The select input code determines to which output the DATA input will be transmitted. In other words, the demultiplexer takes one input data source and selectively distributes it to 1 of N output channels just like a multiposition switch.

FIGURE 9-28 General demultiplexer.

1-Line-to-8-Line Demultiplexer

Figure 9-29 shows the logic diagram for a demultiplexer that distributes one input line to eight output lines. The single data input line I is connected to all eight AND gates, but only one of these gates will be enabled by the SELECT input lines. For example, with $S_2S_1S_0 = 000$, only AND gate 0 will be enabled, and data input I will appear at output O_0. Other SELECT codes cause input I to reach the other outputs. The truth table summarizes the operation.

The demultiplexer circuit of Figure 9-29 is very similar to the 3-line-to-8-line decoder circuit in Figure 9-2 except that a fourth input (I) has been added to each gate. It was pointed out earlier that many IC decoders have an ENABLE input, which is an extra input added to the decoder gates. This type of decoder chip can therefore be used as a demultiplexer, with the binary

FIGURE 9-50
Bidirectional register connected to data bus.

(a)

(b)

FIGURE 9-51

(Figure showing bidirectional register connected to Data Bus, 4-wire)

1. What will happen if $\overline{OE}_A = \overline{OE}_B = $ LOW in Figure 9-44?
2. What logic level is on a data bus line when all devices tied to the bus are disabled?
3. What is the function of a bus driver?
4. What are the reasons for having registers with common I/O lines?
5. Redraw Figure 9-50(a) using the bundled line representation. (The answer is shown in Figure 9-51 above.)

9-15 DECODERS USING HDL

Section 9-1 introduced the decoder as a device that can recognize a binary number on its input and activate a corresponding output. Specifically, the 74138 1-of-8 decoder was presented. It uses three binary inputs to activate one of the eight outputs when the chip is enabled. In order to study HDL methods for implementing the types of digital devices that are covered in this chapter, we will focus primarily on conventional MSI parts, which have been discussed earlier. Not only is the operation of these devices already described in this book, but further reference material is readily available in logic data books. In all of these cases, it is vital that you understand what the device is supposed to do before trying to dissect the HDL code that describes it.

In actual practice, we are not recommending, for example, that new code be written to perform the task of a 74138. After all, there is a macrofunction already available that works exactly like this standard part. Using these devices as examples and showing the HDL techniques used to create them opens the door for embellishment of these devices so that a circuit that will uniquely fit the application at hand can be described. In some instances, we will add our own embellishments to a circuit that has been described; in other instances, we will describe a simpler version of a part in order to focus on the core principle in HDL and avoid other confusing features.

The methods used to define the inputs and outputs should take into consideration the purpose of these signals. In the case of a 1-of-8 decoder such as the 74138 described in Figure 9-3, there are three enable inputs (E_1, $\overline{E_2}$, and $\overline{E_3}$) that should be described as individual inputs to the device. On the other hand, the binary inputs that are to be decoded (A_2, A_1, A_0) should be described as three-bit numbers. The outputs can be described as eight individual bits. They can also be described as an array of eight bits, with output 0 represented by element 0 in the array, and so on, to output 7 represented by element 7. Depending on the way the code is written, one strategy may be easier to write than the other. Generally, using individual names can make the purpose of each I/O bit clearer, and using bit arrays makes it easier to write the code.

When an application such as a decoder calls for a unique response from the circuit corresponding to each combination of its input variables, the two methods that best serve this purpose are the CASE construct and the truth TABLE. The interesting aspect of this decoder is that the output response should happen only when *all* the enables are activated. If any of the enables are not in their active state, it should cause all the outputs to go HIGH. Each of the examples that follow will demonstrate ways to decode the input number only when *all* of the enables are activated.

AHDL DECODERS

The first illustration of an AHDL decoder, shown in Figure 9-52, is intended to demonstrate the use of a CASE construct that is evaluated only under the condition that all enables are active. The outputs must all revert back to HIGH as soon as any enable is deactivated. This example also illustrates a way to accomplish this without explicitly assigning a value to each output for each case, and it uses individually named output bits.

Line 3 defines the three-bit binary number that will be decoded. Line 4 defines the three enable inputs, and line 5 specifically names each output. The unique property of this solution is the use of the **DEFAULTS** keyword in AHDL (lines 10 to 13) to establish a value for variables that are not specified elsewhere in the code. This maneuver allows each case to force one bit LOW without specifically stating that the others must go HIGH.

The next illustration, in Figure 9-53, is intended to demonstrate the same decoder using the truth table approach. Notice that the outputs are defined as bit arrays but are still numbered *y[7]* down to *y[0]*. The unique aspect of this code is the use of the don't-care values in the truth table. Line 11 is used to concatenate the six input bits into a single variable (bit array) named *inputs[ ]*. Notice that in lines 14, 15, and 16 of the table, only one bit value is specified as 1 or 0. The others are all in the don't-care state (X). Line 14 says, "As long as *e1*

FIGURE 9-52 AHDL equivalent to the 74138 decoder.

```
1   SUBDESIGN fig9_52
2   (
3       a[2..0]                     :INPUT;     -- binary inputs
4       e1, e2bar, e3bar            :INPUT;     -- enable inputs
5       y7,y6,y5,y4,y3,y2,y1,y0     :OUTPUT;    -- decoded outputs
6   )
7   VARIABLE
8       enable                      :NODE;
9   BEGIN
10      DEFAULTS
11          y7=VCC;y6=VCC;y5=VCC;y4=VCC;
12          y3=VCC;y2=VCC;y1=VCC;y0=VCC;     -- defaults all HIGH out
13      END DEFAULTS;
14      enable = e1 & !e2bar & !e3bar;     -- all enables activated
15      IF enable    THEN
16          CASE a[] IS
17              WHEN 0    =>    y0 = GND;
18              WHEN 1    =>    y1 = GND;
19              WHEN 2    =>    y2 = GND;
20              WHEN 3    =>    y3 = GND;
21              WHEN 4    =>    y4 = GND;
22              WHEN 5    =>    y5 = GND;
23              WHEN 6    =>    y6 = GND;
24              WHEN 7    =>    y7 = GND;
25          END CASE;
26      END IF;
27  END;
```

FIGURE 9-53 AHDL decoder using a TABLE.

```
1   SUBDESIGN fig9_53
2   (
3       a[2..0]             :INPUT;     -- decoder inputs
4       e1, e2bar, e3bar    :INPUT;     -- enable inputs
5       y[7..0]             :OUTPUT;    -- decoded outputs
6   )
7   VARIABLE
8       inputs[5..0]        :NODE;      -- all 6 inputs combined
9
10  BEGIN
11      inputs[] = (e1, e2bar, e3bar, a[]); -- concatenate the inputs
12      TABLE
13          inputs[]        =>      y[];
14          B"0XXXXX"       =>      B"11111111";    -- e1 not enabled
15          B"X1XXXX"       =>      B"11111111";    -- e2bar disabled
16          B"XX1XXX"       =>      B"11111111";    -- e3bar disabled
17          B"100000"       =>      B"11111110";    -- Y0 active
18          B"100001"       =>      B"11111101";    -- Y1 active
19          B"100010"       =>      B"11111011";    -- Y2 active
20          B"100011"       =>      B"11110111";    -- Y3 active
21          B"100100"       =>      B"11101111";    -- Y4 active
22          B"100101"       =>      B"11011111";    -- Y5 active
23          B"100110"       =>      B"10111111";    -- Y6 active
24          B"100111"       =>      B"01111111";    -- Y7 active
25      END TABLE;
26  END;
```

is *not* enabled, it does not matter what the other inputs are doing; the outputs will be HIGH." Lines 15 and 16 do the same thing, making sure that if *e2bar* or *e3bar* is HIGH (disabled), the outputs will be HIGH. Lines 17 through 24 state that as long as the first three bits (enables) are "100," the proper decoder output will be activated to correspond with the lower three bits of *inputs[ ]*.

VHDL DECODERS

The VHDL solution presented in Figure 9-54 essentially uses a truth table approach. The key strategy in this solution involves the concatenation of the three enable bits (*e1, e2bar, e3bar*) with the binary input *a* on line 11. The VHDL selected signal assignment is used to assign a value to a signal when a specific combination of inputs is present. Line 12 (WITH inputs SELECT) indicates that we are using the value of the intermediate signal *inputs* to determine which value is assigned to *y*. Each of the *y* outputs is listed on lines 13–20. Notice that only combinations that begin with 100 follow the WHEN clause on lines 13–20. This combination of *e1, e2bar*, and *e3bar* is necessary to make each of the enables active. Line 12 assigns a disabled state to each output when any combination other than 100 is present on the enable inputs.

```
1    ENTITY fig9_54 IS
2    PORT(
3        a                 :IN BIT_VECTOR (2 DOWNTO 0);
4        e1, e2bar, e3bar  :IN BIT;
5        y                 :OUT BIT_VECTOR (7 DOWNTO 0)
6        );
7    END fig9_54 ;
8    ARCHITECTURE truth OF fig9_54 IS
9    SIGNAL inputs:  BIT_VECTOR (5 DOWNTO 0); --combine enables w/ binary in
10      BEGIN
11        inputs <= e1 & e2bar & e3bar & a;
12        WITH inputs SELECT
13          y  <= "11111110"  WHEN "100000",  --Y0 active
14              "11111101"  WHEN "100001",  --Y1 active
15              "11111011"  WHEN "100010",  --Y2 active
16              "11110111"  WHEN "100011",  --Y3 active
17              "11101111"  WHEN "100100",  --Y4 active
18              "11011111"  WHEN "100101",  --Y5 active
19              "10111111"  WHEN "100110",  --Y6 active
20              "01111111"  WHEN "100111",  --Y7 active
21              "11111111"  WHEN OTHERS;    --disabled
22      END truth;
```

FIGURE 9-54   VHDL equivalent to the 74138 decoder.

REVIEW QUESTIONS

1. What is the purpose of the three inputs *e1, e2bar*, and *e3bar*?
2. Name two AHDL methods to describe a decoder's operation.
3. Name two VHDL methods to describe a decoder's operation.

9-16 THE HDL 7-SEGMENT DECODER/DRIVER

Section 9-2 described a BCD-to-7-segment decoder/driver. The standard part number for the circuit described is a 7447. In this section, we look into the HDL code necessary to produce a device that meets the same criteria as the 7447. Recall that the $\overline{BI}$ (blanking input) is the overriding control that turns all segments off regardless of other input levels. The $\overline{LT}$ (lamp test) input is used to test all the segments on the display by lighting them up. The $\overline{RBO}$ (ripple blanking output) is designed to go LOW when $\overline{RBI}$ (ripple blanking input) is LOW and the BCD input value is 0. Typically, in multiple-digit display applications, each $\overline{RBO}$ pin is connected to the $\overline{RBI}$ pin of the next digit to the right. This setup creates the feature of blanking all leading zeros in a display value without blanking zeros in the middle of a number. For example, the number 2002 would display as 2002, but the number 0002 would *not* display as 0002, but rather _ _ _ 2. One feature of the 7447 that would be difficult to replicate in HDL is the combination input/output pin named $\overline{BI}/\overline{RBO}$. Rather than complicate the code, we have decided to create a separate input ($\overline{BI}$) and an output ($\overline{RBO}$) on two different pins. This discussion also makes no attempt to replicate the non-BCD display characters of a 7447 but simply blanks all segments for values greater than 9.

Several decisions must be made when designing a circuit such as this one. The first involves the type of display we intend to use. If it is a common cathode, then a logic 1 lights the LED segment. If it is a common anode, then a logic 0 is required to turn on a segment. Next, we must decide on the type of inputs, outputs, and intermediate variables. We have decided that the outputs for each individual segment should be assigned a bit name (*a–g*) rather than using a bit array. This arrangement will make it clearer when connecting the display to the IC. These individual bits can be grouped as a set of bits and assigned binary values, as we have done in AHDL, or an intermediate variable bit array can be used to make it convenient when assigning all seven bit levels in a single statement, as we have done in VHDL. The BCD inputs are treated as a four-bit number, and the blanking controls are individual bits. The other issue that greatly affects the bit patterns in the HDL code is the arbitrary decision of the order of the segment names *a–g*. In this discussion, we have assigned segment *a* to the leftmost bit in the binary bit pattern, with the bits moving alphabetically left to right.

Some of the controls must have precedence over other controls. For example, the $\overline{LT}$ (lamp test) should override any regular digit display, and the $\overline{BI}$ (blanking input) should override even the lamp test input. In these illustrations, the IF/ELSE control structure is used to establish precedence. The first condition that is evaluated as true will determine the resulting output, regardless of the other input levels. Subsequent ELSE statements will have no effect, which is why the code tests first for $\overline{BI}$, then $\overline{LT}$, then $\overline{RBI}$, and finally determines the correct segment pattern.

AHDL DECODER/DRIVER

The AHDL code for this circuit is shown in Figure 9-55. AHDL allows output bits to be grouped in a set by separating the bits with commas and enclosing them in parentheses. A group of binary states can be assigned directly to these bit sets, as shown on lines 9, 11, 13, and 15. This convention avoids the need for an intermediate variable and is much shorter than eight separate assignment statements. The TABLE feature of AHDL is useful in this application to correlate an input BCD value to a 7-segment bit pattern.

```
1    SUBDESIGN fig9_55
2    (
3       bcd[3..0]            :INPUT;          -- 4-bit number
4       lt, bi, rbi         :INPUT;          -- 3 independent controls
5       a,b,c,d,e,f,g,rbo   :OUTPUT;         -- individual outputs
6    )
7    BEGIN
8        IF !bi THEN
9            (a,b,c,d,e,f,g,rbo) = (1,1,1,1,1,1,1,0);      % blank all %
10       ELSIF    !lt THEN
11           (a,b,c,d,e,f,g,rbo) = (0,0,0,0,0,0,0,1);      % test segments %
12       ELSIF !rbi & bcd[] == 0  THEN
13           (a,b,c,d,e,f,g,rbo) = (1,1,1,1,1,1,1,0);      % blank leading 0's %
14       ELSIF bcd[] > 9 THEN
15           (a,b,c,d,e,f,g,rbo) = (1,1,1,1,1,1,1,1);      % blank non BCD input %
16       ELSE
17           TABLE                              % display 7 segment Common Anode pattern %
18           bcd[]    =>    a,b,c,d,e,f,g,rbo;
19           0        =>    0,0,0,0,0,0,1,1;
20           1        =>    1,0,0,1,1,1,1,1;
21           2        =>    0,0,1,0,0,1,0,1;
22           3        =>    0,0,0,0,1,1,0,1;
23           4        =>    1,0,0,1,1,0,0,1;
24           5        =>    0,1,0,0,1,0,0,1;
25           6        =>    1,1,0,0,0,0,0,1;
26           7        =>    0,0,0,1,1,1,1,1;
27           8        =>    0,0,0,0,0,0,0,1;
28           9        =>    0,0,0,1,1,0,0,1;
29           END TABLE;
30       END IF;
31   END;
```

FIGURE 9-55   AHDL 7-segment BCD display decoder.

VHDL DECODER/DRIVER

The VHDL code for this circuit is shown in Figure 9-56. This illustration demonstrates the use of a VARIABLE as opposed to a SIGNAL. A VARIABLE can be thought of as a piece of scrap paper used to write down some numbers that will be needed later. A SIGNAL, on the other hand, is usually thought of as a wire connecting two points in the circuit. In line 12, the keyword VARIABLE is used to declare *segments* as a bit vector with seven bits. Take special note of the order of the indices for this variable. They are declared as 0 TO 6. In VHDL, this means that element 0 appears on the left end of the binary bit pattern and element 6 appears on the right end. This is exactly opposite of the way most examples in this text have presented variables, but it is important to realize the significance of the declaration statement in VHDL. For this illustration, segment *a* is bit 0 (on the left), segment *b* is bit 1 (moving to the right), and so on.

Notice that on line 3, the BCD input is declared as an INTEGER. This allows us to refer to it by its numeric value in decimal rather than being limited to bit pattern references. A PROCESS is employed here in order to allow us to use the IF/ELSE constructs to establish the precedence of one input over the other. Notice that the sensitivity list contains all the inputs. The code within the

```
1    ENTITY fig9_56 IS
2    PORT  (
3            bcd                 :IN INTEGER RANGE 0 TO 15;
4            lt, bi, rbi        :IN BIT;
5            a,b,c,d,e,f,g,rbo :OUT BIT
6          );
7    END fig9_56 ;
8
9    ARCHITECTURE vhdl OF fig9_56 IS
10   BEGIN
11   PROCESS  (bcd, lt, bi, rbi)
12   VARIABLE  segments        :BIT_VECTOR (0 TO 6);
13      BEGIN
14        IF  bi = '0' THEN
15           segments := "1111111";  rbo <= '0'; -- blank all
16        ELSIF lt = '0' THEN
17           segments := "0000000";  rbo <= '1'; -- test segments
18        ELSIF (rbi = '0' AND bcd = 0) THEN
19           segments := "1111111";  rbo <= '0'; -- blank leading 0's
20        ELSE
21           rbo <= '1';
22           CASE bcd IS     -- display 7 segment Common Anode pattern
23              WHEN 0      => segments := "0000001";
24              WHEN 1      => segments := "1001111";
25              WHEN 2      => segments := "0010010";
26              WHEN 3      => segments := "0000110";
27              WHEN 4      => segments := "1001100";
28              WHEN 5      => segments := "0100100";
29              WHEN 6      => segments := "1100000";
30              WHEN 7      => segments := "0001111";
31              WHEN 8      => segments := "0000000";
32              WHEN 9      => segments := "0001100";
33              WHEN OTHERS => segments := "1111111";
34           END CASE;
35        END IF;
36      a <= segments(0); --assign bits of array to output pins
37      b <= segments(1);
38      c <= segments(2);
39      d <= segments(3);
40      e <= segments(4);
41      f <= segments(5);
42      g <= segments(6);
43        END PROCESS;
44   END vhdl;
```

FIGURE 9-56   VHDL 7-segment BCD display decoder.

PROCESS describes the behavioral operation of the circuit that is necessary whenever any of the inputs in the sensitivity list changes state. Another very important point in this illustration is the assignment operator for variables. Notice in line 15, for example, the statement *segments := "1111111"*. The variable assignment operator := is used for variables in place of the <= operator that was used for signal assignments. In lines 36–42, the individual bits that were established in the IF/ELSE decisions are assigned to the proper output bits.

1. What feature of a 7447 is very difficult to duplicate in PLD hardware and HDL code?
2. Are these illustrations intended to drive common anode or common cathode 7-segment displays?
3. How are certain inputs (e.g., lamp test) given precedence over other inputs (e.g., RBI) in the HDL code in this section?

9-17 ENCODERS USING HDL

In Section 9-4, we discussed encoders and priority encoders. Similarities exist, of course, between decoders and encoders. Decoders take a binary number and activate one output that corresponds to that number. An encoder works in the other direction by monitoring one of its several inputs; when one of the inputs is activated, it produces a binary number corresponding to that input. If more than one of its inputs is activated at the same time, a priority encoder ignores the input of lower significance and produces the binary value that corresponds to the most significant input. In other words, it gives more significant inputs priority over less significant inputs. This section focuses on the methods that can be used in HDL to describe circuits that have this characteristic of priority for some inputs over others.

Another very important concept, which was presented in Chapter 8, was the tristate output circuit. Devices with tristate outputs can produce a logic HIGH or a logic LOW, just like a normal circuit, when their output is enabled. However, these devices can have their outputs disabled, which puts them in a "disconnected" or a high-impedance state. This is very important for devices connected to common buses, as described in Section 9-12. The next logical question is, "How do we describe tristate outputs using HDL?" This section incorporates tristate outputs in the encoder design to address this issue. In order to keep the discussion focused on the essentials, we create a circuit that emulates the 74147 priority encoder, with one added feature of having active-HIGH tristate outputs. Other features like cascading inputs and outputs (such as those found on a 74148) are left for you to try later. A symbol for the circuit we are describing is shown in Figure 9-57. Because the inputs are all labeled in a manner very similar to bit array notation, it makes sense to use a bit array to describe the encoder inputs. The tristate enable must be a single bit, and the encoded outputs can be described as an integer numeric value.

FIGURE 9-57 Graphic description of an encoder with tristate outputs.

AHDL ENCODER

The most important point to be made from Figure 9-58 is the method of establishing priority, but also note the I/O assignments. The AHDL input/output descriptions do not provide a separate type for integers but allow a bit array to be referred to as an integer. Consequently, line 4 describes the outputs as a bit array. In this illustration, a TABLE is used that is very similar to the tables often found in data books describing this circuit's operation. The key to this table is the use of the don't-care state (X) on inputs. The priority is described by the way we position these don't-care states in the truth table. Reading line 15, for instance, we see that as soon as we encounter an active input (LOW on input *a[4]*), the lower order input bits do not matter. The output has been determined to be 4. The tristate outputs are made possible by using the built-in primitive function :TRI on line 6. This line assigns the attributes of a tristate buffer to the variable that has been named *buffer*. Recall that this is the same way a flip-flop is described in AHDL. The ports of a tristate buffer are quite straightforward. They represent the input (*in*), the output (*out*), and the tristate output enable (*oe*).

FIGURE 9-58 AHDL priority encoder with tristate outputs.

```
1    SUBDESIGN fig9_58
2    (
3        a[9..0], oe            :INPUT;
4        d[3..0]                :OUTPUT;
5    )
6    VARIABLE buffer[3..0]    :TRI;
7    BEGIN
8        TABLE
9           a[]                   => buffer[].in;
10          B"1111111111"  => B"1111";     -- no input active
11          B"1111111110"  => B"0000";     -- 0
12          B"111111110X"  => B"0001";     -- 1
13          B"11111110XX"  => B"0010";     -- 2
14          B"1111110XXX"  => B"0011";     -- 3
15          B"111110XXXX"  => B"0100";     -- 4
16          B"11110XXXXX"  => B"0101";     -- 5
17          B"1110XXXXXX"  => B"0110";     -- 6
18          B"110XXXXXXX"  => B"0111";     -- 7
19          B"10XXXXXXXX"  => B"1000";     -- 8
20          B"0XXXXXXXXX"  => B"1001";     -- 9
21       END TABLE;
22       buffer[].oe = oe;      -- hook up enable line
23       d[] = buffer[].out;    -- hook up outputs
24    END;
```

The next illustration (Figure 9-59) uses the IF/ELSE construct to establish priority, very much like the method demonstrated in the 7-segment decoder example. The first IF condition that evaluates TRUE will THEN cause the corresponding value to be applied to the tristate buffer inputs. The priority is established by the order in which we list the IF conditions. Notice that they start with input 9, the highest order input. This illustration adds another feature of putting the outputs into the high-impedance state when no input is being activated. Line 20 shows that the output enables will be acti-

```
1    SUBDESIGN fig9_59
2    (
3       sw[9..0], oe    :INPUT;
4       d[3..0]         :OUTPUT;
5    )
6    VARIABLE
7       buffers[3..0]   :TRI;
8    BEGIN
9       IF   !sw[9]   THEN   buffers[].in = 9;
10      ELSIF !sw[8]   THEN   buffers[].in = 8;
11      ELSIF !sw[7]   THEN   buffers[].in = 7;
12      ELSIF !sw[6]   THEN   buffers[].in = 6;
13      ELSIF !sw[5]   THEN   buffers[].in = 5;
14      ELSIF !sw[4]   THEN   buffers[].in = 4;
15      ELSIF !sw[3]   THEN   buffers[].in = 3;
16      ELSIF !sw[2]   THEN   buffers[].in = 2;
17      ELSIF !sw[1]   THEN   buffers[].in = 1;
18      ELSE                  buffers[].in = 0;
19      END IF;
20      buffers[].oe = oe & sw[]!=b"1111111111";   -- enable on any input
21      d[] = buffers[].out;                        -- connect to outputs
22    END;
```

FIGURE 9-59   AHDL priority encoder using IF/ELSE.

vated only when the *oe* pin is activated and one of the inputs is activated. Another item of interest in this illustration is the use of bit array notation to describe individual inputs. For example, line 9 states that IF switch input 9 is activated (LOW), THEN the inputs to the tristate buffer will be assigned the value 9 (in binary, of course).

VHDL ENCODER

Two very important VHDL techniques are demonstrated in this description of a priority encoder. The first is the use of tristate outputs in VHDL, and the second is a new method of describing priority. Figure 9-60 shows the input/output definitions for this encoder circuit. Notice on line 6 that the input switches are defined as bit vectors with indices from 9 to 0. Also note that the *d* output is defined as an IEEE standard bit array (std_logic_vector type). This definition is necessary to allow the use of high-impedance states (tristate) on the outputs and also explains the need for the LIBRARY and USE statements on lines 1 and 2. As we mentioned, a very important point of this illustration is the method of describing precedence for the inputs. This code uses the **conditional signal assignment statement** starting on line 14 and continuing through line 24. On line 14, it assigns the value listed to the right of <= to the variable *d* on the left, assuming the condition following WHEN is true. If this clause is not true, the clauses following ELSE are evaluated one at a time until one that is true is found. The value preceding WHEN will then be assigned to *d*. A very important attribute of the conditional signal assignment statement is the sequential evaluation. The precedence of these statements is established by the order in which they are listed. Notice that in this illustration, the first condition being tested (line 14) is the enabling of the tristate outputs. Recall from Chapter 8 that the three states of a

FIGURE 9-60
VHDL priority encoder using Conditional Signal Assignment.

```
1   LIBRARY    ieee;
2   USE ieee.std_logic_1164.ALL;
3
4   ENTITY fig9_60 IS
5   PORT(
6       sw     :IN BIT_VECTOR (9 DOWNTO 0);  -- standard logic not needed
7       oe     :IN BIT;                       -- standard logic not needed
8       d      :OUT STD_LOGIC_VECTOR (3 DOWNTO 0)  -- std logic for hi-Z
9       );
10  END fig9_60;
11
12  ARCHITECTURE a OF fig9_60 IS
13     BEGIN
14      d  <= "ZZZZ"  WHEN  ((oe = '0') OR (sw = "1111111111")) ELSE
15            "1001"  WHEN  sw(9) = '0' ELSE
16            "1000"  WHEN  sw(8) = '0' ELSE
17            "0111"  WHEN  sw(7) = '0' ELSE
18            "0110"  WHEN  sw(6) = '0' ELSE
19            "0101"  WHEN  sw(5) = '0' ELSE
20            "0100"  WHEN  sw(4) = '0' ELSE
21            "0011"  WHEN  sw(3) = '0' ELSE
22            "0010"  WHEN  sw(2) = '0' ELSE
23            "0001"  WHEN  sw(1) = '0' ELSE
24            "0000"  WHEN  sw(0) = '0';
25  END a;
```

tristate output are HIGH, LOW, and high impedance, which is referred to as high Z. When the value "ZZZZ" is assigned to the output, each output is in the high-impedance state. If the outputs are to be disabled (high Z), then none of the other encoding matters. Line 15 tests the highest priority input, which is bit 9 of the *sw* input array. If it is active (LOW), then a value of 9 is output regardless of whether other inputs are being activated at the same time.

REVIEW QUESTIONS

1. Name two methods in AHDL for giving priority to some inputs over others.
2. Name two methods in VHDL for giving priority to some inputs over others.
3. In AHDL, how are tristate outputs implemented?
4. In VHDL, how are tristate outputs implemented?

9-18 HDL MULTIPLEXERS AND DEMULTIPLEXERS

A multiplexer is a device that acts like a selector switch for digital signals. The select inputs are used to specify the input channel that is to be "connected" to the output pins. A demultiplexer works in the opposite direction by taking a digital signal as an input and distributing it to one of its outputs. Figure 9-61 shows a multiplexer/demultiplexer system with four data input channels. Each input is a four-bit number. These devices are not exactly like any of the multiplexers or demultiplexers described earlier in this chapter, but they operate in the same way. The system in this illustration allows the four digital signals to share a common "pipeline" in order to get data from

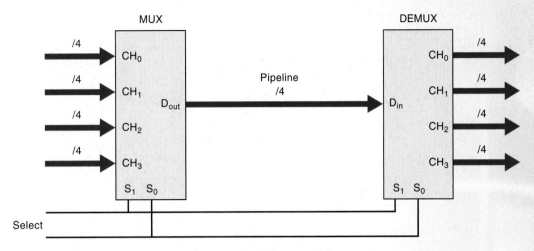

FIGURE 9-61 Four channels of data sharing a common data path.

one point to the other. The select inputs are used to decide which signal is going through the pipeline at any time.

In this section, we examine some code that implements both the multiplexer and the demultiplexer. The key HDL issue in both the MUX and DEMUX is assigning signals under certain conditions. For the demux, another issue is assigning a state to whichever outputs are not currently selected to distribute data. In other words, when an output is not being used for data (not selected), do we want it to have all bits HIGH, all bits LOW, or the tristate disabled? In the following descriptions, we have chosen to make them all HIGH when not selected, but with the structure shown, it would be a simple matter to change to one of the other possibilities.

AHDL MUX AND DEMUX

We will implement the multiplexer first. Figure 9-62 describes a multiplexer with four inputs of four bits each. Each input channel is named in a way that identifies its channel number. In this figure, each input is described as a four-bit array. The select input (*s[ ]*) requires two bits to specify the four channel numbers (0–3). A CASE construct is used here to assign an input channel conditionally to the output pins. Line 9, for example, states that in the case when

FIGURE 9-62
Four-bit × four-channel MUX in AHDL.

```
1   SUBDESIGN fig9_62
2   (
3     ch0[3..0], ch1[3..0], ch2[3..0], ch3[3..0]:INPUT;
4     s[1..0]                              :INPUT; -- select inputs
5     dout[3..0]                           :OUTPUT;
6   )
7   BEGIN
8     CASE s[] IS
9         WHEN 0 =>    dout[] = ch0[];
10        WHEN 1 =>    dout[] = ch1[];
11        WHEN 2 =>    dout[] = ch2[];
12        WHEN 3 =>    dout[] = ch3[];
13    END CASE;
14  END;
```

FIGURE 9-63 Four-bit ×
four-channel DEMUX in
AHDL.

```
1    SUBDESIGN fig9_63
2    (
3        ch0[3..0], ch1[3..0], ch2[3..0], ch3[3..0]   :OUTPUT;
4        s[1..0]                                       :INPUT;
5        din[3..0]                                     :INPUT;
6    )
7    BEGIN
8        DEFAULTS
9            ch0[] = B"1111";
10           ch1[] = B"1111";
11           ch2[] = B"1111";
12           ch3[] = B"1111";
13       END DEFAULTS;
14
15       CASE S[] IS
16            WHEN 0 =>      ch0[] = din[];
17            WHEN 1 =>      ch1[] = din[];
18            WHEN 2 =>      ch2[] = din[];
19            WHEN 3 =>      ch3[] = din[];
20       END CASE;
21   END;
```

the select inputs (*s[ ]*) are set to 0 (that is, binary 00), the circuit should connect the channel 0 input to the data output. Notice that when assigning connections, the destination (output) of the signal is on the left of the = sign and the source (input) is on the right.

The demultiplexer code works in a similar way but has only one input channel and four output channels. It must also ensure that the outputs are all HIGH when they are not selected. In Figure 9-63, the inputs and outputs are declared as usual on lines 3–5. The default condition for each channel is specified after the keyword DEFAULTS, which tells the compiler to generate a circuit that will have HIGHs on the outputs unless specifically assigned a value elsewhere in the code. If this default section is not specified, the output values would default automatically to all LOW. Notice on lines 16–19 that the input signal is assigned conditionally to one of the output channels. Consequently, the output channel is on the left of the = sign and the input signal is on the right.

VHDL MUX AND DEMUX

Figure 9-64 shows the code that creates a four-channel MUX with four bits per channel. The inputs are declared as bit arrays on line 3. They could have been declared just as easily as integers ranging from 0 to 15. Whichever way the inputs are declared, the outputs must be of the same type. Notice on line 4 that the select input (*s*) is declared as a decimal integer from 0 to 3 (equivalent to binary 00 to 11). This allows us to refer to it by its decimal channel number in the code, making it easier to understand. Lines 11–15 use the selected signal assignment statement to "connect" the appropriate input to the output, depending on the value on the select inputs. For example, line 15 states that channel 3 should be selected to connect to the data outputs when the select inputs are set to 3.

The demultiplexer code works in a similar way but has only one input channel and four output channels. In Figure 9-65, the inputs and outputs are declared as usual on lines 3–5. Notice that in line 3, the select input(s) is typed as an integer, just like the MUX code in Figure 9-64. The operation of

FIGURE 9-64 Four-bit ×
four-channel MUX in
VHDL.

```
1    ENTITY fig9_64 IS
2    PORT    (
3               ch0, ch1, ch2, ch3   :IN BIT_VECTOR (3 DOWNTO 0);
4               s                    :IN INTEGER RANGE 0 TO 3;
5               dout                 :OUT BIT_VECTOR (3 DOWNTO 0)
6            );
7    END fig9_64;
8
9    ARCHITECTURE selecter OF fig9_64 IS
10       BEGIN
11          WITH s SELECT
12          dout  <=  ch0 WHEN 0,  -- switch channel 0 to output
13                    ch1 WHEN 1,  -- switch channel 1 to output
14                    ch2 WHEN 2,  -- switch channel 2 to output
15                    ch3 WHEN 3;  -- switch channel 3 to output
16       END selecter;
```

FIGURE 9-65 Four-bit ×
four-channel DEMUX in
VHDL.

```
1    ENTITY fig9_65 IS
2    PORT    (
3               s                    :IN INTEGER RANGE 0 TO 3;
4               din                  :IN BIT_VECTOR (3 DOWNTO 0);
5               ch0, ch1, ch2, ch3   :OUT BIT_VECTOR(3 DOWNTO 0)
6            );
7    END fig9_65;
8
9    ARCHITECTURE selecter OF fig9_65 IS
10       BEGIN
11          ch0 <= din WHEN s = 0  ELSE "1111";
12          ch1 <= din WHEN s = 1  ELSE "1111";
13          ch2 <= din WHEN s = 2  ELSE "1111";
14          ch3 <= din WHEN s = 3  ELSE "1111";
15       END selecter;
```

a DEMUX is described most easily using several conditional signal assignment statements, as shown in lines 11–14. We decided earlier that the code for this DEMUX must ensure that the outputs are all HIGH when they are not selected. This is accomplished with the ELSE clause of each conditional signal assignment. If the ELSE clause is not used, the output values would default automatically to all LOW. For example, line 13 states that channel 2 will be connected to the data inputs whenever the select inputs are set to 2. If s is set to any other value, then channel 2 will be forced to have all bits HIGH.

REVIEW QUESTIONS

1. For the four-bit by four-channel MUX, name the data inputs, the data outputs, and the control inputs that choose one channel of the four.
2. For the four-bit by four-channel DEMUX, name the data inputs, the data outputs, and the control inputs that choose one channel of the four.
3. In the AHDL example, how are the logic states determined for the channels that are not selected?
4. In the VHDL example, how are the logic states determined for the channels that are not selected?

9-19 HDL MAGNITUDE COMPARATORS

In Section 9-10, we studied a 7485 magnitude comparator chip. As the name implies, this device compares the magnitude of two binary numbers and indicates the relationship between the two (greater than, less than, equal to). Control inputs are provided for the purpose of cascading these chips. These chips are interconnected so that the chip comparing the lower-order bits has its outputs connected to the control inputs of the next higher-order chip, as shown in Figure 9-37. When the highest order stage detects that its data inputs have equal magnitude, it will look to the next lower stage and use these control inputs to make the final decision. The way we can describe the operation of each stage is:

If A > B, then indicate that A > B.

If A < B, then indicate that A < B.

If A = B, then use the control (cascade) inputs to indicate the relationship.

It is best to use an IF/ELSE construct because each IF clause evaluates a relationship between two values, as opposed to looking for the single value of a variable, like a CASE. The two inputs being compared should definitely be declared as numerical values. The three cascading control inputs and three comparator outputs should be declared as individual bits in order to label each bit's purpose clearly.

AHDL COMPARATOR

The AHDL code in Figure 9-66 follows the algorithm we have described using IF/ELSE constructs. Notice in line 3 that the data values are declared as four-bit numbers, but the control inputs and outputs are individual bits. Also note in lines 11, 13, and 14 that several statements can be used to specify the circuit's operation when the IF clause is true. Each statement is used to set the level on one of the outputs. These three statements are considered concurrent, and the order in which they are listed makes no difference. For example, in line 11, when *A* is greater than *B*, the *agtb* output will go HIGH at the same time the other two outputs (*altb, aeqb*) go LOW.

FIGURE 9-66
Magnitude comparator in AHDL.

```
1    SUBDESIGN fig9_66
2    (
3       a[3..0], b[3..0]   :INPUT;
4       gtin, ltin, eqin   :INPUT;        % cascade inputs %
5       agtb, altb, aeqb   :OUTPUT;
6    )
7       % standard cascade inputs:   gtin = ltin = GND eqin = VCC %
8
9    BEGIN
10      IF     a[] > b[] THEN
11              agtb = VCC;      altb = GND;         aeqb = GND;
12      ELSIF a[] < b[] THEN
13              agtb = GND;      altb = VCC;         aeqb = GND;
14      ELSE    agtb = gtin;     altb = ltin;        aeqb = eqin;
15      END IF;
16   END;
```

VHDL

VHDL COMPARATOR

The VHDL code in Figure 9-67 follows the algorithm we have described using IF/ELSE constructs. Notice in line 2 that the data values are declared as four-bit integers, but the control inputs and outputs are individual bits. Remember, in VHDL, the IF/ELSE constructs can be used only inside a PROCESS. In this case, we want to evaluate the PROCESS whenever any of the inputs (data or control) change state. Consequently, each input is listed in the sensitivity list within the parentheses. Also note in lines 12, 13, and 14 that several statements can be used to specify the circuit's operation when the IF clause is true. Each statement is used to set the level on one of the outputs. These three statements are considered concurrent, and the order in which they are listed makes no difference. For example, on line 13, when A is greater than B, the *agtb* output will go HIGH at the same time the other two outputs (*altb*, *aeqb*) go LOW.

```
1    ENTITY fig9_67 IS
2    PORT (   a, b              : IN INTEGER RANGE 0 TO 15;
3               gtin, ltin, eqin  : IN BIT;                    -- cascade inputs
4               agtb, altb, aeqb  : OUT BIT);
5    END fig9_67;
6      -- standard cascade inputs:    gtin = ltin = '0' eqin = '1'
7
8    ARCHITECTURE vhdl OF fig9_67 IS
9    BEGIN
10     PROCESS (a, b, gtin, ltin, eqin)
11     BEGIN
12        IF    a < b   THEN    altb <= '1';   agtb <= '0';   aeqb <= '0';
13        ELSIF a > b   THEN    altb <= '0';   agtb <= '1';   aeqb <= '0';
14        ELSE                  altb <= ltin;  agtb <= gtin;  aeqb <= eqin;
15        END IF;
16     END PROCESS;
16   END vhdl;
```

FIGURE 9-67   Magnitude comparator in VHDL.

REVIEW QUESTIONS

1. What type of data objects must be declared for data inputs to a comparator?
2. What is the key control structure used to describe a comparator?
3. What are the key operators used?

9-20 HDL CODE CONVERTERS

Section 9-11 demonstrated some methods using adder circuits in an interesting but not at all intuitive way to create a BCD-to-binary conversion. In Chapter 6, we discussed adder circuits, and the circuit of Figure 9-40 can certainly be implemented using HDL and 7483 macrofunctions or adder descriptions that we know how to write. However, this is an excellent opportunity to point out the tremendous advantage that HDL can offer because it allows a circuit to be described in a way that makes the most sense. In the case of BCD-to-binary conversion, the sensible method of conversion is to use the concepts that we all learned in the third grade about the decimal number system. You were once taught that the number 275 was actually:

$$
\begin{array}{rcccl}
 & 2 & \times & 100 & = & 200 \\
+ & 7 & \times & 10 & = & 70 \\
+ & 5 & \times & 1 & = & \underline{5} \\
 & & & & & 275
\end{array}
$$

Now we have studied the BCD number system and realize that 275 is represented in BCD as 0010 0111 0101. Each digit is simply represented in binary. If we could multiply these binary digits by the decimal weight (represented in binary) and add them, we would have a binary answer that is equivalent to the BCD quantity. For example, let's try using the BCD representation for 275:

BCD	Decimal Weight (in binary)	Partial Product (in binary)
0010 ×	1100100 =	11001000
+ 0111 ×	1010 =	01000110
+ 0101 ×	1 =	0101

$$100010011 \implies 275_{10}$$

The solution presented here for our eight-bit (two BCD digits) HDL code converter will use the following strategy:

Take the most significant BCD digit (the tens place) and multiply it by 10.
Add this product to the least significant BCD digit (the ones place).

The answer will be a binary number representing the BCD quantity. It is important to realize that the HDL compiler does not necessarily try to implement an actual multiplier circuit in its solution. It will create the most efficient circuit that will do the job, which allows the designer to describe its behavior in the most sensible way.

AHDL BCD-TO-BINARY CODE CONVERTER

The key to this strategy is in being able to multiply by 10. AHDL does not offer a multiplication operator, so in order to use this overall strategy, we need some math tricks. We will use the shifting of bits to perform multiplication and then employ the distributive property from algebra to multiply by 10. In the same way that we can shift a decimal number left by one digit, thus multiplying it by 10, we can likewise shift a binary number one place to the left and multiply it by 2. Shifting two places multiplies a binary number by 4, and shifting three places multiplies by 8. The distributive property tells us that:

$$\text{num} \times 10 = \text{num} \times (8 + 2) = (\text{num} \times 8) + (\text{num} \times 2)$$

If we can take the BCD tens digit and shift it left three bit positions (i.e., multiply it by 8), then take the same number and shift it left one place (i.e., multiply it by 2), and then add them together, the result will be the same as multiplying the BCD digit by 10. This value is then added to the BCD ones digit to produce the binary equivalent of the two-digit BCD input.

The next challenge is to shift the BCD digit left using AHDL. Because AHDL allows us to make up sets of variables, we can shift the bits by appending zeros to the right end of the array. For example, if we have the number 5 in BCD (0101) and we want to shift it three places, we can concatenate the number 0101 with the number 000 in a set, as follows:

$$(B"0101", B"000") = B"0101000"$$

The AHDL code in Figure 9-68 begins by declaring inputs for the BCD ones and tens digits. The binary output must be able to represent 99_{10}, which requires seven bits. We also need a variable to hold the product of the BCD digit multiplied by 10. Line 5 declares this variable as a seven-bit number. Line 8 performs the shifting of the *tens[ ]* array three times and adds it to the *tens[ ]* array shifted one place to the left. Notice that this latter set must have seven bits in order to be added to the first set, thus the need to concatenate B"00" on the left end. Finally, in line 10, the result from line 8 is added to the BCD ones digit with leading zero extensions (to make seven bits) to form the binary output.

```
1    SUBDESIGN fig9_68
2    (  ones[3..0], tens[3..0]     :INPUT;
3       binary[6..0]               :OUTPUT;  )
4
5    VARIABLE times10[6..0]         :NODE;    % variable for tens digit times 10%
6
7    BEGIN
8        times10[] = (tens[],B"000") + (B"00",tens[],B"0");
9            % shift left 3X (times 8) + shift left 1X (times 2) %
10       binary[] = times10[] + (B"000",ones[]);
11           % tens digit times 10 + ones digit %
12   END;
```

FIGURE 9-68   BCD to binary code converter in AHDL.

VHDL BCD-TO-BINARY CODE CONVERTER

The VHDL solution in Figure 9-69 is very simple due to the powerful math operations available in the language. The inputs and outputs must be declared as integers because we intend to perform arithmetic operations on them. Notice that the range is specified based on the largest valid BCD number using two digits. In line 9, the tens digit is multiplied by ten, and in line 10, the ones digit is added to form the binary equivalent of the BCD input.

FIGURE 9-69   BCD to binary code converter in VHDL.

```
1    ENTITY fig9_69 IS
2    PORT (   ones, tens  :IN INTEGER RANGE 0 TO 9;
3             binary      :OUT INTEGER RANGE 0 TO 99);
4    END fig9_69;
5
6    ARCHITECTURE vhdl OF fig9_69 IS
7    SIGNAL times10      :INTEGER RANGE 0 TO 90;
8    BEGIN
9        times10 <= tens * 10;
10       binary <= times10 + ones;
11   END vhdl;
```

1. For a two-digit BCD (eight-bit) number, what is the decimal weight of the most significant digit?
2. In AHDL, how is multiplication by 10 accomplished?
3. In VHDL, how is multiplication by 10 accomplished?

SUMMARY

1. A decoder is a device whose output is activated only when a unique binary combination (code) is presented on its inputs. Many MSI decoders have several outputs, each one corresponding to only one of the many possible input combinations.

2. Digital systems often need to display decimal numbers. This is done using 7-segment displays that are driven by special chips that decode the binary number and translate it into segment patterns that represent decimal numbers to people. The segment elements can be light-emitting diodes, liquid crystals, or glowing electrodes surrounded by neon gas.

3. Graphical LCDs use a matrix of picture elements called pixels to create an image on a large screen. Each pixel is controlled by activating the row and column that have that pixel in common. The brightness level of each pixel is stored as a binary number in the video memory. A fairly complex digital circuit must scan through the video memory and all the row/column combinations, controlling the amount of light that can pass through each pixel.

4. An encoder is a device that generates a unique binary code in response to the activation of each individual input.

5. Troubleshooting a digital system involves *observation/analysis* to identify the possible causes, and a process of elimination called *divide-and-conquer* to isolate and identify the cause.

6. Multiplexers act like digitally controlled switches that select and connect one logic input at a time to the output pin. By taking turns, many different data signals can share the same data path using multiplexers. Demultiplexers are used at the other end of the data path to separate the signals that are sharing a data path and distribute them to their respective destinations.

7. Magnitude comparators serve as an indicator of the relationship between two binary numbers, with outputs that show >, <, and =.

8. It is often necessary to translate between and among various methods of representing quantities with binary numbers. Code converters are devices that take in one form of binary representation and convert it to another form.

9. In digital systems, many devices must often share the same data path. This data path is often called a *data bus*. Even though many devices can be "riding" on the bus, there can be only one bus "driver" at any one time. Thus, devices must take turns applying logic signals to the data bus.

10. In order to take turns, the devices must have *tristate outputs* that can be disabled when another device is driving the bus. In the disabled state, the device's output is essentially electrically disconnected from the bus by going into a state that offers a high-impedance path to both ground and the positive power supply. Devices designed to interface to a bus have outputs that can be HIGH, LOW, or disabled (high impedance).

11. PLDs offer an alternative to the use of MSI circuits to implement digital systems. Boolean equations can be used to describe the operation of these circuits, but HDLs also offer high-level language constructs.

12. HDL macrofunctions are available for many MSI standard parts described in this chapter.

13. Custom code can be written in HDL to describe each of the common logic functions presented in this chapter.

14. Priority and precedence can be established in AHDL using don't-care entries in truth tables and using IF/ELSE decisions. Priority and precedence can be established in VHDL using conditional signal assignments or using a PROCESS containing IF/ELSE or CASE decisions.

15. Tristate outputs can be created in HDL. AHDL uses :TRI primitives that drive the outputs. VHDL assigns Z (high impedance) as a valid state for STD_LOGIC outputs.

16. The DEFAULTS statement in AHDL can be used to define the proper level for outputs that are not explicitly defined in the code.

17. The ELSE clause in the conditional signal assignment statement of VHDL can be used to define the default state of an output.

IMPORTANT TERMS

decoder
BCD-to-decimal
  decoder
driver
BCD-to-7-segment
  decoder/driver
common anode
common cathode
LCD
backplane
pixel

encoding
encoder
priority encoder
observation/analysis
divide-and-conquer
multiplexer (MUX)
multiplexing
parallel-to-serial
  conversion
demultiplexer
  (DEMUX)

data bus
floating bus
word
bus driver
bidirectional data
  lines
DEFAULTS
conditional signal
  assignment
  statement

PROBLEMS

SECTION 9-1

B   9-1. Refer to Figure 9-3. Determine the levels at each decoder output for the following sets of input conditions.

(a) All inputs LOW

(b) All inputs LOW except E_3 = HIGH

(c) All inputs HIGH except $\overline{E}_1 = \overline{E}_2$ = LOW

(d) All inputs HIGH

B   9-2. What is the number of inputs and outputs of a decoder that accepts 64 different input combinations?

B   9-3. For a 74ALS138, what input conditions will produce the following outputs:

(a) LOW at $\overline{O}_6$

(b) LOW at $\overline{O}_3$

(c) LOW at $\overline{O}_5$

(d) LOW at $\overline{O}_0$ and $\overline{O}_7$, simultaneously

D   9-4. Show how to use 74LS138s to form a 1-of-16 decoder.

9-5. Figure 9-70 shows how a decoder can be used in the generation of control signals. Assume that a RESET pulse has occurred at time t_0, and determine the CONTROL waveform for 32 clock pulses.

FIGURE 9-70   Problems 9-5 and 9-6.

D   9-6. Modify the circuit of Figure 9-70 to generate a CONTROL waveform that goes LOW from t_{20} to t_{24}. (*Hint:* The modification does not require additional logic.)

9-7. The 7442 decoder of Figure 9-5 does not have an ENABLE input. However, we can operate it as a 1-of-8 decoder by not using outputs $\overline{O}_8$ and $\overline{O}_9$ and by using the D input as an ENABLE. This is illustrated in Figure 9-71. Describe how this arrangement works as an enabled 1-of-8 decoder, and state how the level on D either enables or disables the outputs.

FIGURE 9-71   Problem 9-7.

9-8. Consider the waveforms in Figure 9-72. Apply these signals to the 74LS138 as follows:

$$A \rightarrow A_0 \qquad B \rightarrow A_1 \qquad C \rightarrow A_2 \qquad D \rightarrow E_3$$

Assume that $\overline{E}_1$ and $\overline{E}_2$ are tied LOW, and draw the waveforms for outputs $\overline{O}_0$, $\overline{O}_3$, $\overline{O}_6$, and $\overline{O}_7$.

FIGURE 9-72 Problems 9-8 and 9-41.

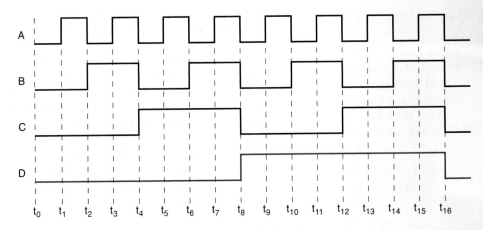

D 9-9. Modify the circuit of Figure 9-6 so that relay K_1 stays energized from t_3 to t_5 and K_2 stays energized from t_6 to t_9. (*Hint:* This modification requires no additional circuitry.)

SECTIONS 9-2 AND 9-3

B, D 9-10. Show how to connect BCD-to-7-segment decoder/drivers and LED 7-segment displays to the clock circuit of Figure 7-47 to display minutes and hours. Assume that each segment is to operate at approximately 10 mA at 2.5 V.

B 9-11. (a) Refer to Figure 9-10 and draw the segment and backplane waveforms relative to ground for CONTROL = 0. Then draw the waveform of segment voltage relative to backplane voltage.

(b) Repeat part (a) for CONTROL = 1.

C, D 9-12. The BCD-to-7-segment decoder/driver of Figure 9-8 contains the logic for activating each segment for the appropriate BCD inputs. Design the logic for activating the g segment.

SECTION 9-4

B 9-13. DRILL QUESTION

For each item, indicate whether it is referring to a decoder or an encoder.

(a) Has more inputs than outputs.

(b) Is used to convert key actuations to a binary code.

(c) Only one output can be activated at one time.

(d) Can be used to interface a BCD input to an LED display.

(e) Often has driver-type outputs to handle large I and V.

9-14. Determine the output levels for the 74147 encoder when $\overline{A}_8 = \overline{A}_4 = 0$ and all other inputs are HIGH.

FIGURE 9-82　Problems 9-60, 9-61, and 9-62.

C　**9-61.** Now that the circuitry of Figure 9-82 has been added to Figure 9-44, a total of five devices are connected to the data bus. The circuit in Figure 9-83(a) will now be used to generate the enable signals needed to perform the different data transfers over the data bus. It uses a 74HC139 chip that contains two identical independent 1-of-4 decoders with an active-LOW enable. The top decoder is used to select the device that will put data on the data bus (output select), and the bottom decoder is used to select the device that is to take the data from the data bus (input select). Assume that the decoder outputs are connected to the corresponding enable inputs of the devices tied to the data bus. Also assume that all registers initially contain 0000 at time t_0, and the switches are in the positions shown in Figure 9-82.

(a) Determine the contents of each register at times t_1, t_2, and t_3 in response to the waveforms in Figure 9-83(b).

(b) Can bus contention ever occur with this circuit? Explain.

9-62. Show how a 74HC541 (Figure 9-47) can be used in the circuit of Figure 9-82.

FIGURE 9-83   Problem 9-61.

MICROCOMPUTER APPLICATIONS

C, N 9-63. Figure 9-84 shows the basic circuitry to interface a microprocessor (MPU) to a memory module. The memory module will contain one or more memory ICs (Chapter 12) that can either receive data from the MPU (a WRITE operation) or send data to the MPU (a READ operation). The data are transferred over the eight-line data bus. The MPU's data lines and the memory's I/O data lines are connected to this common bus. For now we will be concerned with how the MPU controls the selection of the memory module for a READ or WRITE operation. The steps involved are as follows:

1. The MPU places the memory address on its address output lines A_{15} to A_0.

2. The MPU generates the $R/\overline{W}$ signal to inform the memory module which operation is to be performed: $R/\overline{W} = 1$ for READ, $R/\overline{W} = 0$ for WRITE.

3. The upper five bits of the MPU address lines are decoded by the 74ALS138, which controls the ENABLE input of the memory module. This ENABLE input must be active in order for the memory module to do a READ or WRITE operation.

4. The other 11 address bits are connected to the memory module, which uses them to select the specific *internal* memory location being accessed by the MPU, provided that ENABLE is active.

In order to read from or write into the memory module, the MPU must put the correct address on the address lines to enable the memory, and then pulse *CP* to the HIGH state.

(a) Determine which, if any, of these hexadecimal addresses will activate the memory module: 607F, 57FA, 5F00.

(b) Determine what range of hex addresses will activate the memory. (*Hint:* Inputs A_0 to A_{10} to memory can be any combination.)

FIGURE 9-84 Basic microprocessor-to-memory interface circuit for Problem 9-63.

(c) Assume that a second identical memory module is added to the circuit with its address, $R/\overline{W}$, and data I/O lines connected exactly the same as the first module *except* that its ENABLE input is tied to decoder output $\overline{O}_4$. What range of hex addresses will activate this second module?

(d) Is it possible for the MPU to read from or write to both modules at the same time? Explain.

DESIGN PROBLEM

C, D
9-64. The keyboard entry circuit of Figure 9-16 is to be used as part of an electronic digital lock that operates as follows: when activated, an UN-LOCK output goes HIGH. This HIGH is used to energize a solenoid that retracts a bolt and allows a door to be opened. To activate UN-LOCK, the operator must press the CLEAR key and then enter the correct three-key sequence.

(a) Show how 74HC85 comparators and any other needed logic can be added to the keyboard entry circuit to produce the digital lock operation described above for a key sequence of CLEAR-3-5-8.

(b) Modify the circuit to activate an ALARM output if the operator enters something other than the correct three-key sequence.

SECTIONS 9-15 TO 9-20

H, D, N 9-65. Write the HDL code for a BCD-to-decimal decoder (the equivalent of a 7442).

H, D, N 9-66. Write the HDL code for a HEX decoder/driver for a 7-segment display. The first 10 characters should appear as shown in Figure 9-7. The last six characters should appear as shown in Figure 9-85.

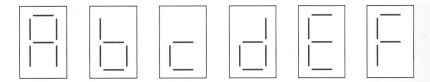

FIGURE 9-85   HEX characters for Problem 9-66.

B, H, N 9-67. Write a low-priority ENCODER description that will always encode the lowest number if two inputs are activated simultaneously.

H, N 9-68. Use an HDL and the techniques covered in Chapter 6 to cascade two 7485 macrofunctions (four-bit comparators) to make an eight-bit comparator.

H, N 9-69. Rewrite the code of the four-bit comparator of Figure 9-66 or 9-67 to make an eight-bit comparator without using macrofunctions.

H, N 9-70. Use HDL to describe a four-bit binary number to a two-digit BCD code converter.

H, N 9-71. Use HDL to describe a three-digit BCD code to eight-bit binary number converter. (Maximum BCD input is 255).

ANSWERS TO SECTION REVIEW QUESTIONS

SECTION 9-1

1. No   2. The enable input controls whether or not the decoder logic responds to the input binary code.   3. The 7445 has open-collector outputs that can handle up to 30 V and 80 mA.   4. 24 pins: 2 enables, 4 inputs, 16 outputs, V_{CC}, and ground

SECTION 9-2

1. *a, b, c, f, g*   2. True

SECTION 9-3

1. LEDs: (a), (e), (f). LCDs: (b), (c), (d), (e)   2. (a) four-bit BCD, (b) seven- or eight-bit ASCII, (c) binary value for pixel intensity

SECTION 9-4

1. An encoder produces an output code corresponding to the activated input. A decoder activates one output corresponding to an applied input code.   2. In a priority encoder, the output code corresponds to the *highest*-numbered input that is activated.   3. Normal BCD = 0110   4. (a) produces a PGT when a key is pressed, (b) converts key actuation to its BCD code, (c) generates bounce-free pulse

to trigger the ring counter, (d) form a ring counter that sequentially clocks output registers, (e) store BCD codes generated by key actuations

SECTION 9-6

1. The binary number at the select inputs determines which data input will pass through to the output.   2. Thirty-two data inputs and five select inputs

SECTION 9-7

1. Parallel-to-serial conversion, data routing, logic-function generation, operations sequencing   2. False; they are applied to the select inputs.   3. Counter

SECTION 9-8

1. A MUX selects one of many input signals to be passed to its output; a DEMUX selects one of many outputs to receive the input signal.   2. True, provided that the decoder has an ENABLE input   3. The LEDs will go on and off in sequence.

SECTION 9-10

1. To provide a means for expanding the compare operations to numbers with more than four bits.   2. $O_{A=B} = 1$; other outputs are 0.

SECTION 9-11

1. A code converter takes input data represented in one type of binary code and converts it to another type of binary code.   2. Three digits can represent decimal values up to 999. To represent 999 in straight binary requires 10 bits.

SECTION 9-12

1. A set of connecting lines to which the inputs and outputs of many different devices can be connected   2. Bus contention occurs when the outputs of more than one device connected to a bus are enabled at the same time. It is prevented by controlling the device enable inputs so that this cannot happen.   3. A condition in which all devices connected to a bus are in the Hi-Z state

SECTION 9-13

1. 1011   2. True   3. 0000

SECTION 9-14

1. Bus contention   2. Floating, Hi-Z   3. Provides tristate low-impedance outputs   4. Reduces the number of IC pins and the number of connections to the data bus   5. See Figure 9-51.

SECTION 9-15

1. They are enable inputs. All must be active for the decoder to work.   2. The CASE and the TABLE   3. The selected signal assignment statement and the CASE

SECTION 9-16

1. The combination input/output pin BI/RBO   2. Common anode. The outputs connect to the cathodes and go LOW to light the segments.   3. The IF/ELSE control structure is evaluated sequentially and gives precedence in the order in which decisions are listed.

SECTION 9-17

1. The don't-care entry in a truth table and the IF/ELSE control structure   2. The IF/ELSE control structure and the conditional signal assignment statement   3. By use of the :TRI primitive and assigning a value to OE   4. By use of the IEEE STD_LOGIC data type that has a possible value of Z

SECTION 9-18

1. Inputs: *ch0, ch1, ch2, ch3*; output: *dout*; control inputs (Select): s   2. Input *din*; outputs: *ch0, ch1, ch2, ch3*; control inputs (Select): *s*   3. DEFAULTS   4. ELSE

SECTION 9-19

1. numerical data objects (e.g., INTEGER in VHDL)   2. IF/ ELSE   3. Relational operators (< , >)

SECTION 9-20

1. 10   2. By multiplying by 8 + 2. Shifting the BCD digit three places left multiplies by 8, and shifting the same BCD digit one place left multiplies by 2. Adding these results produced the BCD digit multiplied by 10.   3. VHDL simply uses the ⋆ operator to multiply.

DIGITAL SYSTEM PROJECTS USING HDL

■ OUTLINE

■ OBJECTIVES

Upon completion of this chapter, you will be able to:

■ Analyze the operation of systems made of several components that have been covered earlier in this textbook.

■ Describe an entire project with one HDL file.

■ Describe the process of hierarchical project management.

■ Understand how to break a project into manageable pieces.

■ Use MAX+PLUS II software tools to implement a hierarchical modular project.

■ Plan ways to test the operation of the circuits you build.

■ INTRODUCTION

Throughout the first nine chapters of this book, we have explained the fundamental building blocks of digital systems. Now that we have taken out each block and looked it over, we do not want to put them all away and forget them; it is time to build something with the blocks. Some of the examples we have used to demonstrate the operation of individual circuits are really digital systems in their own right, and we have studied how they work. In this chapter, we want to focus more on the building process.

Surveys of graduates show us that most of the professionals in the electrical and computer engineering and technology field have the responsibility of project management. Experience with students has also shown us that the most efficient way to manage a project is not intuitively obvious to everyone, which explains why so many of us end up attending the school of hard knocks (learning through trial and error). This chapter is intended to give you a strategic plan for managing projects while learning about digital systems and the modern tools used to develop them. The principles here are not limited to digital or even electronic projects in general. They could apply to building a house or building your own business. They will definitely improve your success rate and reduce the frustration factor.

Hardware description languages were really created for the purpose of managing large digital systems: for documentation, simulation testing, and the synthesis of working circuits. Likewise, the Altera software tools are specifically designed to work with managing projects that go far beyond the scope of this text. We will describe some of the features of the MAX+PLUS II software package as we go through the steps of developing these small projects. This concept of modular project development, which was introduced in Chapter 4, will be demonstrated here through a series of examples.

10-1   SMALL-PROJECT MANAGEMENT

The first projects described here are relatively small systems that consist of a small number of building blocks. These projects can be developed in separate modules, but this approach would only add to the complexity. They are small enough that it makes sense to implement the entire project in a single HDL design file. This does not mean, however, that a structured process should *not* be followed to complete the project. In fact, most of the same steps that should be employed in a large modular project are also applicable in these examples. The steps that should be followed are (1) overall definition, (2) strategic planning to break the project into small pieces, (3) synthesis and testing of each piece, and (4) system integration and testing.

Definition

The first step in any project is the thorough definition of its scope. In this step, the following issues should be determined:

- How many bits of data are needed?
- How many devices are controlled by the outputs?
- What are the names of each input and output?
- Are the inputs and outputs active-HIGH or active-LOW?
- What are the speed requirements?
- Do I understand fully how this device should operate?
- What will define successful completion of this project?

From this step should come a complete and thorough description of the overall project's operation, a definition of its inputs and outputs, and complete numeric specifications that define its capabilities and limitations.

Strategic Planning

The second step involves developing a strategy for dividing this overall project into manageable pieces. The requirements of the pieces are:

- A way to test each piece must be developed.
- Each piece must fit together to make up the whole system.
- We must know the nature of all the signals that connect the pieces.
- The exact operation of each block must be thoroughly defined and understood.
- We must have a clear vision of how to make each block work.

This last requirement might seem obvious, but it is amazing how many projects are planned around one central block that involves a not-yet-discovered technical miracle or violates silly little laws like conservation of energy. In this stage, each subsystem (section block) becomes somewhat of a project in and of itself, with the possibility of additional subsystems defined within its boundaries. This is the concept of hierarchical design.

Synthesis and Testing

Each subsystem should be built starting at the simplest level. In the case of a digital system designed using HDL, it means writing pieces of code. It also means developing a plan for testing that code to make sure it meets all the criteria. This is often accomplished through some sort of simulation. When a circuit is simulated on a computer, the designer must create all the different scenarios that will be experienced by the actual circuit and must also know what the proper response to those inputs should be. This testing often takes a great deal of thought and is not an area that should be overlooked. The worst mistake you can make is to conclude that a fundamental block works perfectly, only to find later those few situations where it fails. This predicament often forces you to re-think many of the other blocks, thus nullifying much of your work.

System Integration and Testing

The last step is to put the blocks together and test them as a unit. Blocks are added and tested at each stage until the entire project is working. This area is often trivialized but rarely goes smoothly. Even if you took care of all the details you thought about, there are always the "gotchas" that nobody thought about.

Some aspects of project planning and management go beyond the scope of this text. One is the selection of a hardware platform that will best fit your application. In Chapter 13, we will explore the broad field of digital systems and look specifically at the capabilities and limitations of PLDs in various categories. Another very critical dimension in project management is time. Your boss will give you only a certain amount of time to complete your project, and you must plan your work (and effort) to meet this deadline. We will not be able to cover time management in this text, but as a general rule you will find that most facets of the project will actually take two to three times longer than you think they will when you begin.

REVIEW QUESTIONS

1. Name the steps of project management.
2. At what stage should you decide how to measure success?

10-2 STEPPER MOTOR DRIVER PROJECT

The purpose of this section is to demonstrate a typical application of counters combined with decoding circuits. A digital system often contains a counter that cycles through a specified sequence and whose output states are decoded by a combinational logic circuit, which in turn controls the operation of the system. Many applications also have external inputs that are used to put the system into various modes of operation. This section discusses all these features to control a stepper motor.

In a real project, the first step of definition often involves some research on the part of the project manager. In this section (or project), it is vital that we understand what a stepper motor is and how it works before we try to create a circuit that is supposed to control it. In Section 7-14, we showed you how to design a simple synchronous counter that could be used to drive a stepper motor. The sequence demonstrated in that section is called the *full-step sequence*. As you recall, it involved two flip-flops and their Q and $\overline{Q}$ outputs driving the four coils of the motor. The full-step sequence always has two coils of the stepper motor energized in any state of the sequence and typically causes 15° of shaft rotation per step. Other sequences, however, will also cause a stepper to rotate. If you look at the full-step sequence, you will notice that each state transition involves turning off one coil and simultaneously turning on another coil. For example, look at the first state (1010) in the full-step sequence of Table 10-1. When it switches to the second state in the sequence, *coil 1* is turned off and *coil 0* is turned on. The *half-step sequence* is created by inserting a state with only one coil energized between full steps, as shown in the middle column of Table 10-1. In this sequence, one coil is de-energized before the other is energized. The first state is 1010 and the second state is 1000, meaning that *coil 1* is turned off for one state before *coil 0* is turned on. This intermediate state causes the stepper shaft to rotate half as far (7.5°) as it would in the full-step sequence (15°). The half-step sequence is used when smaller steps are desirable and more steps per revolution are acceptable. As it turns out, the stepper motor will rotate in a manner similar to the full-step sequence (15° per step) if you apply only the sequence of intermediate states with one coil energized at a time. This sequence, called the *wave-drive sequence*, has less torque but operates more smoothly than the full-step sequence at moderate speeds. The wave-drive sequence is shown in the right-hand column of Table 10-1.

TABLE 10-1 Stepper motor coil drive sequences.

Full-Step	Half-Step	Wave-Drive
1010	1010	
	1000	1000
1001	1001	
	0001	0001
0101	0101	
	0100	0100
0110	0110	
	0010	0010

Problem Statement

A microprocessor laboratory needs a universal interface to drive a stepper motor. In order to experiment with microcontrollers driving stepper motors, it would be useful to have a single universal interface IC wired to the stepper motor. This circuit needs to accept any of the typical forms of stepper drive signals from a microcontroller and activate the windings of the motor to make it move in the desired manner. The interface needs to operate in one of four modes: decoded full-step, decoded half-step, decoded wave-drive, or nondecoded direct drive. The mode is selected by controlling the logic levels on the *M1, M0* input pins. In the first three modes, the interface receives just two control bits—a step pulse and a direction control bit—from the microcontroller. Each time it sees a *rising* edge on the step input, the circuit must cause the motor to move one increment of motion clockwise or counterclockwise, depending on the level present on the direction bit. Depending on the mode that the IC is in, the outputs will respond to each step pulse by changing state according to the sequences shown in Table 10-1. The fourth mode of operation of this circuit must allow the microcontroller to control each winding of the motor directly. In this mode, the circuit accepts four control bits from the microcontroller and passes these logic levels directly to its outputs, which are used to energize the stepper coils. The four modes are summarized in Table 10-2.

TABLE 10-2

Mode	M1 M0	Input Signals	Output
0	0 0	Step, direction	Full-step count sequence
1	0 1	Step, direction	Wave-drive count sequence
2	1 0	Step, direction	Half-step count sequence
3	1 1	Four control inputs	Direct drive from control inputs

In modes 0, 1, and 2, the outputs count through the corresponding count sequence on each rising edge of the step input. The direction input determines whether the sequence moves forward or backward through the states in Table 10-1, thus moving the motor clockwise or counterclockwise. From this description, we can make some decisions about the project.

Inputs

step: rising edge trigger

direction: 0 = backward through table, 1 = forward through table

cin0, cin1, cin2, cin3, m1, m0: active-HIGH control inputs

Outputs

cout0, cout1, cout2, cout3: active-HIGH control outputs

Strategic Planning

This project has two key requirements. It requires a sequential counter circuit that will control the outputs in three of the modes. In the last mode, the output does not follow a counter but rather follows the control inputs. While there are many ways to divide this project and still fulfill these requirements, we will choose to have a simple up/down binary counter that responds to the step and direction inputs. A separate combinational logic circuit will translate (decode) the binary count into the appropriate output state, depending on the mode input setting. This circuit will also ignore the counter inputs and pass the control inputs directly to the outputs when the mode is set to 3. The circuit diagram is shown in Figure 10-1.

Breaking this problem into manageable pieces is also fairly straightforward. The first step is to build an up/down counter. This counter should be tested on a simulator using only the direction and step inputs. Next, try to make each decoded sequence work individually with the counter. Then try to get the mode inputs to select one of the decoder sequences and add the direct-drive option (which is fairly trivial). When the circuit can follow the states shown in Table 10-1 in either direction, for each mode sequence, and pass the four *cin* signals directly to *cout* in mode 3, we will be successful.

Synthesis and Testing

The code in Figures 10-2 and 10-3 shows the first stage of development: designing and testing an up/down counter. We will use an intermediate integer variable for the counter value and test it by outputting the count directly to *q*. To test this part of the design, we simply need to make sure it can count up and down through the eight states. Figure 10-4 shows the simulation results. We only need to provide the clock pulses and make up a direction control signal, and the simulator demonstrates the counter's response.

FIGURE 10-1   A universal stepper motor interface circuit.

```
SUBDESIGN  fig10_2
(
   step, dir     :INPUT;
   q[2..0]       :OUTPUT;
)
VARIABLE
count[2..0]      : DFF;

BEGIN
   count[].clk = step;
   IF dir THEN count[].d = count[].q + 1;
   ELSE         count[].d = count[].q - 1;
   END IF;
   q[] = count[].q;
END;
```

```
ENTITY fig10_3 IS
PORT( step, dir    :IN BIT;
      q            :OUT INTEGER RANGE 0 TO 7);
END fig10_3;

ARCHITECTURE vhdl OF fig10_3 IS
BEGIN
   PROCESS (step)
   VARIABLE count    :INTEGER RANGE 0 TO 7;
   BEGIN
      IF (step'EVENT AND step = '1') THEN
         IF dir = '1' THEN count := count + 1;
         ELSE               count := count - 1;
         END IF;
      END IF;
      q <= count;
   END PROCESS;
END vhdl;
```

FIGURE 10-2   AHDL MOD-8.   **FIGURE 10-3**   VHDL MOD-8.

FIGURE 10-4   Simulation testing of a basic MOD-8.

The next step is to add one of the decoded outputs and test it, which will require adding the four-bit *cout* output specification. The *q* output bits of the MOD-8 counter are kept for the sake of continuity. Figure 10-5 shows the AHDL code for this stage of testing, and Figure 10-6 shows the VHDL code for the same stage of testing. Notice that a CASE construct is used to decode the counter and drive the outputs. In the VHDL code, the *cout* outputs have been declared as bit_vector type because we now want to assign binary bit patterns to them. Figure 10-7 shows the simulated test of its operation with enough clock cycles included to test an entire counter cycle up and down.

The other count sequences are simply variations of the code we just tested. It is probably not necessary to test each one independently, so now is a good time to bring in the mode selector inputs (*m*) and direct-drive coil control inputs (*cin*). Notice that the new inputs have been defined in Figures 10-8 (AHDL) and 10-9 (VHDL). Because the mode control has four possible states and we want to do something different for each state, another CASE construct works best. In other words, we have chosen to use a CASE structure to select the mode and a CASE

```
SUBDESIGN  fig10_5
(
    step, dir        :INPUT;
    q[2..0]          :OUTPUT;
    cout[3..0]       :OUTPUT;
)
VARIABLE
    count[2..0]      : DFF;

BEGIN
    count[].clk = step;
    IF dir THEN count[].d = count[].q + 1;
    ELSE        count[].d = count[].q - 1;
    END IF;
    q[] = count[].q;
    CASE count[] IS
        WHEN  B"000"   => cout[] = B"1010";
        WHEN  B"001"   => cout[] = B"1001";
        WHEN  B"010"   => cout[] = B"0101";
        WHEN  B"011"   => cout[] = B"0110";
        WHEN  B"100"   => cout[] = B"1010";
        WHEN  B"101"   => cout[] = B"1001";
        WHEN  B"110"   => cout[] = B"0101";
        WHEN  B"111"   => cout[] = B"0110";
    END CASE;
END;
```

FIGURE 10-5 AHDL full-step sequence decoder.

```
ENTITY fig10_6 IS
PORT (  step, dir :IN BIT;
        q           :OUT INTEGER RANGE 0 TO 7;
        cout        :OUT BIT_VECTOR (3 downto 0));
END fig10_6;

ARCHITECTURE vhdl OF fig10_6 IS
BEGIN
    PROCESS (step)
    VARIABLE count    :INTEGER RANGE 0 TO 7;
    BEGIN
        IF (step'EVENT AND step = '1') THEN
            IF dir = '1' THEN count := count + 1;
            ELSE              count := count - 1;
            END IF;
            q <= count;
        END IF;
        CASE count IS
            WHEN 0   => cout <= B"1010";
            WHEN 1   => cout <= B"1001";
            WHEN 2   => cout <= B"0101";
            WHEN 3   => cout <= B"0110";
            WHEN 4   => cout <= B"1010";
            WHEN 5   => cout <= B"1001";
            WHEN 6   => cout <= B"0101";
            WHEN 7   => cout <= B"0110";
        END CASE;
    END PROCESS;
END vhdl;
```

FIGURE 10-6 VHDL full-step sequence decoder.

FIGURE 10-7 Simulation testing of decoded sequence.

digital clock. Notice that it is really made up of several small and relatively simple circuits that are strategically interconnected to make the clock. Recall that in Chapter 4, we mentioned briefly the concept of modular, hierarchical design and development of digital systems. Now it is time to apply these principles to a project that is within your scope of understanding using the MAX+PLUS II development system from Altera. You must understand the operation of the circuits that have just been described before proceeding with the design of this clock using HDL. Take some time to review this material.

Top-Down Hierarchical Design

Top-down design means that we want to start at the highest level of complexity in the hierarchy, or that the entire project is considered to exist in a closed, dark box with inputs and outputs. The details regarding what is in the box are not yet known. We can only say at this point how we want it to behave. The digital clock was chosen because everyone is familiar with the end result of the operation of this device. An important aspect of this stage of the design process is establishing the scope of the project. For example, this digital clock is not going to have a way to set the time, set an alarm time, shut off the alarm, snooze, or incorporate other features that you may find on the clock beside your bed. To add all these features now would only clutter the example with unnecessary complexity for our immediate purpose. We are also not going to include the signal conditioning that transforms a 60-Hz sine wave into a 60-pulse-per-second digital waveform, or the decoder/display circuits. The project we are tackling has the following specifications:

Inputs: 60 pps CMOS compatible waveform (accuracy dependent on line frequency)

Outputs: BCD Hours: 1 bit TENS 4 bits UNITS
         BCD Minutes: 3 bits TENS 4 bits UNITS
         BCD Seconds: 3 bits TENS 4 bits UNITS
         PM indicator

Minutes and Seconds sequence: BCD MOD 60
         00–59 (decimal representation of BCD)
Hours sequence BCD MOD 12
         01–12 (decimal representation of BCD)
Overall range of display
         01:00:00–12:59:59

AM/PM indicator toggles at 12:00:00

A **hierarchy** is a group of objects arranged in rank order of magnitude, importance, or complexity. A block diagram of the overall project (highest level of the hierarchy) is shown in Figure 10-19. Notice that there are four bits for each

FIGURE 10-19 The top level block of the hierarchy.

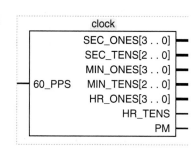

of the BCD units outputs and only three bits for each of the minute and second BCD tens outputs. Because the most significant BCD digit for the tens place is 5 (101_2), only three bits are needed. Notice also that the tens place for the hours (HR_TENS) is only one bit. It will never have a value other than 0 or 1.

The next phase is to break this problem into more manageable sections. First, we need to take the 60-pps input and transform it into a 1-pulse-per-second timing signal. A circuit that divides a reference frequency to a rate required by the system is called a **prescaler**. Next, it makes sense to have individual sections for a seconds counter, minutes counter, and hours counter. So far, the hierarchy diagram looks like Figure 10-20, which shows the project broken into four subsections.

The entire purpose of the frequency prescaler section is to divide the 60-pps input to a frequency of one pulse every second. This requires a MOD-60 counter, and the sequence of the count does not really matter. In this example, the minutes and seconds sections both require MOD-60 counters that count from 00–59 in BCD. Looking for similarities like this is very important in the design process. In this case, we can use the exact same circuit design to implement the frequency prescaler, the minutes counter, and the seconds counters.

A MOD-60 BCD counter can be made quite easily from a MOD-10 (decade) counter cascaded to a MOD-6 BCD counter, as we saw in the diagram of Figure 10-17. This means that inside each of these MOD-60 blocks, we would find a diagram similar to Figure 10-21. The hierarchy of the project now appears as shown in Figure 10-22.

FIGURE 10-20 The section level of the hierarchy.

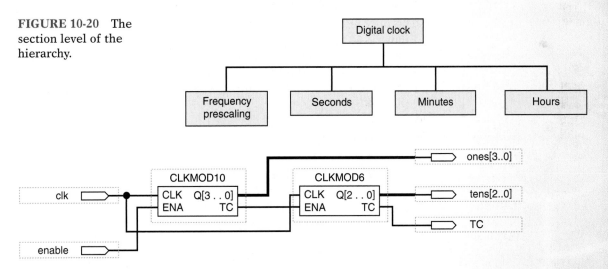

FIGURE 10-21 The blocks inside the MOD-60 section.

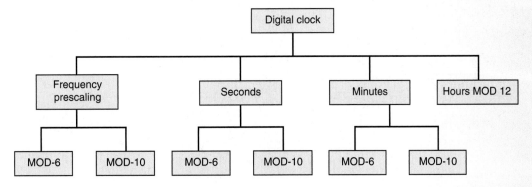

FIGURE 10-22 The complete hierarchy of the clock project.

The final design decision is whether or not to break down the MOD-12 section for Hours into two stages, as shown in Figure 10-18. One option is to connect the macrofunctions of these standard parts from the HDL library, as we have discussed in previous chapters. Because this circuit is rather unique, we have decided instead to describe the MOD-12 hours counter using a single HDL module. We will also describe the MOD-6 and MOD-10 building blocks using HDL. The entire clock circuit can then be built using these three basic circuit descriptions. Of course, even these blocks can be broken down into smaller flip-flop blocks and designed using the schematic entry features of MAX+PLUS II, but it will be much easier using HDL at this level.

Building the Blocks from the Bottom Up

Each of the basic blocks are presented here in both AHDL and VHDL. We present the MOD-6 as a simple modification of the MOD-5 synchronous counter descriptions presented earlier in Chapter 7 (see Figures 7-44 and 7-45). Then we modify this code further to create the MOD-10 counter and finally design the MOD-12 Hours counter from the ground up. We construct the entire clock from these three basic blocks.

MOD-6 COUNTER AHDL

The only additional features that this design needs that are not covered in Figure 7-44 are the count *enable* input and terminal count (*tc*) output shown in Figure 10-23. Notice that the extra input (*enable*, line 3) and output (*tc*, line 4) are included in the I/O definition. A new line (line 11) in the architecture description tests *enable* before deciding how to update the value of *count*

```
1    SUBDESIGN fig10_23
2    (
3       clock, enable      :INPUT;      -- synch clock and enable.
4       q[2..0], tc        :OUTPUT;     -- 3-bit counter
5    )
6    VARIABLE
7       count[2..0]    :DFF;    -- declare a register of D flip-flops.
8
9    BEGIN
10      count[].clk = clock;       -- connect all clocks to synchronous source
11      IF enable THEN
12         IF count[].q < 5 THEN
13            count[].d = count[].q + 1;   -- increment current value by one
14         ELSE count[].d = 0;             -- recycle, force unused states to 0
15         END IF;
16      ELSE count[].d = count[].q;        -- not enabled: hold at this count
17      END IF;
18      tc = enable & count[].q == 5;      -- detect maximum count if enabled
19      q[] = count[].q;                   -- connect register to outputs
20    END;
```

FIGURE 10-23   The MOD-6 design in AHDL.

(lines 12–15). If *enable* is LOW, the same value is held on *count* at every clock edge by the ELSE branch (line 16). Remember always to match an IF with an END IF, as we did on lines 15 and 17. Terminal count (*tc*, line 18) will be HIGH when it is *true* that *count* = = 5 AND *enable* is active. Notice the use of double equal signs (= =) to evaluate equality in AHDL.

MOD-6 COUNTER VHDL

The only additional features that this design needs that are not covered in Figure 7-44 are the count *enable* input and terminal count (*tc*) output shown in Figure 10-24. Notice that the extra input (*enable*, line 2) and output (*tc*, line 4) are included in the I/O definition. A new line (line 15) in the architecture description tests *enable* before deciding how to update the value of *count* (lines 16–20). In the case that *enable* is LOW, the current value is held in the variable *count* and does not count up. Remember always to match an IF with an END IF, as we did on lines 20–22. The terminal count indicator (*tc*, lines 23 and 24) will be HIGH when it is *true* that *count* = 5 AND *enable* is active.

```
1    ENTITY fig10_24 IS
2    PORT( clock, enable      :IN BIT ;
3         q                   :OUT INTEGER RANGE 0 TO 5;
4         tc                  :OUT BIT
5       );
6    END fig10_24;
7
8    ARCHITECTURE a OF fig10_24 IS
9    BEGIN
10      PROCESS (clock)                           -- respond to clock
11      VARIABLE count    :INTEGER RANGE 0 TO 5;
12
13      BEGIN
14        IF (clock = '1' AND clock'event) THEN
15          IF enable = '1' THEN           -- synchronous cascade input
16            IF count < 5 THEN            -- < max (terminal) count?
17                count := count + 1;
18            ELSE
19                count := 0;
20            END IF;
21          END IF;
22        END IF;
23        IF (count = 5) AND (enable = '1') THEN   -- synch cascade output
24            tc <= '1';                           -- indicate terminal ct
25        ELSE  tc <= '0';
26        END IF;
27        q <= count;                              -- update outputs
28      END PROCESS;
29    END a;
```

FIGURE 10-24   The MOD-6 design in VHDL.

FIGURE 10-25   Simulation of the MOD-6 counter.

The simulation testing of the MOD-6 counter in Figure 10-25 verifies that it counts 0–5 and that it responds to the count enable input by ignoring the clock pulses and freezing the count whenever *enable* is LOW. It also generates the *tc* output when it is enabled at its maximum count of 5.

MOD-10 COUNTER AHDL

The MOD-10 counter varies only slightly from the MOD-6 counter that was described in Figure 10-23. The only changes that are necessary involve changing the number of bits in the output port and the register (in the VARIABLE section) along with the maximum value that the counter should reach before rolling over. Figure 10-26 presents the MOD-10 design.

```
1    SUBDESIGN fig10_26
2    (
3       clock, enable      :INPUT;       -- synch clock and enable.
4       q[3..0], tc        :OUTPUT;      -- 4-bit Decade counter
5    )
6    VARIABLE
7       count[3..0] :DFF;                -- declare a register of D flip flops.
8
9    BEGIN
10      count[].clk = clock;        -- connect all clocks to synchronous source
11      IF enable THEN
12         IF count[].q < 9 THEN
13            count[].d = count[].q + 1;  -- increment current value by one
14         ELSE count[].d = 0;            -- recycle,force unused states to 0
15         END IF;
16      ELSE count[].d = count[].q;    -- not enabled: hold at this count
17      END IF;
18      tc = enable & count[].q == 9;   -- detect maximum count
19      q[] = count[].q;                -- connect register to outputs
20   END;
```

FIGURE 10-26   The MOD-10 design in AHDL.

MOD-10 COUNTER VHDL

The MOD-10 counter varies only slightly from the MOD-6 counter that was described in Figure 10-24. The only changes that are necessary involve changing the number of bits in the output port and the variable *count* (using INTEGER RANGE) along with the maximum value that the counter should reach before rolling over. Figure 10-27 presents the MOD-10 design.

```
1    ENTITY fig10_27 IS
2    PORT( clock, enable     :IN BIT ;
3          q                 :OUT INTEGER RANGE 0 TO 9;
4          tc                :OUT BIT
5        );
6    END fig10_27;
7
8    ARCHITECTURE a OF fig10_27 IS
9    BEGIN
10      PROCESS (clock)                          -- respond to clock
11      VARIABLE count   :INTEGER RANGE 0 TO 9;
12
13      BEGIN
14        IF (clock = '1' AND clock'event) THEN
15           IF enable = '1' THEN               -- synchronous cascade input
16              IF count < 9 THEN               -- decade counter
17                 count := count + 1;
18              ELSE
19                 count := 0;
20              END IF;
21           END IF;
22        END IF;
23        IF (count = 9) AND (enable = '1') THEN     -- synch cascade output
24              tc <= '1';
25        ELSE  tc <= '0';
26        END IF;
27        q <= count;                          -- update outputs
28      END PROCESS;
29   END a;
```

FIGURE 10-27   The MOD-10 design in VHDL.

MOD 12 Design

We have already decided that the hours counter is to be implemented as a single design file using HDL. It must be a MOD-12 BCD counter that follows the hours sequence of a clock (1–12) and provides the AM/PM indicator. Recall from the initial design step that the BCD outputs need to be a four-bit array for the low-order digit and a single bit for the high-order digit. To design this counter circuit, consider how it needs to operate. Its sequence is:

01 02 03 04 05 06 07 08 09 10 11 12 01 . . .

By observing this sequence, we can conclude that there are four critical areas that define the operations needed to produce the proper NEXT state:

1. When the value is 01 through 08, increment the low digit and keep the high digit the same.
2. When the value is 09, reset the low digit to 0 and force the high digit to 1.
3. When the value is 10 or 11, increment the low digit and keep the high digit the same.
4. When the value is 12, reset the low digit to 1 and the high digit to 0.

Because these conditions need to evaluate a range of values, it is most appropriate to use an IF/ELSIF construct rather than a CASE construct. There is also a need to identify when it is time to toggle the AM/PM indicator. This time occurs when the hour state is 11 and the enable is HIGH, which means that the lower-order counters are at their maximum (59:59).

MOD-12 COUNTER IN AHDL

The AHDL counter needs a bank of four D flip-flops for the low-order BCD digit and only a single D flip-flop for the high-order BCD digit because its value will always be 0 or 1. A flip-flop is also needed to keep track of A.M. and P.M. These primitives are declared on lines 7–9 of Figure 10-28. Also note that in this design, the same names are used for the output ports. This is a convenient feature of AHDL. When the enable input (*ena*) is active, the circuit evaluates the IF/ELSE

```
1    SUBDESIGN fig10_28
2    (
3       clk, ena           :INPUT;
4       low[3..0], hi, pm  :OUTPUT;
5    )
6    VARIABLE
7       low[3..0]    :DFF;
8       hi           :DFF;
9       am_pm        :JKFF;
10      time         :NODE;
11   BEGIN
12      low[].clk = clk;          -- synchronous clocking
13      hi.clk = clk;
14      am_pm.clk = clk;
15      IF ena THEN               -- use enable to count
16          IF low[].q < 9 & hi.q == 0   THEN
17             low[].d = low[].q + 1; --inc lo digit
18             hi.d = hi.q;   -- hold hi digit
19          ELSIF low[].q == 9 THEN
20             low[].d = 0;
21             hi.d = VCC;
22          ELSIF hi.q == 1 & low[].q < 2 THEN
23                 low[].d = low[].q + 1;
24                 hi.d = hi.q;
25          ELSIF hi.q == 1 & low[].q == 2 THEN
26                 low[].d = 1;
27                 hi.d = GND;
28          END IF;
29      ELSE
30          low[].d = low[].q;
31          hi.d = hi.q;
32      END IF;
33      time = hi.q == 1 & low[3..0].q == 1 & ena;   -- detect 11:59:59
34      am_pm.j = time;        -- toggle am/pm at noon and midnight
35      am_pm.k = time;
36      pm = am_pm.q;
37   END;
```

FIGURE 10-28　The MOD-12 hours counter in AHDL.

statements of lines 16–28 and performs the proper operation on the high and low nibble of the BCD number. Whenever the enable input is LOW, the value remains the same, as shown on lines 30 and 31. Line 33 detects when the count reaches 11 while the counter is enabled. This signal is applied to the *J* and *K* inputs of the am_pm flip-flop to cause it to toggle at 11:59:59.

MOD-12 COUNTER IN VHDL

The VHDL counter of Figure 10-29 needs a four-bit output for the low-order BCD digit and a single output bit for the high-order BCD digit because its

```
1   ENTITY fig10_29 IS
2   PORT( clk, ena        :IN BIT ;
3         low             :OUT INTEGER RANGE 0 TO 9;
4         hi              :OUT INTEGER RANGE 0 TO 1;
5         pm              :OUT BIT              );
6   END fig10_29;
7
8   ARCHITECTURE a OF fig10_29 IS
9   BEGIN
10     PROCESS (clk)                        -- respond to clock
11     VARIABLE am_pm :BIT;
12     VARIABLE ones  :INTEGER RANGE 0 TO 9; -- 4-bit units signal
13     VARIABLE tens  :INTEGER RANGE 0 TO 1; -- 1-bit tens signal
14     BEGIN
15       IF (clk = '1' AND clk'EVENT) THEN
16         IF ena = '1' THEN                -- synchronous cascade input
17           IF (ones = 1) AND (tens = 1) THEN   -- at 11:59:59
18             am_pm := NOT am_pm;               -- toggle am/pm
19           END IF;
20           IF (ones < 9) AND  (tens = 0) THEN  -- states 00-08
21             ones := ones + 1;              -- increment units
22           ELSIF ones = 9 THEN              -- state 09...set to 10:00
23             ones := 0;                     -- units reset to zero
24             tens := 1;                     -- tens bump up to 1
25           ELSIF (tens = 1) AND (ones < 2) THEN-- states 10, 11
26             ones := ones + 1;              -- increment units
27           ELSIF (tens = 1) AND (ones = 2) THEN -- state 12
28             ones := 1;                     -- set to 01:00
29             tens := 0;
30           END IF;
31   -------------------------------------------------------------
32   -- This space is the alternate location for updating am/pm
33   -------------------------------------------------------------
34         END IF;
35       END IF;
36       pm <= am_pm;
37       low <= ones;                         -- update outputs
38       hi <= tens;
39     END PROCESS;
40   END a;
```

FIGURE 10-29 The MOD-12 hours counter in VHDL.

value will always be 0 or 1. These outputs (lines 3 and 4) and also the variables that will produce the outputs (lines 12 and 13) are declared as integers because this makes "counting" possible by simply adding 1 to the variable value. On each active edge of the clock, when the enable input is active, the circuit needs to decide what to do with the BCD units-of-hours counter, the single bit tens-of-hours flip-flop, and also the AM/PM flip-flop.

This example is an excellent opportunity to point out some of the advanced features of VHDL that allow the designer to describe precisely the operation of the final hardware circuit. In previous chapters, we discussed the issue of statements within a PROCESS being evaluated sequentially. Recall that statements outside the PROCESS are considered concurrent, and the order in which they are written in the design file has no effect on the operation of the final circuit. In this example, we must evaluate the current state to decide whether to toggle the AM/PM indicator and also advance the counter to the NEXT state. The issues involved include the following:

1. How do we "remember" the current count value in VHDL?
2. Do we evaluate the current count to see if it is 11 (to determine if we need to toggle the AM/PM flip-flop) and then increment to 12, or do we increment the counter's state from 11 to 12 and then evaluate the count to see if it is 12 (to know we need to toggle the AM/PM flip flop)?

Regarding the first issue, there are two ways to remember the current state of a counter in VHDL. Both SIGNALs and VARIABLEs hold their value until they are updated. Generally, SIGNALs are used to connect nodes in the circuit like wires, and VARIABLEs are used like a register to store data between updates. Consequently, VARIABLEs are generally used to implement counters. The major differences are that VARIABLEs are local to the PROCESS in which they are declared and SIGNALs are global. Also, VARIABLEs are considered to be updated immediately within a sequence of statements in a PROCESS, whereas SIGNALs referred to in a PROCESS are updated when the PROCESS suspends. In this example, we have chosen to use VARIABLEs, which are local to the PROCESS that describes what should happen when the active clock edge occurs.

For the second issue, either of these strategies will work, but how do we describe them using VHDL? If we want the circuit to toggle A.M. and P.M. by detecting 11 prior to the counter updating (like a synchronous cascade), then the test must occur in the code before the VARIABLEs are updated. This test is demonstrated in the design file of Figure 10-29 on lines 17–19. On the other hand, if we want the circuit to toggle A.M. and P.M. by detecting when the hour 12 has arrived after the clock edge (more like a ripple cascade), then the VARIABLEs must be updated prior to testing for the value 12. To modify the design in Figure 10-29 to accomplish this task, the IF construct of lines 17–19 can be moved to the blank area of lines 31–33 and edited as shown in bold below:

```
31    IF (ones = 2) AND (tens = 1) THEN -- at 12:00:00
32      am_pm := NOT am_pm;              -- toggle am/pm
33      END IF;
```

The order of the statements and the value that is decoded makes all the difference in how the circuit operates. On lines 36–38, the *am_pm* VARIABLE is connected to the *pm* port, the units BCD digit is applied to the lower four bits of the output (*low*), and the tens digit (a single-bit variable) is applied to the most significant digit (*hi*) of the output port. Because all these VARIABLEs are local, these statements must occur prior to END PROCESS on line 39.

After the design is compiled, it must be simulated to verify its operation, especially at the critical areas. Figure 10-30 shows an example of a simulation to test this counter. On the left side of the timing diagram, the counter is disabled and is holding the hour 11 because the *hi* digit is at 1 and the *low[ ]* digit is at 1. On the rising edge of the clock, after the enable goes HIGH, the hour goes from 11 to 12 and causes the PM indicator to go HIGH, which means it is noon. The next active edge causes the count to roll over from 12 to 01. On the right half of the timing, the same sequence is simulated, showing that there would actually be many clock pulses between the times the hour increments. On the clock cycle before it must increment, the enable is driven HIGH by the terminal count of the previous stage.

FIGURE 10-30 Simulation of the MOD-12 hours counter.

Combining Blocks Graphically

The building blocks of the project have been defined, created, and individually simulated to verify that they work correctly. Now it is time to combine the blocks to make sections and to combine the sections to make the final product. Altera's MAX+PLUS II software offers several ways to accomplish the integration of all the pieces of a project. Probably the most straightforward and intuitive approach is the use of graphic design files to "connect" the blocks. In Chapter 4, we mentioned that all different types of design files (AHDL, VHDL, VERILOG, Schematic, Waveform) can be combined in a GDF file. This technique is made possible by a feature that allows us to create a "symbol" to represent a particular design file. For example, the MOD-6 counter design file that was written in the VHDL design file fig10_24 can be represented in the software as the circuit block, as shown in Figure 10-31(a). The MAX+PLUS II software creates this symbol at the click of a button. From that point, it will recognize the symbol as operating according to the design specified in the HDL code. The symbol of Figure 10-31(b) was created from the AHDL file for the MOD-10 counter of Figure 10-26, and the symbol of Figure 10-31(c) was created from the VHDL file for the MOD-12 counter of Figure 10-29. (The reason these blocks are named by figure number is simply

FIGURE 10-31 Graphic block symbols generated from HDL design files: (a) MOD-6 from VHDL; (b) MOD-10 from AHDL; (c) MOD-12 from VHDL.

FIGURE 10-32 Graphically combining HDL blocks to make a MOD-60.

FIGURE 10-33 The MOD-60 counter.

to make it easier to locate the design files on the enclosed CD. In a design environment [rather than in a textbook], they should be named according to their purpose, with names like MOD6, MOD10, and CLOCK_HOURS.)

Following the design hierarchy that we established, the next step is to combine the MOD-6 and MOD-10 counters to make a MOD-60 block. These symbols make it simple to draw lines to connect blocks and label the inputs and outputs to the graphic design file. The result is shown in Figure 10-32. This graphic design file can be compiled and used to simulate the operation of the MOD-60 counter. When the design has been verified as working properly, the MAX+PLUS II system allows us to take this circuit and create a block symbol for it, as shown in Figure 10-33.

The MOD-60 symbol can be used repeatedly along with the MOD-12 symbol to create the system level GDF file shown in Figure 10-34. Even this system level diagram can be represented by a block symbol for the entire project, as shown in Figure 10-35.

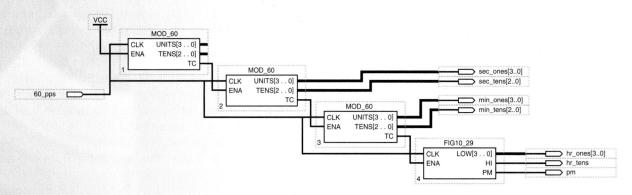

FIGURE 10-34 The complete clock project connected using a graphic design file.

FIGURE 10-35 The entire clock represented by one symbol.

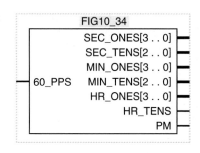

Combining Blocks Using Only HDL

The graphic approach works well as long as it is available and adequate for the purpose at hand. As we mentioned previously, HDL was developed to provide a convenient way to document complex systems and to store the information in a more timeless and software-independent manner. It is reasonable to assume that with AHDL, the option of graphic integration of subdesigns will always be available with the tools from Altera; however, this assumption is not reasonable for users of VHDL. Many VHDL development systems do not offer any equivalent to the graphic design file of Altera, which is why it is important to address the same concept of modular, hierarchical development and project integration using only text-based language tools. Our coverage of AHDL integration will not be as in-depth as our coverage of VHDL because it is less critical in the flexible Altera MAX+PLUS II environment.

AHDL MODULE INTEGRATION

Let's go back to the two AHDL files for the MOD-6 and MOD-10 counters. How do we combine these files into a MOD-60 counter using only text-based AHDL? The method is really very similar to that of graphic integration. Instead of creating a "symbol" representation of the MOD-6 and MOD-10 files, a new kind of file called an "INCLUDE" file is created. It contains all the important information about the AHDL file it represents. To describe a MOD-60 counter, a new TDF file, shown in Figure 10-36, is opened. The building block files are "included" at the top, as shown on lines 1 and 2. Next, the names that were used for the building blocks are used like library components or primitives to define the nature of a variable. On line 10, the variable *mod10* is now used to represent the MOD-10 counter in the other module (fig10_26). *MOD10* now has all the attributes (inputs, outputs, functional operation) described in fig10_26.tdf. Likewise, on line 11, the variable *mod6* is given the attributes of the MOD-6 counter of fig10_23.tdf. Lines 13–19 accomplish the exact same task as drawing lines on the GDF file to connect the components to one another and to the input/output ports.

```
1    INCLUDE "fig10_26.inc";        -- mod-10 counter module
2    INCLUDE "fig10_23.inc";        -- mod-6 counter module
3
4    SUBDESIGN fig10_36
5    (
6        clk, ena                   :INPUT;
7        ones[3..0], tens[2..0], tc :OUTPUT;
8    )
9    VARIABLE
10       mod10              :fig10_26;    -- mod-10 for units
11       mod6               :fig10_23;    -- mod-6 for tens
12   BEGIN
13       mod10.clock = clk;             -- synchronous clocking
14       mod6.clock = clk;
15       mod10.enable = ena;
16       mod6.enable = mod10.tc;        -- cascade
17       ones[3..0] = mod10.q[3..0];    -- 1s
18       tens[2..0] = mod6.q[2..0];     -- 10s
19       tc = mod6.tc;                  -- Make terminal count at 59
20   END;
```

FIGURE 10-36   The MOD-60 made from MOD-10 and MOD-6 in AHDL.

This file (FIG10_36.TDF) can be translated into an "include" file (fig10_36.inc) by the compiler and then used in another tdf file that describes the interconnection of major sections to make up the system. Each level of the hierarchy refers back to the constituent modules of the lower levels.

VHDL MODULE INTEGRATION

Let's go back to the two VHDL files for the MOD-6 and MOD-10 counters, which were shown in Figures 10-24 and 10-27, respectively. How do we combine these files into a MOD-60 counter using only text-based VHDL? The method is really very similar to that of graphic integration. Instead of creating a "symbol" representation of the MOD-6 and MOD-10 files, these design files are described as a COMPONENT, like we studied in Chapter 5. It contains all the important information about the VHDL file it represents. To describe a MOD-60 counter, a new VHDL file, shown in Figure 10-37, is opened. The building block files are described as "components," as shown on lines

```
1   ENTITY fig10_37 IS
2   PORT( clk, ena      :IN BIT ;
3         tens          :OUT INTEGER RANGE 0 TO 5;
4         ones          :OUT INTEGER RANGE 0 TO 9;
5         tc            :OUT BIT              );
6   END fig10_37;
7
8   ARCHITECTURE a OF fig10_37 IS
9   SIGNAL cascade_wire  :BIT;
10  COMPONENT fig10_24                          -- MOD-6 module
11  PORT( clock, enable  :IN BIT ;
12        q              :OUT INTEGER RANGE 0 TO 5;
13        tc             :OUT BIT);
14  END COMPONENT;
15  COMPONENT fig10_27                          -- MOD-10 module
16  PORT( clock, enable  :IN BIT ;
17        q              :OUT INTEGER RANGE 0 TO 9;
18        tc             :OUT BIT);
19  END COMPONENT;
20
21  BEGIN
22      mod10:fig10_27
23          PORT MAP(  clock => clk,
24                     enable => ena,
25                     q  => ones,
26                     tc => cascade_wire);
27
28      mod6:fig10_24
29          PORT MAP(  clock => clk,
30                     enable => cascade_wire,
31                     q  => tens,
32                     tc => tc);
33  END a;
```

FIGURE 10-37 The MOD-60 made from MOD-10 and MOD-6 in VHDL.

```
1    ENTITY fig10_38 IS
2    PORT( pps_60                        :IN BIT ;
3         hour_tens                      :OUT INTEGER RANGE 0 TO 1;
4         hour_ones, min_ones, sec_ones  :OUT INTEGER RANGE 0 TO 9;
5         min_tens, sec_tens             :OUT INTEGER RANGE 0 to 5;
6         pm                             :OUT BIT              );
7    END fig10_38;
8
9    ARCHITECTURE a OF fig10_38 IS
10   SIGNAL cascade_wire1, cascade_wire2, cascade_wire3    :BIT;
11   SIGNAL enabled                                        :BIT;
12   COMPONENT fig10_37    -- MOD-60
13   PORT( clk, ena       :IN BIT ;
14        tens            :OUT INTEGER RANGE 0 TO 5;
15        ones            :OUT INTEGER RANGE 0 TO 9;
16        tc              :OUT BIT        );
17   END COMPONENT;
18   COMPONENT fig10_29    -- MOD-12
19   PORT( clk, ena       :IN BIT ;
20        low             :OUT INTEGER RANGE 0 TO 9;
21        hi              :OUT INTEGER RANGE 0 TO 1;
22        pm              :OUT BIT        );
23   END COMPONENT;
24   BEGIN
25      enabled <= '1';
26
27      prescale:fig10_37         -- MOD-60 prescaler
28         PORT MAP(   clk  => pps_60,
29                     ena  => enabled,
30                     tc   => cascade_wire1);
31
32      second:fig10_37           -- MOD-60 seconds counter
33         PORT MAP(   clk  => pps_60,
34                     ena  => cascade_wire1,
35                     ones => sec_ones,
36                     tens => sec_tens,
37                     tc   => cascade_wire2);
38
39      minute:fig10_37           -- MOD-60 minutes counter
40         PORT MAP(   clk  => pps_60,
41                     ena  => cascade_wire2,
42                     ones => min_ones,
43                     tens => min_tens,
44                     tc   => cascade_wire3);
45
46      hour:fig10_29             -- MOD12 Hours Counter
47         PORT MAP(   clk  => pps_60,
48                     ena  => cascade_wire3,
49                     low  => hour_ones,
50                     hi   => hour_tens,
51                     pm   => pm);
52   END a;
```

FIGURE 10-38   The complete clock in VHDL.

10–14 and lines 15–19 in the architecture description. Next, the names that were used for the building blocks (components) are used along with the PORT MAP keywords to describe the interconnection of these components. The information in the PORT MAP sections describes the exact same operations as drawing wires on a schematic diagram in a GDF file.

Finally, the VHDL file that represents the block at the top of the hierarchy is created using components from Figure 10-37 (MOD-60) and Figure 10-29 (MOD-12). This file is shown in Figure 10-38 (on page 715). Notice that the general form is as follows:

Define I/O: lines 1–7

Define signals: lines 10–11

Define components: lines 12–23

Instantiate components and connect them together: lines 27–52

REVIEW QUESTIONS

1. What is being defined at the top level of a hierarchical design?
2. Where does the design process start?
3. Where does the building process start?
4. At which stage(s) should simulation testing be done?

10-5 FREQUENCY COUNTER PROJECT

The project in this section demonstrates the use of counters and other standard logic functions to implement a system called a frequency counter, which is similar to the piece of test equipment that you have probably used in the laboratory. The theory of operation will be described in terms of conventional MSI logic devices and then related to the building blocks that can be developed using HDL. As with most projects, this example consists of several circuits that we have studied in earlier chapters. They are combined here to form a digital system with a unique purpose. First, let us define a frequency counter.

A **frequency counter** is a circuit that can measure and display the frequency of a signal. As you know, the frequency of a periodic waveform is simply the number of cycles per second. Shaping each cycle of the unknown frequency into a digital pulse allows us to use a digital circuit to count the cycles. The general idea behind measuring frequency involves enabling a counter to count the number of cycles (pulses) of the incoming waveform during a precisely specified period of time called the **sampling interval**. The length of the sampling interval determines the range of frequencies that can be measured. A longer interval provides improved precision for low frequencies but will overflow the counter at high frequencies. A shorter sample interval provides a less precise measurement of low frequencies but can measure a much higher maximum frequency without exceeding the upper limit of the counter.

EXAMPLE 10-1

Assume that a frequency counter uses a four-digit BCD counter. Determine the maximum frequency that can be measured using each of the following sample intervals:

(a) 1 second   (b) 0.1 second   (c) 0.01 second

Solution

(a) With a sampling interval of 1 second, the four-digit counter can count up to 9999 pulses. The frequency is 9999 pulses per second or 9.999 kHz.

(b) The counter can count up to 9999 pulses within the sampling interval of 0.1 seconds. This translates into a frequency of 99,990 pulses per second or 99.99 kHz.

(c) The counter can count up to 9999 pulses within the sampling interval of 0.01 seconds. This translates into a frequency of 999,900 pulses per second or 999.9 kHz.

EXAMPLE 10-2

If a frequency of 3792 pps is applied to the input of the frequency counter, what will the counter read under each of the following sample intervals?

(a) 1 second   (b) 0.1 second   (c) 10 ms

Solution

(a) During a sampling interval of 1 second, the counter will count 3792 cycles. The frequency will read 3.792 kpps.

(b) During a sampling interval of 0.1 second, the number of pulses that will be counted is 379 or 380 cycles, depending on where the sample interval begins. The frequency will read 03.79 kpps or 03.80 kpps.

(c) During a sampling interval of 0.01 second, the number of pulses that will be counted is 37 or 38 cycles, depending on where the sample interval begins. The frequency will read 003.7 kpps or 003.8 kpps.

One of the most straightforward methods for constructing a frequency counter is shown as a block diagram in Figure 10-39. The major blocks are the counter, the display register, the decoder/display, and the timing and control unit. The counter block contains several cascaded BCD counters that are used to count the number of pulses produced by the unknown signal applied to the clock input. The counter block has count enable and clear controls. The time period for counting (sample interval) is controlled by an enable signal that is produced by the timing and control block. The length of time for the BCD counters to be enabled can be selected with the range select input to the timing

FIGURE 10-39 Basic frequency counter block diagram.

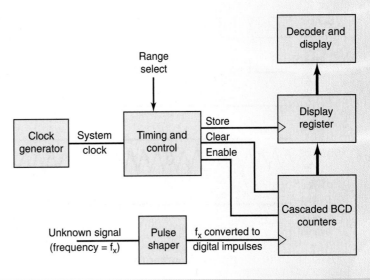

and control block. This allows the user to select the desired frequency range to be measured and effectively determines the location of the decimal point in the digital readout. The pulse width of the enable signal (sample interval) is critical for taking an accurate frequency measurement. The counter must be cleared before it is enabled for a new frequency measurement of the un-known signal. After a new count has been taken, the counter is disabled, and the most recent frequency measurement is stored in the display register. The output of the display register is input to the decoder and display block, where the BCD values are converted into decimal for the display readout. Using a separate display register allows the frequency counter to take a new meas-urement in the background so that the user does not watch the counter while it is totaling the number of pulses for a new reading. The display is instead updated periodically with the last frequency reading.

The accuracy of this frequency counter depends almost entirely on the accuracy of the system clock frequency, which is used to create the proper pulse width for the counter enable signal. A crystal-controlled clock genera-tor is used in Figure 10-39 to produce an accurate system clock for the tim-ing and control block.

A pulse shaper block is needed to ensure that the unknown signal whose frequency is to be measured will be compatible with the clock input for the counter block. A Schmitt-trigger circuit may be used to convert "nonsquare" waveforms (sine, triangle, etc.) as long as the unknown input signal is of sat-isfactory amplitude. If the unknown signal might have a larger or smaller am-plitude than is compatible with a given Schmitt trigger, then additional analog signal conditioning circuitry, such as an automatic gain control, will be required for the pulse shaper block.

The timing diagram for the control of the frequency counter is shown in Figure 10-40. The control clock is derived from the system clock signal by fre-quency dividers contained in the control and timing block. The period of the control clock signal is used to create the desired enable pulse width. A recycling control counter inside the control and timing block is clocked by the control clock signal. It has selected states decoded to produce the repeating control sig-nal sequence (clear, enable, and store). The counter (cascaded BCD stages) is first cleared. Then the counter is enabled for the proper sample interval to

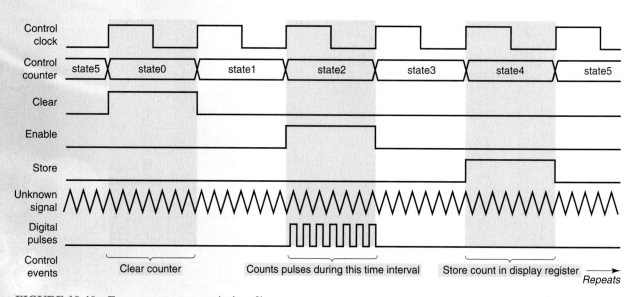

FIGURE 10-40   Frequency counter timing diagram.

count the digital pulses, which have the same frequency as the unknown signal. After disabling the counter, the new count is stored in the display register.

The counter, display register, and decoder/display sections are straight-forward and are not described any further here. The timing and control block provides the "brains" for our frequency counter and deserves a little more discussion to explain its operation. Figure 10-41 shows the sub-blocks within the timing and control block. For our example design, we will assume that the clock generator produces a 100-kHz system clock signal. The system clock frequency is divided by a set of five decade counters (MOD-10). This gives the user six different frequencies that can be selected by the multiplexer for the control clock frequency using the range select control. Because the period of the control clock is the same as the pulse width of the counter enable, this setup allows the frequency counter to have six different frequency measurement ranges. The control counter is a MOD-6 counter that has three selected states decoded by the control signal generator to produce the clear, enable, and store control signals.

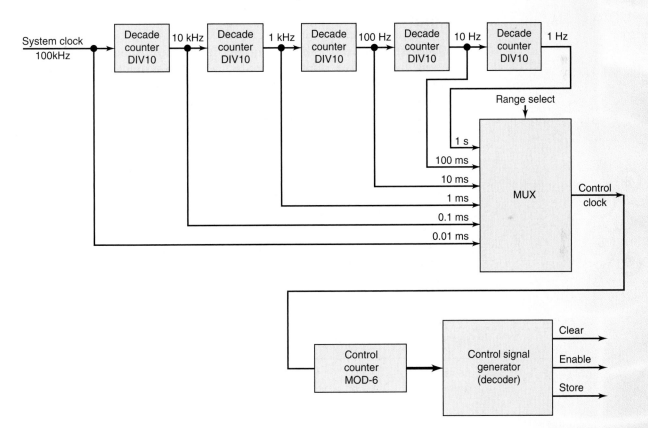

FIGURE 10-41   Timing and control block for frequency counter.

EXAMPLE 10-3

Assume that the BCD counter in Figure 10-39 consists of three cascaded BCD stages and their associated displays. If the unknown frequency is between 1 kpps and 9.99 kpps, which range (sample interval) should be selected using the MUX of Figure 10-41?

Solution

With three BCD counters, the total capacity of the counter is 999. A 9.99-kpps frequency produces a count of 999 if a 0.1-s sample interval were used. Thus, in order to use the full capacity of the counter, the MUX should select the 0.1-s clock period (10 Hz). If a 1-s sampling interval were used, the counter capacity would always be exceeded for frequencies in the specified range. If a shorter sample interval were used, the counter would count only between 1 and 99, which would give a reading to only two significant figures and would be a waste of the counter's capacity.

1. What is the purpose of running the unknown signal through a pulse shaper?
2. What are the units of a frequency measurement?
3. What does the display show during the sample interval?

SUMMARY

1. Successful project management can be accomplished by the following steps: overall project definition; breaking the project into small, strategic pieces; synthesis and testing of each piece; and system integration.
2. Small projects like the stepper motor driver can be completed in a single design file, even though these projects are developed modularly.
3. Projects that consist of several simple building blocks, like the keypad encoder, can produce very useful systems.
4. Larger projects like the digital clock can often take advantage of standard common modules that can be used repeatedly in the overall design.
5. Projects should be built and tested in modules starting at the lowest levels of hierarchy.
6. Preexisting modules can easily be combined with new custom modules using both graphical and text-based description methods.
7. Modules can be combined and represented as a single block in the next higher level of the hierarchy using the Altera design tools.

IMPORTANT TERMS

nesting	prescaler	sampling interval
hierarchy	frequency counter	

PROBLEMS

SECTION 10-1

B 10-1. The security monitoring system of Section 9-8 in Chapter 9 can be developed as a project.
   (a) Write a project definition with specifications for this system.
   (b) Define three major blocks of this project.
   (c) Identify the signals that interconnect the blocks.
   (d) At what frequency must the oscillator run for a 2.5-Hz flash rate?
   (e) Why is it reasonable to use only one current limiting resistor for all eight LEDs?

SECTION 10-2

Problems 10-2 through 10-7 refer to stepper motors described in Section 10-2.

B   10-2. How many full steps must occur for a complete revolution?

B   10-3. How many degrees of rotation result from one compete cycle through the full-step sequence in Table 10-1?

B   10-4. How many degrees of rotation result from one complete cycle through the half-step sequence in Table 10-1?

B   10-5. The *cout* lines of Figure 10-1 started at 1010 and have just progressed through the following sequence: 1010, 1001, 0101, 0110.

   (a) How many degrees has the shaft rotated?

   (b) What sequence will reverse the rotation and return the shaft to its original position?

B   10-6. Describe a method to test the stepper driver in:

   (a) Full-step mode

   (b) Half-step mode

   (c) Wave-drive mode

   (d) Direct-drive mode

D, H   10-7. Rewrite the stepper driver design file of Figure 10-8 or 10-9 without using a CASE statement. Use your favorite HDL.

D, H   10-8. Modify the stepper design file of Figure 10-8 or 10-9 to add an enable input that puts the outputs in the Hi-Z state (tristate) when enable = 0.

SECTION 10-3

B   10-9. Write the state table for the ring counter shown in Figure 10-11 and described in Figure 10-13.

B   10-10. With no keys pressed, what is the value on c[3..0]?

B   10-11. Assume that the ring counter is in state 0111 when someone presses the 7 key. Will the ring counter advance to the NEXT state?

B   10-12. Assume the 9 key is pressed and held until DAV = 1.

   (a) What is the value on the ring counter?

   (b) What is the value encoded by the row encoder?

   (c) What is the value encoded by the column encoder?

   (d) What binary number is on the D[3..0] lines?

B   10-13. In Problem 10-12, will the data be valid on the falling edge of DAV?

B, D   10-14. If you wanted to latch data from the keypad into a 74174 register, which signal from the keypad would you connect to the clock of the register? Draw the circuit.

T   10-15. The keypad is connected to a 74373 octal transparent latch as shown in Figure 10-42. The output is correct as long as a key is held. However, it is unable to latch data between key presses. Why will this circuit *not* work correctly?

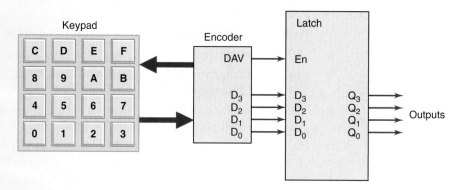

FIGURE 10-42   Problem 10-15.

SECTION 10-4

B   10-16. Assume a 1-Hz clock is applied to the seconds stage of the clock in Figure 10–17. The MOD-10 *units of seconds* counter's terminal count (tc) output is shown in Figure 10-43. Draw a similar diagram showing the number of clock cycles between the tc output pulses of each of the following:

(a) Tens of seconds counter

(b) Units of minutes counter

(c) Tens of minutes counter

FIGURE 10-43   Problem 10-16.

B   10-17. How many cycles of the 60-Hz power line will occur in a 24-hour period? What problem do you think will result if we attempt to simulate the operation of the entire clock circuit?

D   10-18. Many digital clocks are set by simply making them count faster while a push button is held down. Modify the design to add this feature.

D, H   10-19. Modify the hours stage of Figure 10-18 to keep military time (00–23 hours).

SECTION 10-5

B   10-20. Draw the hierarchy diagram for the frequency counter project.

D, H   10-21. Write the HDL code for the MOD-6 control counter and control signal generator in Figure 10-41.

D, H   10-22. Write the HDL code for the MUX of Figure 10-41.

D   10-23. Use graphic design techniques and the BCD counter described in Figure 10-31, the MUX, and the control signal generator design to create the entire timing and control block for the frequency counter project.

D, H 10-24. Write the HDL code for the timing and control section of the frequency counter.

ANSWERS TO SECTION REVIEW QUESTIONS

SECTION 10-1

1. Definition, strategic planning, synthesis and testing, system integration and testing   2. The definition stage

SECTION 10-2

1. Full-step, half-step, wave-drive, and direct-drive   2. cin_0-cin_3 [mode selector switches set to (1,1)]   3. Step, direction [mode selector switches set to (0,1)]
4. Eight states

SECTION 10-3

1. Only one   2. The first one scanned after being pressed (usually the first one pressed)   3. To make DAV go HIGH after the data stabilizes   4. No, it goes HIGH on the next clock after the key is pressed.   5. Whenever OE is LOW or when no keys are pressed

SECTION 10-4

1. The overall operating specifications and the system inputs and outputs. Delete the boundaries and connections between the major sections.   2. At the top of the hierarchy   3. At the bottom, building the simplest blocks first   4. At each stage of modular implementation

SECTION 10-5

1. To change the shape of the analog signal into a digital signal of the same frequency   2. Cycles per second (Hz) or pulses per second (pps)   3. The display shows the frequency measured during the previous sample interval.

CHAPTER 11

INTERFACING WITH THE ANALOG WORLD

■ OUTLINE

■ OBJECTIVES

Upon completion of this chapter, you will be able to:

- Understand the theory of operation and the circuit limitations of several types of digital-to-analog converters (DACs).

- Read and understand the various DAC manufacturer specifications.

- Use different test procedures to troubleshoot DAC circuits.

- Compare the advantages and disadvantages among the digital-ramp analog-to-digital converter (ADC), successive-approximation ADC, and flash ADC.

- Analyze the process by which a computer, in conjunction with an ADC, digitizes an analog signal and then reconstructs that analog signal from the digital data.

- Describe the basic operation of a digital voltmeter.

- Understand the need for using sample-and-hold circuits in conjunction with ADCs.

- Describe the operation of an analog multiplexing system.

- Understand the features and basic operation of a digital storage oscilloscope.

- Understand the basic concepts of digital signal processing.

11-1 REVIEW OF DIGITAL VERSUS ANALOG

A **digital quantity** has a value that is specified as one of two possibilities, such as 0 or 1, LOW or HIGH, true or false, and so on. In practice, a digital quantity such as a voltage may actually have a value that is anywhere within specified ranges, and we define values within a given range to have the same digital value. For example, for TTL logic, we know that

$$0 \text{ V to } 0.8 \text{ V} = \text{logic } 0$$

$$2 \text{ V to } 5 \text{ V} = \text{logic } 1$$

Any voltage falling in the range from 0 to 0.8 V is given the digital value 0, and any voltage in the range 2 to 5 V is assigned the digital value 1. The exact voltage values are not significant because the digital circuits respond in the same way to all voltage values within a given range.

By contrast, an **analog quantity** can take on any value over a continuous range of values and, most important, its exact value is significant. For example, the output of an analog temperature-to-voltage converter might be measured as 2.76 V, which may represent a specific temperature of 27.6°C. If the voltage were measured as something different, such as 2.34 V or 3.78 V, this would represent a completely different temperature. In other words, each possible value of an analog quantity has a different meaning. Another example of this is the output voltage from an audio amplifier into a speaker. This voltage is an analog quantity because each of its possible values produces a different response in the speaker.

Most physical variables are analog in nature and can take on any value within a continuous range of values. Examples include temperature, pressure, light intensity, audio signals, position, rotational speed, and flow rate. Digital systems perform all of their internal operations using digital circuitry and digital operations. Any information that must be input to a digital system must first be put into digital form. Similarly, the outputs from a digital system are always in digital form. When a digital system such as a computer is to be used to monitor and/or control a physical process, we must deal with the difference between the digital nature of the computer and the analog nature of the process variables. Figure 11-1 illustrates the situation. This diagram shows the five elements that are involved when a computer is monitoring and controlling a physical variable that is assumed to be analog:

1. **Transducer.** The physical variable is normally a nonelectrical quantity. A **transducer** is a device that converts the physical variable to an electrical variable. Some common transducers include thermistors, photocells, photodiodes, flow meters, pressure transducers, and tachometers. The electrical output of the transducer is an analog current or voltage that is proportional to the physical variable that it is monitoring. For example, the physical variable could be the temperature of water in a large tank that is being filled from cold and hot water pipes. Let's say that the water temperature varies from 80 to 150°F and that a thermistor and its associated circuitry convert this water temperature to a voltage ranging from 800 to 1500 mV. Note that the transducer's output is directly proportional to temperature such that each 1°F produces a 10-mV output. This proportionality factor was chosen for convenience.

2. **Analog-to-digital converter (ADC).** The transducer's electrical analog output serves as the analog input to the ADC. The ADC converts this analog input to a digital output. This digital output consists of a number of bits that represent the value of the analog input. For example, the ADC might convert the transducer's 800- to 1500-mV analog values to binary values ranging from 01010000 (80) to 10010110 (150). Note that the binary output from the ADC is proportional to the analog input voltage so that each unit of the digital output represents 10 mV.

FIGURE 11-1   Analog-to-digital converter (ADC) and digital-to-analog converter (DAC) are used to interface a computer to the analog world so that the computer can monitor and control a physical variable.

3. **Computer.** The digital representation of the process variable is transmitted from the ADC to the digital computer, which stores the digital value and processes it according to a program of instructions that it is executing. The program might perform calculations or other operations on this digital representation of temperature to come up with a digital output that will eventually be used to control the temperature.

4. **Digital-to-analog converter (DAC).** This digital output from the computer is connected to a DAC, which converts it to a proportional analog voltage or current. For example, the computer might produce a digital output ranging from 00000000 to 11111111, which the DAC converts to a voltage ranging from 0 to 10 V.

5. **Actuator.** The analog signal from the DAC is often connected to some device or circuit that serves as an actuator to control the physical variable. For our water temperature example, the actuator might be an electrically controlled valve that regulates the flow of hot water into the tank in accordance with the analog voltage from the DAC. The flow rate would vary in proportion to this analog voltage, with 0 V producing no flow and 10 V producing the maximum flow.

Thus, we see that ADCs and DACs function as *interfaces* between a completely digital system, such as a computer, and the analog world. This function has become increasingly more important as inexpensive microcomputers have moved into areas of process control where computer control was previously not feasible.

REVIEW QUESTIONS

1. What is the function of a transducer?
2. What is the function of an ADC?
3. What does a computer often do with the data that it receives from an ADC?
4. What function does a DAC perform?
5. What is the function of an actuator?

11-2 DIGITAL-TO-ANALOG CONVERSION

We will now begin our study of digital-to-analog (D/A) and analog-to-digital (A/D) conversion. Many A/D conversion methods utilize the D/A conversion process, so we will examine D/A conversion first.

Basically, D/A *conversion* is the process of taking a value represented in *digital* code (such as straight binary or BCD) and converting it to a voltage or current that is proportional to the digital value. Figure 11-2(a) shows the symbol for a typical four-bit D/A converter. We will not concern ourselves with the internal circuitry until later. For now, we will examine the various input/output relationships.

Notice that there is an input for a voltage reference, V_{ref}. This input is used to determine the **full-scale output** or maximum value that the D/A converter can produce. The digital inputs $D, C, B,$ and A are usually derived from the output register of a digital system. The $2^4 = 16$ different binary numbers represented by these four bits are listed in Figure 11-2(b). For each input number, the D/A converter output voltage is a unique value. In fact, for this case, the analog output voltage V_{OUT} is equal in volts to the binary number. It

D	C	B	A	V_{OUT}	
0	0	0	0	0	Volts
0	0	0	1	1	
0	0	1	0	2	
0	0	1	1	3	
0	1	0	0	4	
0	1	0	1	5	
0	1	1	0	6	
0	1	1	1	7	
1	0	0	0	8	
1	0	0	1	9	
1	0	1	0	10	
1	0	1	1	11	
1	1	0	0	12	
1	1	0	1	13	
1	1	1	0	14	
1	1	1	1	15	Volts

(b)

FIGURE 11-2   Four-bit DAC with voltage output.

could also have been twice the binary number or some other proportionality factor. The same idea would hold true if the D/A output were a current I_{OUT}.

In general,

$$\text{analog output} = K \times \text{digital input} \qquad (11\text{-}1)$$

where K is the proportionality factor and is a constant value for a given DAC connected to a fixed reference voltage. The analog output can, of course, be a voltage or a current. When it is a voltage, K will be in voltage units, and when the output is a current, K will be in current units. For the DAC of Figure 11-2, $K = 1$ V, so that

$$V_{OUT} = (1 \text{ V}) \times \text{digital input}$$

We can use this to calculate V_{OUT} for any value of digital input. For example, with a digital input of $1100_2 = 12_{10}$, we obtain

$$V_{OUT} = 1 \text{ V} \times 12 = 12 \text{ V}$$

EXAMPLE 11-1A

A five-bit DAC has a current output. For a digital input of 10100, an output current of 10 mA is produced. What will I_{OUT} be for a digital input of 11101?

Solution

The digital input 10100_2 is equal to decimal 20. Because $I_{OUT} = 10$ mA for this case, the proportionality factor must be 0.5 mA. Thus, we can find I_{OUT} for any digital input such as $11101_2 = 29_{10}$ as follows:

$$I_{OUT} = (0.5 \text{ mA}) \times 29$$
$$= 14.5 \text{ mA}$$

Remember, the proportionality factor, K, varies from one DAC to another and depends on the reference voltage.

EXAMPLE 11-1B

What is the largest value of output voltage from an eight-bit DAC that produces 1.0 V for a digital input of 00110010?

Solution

$$00110010_2 = 50_{10}$$
$$1.0 \text{ V} = K \times 50$$

Therefore,

$$K = 20 \text{ mV}$$

The largest output will occur for an input of $11111111_2 = 255_{10}$.

$$V_{OUT}(\text{max}) = 20 \text{ mV} \times 255$$
$$= 5.10 \text{ V}$$

Analog Output

The output of a DAC is technically not an analog quantity because it can take on only specific values, such as the 16 possible voltage levels for V_{OUT} in Figure 11-2, as long as V_{ref} is constant. Thus, in that sense, it is actually digital. As we will see, however, the number of different possible output values can be increased and the difference between successive values decreased by increasing the number of input bits. This will allow us to produce an output that is more and more like an analog quantity that varies continuously over a range of values. In other words, the DAC output is a "pseudo-analog" quantity. We will continue to refer to it as analog, keeping in mind that it is an approximation to a pure analog quantity.

Input Weights

For the DAC of Figure 11-2, note that each digital input contributes a different amount to the analog output. This is easily seen if we examine the cases where only one input is HIGH (Table 11-1). The contributions of each digital input are *weighted* according to their position in the binary number. Thus, A, which is the LSB, has a *weight* of 1 V; B has a weight of 2 V; C has a weight of 4 V; and D, the MSB, has the largest weight, 8 V. The weights are successively doubled for each bit, beginning with the LSB. Thus, we can consider V_{OUT} to be the weighted sum of the digital inputs. For instance, to find V_{OUT} for the digital input 0111, we can add the weights of the C, B, and A bits to obtain 4 V + 2 V + 1 V = 7 V.

TABLE 11-1

D	C	B	A		V_{OUT} (V)
0	0	0	1	→	1
0	0	1	0	→	2
0	1	0	0	→	4
1	0	0	0	→	8

EXAMPLE 11-2

A five-bit D/A converter produces $V_{OUT} = 0.2$ V for a digital input of 00001. Find the value of V_{OUT} for an input of 11111.

Solution

Obviously, 0.2 V is the weight of the LSB. Thus, the weights of the other bits must be 0.4 V, 0.8 V, 1.6 V, and 3.2 V, respectively. For a digital input of 11111, then, the value of V_{OUT} will be 3.2 V + 1.6 V + 0.8 V + 0.4 V + 0.2 V = 6.2 V.

Resolution (Step Size)

Resolution of a D/A converter is defined as the smallest change that can occur in the analog output as a result of a change in the digital input. Referring to the table in Figure 11-2, we can see that the resolution is 1 V because V_{OUT} can change by no less than 1 V when the digital input value is changed. The resolution is always equal to the weight of the LSB and is also referred to as the **step size** because it is the amount that V_{OUT} will change as the digital input value is changed from one step to the next. This is illustrated better in Figure 11-3, where the outputs from a four-bit binary counter provide the inputs to our DAC. As the counter is being continually cycled through its 16 states by the clock signal, the DAC output is a **staircase** waveform that goes up 1 V per step. When the counter is at 1111, the DAC output is at its maximum value of 15 V; this is its full-scale output. When the counter recycles to 0000, the DAC output returns to 0 V. The resolution (or step size) is the size of the jumps in the staircase waveform; in this case, each step is 1 V.

Note that the staircase has 16 levels corresponding to the 16 input states, but there are only 15 steps or jumps between the 0-V level and full-scale. In general, for an N-bit DAC the number of different levels will be 2^N, and the number of steps will be $2^N - 1$.

You may have already figured out that resolution (step size) is the same as the proportionality factor in the DAC input/output relationship:

$$\text{analog output} = K \times \text{digital input}$$

FIGURE 11-3   Output waveforms of a DAC as inputs are provided by a binary counter.

A new interpretation of this expression would be that the digital input is equal to the number of steps, K is the amount of voltage (or current) per step, and the analog output is the product of the two. We now have a convenient way of calculating the value of K for the D/A:

$$\text{resolution} = K = \frac{A_{\text{fs}}}{(2^n - 1)} \qquad (11\text{-}2)$$

where A_{fs} is the analog full-scale output and n is the number of bits.

EXAMPLE 11-3A

What is the resolution (step size) of the DAC of Example 11-2? Describe the staircase signal out of this DAC.

Solution

The LSB for this converter has a weight of 0.2 V. This is the resolution or step size. A staircase waveform can be generated by connecting a five-bit counter to the DAC inputs. The staircase will have 32 levels, from 0 V up to a full-scale output of 6.2 V, and 31 steps of 0.2 V each.

EXAMPLE 11-3B

For the DAC of Example 11-2, determine V_{OUT} for a digital input of 10001.

Solution

The step size is 0.2 V, which is the proportionality factor K. The digital input is $10001 = 17_{10}$. Thus, we have

$$\begin{aligned} V_{\text{OUT}} &= (0.2 \text{ V}) \times 17 \\ &= 3.4 \text{ V} \end{aligned}$$

Percentage Resolution

Although resolution can be expressed as the amount of voltage or current per step, it is also useful to express it as a percentage of the *full-scale output*. To illustrate, the DAC of Figure 11-3 has a maximum full-scale output of 15 V (when the digital input is 1111). The step size is 1 V, which gives a percentage resolution of

$$\% \text{ resolution} = \frac{\text{step size}}{\text{full scale (F.S.)}} \times 100\% \qquad (11\text{-}3)$$

$$= \frac{1 \text{ V}}{15 \text{ V}} \times 100\% = 6.67\%$$

EXAMPLE 11-4

A 10-bit DAC has a step size of 10 mV. Determine the full-scale output voltage and the percentage resolution.

Solution

With 10 bits, there will be $2^{10} - 1 = 1023$ steps of 10 mV each. The full-scale output will therefore be 10 mV $\times$ 1023 = 10.23 V, and

$$\% \text{ resolution} = \frac{10 \text{ mV}}{10.23 \text{ V}} \times 100\% \approx 0.1\%$$

Example 11-4 helps to illustrate the fact that the percentage resolution becomes smaller as the number of input bits is increased. In fact, the percentage resolution can also be calculated from

$$\% \text{ resolution} = \frac{1}{\text{total number of steps}} \times 100\% \qquad (11\text{-}4)$$

For an N-bit binary input code, the total number of steps is $2^N - 1$. Thus, for the previous example,

$$\% \text{ resolution} = \frac{1}{2^{10} - 1} \times 100\%$$
$$= \frac{1}{1023} \times 100\%$$
$$\approx 0.1\%$$

This means that it is *only the number of bits* that determines the *percentage* resolution. Increasing the number of bits increases the number of steps to reach full scale, so that each step is a smaller part of the full-scale voltage. Most DAC manufacturers specify resolution as the number of bits.

What Does Resolution Mean?

A DAC cannot produce a continuous range of output values and so, strictly speaking, its output is not truly analog. A DAC produces a finite set of output values. In our water temperature example of Section 11-1, the computer generates a digital output to provide an analog voltage between 0 and 10 V to an electrically controlled valve. The DAC's resolution (number of bits) determines how many possible voltage values the computer can send to the valve. If a six-bit DAC is used, there will be 63 possible steps of 0.159 V between 0 and 10 V. When an eight-bit DAC is used, there will be 255 possible steps of 0.039 V between 0 and 10 V. The greater the number of bits, the finer the resolution (the smaller the step size).

The system designer must decide what resolution is needed on the basis of the required system performance. The resolution limits how close the DAC output can come to a given analog value. Generally, the cost of DACs increases with the number of bits, and so the designer will use only as many bits as necessary.

EXAMPLE 11-5

Figure 11-4 shows a computer controlling the speed of a motor. The 0- to 2-mA analog current from the DAC is amplified to produce motor speeds from 0 to 1000 rpm (revolutions per minute). How many bits should be used if the computer is to be able to produce a motor speed that is within 2 rpm of the desired speed?

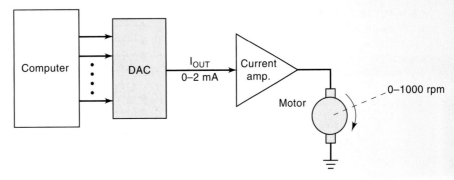

FIGURE 11-4   Example 11-5.

Solution

The motor speed will range from 0 to 1000 rpm as the DAC goes from zero to full scale. Each step in the DAC output will produce a step in the motor speed. We want the step size to be no greater than 2 rpm. Thus, we need at least 500 steps (1000/2). Now we must determine how many bits are required so that there are at least 500 steps from zero to full scale. We know that the number of steps is $2^N - 1$, and so we can say

$$2^N - 1 \geq 500$$

or

$$2^N \geq 501$$

Since $2^8 = 256$ and $2^9 = 512$, the smallest number of bits that will produce at least 500 steps is *nine.* We could use more than nine bits, but this might add to the cost of the DAC.

EXAMPLE 11-6

Using nine bits, how close to 326 rpm can the motor speed be adjusted?

Solution

With nine bits, there will be 511 steps ($2^9 - 1$). Thus, the motor speed will go up in steps of 1000 rpm/511 = 1.957 rpm. The number of steps needed to reach 326 rpm is 326/1.957 = 166.58. This is not a whole number of steps, and so we will round it to 167. The actual motor speed on the 167th step will be $167 \times 1.957 = 326.8$ rpm. Thus, the computer must output the nine-bit binary equivalent of 167_{10} to produce the desired motor speed within the resolution of the system.

In all of our examples, we have assumed that the DACs have been perfectly accurate in producing an analog output that is directly proportional to the binary input, and that the resolution is the only thing that limits how close we can come to a desired analog value. This, of course, is unrealistic because all devices contain inaccuracies. We will examine the causes and effects of DAC inaccuracy in Sections 11-3 and 11-4.

FIGURE 11-5 DAC using BCD input code. This converter accepts a two-digit input and generates 100 possible analog output values.

BCD Input Code

The DACs we have considered thus far have used a binary input code. Some DACs use a BCD input code where four-bit code groups are used for each decimal digit. Figure 11-5 shows the diagram of an eight-bit (two-digit) converter of this type. Each four-bit code group can range from 0000 to 1001, and so the BCD inputs represent decimal numbers from 00 to 99. Within each code group, the weights of the different bits are in the normal binary proportions (1, 2, 4, 8), but the relative weights of each code group are different by a factor of 10. Figure 11-5 shows the relative weights of the various bits. Note that the bits that make up the BCD code for the most significant digit (MSD) have a relative weight that is 10 times that of the corresponding bits of the LSD.

EXAMPLE 11-7A

If the weight of A_0 is 0.1 V in Figure 11-5, find the following values:

(a) Step size

(b) Full-scale output and percentage resolution

(c) V_{OUT} for $D_1C_1B_1A_1 = 0101$ and $D_0C_0B_0A_0 = 1000$

Solution

(a) Step size is the weight of the LSB of the LSD, 0.1 V.

(b) There are 99 steps because there are two BCD digits. Thus, full-scale output is $99 \times 0.1 = 9.9$ V. The resolution [using equation (11-3)] is

$$\frac{\text{step size}}{\text{F.S.}} \times 100\% = \frac{0.1}{9.9} \times 100\% \approx 1\%$$

We could have also used equation (11-4) to calculate percentage resolution because the total number of steps is 99.

(c) The exact weights in volts are listed in Table 11-2. One way to find V_{OUT} for a given input is to add the weights of all bits that are 1s. Thus, for an input of 0101 1000, we have

An easier way is to realize that the BCD input code represents 58_{10} and the step size is 0.1 V, so that

$$V_{OUT} = \overset{C_1}{4\text{ V}} + \overset{A_1}{1\text{ V}} + \overset{D_0}{0.8\text{ V}} = 5.8 \text{ V}$$

TABLE 11-2

	MSD				LSD		
D_1	C_1	B_1	A_1	D_0	C_0	B_0	A_0
8.0	4.0	2.0	1.0	0.8	0.4	0.2	0.1

$$V_{OUT} = (0.1 \text{ V}) \times 58 = 5.8 \text{ V}$$

EXAMPLE 11-7B

A certain 12-bit BCD digital-to-analog converter has a full-scale output of 9.99 V.

(a) Determine the percentage resolution.

(b) Determine the converter's step size.

Solution

(a) Twelve bits correspond to three decimal digits, that is, decimal numbers from 000 to 999. Therefore, the output of this DAC has 999 possible steps from 0 to 9.99 V. Thus, we have

$$\% \text{ resolution} = \frac{1}{\text{number of steps}} \times 100\%$$

$$= \frac{1}{999} \times 100\%$$

$$\approx 0.1\%$$

(b)
$$\text{step size} = \frac{\text{F.S.}}{\text{number of steps}}$$

$$= 9.99 \text{ V}/999$$

$$= \mathbf{0.01 \text{ V}}$$

Bipolar DACs

Up to this point, we have assumed that the binary input to a DAC has been an unsigned number and that the DAC output has been a positive voltage or current. Many DACs are designed to produce both positive and negative values, such as -10 to $+10$ V. This is generally done by using the binary input as a signed number with the MSB as the sign bit (0 for + and 1 for $-$). Negative input values are often represented in 2's-complement form, although the true-magnitude form is also used by some DACs. For example, suppose that we have a six-bit bipolar DAC that uses the 2's-complement system and has a resolution of 0.2 V. The binary input values range from 100000 (-32) to 011111 ($+31$) to produce analog outputs in the range from -6.4 to $+6.2$ V. There are 63 steps ($2^6 - 1$) of 0.2 V between these negative and positive limits.

1. An eight-bit DAC has an output of 3.92 mA for an input of 01100010. What are the DAC's resolution and full-scale output?

2. What is the weight of the MSB of the DAC of question 1?

3. What is the percentage resolution of an eight-bit DAC?

4. How many different output voltages can a 12-bit DAC produce?

5. For the system of Figure 11-4, how many bits should be used if the computer is to control the motor speed within 0.4 rpm?

6. Consider a 12-bit DAC with BCD inputs and a resolution of 10 mV. What will its output be for an input of 100001110011?

7. *True or false:* The percentage resolution of a DAC depends *only* on the number of bits.

8. What is the advantage of a smaller (finer) resolution?

11-3   D/A-CONVERTER CIRCUITRY

There are several methods and circuits for producing the D/A operation that has been described. We shall examine several of the basic schemes to gain an insight into the ideas used. It is not important to be familiar with all of the various circuit schemes because D/A converters are available as ICs or as encapsulated packages that do not require any circuit knowledge. Instead, it is important to know the significant performance characteristics of DACs, in general, so that they can be used intelligently. These will be covered in Section 11-4.

Figure 11-6(a) shows the basic circuit for one type of four-bit DAC. The inputs *A*, *B*, *C*, and *D* are binary inputs that are assumed to have values of either 0 or 5 V. The *operational amplifier* is employed as a summing amplifier,

Input code				
D	C	B	A	V_{OUT} (volts)
0	0	0	0	0
0	0	0	1	−0.625 ← LSB
0	0	1	0	−1.250
0	0	1	1	−1.875
0	1	0	0	−2.500
0	1	0	1	−3.125
0	1	1	0	−3.750
0	1	1	1	−4.375
1	0	0	0	−5.000
1	0	0	1	−5.625
1	0	1	0	−6.250
1	0	1	1	−6.875
1	1	0	0	−7.500
1	1	0	1	−8.125
1	1	1	0	−8.750
1	1	1	1	−9.375 ← Full-scale

(b)

FIGURE 11-6 Simple DAC using an op-amp summing amplifier with binary-weighted resistors.

which produces the weighted sum of these input voltages. Recall that the summing amplifier multiplies each input voltage by the ratio of the feedback resistor R_F to the corresponding input resistor R_{IN}. In this circuit $R_F = 1$ kΩ, and the input resistors range from 1 to 8 kΩ. The D input has $R_{IN} = 1$ kΩ, so the summing amplifier passes the voltage at D with no attenuation. The C input has $R_{IN} = 2$ kΩ, so that it will be attenuated by $1/2$. Similarly, the B input will be attenuated by $1/4$, and the A input by $1/8$. The amplifier output can thus be expressed as

$$V_{OUT} = -(V_D + \tfrac{1}{2}V_C + \tfrac{1}{4}V_B + \tfrac{1}{8}V_A) \tag{11-5}$$

The negative sign is present because the summing amplifier is a polarity-inverting amplifier, but it will not concern us here.

Clearly, the summing amplifier output is an analog voltage that represents a weighted sum of the digital inputs, as shown by the table in Figure 11-6(b). This table lists all of the possible input conditions and the resultant amplifier output voltage. The output is evaluated for any input condition by setting the appropriate inputs to either 0 or 5 V. For example, if the digital input is 1010, then $V_D = V_B = 5$ V and $V_C = V_A = 0$ V. Thus, using equation (11-5),

$$V_{OUT} = -(5\text{ V} + 0\text{ V} + \tfrac{1}{4} \times 5\text{ V} + 0\text{ V})$$
$$= -6.25\text{ V}$$

The resolution of this D/A converter is equal to the weighting of the LSB, which is $\tfrac{1}{8} \times 5$ V $= 0.625$ V. As shown in the table, the analog output increases by 0.625 V as the binary input number advances one step.

EXAMPLE 11-8

(a) Determine the weight of each input bit of Figure 11-6(a).

(b) Change R_F to 250 Ω and determine the full-scale output.

Solution

(a) The MSB passes with gain = 1, so its weight in the output is 5 V. Thus,

$$\text{MSB} \rightarrow 5\text{ V}$$
$$\text{2nd MSB} \rightarrow 2.5\text{ V}$$
$$\text{3rd MSB} \rightarrow 1.25\text{ V}$$
$$\text{4th MSB} = \text{LSB} \rightarrow 0.625\text{ V}$$

(b) If R_F is reduced by a factor of 4, to 250 Ω, each input weight will be four times *smaller* than the values above. Thus, the full-scale output will be reduced by this same factor and becomes $-9.375/4 = -2.344$ V.

If we look at the input resistor values in Figure 11-6, it should come as no surprise that they are *binarily weighted*. In other words, starting with the MSB resistor, the resistor values increase by a factor of 2. This, of course, produces the desired weighting in the voltage output.

Conversion Accuracy

The table in Figure 11-6(b) gives the *ideal* values of V_{OUT} for the various input cases. How close the circuit comes to producing these values depends primarily assume that the DAC of Figure 11-6 has an accuracy of $\pm0.01\%$ F.S. Because this converter has a full-scale output of 9.375 V, this percentage converts to

$$\pm0.01\% \times 9.375 \text{ V} = \pm0.9375 \text{ mV}$$

This means that the output of this DAC can, at any time, be off by as much as 0.9375 mV from its expected value.

Linearity error is the maximum deviation in step size from the ideal step size. For example, the DAC of Figure 11-6 has an expected step size of 0.625 V. If this converter has a linearity error of $\pm0.01\%$ F.S., this would mean that the actual *step size* could be off by as much as 0.9375 mV.

It is important to understand that accuracy and resolution of a DAC must be compatible. It is illogical to have a resolution of, say, 1 percent and an accuracy of 0.1 percent, or vice versa. To illustrate, a DAC with a resolution of 1 percent and an F.S. output of 10 V can produce an output analog voltage within 0.1 V of any desired value, assuming perfect accuracy. It makes no sense to have a costly accuracy of 0.01% F.S. (or 1 mV) if the resolution already limits the closeness of the desired value to 0.1 V. The same can be said for having a resolution that is very small (many bits) while the accuracy is poor; it is a waste of input bits.

EXAMPLE 11-11

A certain eight-bit DAC has a full-scale output of 2 mA and a full-scale error of $\pm0.5\%$ F.S. What is the range of possible outputs for an input of 10000000?

Solution

The step size is 2 mA/255 = 7.84 μA. Since $10000000 = 128_{10}$, the ideal output should be 128 $\times$ 7.84 μA = 1004 μA. The error can be as much as

$$\pm0.5\% \times 2 \text{ mA} = \pm10 \text{ }\mu\text{A}$$

Thus, the actual output can deviate by this amount from the ideal 1004 μA, so the actual output can be anywhere from 994 to 1014 μA.

Offset Error

Ideally, the output of a DAC will be zero volts when the binary input is all 0s. In practice, however, there will be a very small output voltage for this situation; this is called **offset error**. This offset error, if not corrected, will be added to the expected DAC output for *all* input cases. For example, let's say that a four-bit DAC has an offset error of +2 mV and a *perfect* step size of 100 mV. Table 11-3 shows the ideal and the actual DAC output for several input

TABLE 11-3

Input Code	Ideal Output (mV)	Actual Output (mV)
0000	0	2
0001	100	102
1000	800	802
1111	1500	1502

cases. Note that the actual output is 2 mV greater than expected; this is due to the offset error. Offset error can be negative as well as positive.

Many DACs have an external offset adjustment that allows you to zero the offset. This is usually accomplished by applying all 0s to the DAC input and monitoring the output while an *offset adjustment potentiometer* is adjusted until the output is as close to 0 V as required.

Settling Time

The operating speed of a DAC is usually specified by giving its **settling time,** which is the time required for the DAC output to go from zero to full scale as the binary input is changed from all 0s to all 1s. Actually, the settling time is measured as the time for the DAC output to settle within $\pm\frac{1}{2}$ step size (resolution) of its final value. For example, if a DAC has a resolution of 10 mV, settling time is measured as the time it takes the output to settle within 5 mV of its full-scale value.

Typical values for settling time range from 50 ns to 10 μs. Generally speaking, DACs with a current output will have shorter settling times than those with voltage outputs. The main reason for this difference is the response time of the op-amp that is used as the current-to-voltage converter.

Monotonicity

A DAC is **monotonic** if its output increases as the binary input is incremented from one value to the next. Another way to describe this is that the staircase output will have no downward steps as the binary input is incremented from zero to full scale.

REVIEW QUESTIONS

1. Define *full-scale error.*
2. What is *settling time*?
3. Describe offset error and its effect on a DAC output.
4. Why are voltage DACs generally slower than current DACs?

11-5   AN INTEGRATED-CIRCUIT DAC

The AD7524, a CMOS IC available from several IC manufacturers, is an eight-bit D/A converter that uses an $R/2R$ ladder network. Its block symbol is given in Figure 11-10(a). This DAC has an eight-bit input that can be latched internally under the control of the Chip Select ($\overline{CS}$) and WRITE ($\overline{WR}$) inputs. When both of these control inputs are LOW, the digital data inputs D_7–D_0 produce the analog output current *OUT 1* (the *OUT 2* terminal is normally grounded). When either control input goes HIGH, the digital input data are latched, and the analog output remains at the level corresponding to that latched digital data. Subsequent changes in the digital inputs will have no effect on *OUT 1* in this latched state.

The maximum settling time for the AD7524 is typically 100 ns, and its full-range accuracy is rated at $\pm0.2\%$ F.S. The V_{REF} can range over both negative and positive voltages from 0 to 25 V, so that analog output currents of both polarities can be produced. The output current can be converted to a

FIGURE 11-10 (a) AD7524 eight-bit DAC with latched inputs; (b) AD7524 connected to produce analog output voltage ranging from 0 V to approximately −10 V.

voltage using an op-amp connected as in Figure 11-10(b). Note that the op-amp's feedback resistor is already on the DAC chip.

11-6 DAC APPLICATIONS

DACs are used whenever the output of a digital circuit must provide an analog voltage or current to drive an analog device. Some of the most common applications are described in the following paragraphs.

Control

The digital output from a computer can be converted to an analog control signal to adjust the speed of a motor or the temperature of a furnace, or to control almost any physical variable.

Automatic Testing

Computers can be programmed to generate the analog signals (through a DAC) needed to test analog circuitry. The test circuit's analog output response will normally be converted to a digital value by an ADC and fed into the computer to be stored, displayed, and sometimes analyzed.

Signal Reconstruction

In many applications, an analog signal is **digitized**; that is, successive points on the signal are converted to their digital equivalents and stored in memory. This conversion is performed by an analog-to-digital converter (ADC). A DAC can then be used to convert the stored digitized data back to analog—one point at a time—thereby reconstructing the original signal. This combination of digitizing and reconstructing is used in digital storage oscilloscopes, audio compact disk systems, and digital audio and video recording. We will look at this further after we learn about ADCs.

A/D Conversion

Several types of ADCs use DACs as part of their circuitry, as we shall see in Section 11-8.

Serial DACs

Many of these DAC applications involve a microprocessor. The main problem with using the parallel-data DACs that have been described thus far is that they occupy so many port bits of the microcomputer. In cases where speed of data transfer is of little concern, a microprocessor can output the digital value to a DAC over a serial interface. Serial DACs are now readily available with a built-in serial in/parallel out shift register. Many of these devices have more than one DAC on the same chip. The digital data, along with a code that specifies which DAC you want, are sent to the chip, one bit at a time. As each bit is presented on the DAC input, a pulse is applied to the serial clock input to shift the bit in. After the proper number of clock pulses, the data value is latched and converted to its analog value.

11-7   TROUBLESHOOTING DACs

DACs are both digital and analog. Logic probes and pulsers can be used on the digital inputs, but a meter or an oscilloscope must be used on the analog output. There are basically two ways to test a DAC's operation: a *static accuracy test* and a *staircase test.*

The static test involves setting the binary input to a fixed value and measuring the analog output with a high-accuracy meter. This test is used to check that the output value falls within the expected range consistent with the DAC's specified accuracy. If it does not, there can be several possible causes. Here are some of them:

- Drift in the DAC's internal component values (e.g., resistor values) caused by temperature, aging, or some other factors. This condition can easily produce output values outside the expected accuracy range.

- Open connections or shorts in any of the binary inputs. This could either prevent an input from adding its weight to the analog output or cause its weight to be permanently present in the output. This situation is especially hard to detect when the fault is in the less significant inputs.

- A faulty voltage reference. Because the analog output depends directly on V_{REF}, this could produce results that are way off. For DACs that use external reference sources, the reference voltage can be checked easily with a digital voltmeter. Many DACs have internal reference voltages that cannot be checked, except on some DACs that bring the reference voltage out to a pin of the IC.

- Excessive offset error caused by component aging or temperature. This would produce outputs that are off by a fixed amount. If the DAC has an external offset adjustment capability, this type of error can initially be zeroed out, but changes in operating temperature can cause the offset error to reappear.

The staircase test is used to check the monotonicity of the DAC; that is, it checks to see that the output increases step by step as the binary input is incremented as in Figure 11-3. The steps on the staircase must be of the same size, and there should be no missing steps or downward steps until full scale is reached. This test can help detect internal or external faults that cause an input to have either no contribution or a permanent contribution to the analog output. The following example will illustrate.

EXAMPLE 11-12

How would the staircase waveform appear if the C input to the DAC of Figure 11-3 is open? Assume that the DAC inputs are TTL-compatible.

Solution

An open connection at C will be interpreted as a constant logic 1 by the DAC. Thus, this will contribute a constant 4 V to the DAC output so that the DAC output waveform will appear as shown in Figure 11-11. The dotted lines are the staircase as it would appear if the DAC were working correctly. Note that the faulty output waveform matches the correct one during those times when the bit C input would normally be HIGH.

FIGURE 11-11
Example 11-12.

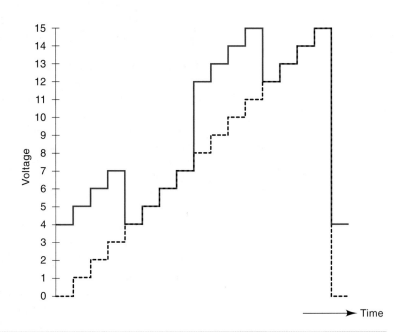

11-8   ANALOG-TO-DIGITAL CONVERSION

An **analog-to-digital converter** takes an analog input voltage and, after a certain amount of time, produces a digital output code that represents the analog input. The A/D conversion process is generally more complex and time-consuming than the D/A process, and many different methods have been developed and used. We shall examine several of these methods in detail, even though it may never be necessary to design or construct ADCs (they are available as completely packaged units). However, the techniques that are used provide an insight into what factors determine an ADC's performance.

Several important types of ADCs utilize a DAC as part of their circuitry. Figure 11-12 is a general block diagram for this class of ADC. The timing for

FIGURE 11-12 General diagram of one class of ADCs.

the operation is provided by the input clock signal. The control unit contains the logic circuitry for generating the proper sequence of operations in response to the START COMMAND, which initiates the conversion process. The op-amp comparator has two *analog* inputs and a *digital* output that switches states, depending on which analog input is greater.

The basic operation of ADCs of this type consists of the following steps:

1. The START COMMAND pulse initiates the operation.
2. At a rate determined by the clock, the control unit continually modifies the binary number that is stored in the register.
3. The binary number in the register is converted to an analog voltage, V_{AX}, by the DAC.
4. The comparator compares V_{AX} with the analog input V_A. As long as $V_{AX} < V_A$, the comparator output stays HIGH. When V_{AX} exceeds V_A by at least an amount equal to V_T (threshold voltage), the comparator output goes LOW and stops the process of modifying the register number. At this point, V_{AX} is a close approximation to V_A. The digital number in the register, which is the digital equivalent of V_{AX}, is also the approximate digital equivalent of V_A, within the resolution and accuracy of the system.
5. The control logic activates the end-of-conversion signal, *EOC*, when the conversion is complete.

The several variations of this A/D conversion scheme differ mainly in the manner in which the control section continually modifies the numbers in the register. Otherwise, the basic idea is the same, with the register holding the required digital output when the conversion process is complete.

REVIEW QUESTIONS

1. What is the function of the comparator in the ADC?
2. Where is the approximate digital equivalent of V_A when the conversion is complete?
3. What is the function of the *EOC* signal?

11-9   DIGITAL-RAMP ADC

One of the simplest versions of the general ADC of Figure 11-12 uses a binary counter as the register and allows the clock to increment the counter one step at a time until $V_{AX} \geq V_A$. It is called a **digital-ramp ADC** because the waveform at V_{AX} is a step-by-step ramp (actually a staircase) like the one shown in Figure 11-3. It is also referred to as a *counter-type* ADC.

Figure 11-13 is the diagram for a digital-ramp ADC. It contains a counter, a DAC, an analog comparator, and a control AND gate. The comparator output serves as the active-LOW end-of-conversion signal $\overline{EOC}$. If we assume that V_A, the analog voltage to be converted, is positive, the operation proceeds as follows:

1. A START pulse is applied to reset the counter to 0. The HIGH at START also inhibits clock pulses from passing through the AND gate into the counter.
2. With all 0s at its input, the DAC's output will be $V_{AX} = 0$ V.
3. Because $V_A > V_{AX}$, the comparator output, $\overline{EOC}$, will be HIGH.
4. When START returns LOW, the AND gate is enabled and clock pulses get through to the counter.
5. As the counter advances, the DAC output, V_{AX}, increases one step at a time, as shown in Figure 11-13(b).
6. This process continues until V_{AX} reaches a step that exceeds V_A by an amount equal to or greater than V_T (typically 10 to 100 μV). At this point, $\overline{EOC}$ will go LOW and inhibit the flow of pulses into the counter, and the counter will stop counting.
7. The conversion process is now complete, as signaled by the HIGH-to-LOW transition at $\overline{EOC}$, and the contents of the counter are the digital representation of V_A.
8. The counter will hold the digital value until the next START pulse initiates a new conversion.

FIGURE 11-13   Digital-ramp ADC.

EXAMPLE 11-13A

Assume the following values for the ADC of Figure 11-13: clock frequency = 1 MHz; V_T = 0.1 mV; DAC has F.S. output = 10.23 V and a 10-bit input. Determine the following values.

(a) The digital equivalent obtained for V_A = 3.728 V

(b) The conversion time

(c) The resolution of this converter

Solution

(a) The DAC has a 10-bit input and a 10.23-V F.S. output. Thus, the number of total possible steps is $2^{10} - 1 = 1023$, and so the step size is

$$\frac{10.23 \text{ V}}{1023} = 10 \text{ mV}$$

This means that V_{AX} increases in steps of 10 mV as the counter counts up from 0. Because V_A = 3.728 V and V_T = 0.1 mV, V_{AX} must reach 3.7281 V or more before the comparator switches LOW. This will require

$$\frac{3.7281 \text{ V}}{10 \text{ mV}} = 372.81 = 373 \text{ steps}$$

At the end of the conversion, then, the counter will hold the binary equivalent of 373, which is 0101110101. This is the desired digital equivalent of V_A = 3.728 V, as produced by this ADC.

(b) Three hundred seventy-three steps were required to complete the conversion. Thus, 373 clock pulses occurred at the rate of one per microsecond. This gives a total conversion time of 373 μs.

(c) The resolution of this converter is equal to the step size of the DAC, which is 10 mV. Expressed as a percentage, it is $1/1023 \times 100\% \approx 0.1\%$.

EXAMPLE 11-13B

For the same ADC of Example 11-13A, determine the approximate range of analog input voltages that will produce the same digital result of $0101110101_2 = 373_{10}$.

TABLE 11-4

Step	V_{AX} (V)
371	3.71
372	3.72
373	3.73
374	3.74
375	3.75

Solution

Table 11-4 shows the ideal DAC output voltage, V_{AX}, for several of the steps on and around the 373rd. If V_A is slightly smaller than 3.72 V (by an amount < V_T), then $\overline{EOC}$ won't go LOW when V_{AX} reaches the 3.72-V step, but it will go LOW on the 3.73-V step. If V_A is slightly smaller than 3.73 V (by an amount < V_T), then $\overline{EOC}$ won't go LOW until V_{AX} reaches the 3.74-V step. Thus, as long as V_A is between approximately 3.72 and 3.73 V, $\overline{EOC}$ will go LOW when V_{AX} reaches the 3.73-V step. The exact range of V_A values is

$$3.72 \text{ V} - V_T \qquad \text{to} \qquad 3.73 \text{ V} - V_T$$

but because V_T is so small, we can simply say that the range is approximately 3.72 to 3.73 V—a range equal to 10 mV, the DAC's resolution. This is illustrated in Figure 11-14.

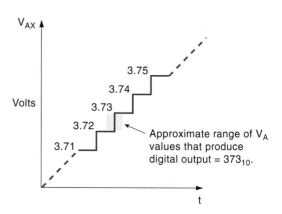

FIGURE 11-14
Example 11-13B.

A/D Resolution and Accuracy

It is very important to understand the errors associated with making any kind of measurements. An unavoidable source of error in the digital-ramp method is that the step size or resolution of the internal DAC is the smallest unit of measure. Imagine trying to measure basketball players' heights by standing them next to a staircase with 12-inch steps and assigning them the height of the first step higher than their head. Anyone over 6 feet would be measured as 7 feet tall! Likewise, the output voltage V_{AX} is a staircase waveform that goes up in discrete steps until it exceeds the input voltage, V_A. By making the step size smaller, we can reduce the potential error, but there will always be a difference between the actual (analog) quantity and the digital value assigned to it. This is called **quantization error**. Thus, V_{AX} is an approximation to the value of V_A, and the best we can expect is that V_{AX} is within 10 mV of V_A if the resolution (step size) is 10 mV. This quantization error, which can be reduced by increasing the number of bits in the counter and the DAC, is sometimes specified as an error of +1 LSB, indicating that the result could be off by as much as the weight of the LSB. In Problem 11-28, we will see how this quantization error can be modified so that it is $\pm\frac{1}{2}$ LSB, which is a more common situation.

   Looking at it from a different aspect, the input V_A can take on an *infinite* number of values from 0 V to F.S. The approximation V_{AX}, however, can take on only a finite number of discrete values. This means that a small range of V_A values will have the same digital representation. In Example 11-13B, we saw that values of V_A from ≈3.72 V to ≈3.73 V required 373 steps, thereby resulting in the same digital representation. In other words, V_A must change by 10 mV (the resolution) to produce a change in digital output.

   As in the DAC, *accuracy* is not related to the resolution but depends on the accuracy of the circuit components, such as the comparator, the DAC's precision resistors and current switches, the reference supplies, and so on. An error specification of 0.01% F.S. indicates that the ADC result may be off by 0.01% F.S., owing to nonideal components. This error is *in addition* to the quantization error due to the resolution. These two sources of error are usually of the same order of magnitude for a given ADC.

EXAMPLE 11-14

A certain eight-bit ADC has a full-scale input of 2.55 V (i.e., $V_A = 2.55$ V produces a digital output of 11111111). It has a specified error of 0.1% F.S. Determine the maximum amount by which the V_{AX} output can differ from the analog input.

Solution

The step size is 2.55 V/$(2^8 - 1)$, which is exactly 10 mV. This means that even if the DAC has no inaccuracies, the V_{AX} output could be off by as much as 10 mV because V_{AX} can change only in 10-mV steps; this is the quantization error. The specified error of 0.1% F.S. is 0.1% $\times$ 2.55 V = 2.55 mV. This means that the V_{AX} value can be off by as much as 2.55 mV because of component inaccuracies. Thus, the total possible error could be as much as 10 mV + 2.55 mV = 12.55 mV.

For example, suppose that the analog input was 1.268 V. If the DAC output were perfectly accurate, the staircase would stop at the 127th step (1.27 V). But let's say that V_{AX} was off by -2 mV, so it was 1.268 V at the 127th step. This would not be large enough to stop the conversion; it would stop at the 128th step. Thus, the digital output would be $10000000_2 = 128_{10}$ for an analog input of 1.268 V, an error of 12 mV.

Conversion Time, t_C

The conversion time is shown in Figure 11-13(b) as the time interval between the end of the START pulse and the activation of the $\overline{EOC}$ output. The counter starts counting from 0 and counts up until V_{AX} exceeds V_A, at which point $\overline{EOC}$ goes LOW to end the conversion process. It should be clear that the value of the conversion time, t_C, depends on V_A. A larger value will require more steps before the staircase voltage exceeds V_A.

The maximum conversion time will occur when V_A is just below full scale so that V_{AX} must go to the last step to activate $\overline{EOC}$. For an N-bit converter, this will be

$$t_C(\text{max}) = (2^N - 1) \text{ clock cycles}$$

For example, the ADC in Example 11-13A would have a maximum conversion time of

$$t_C(\text{max}) = (2^{10} - 1) \times 1 \ \mu s = 1023 \ \mu s$$

Sometimes, average conversion time is specified; it is half of the maximum conversion time. For the digital-ramp converter, this would be

$$t_C(\text{avg}) = \frac{t_C(\text{max})}{2} \approx 2^{N-1} \text{ clock cycles}$$

The major disadvantage of the digital-ramp method is that conversion time essentially doubles for each bit that is added to the counter, so that resolution can be improved only at the cost of a longer t_C. This makes this type of ADC unsuitable for applications where repetitive A/D conversions of a fast-changing analog signal must be made. For low-speed applications, however, the relative simplicity of the digital-ramp converter is an advantage over the more complex, higher-speed ADCs.

EXAMPLE 11-15

What will happen to the operation of a digital-ramp ADC if the analog input V_A is greater than the full-scale value?

Solution

From Figure 11-13, it should be clear that the comparator output will never go LOW because the staircase voltage can never exceed V_A. Thus, pulses will be continually applied to the counter, so that the counter will repetitively count up from 0 to maximum, recycle back to 0, count up, and so on. This will produce repetitive staircase waveforms at V_{AX} going from 0 to full scale, and this will continue until V_A is decreased below full scale.

REVIEW QUESTIONS

1. Describe the basic operation of the digital-ramp ADC.
2. Explain *quantization error.*
3. Why does conversion time increase with the value of the analog input voltage?
4. *True or false:* Everything else being equal, a 10-bit digital-ramp ADC will have a better resolution, but a longer conversion time, than an eight-bit ADC.
5. Give one advantage and one disadvantage of a digital-ramp ADC.
6. For the converter of Example 11-13, determine the digital output for $V_A = 1.345\,\text{V}$. Repeat for $V_A = 1.342\,\text{V}$.

11-10 DATA ACQUISITION

There are many applications in which analog data must be *digitized* (converted to digital) and transferred into a computer's memory. The process by which the computer acquires these digitized analog data is referred to as *data acquisition.* Acquiring a single data point's value is referred to as **sampling** the analog signal, and that data point is often called a *sample.* The computer can do several different things with the data, depending on the application. In a storage application, such as digital audio recording, video recording, or a digital oscilloscope, the internal microcomputer will store the data and then transfer them to a DAC at a later time to reproduce the original analog signal. In a process control application, the computer can examine the data or perform computations on them to determine what control outputs to generate.

Figure 11-15(a) shows how a microcomputer is connected to a digital-ramp ADC for the purpose of data acquisition. The computer generates the START pulses that initiate each new A/D conversion. The $\overline{EOC}$ (end-of-conversion) signal from the ADC is fed to the computer. The computer monitors $\overline{EOC}$ to find out when the current A/D conversion is complete; then it transfers the digital data from the ADC output into its memory.

The waveforms in Figure 11-15(b) illustrate how the computer acquires a digital version of the analog signal, V_A. The V_{AX} staircase waveform that is generated internal to the ADC is shown superimposed on the V_A waveform for purposes of illustration. The process begins at t_0, when the computer generates a START pulse to start an A/D conversion cycle. The conversion is completed at t_1, when the staircase first exceeds V_A, and $\overline{EOC}$ goes LOW. This NGT at $\overline{EOC}$ signals the computer that the ADC has a digital output that now represents the value of V_A at point *a*, and the computer will load these data into its memory.

FIGURE 11-15 (a) Typical computer data acquisition system; (b) waveforms showing how the computer initiates each new conversion cycle and then loads the digital data into memory at the end of conversion.

The computer generates a new START pulse shortly after t_1 to initiate a second conversion cycle. Note that this resets the staircase to 0 and $\overline{EOC}$ back HIGH because the START pulse resets the counter in the ADC. The second conversion ends at t_2 when the staircase again exceeds V_A. The computer then loads the digital data corresponding to point b into its memory. These steps are repeated at t_3, t_4, and so on.

The process whereby the computer generates a START pulse, monitors $\overline{EOC}$, and loads ADC data into memory is done under the control of a program that the computer is executing. This data acquisition program will determine how many data points from the analog signal will be stored in the computer memory.

Reconstructing a Digitized Signal

In Figure 11-15(b), the ADC is operating at its maximum speed because a new START pulse is generated immediately after the computer acquires the ADC output data from the previous conversion. Note that the conversion times are not constant because the analog input value is changing. The problem with this method of storing a waveform is that in order to reconstruct the waveform, we would need to know the point in time that each data value is to be plotted. Normally, when storing a digitized waveform, the samples are taken at fixed intervals at a rate that is at least two times greater than the highest frequency in the analog signal. The digital system will store the waveform as a list of sample data values. Table 11-5 shows the list of data that would be stored if the signal in Figure 11-16(a) were digitized.

In Figure 11-16(a), we see how the ADC continually performs conversions to digitize the analog signal at points *a, b, c, d,* and so on. If these digital data are used to reconstruct the signal, the result will look like that in Figure

TABLE 11-5 Digitized data samples.

Point	Actual Voltage (V)	Digital Equivalent
a	1.22	01111010
b	1.47	10010011
c	1.74	10101110
d	1.70	10101010
e	1.35	10000111
f	1.12	01110000
g	0.91	01011011
h	0.82	01010010

FIGURE 11-16 (a) Digitizing an analog signal; (b) reconstructing the signal from the digital data.

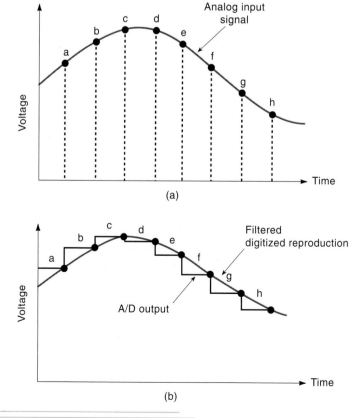

11-16(b). The black line represents the voltage waveform that would actually come out of the D/A converter. The red line would be the result of passing the signal through a simple low-pass RC filter. We can see that it is a fairly good reproduction of the original analog signal because the analog signal does not make any rapid changes between digitized points. If the analog signal contained higher-frequency variations, the ADC would not be able to follow the variations, and the reproduced version would be much less accurate.

Aliasing

The obvious goal in signal reconstruction is to make the reconstruction nearly identical to the original analog signal. In order to avoid loss of information, as has been proven by a man named Harry Nyquist, the incoming signal must be sampled at a rate greater than two times the highest-frequency component in the incoming signal. For example, if you are pretty sure that the highest frequency in an audio system will be less than 10 kHz, you must sample the audio signal at 20,000 samples per second in order to be able to reconstruct the signal. The frequency at which samples are taken is referred to as the **sampling frequency, F_S**. What do you think would happen if for some reason a 12-kHz tone is present in the input signal? Unfortunately, the system would *not* simply ignore it because it is too high! Rather, a phenomenon called *aliasing* would occur. A signal **alias** is produced by sampling the signal at a rate less than the minimum rate identified by Nyquist (twice the highest incoming frequency). In this case, any frequency over 10 kHz will produce an alias frequency. The alias frequency is always the difference between any integer multiple of the sampling frequency F_S (20 kHz) and the incoming frequency that is being digitized (12 kHz). Instead of hearing a 12-kHz tone in the reconstructed signal, you would hear an 8-kHz tone that was not in the original signal.

To see how aliasing can happen, consider the sine wave in Figure 11-17. Its frequency is 1.9 kHz. The dots show where the waveform is sampled every 500 μs (F_S = 2 kHz). If we connect the dots that make up the sampled waveform, we discover that they form a cosine wave that has a period of 10 ms and a frequency of 100 Hz. This demonstrates that the alias frequency is equal to the difference between the sample frequency and the incoming frequency. If we could hear the output that results from this data acquisition, it would not sound like 1.9 kHz; it would sound like 100 Hz.

The problem with **undersampling** ($F_S < 2 F_{in}$ max) is that the digital system has no idea that there was actually a higher frequency at the input. It simply samples the input and stores the data. When it reconstructs the signal, the alias frequency (100 Hz) is present, the original 1.9 kHz is missing, and the reconstructed signal does not sound the same. This is why a data acquisition system must not allow frequencies greater than half of F_S to be placed on the input.

FIGURE 11-17 An alias signal due to under-sampling.

REVIEW QUESTIONS

1. What is *digitizing a signal*?
2. Describe the steps in a computer data acquisition process.
3. What is the minimum sample frequency needed to reconstruct an analog signal?
4. What occurs if the signal is sampled at less than the minimum frequency determined in question 3?

11-11 SUCCESSIVE-APPROXIMATION ADC

The **successive-approximation converter** is one of the most widely used types of ADC. It has more complex circuitry than the digital-ramp ADC but a much shorter conversion time. In addition, successive-approximation converters (SACs) have a fixed value of conversion time that is not dependent on the value of the analog input.

The basic arrangement, shown in Figure 11-18(a), is similar to that of the digital-ramp ADC. The SAC, however, does not use a counter to provide the input to the DAC block but uses a register instead. The control logic modifies the contents of the register bit by bit until the register data are the digital equivalent of the analog input V_A within the resolution of the converter. The basic sequence of operation is given by the flowchart in Figure 11-18(b). We will follow this flowchart as we go through the example illustrated in Figure 11-19.

For this example, we have chosen a simple four-bit converter with a step size of 1 V. Even though most practical SACs would have more bits and smaller resolution than our example, the operation will be exactly the same. At this point, you should be able to determine that the four register bits feeding the DAC have weights of 8, 4, 2, and 1 V, respectively.

Let's assume that the analog input is $V_A = 10.4$ V. The operation begins with the control logic clearing all of the register bits to 0 so that $Q_3 = Q_2 = Q_1 = Q_0 = 0$. We will express this as $[Q] = 0000$. This makes the DAC output $V_{AX} = 0$ V, as indicated at time t_0 on the timing diagram in Figure 11-19. With $V_{AX} < V_A$, the comparator output is HIGH.

At the next step (time t_1), the control logic sets the MSB of the register to 1 so that $[Q] = 1000$. This produces $V_{AX} = 8$ V. Because $V_{AX} < V_A$, the COMP output is still HIGH. This HIGH tells the control logic that the setting of the MSB did not make V_{AX} exceed V_A, so that the MSB is kept at 1.

The control logic now proceeds to the next lower bit, Q_2. It sets Q_2 to 1 to produce $[Q] = 1100$ and $V_{AX} = 12$ V at time t_2. Because $V_{AX} > V_A$, the COMP output goes LOW. This LOW signals the control logic that the value of V_{AX} is too large, and the control logic then clears Q_2 back to 0 at t_3. Thus, at t_3, the register contents are back to 1000 and V_{AX} is back to 8 V.

The next step occurs at t_4, where the control logic sets the next lower bit Q_1 so that $[Q] = 1010$ and $V_{AX} = 10$ V. With $V_{AX} < V_A$, COMP is HIGH and tells the control logic to keep Q_1 set at 1.

The final step occurs at t_5, where the control logic sets the next lower bit Q_0 so that $[Q] = 1011$ and $V_{AX} = 11$ V. Because $V_{AX} > V_A$, COMP goes LOW to signal that V_{AX} is too large, and the control logic clears Q_0 back to 0 at t_6.

At this point, all of the register bits have been processed, the conversion is complete, and the control logic activates its $\overline{EOC}$ output to signal that the digital equivalent of V_A is now in the register. For this example, digital output for $V_A = 10.4$ V is $[Q] = 1010$. Notice that 1010 is actually equivalent to 10 V, which is *less than* the analog input; this is a characteristic of the

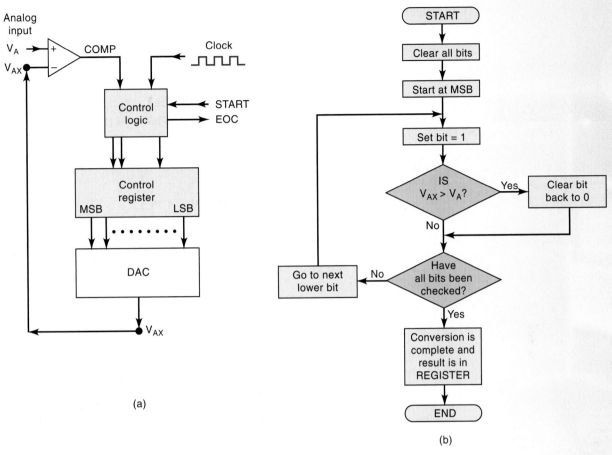

FIGURE 11-18 Successive-approximation ADC: (a) simplified block diagram; (b) flowchart of operation.

FIGURE 11-19 Illustration of four-bit SAC operation using a DAC step size of 1 V and $V_A = 10.4$ V.

successive-approximation method. Recall that in the digital-ramp method, the digital output was always equivalent to a voltage that was on the step above V_A.

EXAMPLE 11-16

An eight-bit SAC has a resolution of 20 mV. What will its digital output be for an analog input of 2.17 V?

Solution

$$2.17 \text{ V}/20 \text{ mV} = 108.5$$

so that step 108 would produce $V_{AX} = 2.16$ V and step 109 would produce 2.18 V. The SAC always produces a final V_{AX} that is at the step *below* V_A. Therefore, for the case of $V_A = 2.17$ V, the digital result would be $108_{10} = 01101100_2$.

Conversion Time

In the operation just described, the control logic goes to each register bit, sets it to 1, decides whether or not to keep it at 1, and goes on to the next bit. The processing of each bit takes one clock cycle, so that the total conversion time for an N-bit SAC will be N clock cycles. That is,

$$t_C \text{ for SAC} = N \times 1 \text{ clock cycle}$$

This conversion time will be the same *regardless of the value of V_A* because the control logic must process each bit to see whether or not a 1 is needed.

EXAMPLE 11-17

Compare the maximum conversion times of a 10-bit digital-ramp ADC and a 10-bit successive-approximation ADC if both utilize a 500-kHz clock frequency.

Solution

For the digital-ramp converter, the maximum conversion time is

$$(2^N - 1) \times (1 \text{ clock cycle}) = 1023 \times 2 \text{ } \mu s = 2046 \text{ } \mu s$$

For a 10-bit successive-approximation converter, the conversion time is always 10 clock periods or

$$10 \times 2 \text{ } \mu s = 20 \text{ } \mu s$$

Thus, it is about 100 times faster than the digital-ramp converter.

Because SACs have relatively fast conversion times, their use in data acquisition applications will permit more data values to be acquired in a given time interval. This feature can be very important when the analog data are changing at a relatively fast rate.

Because many SACs are available as ICs, it is rarely necessary to design the control logic circuitry, and so we will not cover it here. For those who are interested in the details of the control logic, many manufacturers' data books should provide sufficient detail.

An Actual IC: The ADC0804 Successive-Approximation ADC

ADCs are available from several IC manufacturers with a wide range of operating characteristics and features. We will take a look at one of the more popular devices to get an idea of what is actually used in system applications. Figure 11-20 is the pin layout for the ADC0804, which is a 20-pin CMOS IC that performs A/D conversion using the successive-approximation method. Some of its important characteristics are as follows:

- It has two analog inputs, $V_{IN}(+)$ and $V_{IN}(-)$, to allow **differential inputs**. In other words, the actual analog input, V_{IN}, is the difference in the voltages applied to these pins [analog $V_{IN} = V_{IN}(+) - V_{IN}(-)$]. In single-ended measurements, the analog input is applied to $V_{IN}(+)$, while $V_{IN}(-)$ is connected to analog ground. During normal operation, the converter uses $V_{CC} = +5\,V$ as its reference voltage, and the analog input can range from 0 to 5 V full scale.

- It converts the analog input voltage to an eight-bit digital output. The digital outputs are tristate buffered so that they can be easily connected in a data bus arrangement. With eight bits, the resolution is $5\,V/255 = 19.6\,mV$.

- It has an internal clock generator circuit that produces a frequency of $f = 1/(1.1RC)$, where R and C are values of externally connected components. A typical clock frequency is 606 kHz using $R = 10\,k\Omega$ and $C = 150\,pF$. An external clock signal can be used, if desired, by connecting it to the CLK IN pin.

- Using a 606-kHz clock frequency, the conversion time is approximately 100 μs.

- It has separate ground connections for digital and analog voltages. Pin 8 is the analog ground that is connected to the common reference point of the analog circuit that is generating the analog voltage. Pin 10 is the digital

FIGURE 11-20 ADC0804 eight-bit successive-approximation ADC with tristate outputs. The numbers in parentheses are the IC's pin numbers.

ground that is the one used by all of the digital devices in the system. (Note the different symbols used for the different grounds.) The digital ground is inherently noisy because of the rapid current changes that occur as digital devices change states. Although it is not necessary to use a separate analog ground, doing so ensures that the noise from digital ground is prevented from causing premature switching of the analog comparator inside the ADC.

This IC is designed to be easily interfaced to a microprocessor data bus. For this reason, the names of some of the ADC0804 inputs and outputs are based on functions that are common to microprocessor-based systems. The functions of these inputs and outputs are defined as follows:

- $\overline{CS}$ (Chip Select). This input must be in its active-LOW state for $\overline{RD}$ or $\overline{WR}$ inputs to have any effect. With $\overline{CS}$ HIGH, the digital outputs are in the Hi-Z state, and no conversions can take place.
- $\overline{RD}$ (READ). This input is used to enable the digital output buffers. With $\overline{CS} = \overline{RD} = $ LOW, the digital output pins will have logic levels representing the results of the *last* A/D conversion. The microcomputer can then *read* (fetch) this digital data value over the system data bus.
- $\overline{WR}$ (WRITE). A LOW pulse is applied to this input to signal the start of a new conversion. This is actually a start conversion input. It is called a **WRITE** input because in a typical application, the microcomputer generates a WRITE pulse (similar to one used for writing to memory) that drives this input.
- $\overline{INTR}$ (INTERRUPT). This output signal will go HIGH at the start of a conversion and will return LOW to signal the end of conversion. This is actually an end-of-conversion output signal, but it is called INTERRUPT because in a typical situation, it is sent to a microprocessor's interrupt input to get the microprocessor's attention and let it know that the ADC's data are ready to be read.
- $V_{ref}/2$. This is an optional input that can be used to reduce the internal reference voltage and thereby change the analog input range that the converter can handle. When this input is unconnected, it sits at 2.5 V ($V_{CC}/2$) because V_{CC} is being used as the reference voltage. By connecting an external voltage to this pin, the internal reference is changed to twice that voltage, and the analog input range is changed accordingly (see Table 11-6).
- CLK OUT. A resistor is connected to this pin to use the internal clock. The clock signal appears on this pin.
- CLK IN. Used for external clock input, or for a capacitor connection when the internal clock is used.

Figure 11-21(a) shows a typical connection of the ADC0804 to a microcomputer in a data acquisition application. The microcomputer controls

TABLE 11-6

$V_{ref}/2$	Analog Input Range (V)	Resolution (mV)
Open	0–5	19.6
2.25	0–4.5	17.6
2.0	0–4	15.7
1.5	0–3	11.8

(a)

(b)

FIGURE 11-21 (a) An application of an ADC0804; (b) typical timing signals during data acquisition.

when a conversion is to take place by generating the $\overline{CS}$ and $\overline{WR}$ signals. It then acquires the ADC output data by generating the $\overline{CS}$ and $\overline{RD}$ signals after detecting an NGT at $\overline{INTR}$, indicating the end of conversion. The waveforms in Figure 11-21(b) show the signal activity during the data acquisition process. Note that $\overline{INTR}$ goes HIGH when $\overline{CS}$ and $\overline{WR}$ are LOW, but the conversion process does not begin until $\overline{WR}$ returns HIGH. Also note that the ADC output data lines are in their Hi-Z state until the microcomputer activates $\overline{CS}$ and $\overline{RD}$; at that point the ADC's data buffers are enabled so that the

ADC data are sent to the microcomputer over the data bus. The data lines return to the Hi-Z state when either $\overline{CS}$ or $\overline{RD}$ is returned HIGH.

In this application of the ADC0804, the input signal is varying over a range of 0.5 to 3.5 V. In order to make full use of the eight-bit resolution, the A/D must be matched to the analog signal specifications. In this case, the full-scale range is 3.0 V. However, it is offset from ground by 0.5 V. The offset of 0.5 V is applied to the negative input $V_{IN}(-)$, establishing this as the 0 value reference. The range of 3.0 V is set by applying 1.5 V to $V_{ref}/2$, which establishes V_{ref} as 3.0 V. An input of 0.5 V will produce a digital value of 00000000, and an input of 3.5 V will produce 11111111.

Another major concern when interfacing digital and analog signals is *noise.* Notice that the digital and analog ground paths are separated. The two grounds are tied together at a point that is very close to the A/D converter. A very low-resistance path ties this point directly to the negative terminal of the power supply. It is also wise to route the positive supply lines separately to digital and analog devices and make extensive use of decoupling capacitors $(0.01\ \mu F)$ from very near each chip's supply connection to ground.

REVIEW QUESTIONS

1. What is the main advantage of a SAC over a digital-ramp ADC?
2. What is its principal disadvantage compared with the digital-ramp converter?
3. *True or false:* The conversion time for a SAC increases as the analog voltage increases.
4. Answer the following concerning the ADC0804.
   (a) What is its resolution in bits?
   (b) What is the normal analog input voltage range?
   (c) Describe the functions of the $\overline{CS}$, $\overline{WR}$, and $\overline{RD}$ inputs.
   (d) What is the function of the $\overline{INTR}$ output?
   (e) Why does it have two separate grounds?
   (f) What is the purpose of $V_{IN}(-)$?

11-12 FLASH ADCs

The **flash converter** is the highest-speed ADC available, but it requires much more circuitry than the other types. For example, a six-bit flash ADC requires 63 analog comparators, while an eight-bit unit requires 255 comparators, and a ten-bit converter requires 1023 comparators. The large number of comparators has limited the size of flash converters. IC flash converters are commonly available in two- to eight-bit units, and most manufacturers offer nine- and ten-bit units as well.

The principle of operation will be described for a three-bit flash converter in order to keep the circuitry at a workable level. Once the three-bit converter is understood, it should be easy to extend the basic idea to higher-bit flash converters.

The flash converter in Figure 11-22(a) has a three-bit resolution and a step size of 1 V. The voltage divider sets up reference levels for each comparator so that there are seven levels corresponding to 1 V (weight of LSB), 2V, 3V, . . . , and 7 V (full scale). The analog input, V_A, is connected to the other input of each comparator.

With $V_A < 1$ V, all of the comparator outputs C_1 through C_7 will be HIGH. With $V_A > 1$ V, one or more of the comparator outputs will be LOW. The comparator outputs are fed into an active-LOW priority encoder that generates a binary output corresponding to the highest-numbered comparator output that is LOW. For example, when V_A is between 3 and 4 V, outputs C_1, C_2, and C_3 will be LOW and all others will be HIGH. The priority encoder will respond only to the LOW at C_3 and will produce a binary output $CBA = 011$, which represents the digital equivalent of V_A, within the resolution of 1 V. When V_A is greater than 7 V, C_1 to C_7 will all be LOW, and the encoder will produce $CBA = 111$ as the digital equivalent of V_A. The table in Figure 11-22(b) shows the responses for all possible values of analog input.

The flash ADC of Figure 11-22 has a resolution of 1 V because the analog input must change by 1 V in order to bring the digital output to its next step.

(a)

Analog in	Comparator outputs							Digital outputs		
V_A	C_1	C_2	C_3	C_4	C_5	C_6	C_7	C	B	A
0–1 V	1	1	1	1	1	1	1	0	0	0
1–2 V	0	1	1	1	1	1	1	0	0	1
2–3 V	0	0	1	1	1	1	1	0	1	0
3–4 V	0	0	0	1	1	1	1	0	1	1
4–5 V	0	0	0	0	1	1	1	1	0	0
5–6 V	0	0	0	0	0	1	1	1	0	1
6–7 V	0	0	0	0	0	0	1	1	1	0
> 7 V	0	0	0	0	0	0	0	1	1	1

(b)

FIGURE 11-22   (a) Three-bit flash ADC; (b) truth table.

To achieve finer resolutions, we would have to increase the number of input voltage levels (i.e., use more voltage-divider resistors) and the number of comparators. For example, an eight-bit flash converter would require $2^8 = 256$ voltage levels, including 0 V. This would require 256 resistors and 255 comparators (there is no comparator for the 0-V level). The 255 comparator outputs would feed a priority encoder circuit that would produce an eight-bit code corresponding to the highest-order comparator output that is LOW. In general, an N-bit flash converter would require $2^N - 1$ comparators, 2^N resistors, and the necessary encoder logic.

Conversion Time

The flash converter uses no clock signal because no timing or sequencing is required. The conversion takes place continuously. When the value of analog input changes, the comparator outputs will change, thereby causing the encoder outputs to change. The conversion time is the time it takes for a new digital output to appear in response to a change in V_A, and it depends only on the propagation delays of the comparators and encoder logic. For this reason, flash converters have extremely short conversion times. For example, the Analog Devices AD9020 is a 10-bit flash converter with a conversion time under 17 ns.

REVIEW QUESTIONS

1. *True or false:* A flash ADC does not contain a DAC.
2. How many comparators would a 12-bit flash converter require? How many resistors?
3. State the major advantage and disadvantage of a flash converter.

11-13 OTHER A/D CONVERSION METHODS

Several other methods of A/D conversion have been in use for some time, each with its relative advantages and disadvantages. We will briefly describe some of them now.

Up/Down Digital-Ramp ADC (Tracking ADC)

As we have seen, the digital-ramp ADC is relatively slow because the counter is reset to 0 at the start of each new conversion. The staircase always begins at 0 V and steps its way up to the "switching point" where V_{AX} exceeds V_A and the comparator output switches LOW. The time it takes the staircase to reset to 0 and step back up to the new switching point is really wasted. The **up/down digital-ramp ADC** uses an up/down counter to reduce this wasted time.

The up/down counter replaces the up counter that feeds the DAC. It is designed to count up whenever the comparator output indicates that $V_{AX} < V_A$ and to count down whenever $V_{AX} > V_A$. Thus, the DAC output is always being stepped in the direction of the V_A value. Each time the comparator output switches states, it indicates that V_{AX} has "crossed" the V_A value, the digital equivalent of V_A is in the counter, and the conversion is complete.

When a new conversion is to begin, the counter is *not reset to 0* but begins counting either up or down from its last value, depending on the comparator output. It will count until the staircase crosses V_A again to end the conversion. The V_{AX} waveform, then, will contain both positive-going and negative-going staircase signals that "track" the V_A signal. For this reason, this ADC is often called a *tracking ADC*.

Clearly, the conversion times will generally be reduced because the counter does not start over from 0 each time but simply counts up or down from its previous value. Of course, the value of t_C will still depend on the value of V_A, and so it will not be constant.

Dual-Slope Integrating ADC

The **dual-slope converter** has one of the slowest conversion times (typically 10 to 100 ms) but has the advantage of relatively low cost because it does not require precision components such as a DAC or a VCO. The basic operation of this converter involves the *linear* charging and discharging of a capacitor using constant currents. First, the capacitor is charged up for a fixed time interval from a constant current derived from the analog input voltage, V_A. Thus, at the end of this fixed charging interval, the capacitor voltage will be proportional to V_A. At that point, the capacitor is linearly discharged from a constant current derived from a precise reference voltage, V_{ref}. When the capacitor voltage reaches 0, the linear discharging is terminated. During the discharge interval, a digital reference frequency is fed to a counter and counted. The duration of the discharge interval will be proportional to the initial capacitor voltage. Thus, at the end of the discharge interval, the counter will hold a count proportional to the initial capacitor voltage, which, as we said, is proportional to V_A.

In addition to its low cost, another advantage of the dual-slope ADC is its low sensitivity to noise and to variations in its component values caused by temperature changes. Because of its slow conversion times, the dual-slope ADC is not used in any data acquisition applications. The slow conversion times, however, are not a problem in applications such as digital voltmeters or multimeters, and this is where they find their major application.

Voltage-to-Frequency ADC

The **voltage-to-frequency ADC** is simpler than other ADCs because it does not use a DAC. Instead it uses a *linear voltage-controlled oscillator (VCO)* that produces an output frequency proportional to its input voltage. The analog voltage that is to be converted is applied to the VCO to generate an output frequency. This frequency is fed to a counter to be counted for a fixed time interval. The final count is proportional to the value of the analog voltage.

To illustrate, suppose that the VCO generates a frequency of 10 kHz for each volt of input (i.e., 1 V produces 10 kHz, 1.5 V produces 15 kHz, 2.73 V produces 27.3 kHz). If the analog input voltage is 4.54 V, then the VCO output will be a 45.4-kHz signal that clocks a counter for, say, 10 ms. After the 10-ms counting interval, the counter will hold the count of 454, which is the digital representation of 4.54 V.

Although this is a simple method of conversion, it is difficult to achieve a high degree of accuracy because of the difficulty in designing VCOs with accuracies of better than 0.1 percent.

One of the main applications of this type of converter is in noisy industrial environments where small analog signals must be transmitted from transducer circuits to a control computer. The small analog signals can be drastically affected by noise if they are directly transmitted to the control computer. A better approach is to feed the analog signal to a VCO, which generates a digital signal whose output frequency changes according to the analog input. This digital signal is transmitted to the computer and will be much less affected by noise. Circuitry in the control computer will count the digital pulses (i.e., perform a frequency-counting function) to produce a digital value equivalent to the original analog input.

Sigma/Delta Modulation*

Another approach to representing analog information in digital form is called **sigma/delta modulation**. A sigma/delta A/D converter is an oversampling device, which means that it effectively samples the analog information more often than the minimum sample rate. The minimum sample rate is two times higher than the highest frequency in the incoming analog wave. The sigma/delta approach, like the voltage-to-frequency approach, does not directly produce a multibit number for each sample. Instead, it represents the analog voltage by varying the density of logic 1s in a single-bit stream of serial data. To represent the positive portions of a waveform, a stream of bits with a high density of 1s is generated by the modulator (e.g., 01111101111110111110111). To represent the negative portions, a lower density of 1s (i.e., a higher density of 0s) is generated (e.g., 00010001000010001000).

Sigma/delta modulation is used in A/D as well as D/A conversion. One form of a sigma/delta modulator circuit is designed to convert a continuous analog signal into a modulated bit stream (A/D). The other form converts a sequence of digital samples into the modulated bit stream (D/A). We are coming from the perspective of digital systems, so it is easiest to understand the latter of these two circuits because it consists of all digital components that we have studied. Figure 11-23 shows a circuit that takes a five-bit signed digital value as its in-

FIGURE 11-23   Sigma/delta modulator in a D/A converter.

---

*An excellent article published on the web by Jim Thompson, University of Washington, served as a basis for this description. Visit the *Digital Systems: Principles and Applications* Companion Web Site at http://www.prenhall.com/Tocci for the link to this article.

put and converts it into a sigma/delta bit stream. We will assume that the numbers that can be placed on this circuit's input range from -8 to $+8$. The first component is simply a subtractor (the delta section) similar to the one studied in Figure 6-13. The subtractor determines how far the input number is from its maximum or minimum value. This difference is often called the error signal. The second two components (the adder and the D register) form an accumulator very similar to the circuit in Figure 6-9 (the sigma section). For each sample that comes in, the accumulator adds the difference (error signal) to the running total. When the error is small, this running total (sigma) changes by small increments. When the error is large, the sigma changes by large increments. The last component compares the running total from the accumulator with a fixed threshold, which in this case is zero. In other words, it is simply determining if the total is positive or negative. This is accomplished by using the MSB (sign bit) of sigma. As soon as the total goes positive, the MSB goes LOW and feeds back to the delta section the maximum positive value $(+8)$. When the MSB of sigma goes negative, it feeds back the maximum negative value (-8).

A spreadsheet is an excellent way to analyze a circuit like this. The tables in this section are taken from the spreadsheet that is included on the CD at the back of this book. Table 11-7 shows the operation of the converter when a value of zero is the input. Notice that the bit stream output alternates between 1 and 0, and the average value of the analog output is 0 volts. Table 11-8 shows what happens when the digital input is 4. If we assume that 8 is full scale, this represents $\frac{4}{8} = 0.5$. The output is HIGH for three samples and LOW for one sample, a pattern that repeats every four samples. The average value of the analog output is $(1 + 1 + 1 - 1)/4 = 0.5\,\text{V}$.

As a final example, let's use an input of -5, which represents $-5/8 = -0.625$. Table 11-9 shows the resulting output. The pattern in the bit stream is not periodic. From the sigma column, we can see that it takes 16 samples for the pattern to repeat. If we take the overall bit density, however, and calculate the average value of the analog output over 16 samples, we will find that it is

TABLE 11-7
Sigma/delta modulator with an input of 0.

Sample (n)	Digital IN	Delta	Sigma	Bit Stream Out	Analog OUT	Feedback
1	0	−8	0	1	1	8
2	0	8	−8	0	−1	−8
3	0	−8	0	1	1	8
4	0	8	−8	0	−1	−8
5	0	−8	0	1	1	8
6	0	8	−8	0	−1	−8
7	0	−8	0	1	1	8
8	0	8	−8	0	−1	−8

TABLE 11-8
Sigma/delta modulator with an input of 4.

Sample (n)	Digital IN	Delta	Sigma	Bit Stream Out	Analog OUT	Feedback
1	4	−4	4	1	1	8
2	4	−4	0	1	1	8
3	4	12	−4	0	−1	−8
4	4	−4	8	1	1	8
5	4	−4	4	1	1	8
6	4	−4	0	1	1	8
7	4	12	−4	0	−1	−8
8	4	−4	8	1	1	8

equal to -0.625. Your CD player probably uses a sigma/delta D/A converter that operates in this fashion. The 16-bit digital numbers come off the CD serially; then they are formatted into parallel data patterns and clocked into a converter. As the changing numbers come into the converter, the average value of the analog out changes accordingly. Next, the analog output goes through a circuit called a low-pass filter that smoothes out the sudden changes and produces a smoothly changing voltage that is the average value of the bit stream. In your headphones, this changing analog signal sounds just like the original recording. A sigma/delta A/D converter works in a very similar way but converts the analog voltage into the modulated bit stream. To store the digitized data as a list of N-bit binary numbers, the average bit density of 2^N bit-stream samples is calculated and stored.

TABLE 11-9
Sigma/delta modulator
with an input of -5.

Sample (n)	Digital IN	Delta	Sigma	Bit Stream Out	Analog OUT	Feedback
1	-5	3	-5	0	-1	-8
2	-5	3	-2	0	-1	-8
3	-5	-13	1	1	1	8
4	-5	3	-12	0	-1	-8
5	-5	3	-9	0	-1	-8
6	-5	3	-6	0	-1	-8
7	-5	3	-3	0	-1	-8
8	-5	-13	0	1	1	8
9	-5	3	-13	0	-1	-8
10	-5	3	-10	0	-1	-8
11	-5	3	-7	0	-1	-8
12	-5	3	-4	0	-1	-8
13	-5	3	-1	0	-1	-8
14	-5	-13	2	1	1	8
15	-5	3	-11	0	-1	-8
16	-5	3	-8	0	-1	-8
17	-5	3	-5	0	-1	-8
18	-5	3	-2	0	-1	-8

REVIEW QUESTIONS

1. How does the up/down digital-ramp ADC improve on the digital-ramp ADC?
2. What is the main element of a voltage-to-frequency ADC?
3. Cite two advantages and one disadvantage of the dual-slope ADC.
4. Name three types of ADCs that do not use a DAC.
5. How many output data bits does a sigma/delta modulator use?

11-14 DIGITAL VOLTMETER

A digital voltmeter (DVM) converts an analog voltage to its BCD-code representation, which is then decoded and displayed on some type of readout. Figure 11-24 shows a three-digit DVM circuit that uses a digital-ramp ADC

FIGURE 11-24 Continuous-conversion DVM using a digital-ramp ADC.

(shown inside the dashed lines). Three cascaded BCD counters provide the inputs to a three-digit BCD-to-analog converter that has a step size of 10 mV and a full-scale output of 9.99 V. Each BCD counter stage also drives a four-bit register that feeds a decoder/driver and a display. The contents of the BCD counters are transferred to the registers at the end of each conversion cycle, so that the displays do not show the counters resetting and counting, but only the final count that represents the unknown voltage.

As long as $V_{AX} < V_A$, the COMP output stays HIGH, and clock pulses pass through the AND gate to the counter. As the counter increments, the V_{AX} signal increases at 10 mV per step until it exceeds V_A (see Figure 11-25). At that point, COMP goes LOW to disable the AND gate and stop the counter, thereby ending the conversion. The NGT of COMP also triggers one-shot 1 (OS1) to produce a 1-μs pulse at Q_1. The PGT of Q_1 transfers the outputs of the BCD counter stages to their respective registers to be stored and displayed. The NGT of Q_1 triggers a second one-shot, OS2, to produce a pulse that resets all the counters to 0. This brings V_{AX} back to 0, and COMP returns HIGH, allowing pulses into the counter to begin a new conversion cycle.

FIGURE 11-25   Waveforms for a DVM.

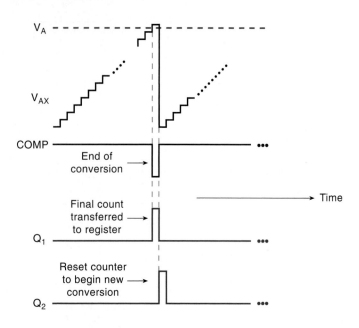

Thus, this DVM will perform one conversion right after another. Of course, the storage registers will keep the displays from showing the conversion process. The display readings will change only if V_A changes, so that different counter contents are transferred to the registers at the end of the conversion cycle.

A numerical example will help illustrate this circuit's operation. Assume that V_A is 6.372 V. In order for the COMP output to switch LOW, V_A must exceed 6.3721 V. Because the DAC output increases by 10 mV/step, this requires

$$\frac{6.3721 \text{ V}}{10 \text{ mV}} = 637.21 \rightarrow 638 \text{ steps}$$

Thus, the counters will count up to 638, which will be transferred to the registers and displayed. A small LED can be used to display a decimal point so that the operator sees 6.38 V.

EXAMPLE 11-18

What will happen if $V_A > 9.99$ V?

Solution

With $V_A > 9.99$ V, the comparator output will stay HIGH, allowing clock pulses into the counter continuously. The counter will repetitively count up to 999 and recycle to 000. One-shot Q_1 will *never* be triggered, and the register contents will not change from their previous value. A well-designed DVM would have some means of detecting this *over-range* condition and activating some type of over-range indicator. One possible method would simply add a thirteenth bit to the counter string. This bit would go HIGH only if the counter recycled from 999 to 000, and then it would indicate the over-range condition.

The DVM can be modified to read input voltages over several ranges by using a suitable amplifier or attenuator between V_A and the comparator. For example, if V_A were 63.72 V, it could be attenuated by a factor of 10, so that the comparator would receive 6.372 V at its + input and the counters would display 638 at the end of the conversion. The decimal-point indicator would be placed in front of the LSD, so that the display would read 63.8 V.

The DVM can be modified to operate as a DMM that can measure current and resistance as well as voltage. To measure current, the unknown current is made to flow through a fixed reference resistor to produce a voltage. To measure resistance, a fixed reference current is made to flow through the unknown resistance to develop a voltage. In each case, the resulting voltage is fed to a circuit like that of Figure 11-24 to be measured and displayed. The circuit may require an attenuation scaling factor so that the displayed value correctly represents the quantity (current or resistance) being measured.

AC voltages can be measured by this same DVM if they are first converted to a dc voltage. This can be done by using a circuit that converts the true rms value of a signal (pure ac or otherwise) to a dc value. The AD536A from Analog Devices is one such IC that performs this true rms-to-dc conversion.

REVIEW QUESTIONS

1. What is the purpose of the registers in the DVM circuit?
2. What is the purpose of the two one-shots?
3. How would the performance of the DVM change if a SAC were used?

11-15 SAMPLE-AND-HOLD CIRCUITS

When an analog voltage is connected directly to the input of an ADC, the conversion process can be adversely affected if the analog voltage is changing during the conversion time. The stability of the conversion process can be improved by using a **sample-and-hold (S/H) circuit** to hold the analog voltage constant while the A/D conversion is taking place. A simplified diagram of a sample-and-hold (S/H) circuit is shown in Figure 11-26.

The S/H circuit contains a unity-gain buffer amplifier A_1 that presents a high impedance to the analog signal and has a low output impedance that can rapidly charge the hold capacitor, C_h. The capacitor will be connected to the output of A_1 when the digitally controlled switch is closed. This is called the *sample* operation. The switch will be closed long enough for C_h to charge

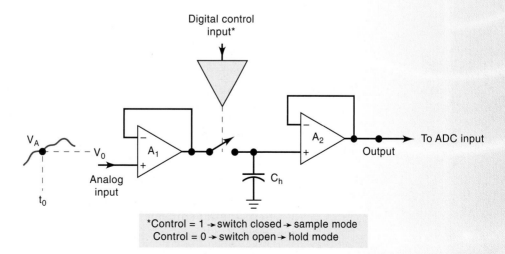

FIGURE 11-26 Simplified diagram of a sample-and-hold circuit.

to the present value of the analog input. For example, if the switch is closed at time t_0, the A_1 output will quickly charge C_h up to a voltage V_0. When the switch opens, C_h will *hold* this voltage so that the output of A_2 will apply this voltage to the ADC. The unity-gain buffer amplifier A_2 presents a high input impedance that will not discharge the capacitor voltage appreciably during the conversion time of the ADC, and so the ADC will essentially receive a dc input voltage V_0.

In a computer-controlled data acquisition system such as the one discussed earlier, the sample-and-hold switch would be controlled by a digital signal from the computer. The computer signal would close the switch in order to charge C_h to a new sample of the analog voltage; the amount of time the switch would have to remain closed is called the **acquisition time**, and it depends on the value of C_h and the characteristics of the S/H circuit. The computer signal would then open the switch to allow C_h to hold its value and provide a relatively constant analog voltage at the A_2 output.

The AD1154 is a sample-and-hold integrated circuit that has an internal hold capacitor with an acquisition time of 3.5 μs. During the hold time, the capacitor voltage will droop (discharge) at a rate of only 0.1 μV/μs. The voltage droop within the sampling interval should be less than the weight of the LSB. For example, a 10-bit converter with a full-scale range of 10 V would have an LSB weight of approximately 10 mV. It would take 100 ms before the capacitor droop would equal the weight of the ADC's LSB. It is not likely, however, that it would ever be necessary to hold the sample for such a long time in the conversion process.

REVIEW QUESTIONS

1. Describe the function of a sample-and-hold circuit.
2. *True or false:* The amplifiers in an S/H circuit are used to provide voltage amplification.

11-16 MULTIPLEXING

When analog inputs from several sources are to be converted, a multiplexing technique can be used so that one ADC may be time-shared. The basic scheme is illustrated in Figure 11-27 for a four-channel acquisition system. Rotary switch S is used to switch each analog signal to the input of the ADC, one at a time in sequence. The control circuitry controls the switch position according to the *select address* bits, A_1, A_0, from the MOD-4 counter. For example, with $A_1A_0 = 00$, the switch connects V_{A0} to the ADC input; $A_1A_0 = 01$ connects V_{A1} to the ADC input; and so on. Each input channel has a specific address code that, when present, connects that channel to the ADC.

The operation proceeds as follows:

1. With select address = 00, V_{A0} is connected to the ADC input.
2. The control circuit generates a START pulse to initiate the conversion of V_{A0} to its digital equivalent.
3. When the conversion is complete, *EOC* signals that the ADC output data are ready. Typically, these data will be transferred to a computer over a data bus.

FIGURE 11-27 Conversion of four analog inputs by multiplexing through one ADC.

4. The multiplexing clock increments the select address to 01, which connects V_{A1} to the ADC.

5. Steps 2 and 3 are repeated with the digital equivalent of V_{A1} now present at the ADC outputs.

6. The multiplexing clock increments the select address to 10, and V_{A2} is connected to the ADC.

7. Steps 2 and 3 are repeated with the digital equivalent of V_{A2} now present at the ADC outputs.

8. The multiplexing clock increments the select address to 11, and V_{A3} is connected to the ADC.

9. Steps 2 and 3 are repeated with the digital equivalent of V_{A3} now present at the ADC outputs.

The multiplexing clock controls the rate at which the analog signals are sequentially switched into the ADC. The maximum rate is determined by the delay time of the switches and the conversion time of the ADC. The switch delay time can be minimized by using semiconductor switches such as the CMOS bilateral switch described in Chapter 8. It may be necessary to connect a sample-and-hold circuit at the input of the ADC if the analog inputs will change significantly during the ADC conversion time.

Many integrated ADCs contain the multiplexing circuitry on the same chip as the ADC. The ADC0808, for example, can multiplex eight different analog inputs into one ADC. It uses a three-bit select input code to determine which analog input is connected to the ADC.

REVIEW QUESTIONS

1. What is the advantage of this multiplexing scheme?
2. How would the address counter be changed if there were eight analog inputs?

11-17   DIGITAL STORAGE OSCILLOSCOPE

As a final example of the application of D/A and A/D converters, we will take a brief look at the digital storage oscilloscope (DSO). The DSO uses both of these devices to digitize, store, and display analog waveforms. A DSO has many advantages over the conventional storage oscilloscope (CSO), which stores the waveform image as electrical charges on a mesh screen between the CRT's electron gun and the phosphor-coated screen. Here are some of the advantages of a DSO:

■ The DSO can store the waveforms for an indefinite time because the digital values are stored in digital memory (i.e., flip-flops). In a CSO, the image gradually fades as the charges on the mesh screen are slowly neutralized.

■ In many DSOs, the stored waveform can be positioned anywhere on the CRT screen, and its vertical and horizontal scales can be changed to suit the measurement situation. This is not possible with a CSO.

■ With a DSO, it is possible to store and display portions of the waveform that occur *before* the occurrence of the trigger point. Again, this cannot be done with a CSO.

■ Many DSOs can store several waveforms and print them out on a standard printer.

A block diagram of a DSO is shown in Figure 11-28. The overall operation is controlled and synchronized by the circuits in the CONTROL block, which usually contains a microprocessor executing a control program stored in ROM (read-only memory). The data acquisition portion of the system contains a sample-and-hold (S/H) and an ADC that repetitively samples and digitizes the input signal at a rate determined by the SAMPLE CLOCK and then sends the digitized data to memory for storage. The CONTROL block makes sure that successive data points are stored in successive memory locations by continually updating the memory's ADDRESS COUNTER.

FIGURE 11-28   Block diagram of a digital storage oscilloscope.

When memory is full, the next data point from the ADC is stored in the first memory location, writing over the old data, and so on, for successive data points. This data acquisition and storage process continues until the CONTROL block receives a trigger signal from either the input waveform (INTERNAL trigger) or an EXTERNAL trigger source. When the trigger occurs, the system stops acquiring new data and enters the display mode of operation, in which all or part of the memory data is repetitively displayed on the CRT.

The display operation uses two DACs to provide the vertical and horizontal deflection voltages for the CRT. Data from memory produce the vertical deflection of the electron beam, while the TIME-BASE COUNTER provides the horizontal deflection in the form of a staircase sweep signal. The CONTROL block synchronizes the display operation by incrementing the memory ADDRESS COUNTER and the TIME-BASE COUNTER at the same time so that each horizontal step of the electron beam is accompanied by a new data value from memory to the vertical DAC. The counters are continuously recycled so that the stored data points are repetitively replotted on the CRT screen. The screen display consists of discrete dots representing the various data points, but the number of dots is usually so large (typically 1000 or more) that they tend to blend together and appear to be a smooth, continuous waveform. The display operation is terminated when the operator presses a front-panel button that commands the DSO to begin a new data acquisition cycle.

Related Applications

The same sequence of operations performed in a DSO—data acquisition/digitizing/storage/data outputting—is used in other applications of DACs and ADCs. For example, in digital audio recording, the analog audio signal produced by a microphone is digitized (using an ADC) and then stored on some medium such as magnetic tape, magnetic disk, or optical disk. Later the stored data are "played back" by sending them to a DAC to reconstruct the analog signal, which is fed to the amplifier and speaker system to produce the recorded sound.

REVIEW QUESTIONS

1. What are some of the advantages of a DSO over a CSO?
2. Describe the functions of the ADC and DACs that are part of the DSO.

11-18   DIGITAL SIGNAL PROCESSING (DSP)

One of the most dynamic areas of digital systems today is in the field of **digital signal processing (DSP)**. A DSP is a very specialized form of microprocessor that has been optimized to perform repetitive calculations on streams of digitized data. The digitized data are usually being fed to the DSP from an A/D converter. It is beyond the scope of this text to explain the mathematics that allow a DSP to process these data values, but suffice it to say that for each new data point that comes in, a calculation is performed (very quickly). This calculation involves the most recent data point as well as several of the preceding data samples. The result of the calculation produces a new output data point, which is usually sent to a D/A converter. A DSP system is similar to the block diagram shown in Figure 11-1. The main difference is in the specialized hardware contained in the computer section.

A major application for DSP is in filtering and conditioning of analog signals. As a very simple example, a DSP can be programmed to take in an analog waveform, such as the output from an audio preamplifier, and pass to the output only those frequency components that are below a certain frequency. All higher frequencies are attenuated by the filter. Perhaps you recall from your study of analog circuits that the same thing can be accomplished by a simple low-pass filter made from a resistor and capacitor. The advantage of DSP over resistors and capacitors is the flexibility of being able to change the critical frequency without switching any components. Instead, numbers are simply changed in the calculations to adapt the dynamic response of the filter. Have you ever been in an auditorium when the PA system started to squeal? This can be prevented if the degenerative feedback frequency can be filtered out. Unfortunately, the frequency that causes the squeal changes with the number of people in the room, the clothes they wear, and many other factors. With a DSP-based audio equalizer, the oscillation frequency can be detected and the filters dynamically adjusted to tune it out.

Digital Filtering

To help you understand digital filtering, imagine you are buying and selling stock. To decide when to buy and sell, you need to know what the market is doing. You want to ignore sudden, short-term (high-frequency) changes but respond to the overall trends (30-day averages). Every day you read the newspaper, take a sample of the closing price for your stock, and write it down. Then you use a formula to calculate the average of the last 30 days' prices. This average value is plotted as shown in Figure 11-29, and the resulting graph is used to make decisions. This is a way of filtering the digital signal (sequence of data samples) that represents the stock market activity.

Now imagine that instead of sampling stock prices, a digital system is sampling an audio (analog) signal from a microphone using an A/D converter. Instead of taking a sample once a day, it takes a sample 20,000 times each second (every 50 μs). For each sample, a weighted averaging calculation is performed using the last 256 data samples and produces a single output data point. A **weighted average** means that some of the data points are considered more important than others. Each of the samples is multiplied by a fractional number (between 0 and 1) before adding them together. This averaging calculation is processing (filtering) the audio signal. The most difficult part of this form of DSP is determining the correct weighting con-

FIGURE 11-29 Digital filtering of stock market activity.

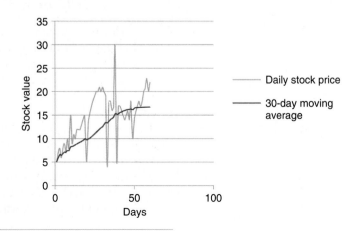

stants for the averaging calculation in order to achieve the desired filter characteristics. Fortunately, there is readily available software for PCs that makes this very easy. The special DSP hardware must perform the following operations:

Read the newest sample (one new number) from A/D.

Replace the oldest sample (of 256) with the new one from A/D.

Multiply each of the 256 samples by their corresponding weight constant.

Add all of these products.

Output the resulting sum of products (1 number) to the D/A.

Figure 11-30 shows the basic architecture of a DSP. The multiply and accumulate (**MAC**) section is central to all DSPs and is used in most applications. Special hardware, like you will study in Chapter 12, is used to implement the memory system that stores the data samples and weight values. The **arithmetic logic unit** and **barrel shifter** (shift register) provide the necessary support to deal with the binary number system while processing signals.

FIGURE 11-30   Digital signal processor architecture.

Another useful application of DSP is called **oversampling** or **interpolation filtering**. As you recall, the reconstructed waveform is always an approximation of the original due to quantization error. The sudden step changes from one data point to the next also introduce high-frequency noise into the reconstructed signal. A DSP can insert interpolated data points into the digital signal. Figure 11-31 shows how 4X oversampling interpolation filtering smoothes out the waveform and makes final filtering possible with simpler analog circuitry. DSP performs this role in your CD player to provide an excellent audio reproduction. The round dots represent the digitally

FIGURE 11-31 Inserting an interpolated data point into a digital signal to reduce noise.

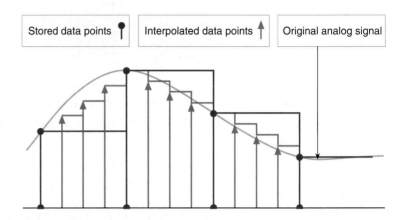

recorded data on your CD. The triangles represent the interpolated data points that the digital filter in your CD player inserts before the final analog output filter.

Many of the important concepts that you need to understand in order to move on to DSP have been presented in this and previous chapters. A/D and D/A conversion methods and hardware along with data acquisition and sampling concepts are vital. Topics such as signed binary number representations (including fractions), signed binary addition and multiplication (covered in Chapter 6), and shift registers (Chapter 7) are necessary to understand the hardware and programming of a DSP. Memory system concepts, which will be presented in the next chapter, will also be important.

DSP is being integrated into many common systems that you are familiar with. CD players use DSP to filter the digital data being read from the disk to minimize the quantization noise that is unavoidably caused by digitizing the music. Telephone systems use DSP to cancel echoes on the phone lines. The high-speed modems that are standard on PCs have been made possible and affordable by DSP. Special effects boxes for guitars and other instruments perform echo, reverb, phasing, and other effects using DSP. Applications of DSP are growing right now at the same rate that microprocessor applications grew in the early 1980s. They provide a digital solution to many traditionally analog problems. Some other examples of applications include speech recognition, telecommunications data encryption, Fast Fourier Transforms, image processing in digital television, ultrasonic beam forming in medical electronics, and noise cancellation in industrial controls. As this trend continues, you can expect to see nearly all electronic systems containing digital signal processing circuitry.

REVIEW QUESTIONS

1. What is a major application of DSP?
2. What is the typical source of digital data for a DSP to process?
3. What advantage does a DSP filter have over an analog filter circuit?
4. What is the central hardware feature of a DSP?
5. How many interpolated data points are inserted between samples when performing 4X oversampled digital filtering? How many for 8X oversampling?

SUMMARY

1. Physical variables that we want to measure, such as temperature, pressure, humidity, distance velocity, and so on, are continuously variable quantities. A transducer can be used to translate these quantities into an electrical signal of voltage or current that fluctuates in proportion to the physical variable. These continuously variable voltage or current signals are called *analog* signals.

2. To measure a physical variable, a digital system must assign a binary number to the analog value that is present at that instant. This is accomplished by an A/D converter. To generate variable voltages or current values that can control physical processes, a digital system must translate binary numbers into a voltage or current magnitude. This is accomplished by a D/A converter.

3. A D/A converter with n bits divides a range of analog values (voltage or current) into $2^n - 1$ pieces. The size or magnitude of each piece is the analog equivalent weight of the least significant bit. This is called the *resolution* or *step size.*

4. Most D/A converters use resistor networks that can cause weighted amounts of current to flow when any of its binary inputs are activated. The amount of current that flows is proportional to the binary weight of each input bit. These weighted currents are summed to create the analog signal out.

5. An A/D converter must assign a binary number to an analog (continuously variable) quantity. The precision with which an A/D converter can perform this conversion depends on how many different numbers it can assign and how wide the analog range is. The smallest change in analog value that an A/D can measure is called its *resolution,* the weight of its least significant bit.

6. By repeatedly sampling the incoming analog signal, converting it to digital, and storing the digital values in a memory device, an analog waveform can be captured. To reconstruct the signal, the digital values are read from the memory device at the same rate at which they were stored, and then they are fed into a D/A converter. The output of the D/A is then filtered to smooth the stair steps and re-create the original waveform. The bandwidth of sampled signals is limited to $\frac{1}{2} F_S$. Incoming frequencies greater than $\frac{1}{2} F_S$ create an *alias* that has a frequency equal to the difference between the nearest integer multiple of F_S and the incoming frequency. This difference will always be less than $\frac{1}{2} F_S$.

7. A digital-ramp A/D is the simplest to understand but it is not often used due to its variable conversion time. A successive-approximation converter has a constant conversion time and is probably the most common general-purpose converter.

8. Flash converters use analog comparators and a priority encoder to assign a digital value to the analog input. These are the fastest converters because the only delays involved are propagation delays.

9. Other popular methods of A/D include up/down tracking, integrating, voltage-to-frequency conversion, and sigma/delta conversion. Each type of converter has its own niche of applications.

10. Any D/A converter can be used with other circuitry such as analog multiplexers that select one of several analog signals to be converted, one at

a time. Sample-and-hold circuits can be used to "freeze" a rapidly changing analog signal while the conversion is taking place.

11. Digital signal processing is an exciting new growth field in electronics. These devices allow calculations to be performed quickly in order to emulate the operation of many analog filter circuits digitally. The primary architectural feature of a DSP is a hardware multiplier and adder circuit that can multiply pairs of numbers together and accumulate the running total (sum) of these products. This circuitry is used to perform efficiently the weighted moving average calculations that are used to implement digital filters and other DSP functions. DSP is responsible for many of the recent advances in high-fidelity audio, high-definition TV, and telecommunications.

IMPORTANT TERMS

digital quantity	digital-ramp ADC	voltage-to-frequency
analog quantity	quantization error	ADC
transducer	sampling	sigma/delta
digital-to-analog	sampling frequency,	modulation
converter (DAC)	F_S	sample-and-hold
full-scale output	alias	(S/H) circuit
resolution	undersampling	acquisition time
step size	successive-	digital signal
staircase	approximation	processing (DSP)
full-scale error	ADC	weighted average
linearity error	differential inputs	MAC
offset error	WRITE	arithmetic logic unit
settling time	flash ADC	barrel shifter
monotonicity	up/down digital-ramp	oversampling
digitization	ADC	interpolation
analog-to-digital	dual-slope ADC	filtering
converter (ADC)		

PROBLEMS

SECTIONS 11-1 AND 11-2

B   **11-1. DRILL QUESTION**

    (a) What is the expression relating the output and inputs of a DAC?

    (b) Define *step size* of a DAC.

    (c) Define *resolution* of a DAC.

    (d) Define *full scale.*

    (e) Define *percentage resolution.*

    (f) *True or false:* A 10-bit DAC will have a smaller resolution than a 12-bit DAC for the same full-scale output.

    (g) *True or false:* A 10-bit DAC with full-scale output of 10 V has a smaller percentage resolution than a 10-bit DAC with 12 V full scale.

B   **11-2.** An eight-bit DAC produces an output voltage of 2.0 V for an input code of 01100100. What will the value of V_{OUT} be for an input code of 10110011?

B   11-3. Determine the weight of each input bit for the DAC of Problem 11-2.

B   11-4. What is the resolution of the DAC of Problem 11-2? Express it in volts and as a percentage.

B   11-5. What is the resolution in volts of a 10-bit DAC whose F.S. output is 5 V?

B   11-6. How many bits are required for a DAC so that its F.S. output is 10 mA and its resolution is less than 40 μA?

B   11-7. What is the percentage resolution of the DAC of Figure 11-32? What is the step size if the top step is 2 V?

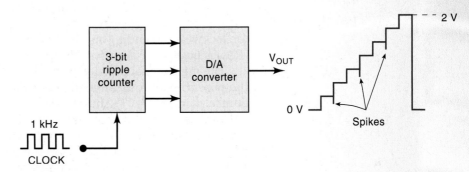

FIGURE 11-32   Problem 11-8.

C   11-8. What is the cause of the negative-going spikes on the V_{OUT} waveform of Figure 11-32? (*Hint:* Note that the counter is a ripple counter and that the spikes occur on every other step.)

B   11-9. Assuming a 12-bit DAC with perfect accuracy, how close to 250 rpm can the motor speed be adjusted in Figure 11-4?

11-10. A 12-bit (three-digit) DAC that uses the BCD input code has a full-scale output of 9.99 V. Determine the step size, the percentage resolution, and the value of V_{OUT} for an input code of 011010010101.

11-11. Compare the step size and the percentage resolution of a DAC with an eight-bit binary input with those of a DAC having an eight-bit BCD input. Assume 990 mV full scale for each DAC.

11-12. Determine the weight of each bit for the BCD DAC of Problem 11-10. What do you think the output would be if the input were 100000100011?

SECTION 11-3

D   11-13. The step size of the DAC of Figure 11-6 can be changed by changing the value of R_F. Determine the required value of R_F for a step size of 0.5 V. Will the new value of R_F change the percentage resolution?

D   11-14. Assume that the output of the DAC in Figure 11-8(a) is connected to the op-amp of Figure 11-8(b).

(a) With $V_{REF} = 5$ V, $R = 20$ kΩ, and $R_F = 10$ kΩ, determine the step size and the full-scale voltage at V_{OUT}.

(b) Change the value of R_F so that the full-scale voltage at V_{OUT} is -2 V.

(c) Use this new value of R_F, and determine the proportionality factor, K, in the relationship $V_{OUT} = K(V_{REF} \times B)$.

11-15. What is the advantage of the DAC of Figure 11-9 over that of Figure 11-8, especially for a larger number of input bits?

SECTIONS 11-4 TO 11-6

11-16. An eight-bit DAC has a full-scale error of 0.2% F.S. If the DAC has a full-scale output of 10 mA, what is the most that it can be in error for any digital input? If the D/A output reads 50 μA for a digital input of 00000001, is this within the specified range of accuracy? (Assume no offset error.)

C, N   11-17. The control of a positioning device may be achieved using a *servomotor*, which is a motor designed to drive a mechanical device as long as an error signal exists. Figure 11-33 shows a simple servo-controlled system that is controlled by a digital input that could be coming directly from a computer or from an output medium such as magnetic tape. The lever arm is moved vertically by the servomotor. The motor rotates clockwise or counterclockwise, depending on whether the voltage from the power amplifier (P.A.) is positive or negative. The motor stops when the P.A. output is 0.

The mechanical position of the lever is converted to a dc voltage by the potentiometer arrangement shown. When the lever is at its 0 reference point, $V_P = 0$ V. The value of V_P increases at the rate of 1 V/inch until the lever is at its highest point (10 inches) and $V_P = 10$ V. The desired position of the lever is provided as a digital code from the computer and is then fed to a DAC, producing V_A. The *difference* between V_P and V_A (called *error*) is produced by the *differential* amplifier and is amplified by the P.A. to drive the motor in the direction that causes the error signal to decrease to 0—that is, moves the lever until $V_P = V_A$.

(a) If the lever must be positioned within a resolution of 0.1 inch, what is the number of bits needed in the digital input code?

(b) In actual operation, the lever arm might oscillate slightly around the desired position, especially if a *wire-wound* potentiometer is used. Can you explain why?

FIGURE 11-33   Problem 11-17.

B 11-18. **DRILL QUESTION**

(a) Define *binary-weighted resistor network.*

(b) Define *R/2R ladder network.*

(c) Define *DAC settling time.*

(d) Define *full-scale error.*

(e) Define *offset error.*

11-19. A particular six-bit DAC has a full-scale output rated at 1.260 V. Its accuracy is specified as ±0.1% F.S., and it has an offset error of ±1 mV. Assume that the offset error has not been zeroed out. Consider the measurements made on this DAC (Table 11-10), and determine which of them are not within the device's specifications. (*Hint:* The offset error is added to the error caused by component inaccuracies.)

TABLE 11-10

Input Code	Output
000010	41.5 mV
000111	140.2 mV
001100	242.5 mV
111111	1.258 V

SECTION 11-7

T 11-20. A certain DAC has the following specifications: eight-bit resolution, full scale = 2.55 V, offset ≤ 2 mV; accuracy = ±0.1% F.S. A static test on this DAC produces the results shown in Table 11-11. What is the probable cause of the malfunction?

TABLE 11-11

Input Code	Output
00000000	8 mV
00000001	18.2 mV
00000010	28.5 mV
00000100	48.3 mV
00001111	158.3 mV
10000000	1.289 V

T 11-21. Repeat Problem 11-20 using the measured data given in Table 11-12.

TABLE 11-12

Input Code	Output
00000000	20.5 mV
00000001	30.5 mV
00000010	20.5 mV
00000100	60.6 mV
00001111	150.6 mV
10000000	1.300 V

T **11-22.** A technician connects a counter to the DAC of Figure 11-3 to perform a staircase test using a 1-kHz clock. The result is shown in Figure 11-34. What is the probable cause of the incorrect staircase signal?

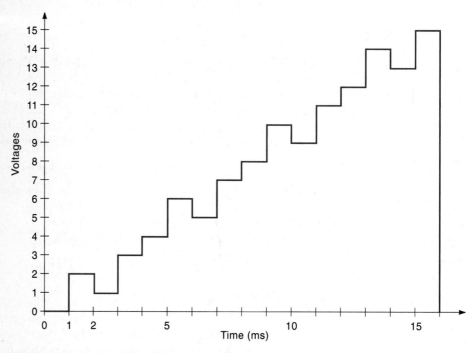

FIGURE 11-34 Problem 11-22.

SECTIONS 11-8 AND 11-9

11-23. DRILL QUESTION

Fill in the blanks in the following description of the ADC of Figure 11-13. Each blank may be one or more words.

A START pulse is applied to _____ the counter and to keep _____ from passing through the AND gate into the _____. At this point, the DAC output, V_{AX}, is _____ and $\overline{EOC}$ is _____.

When START returns _____, the AND gate is _____, and the counter is allowed to _____. The V_{AX} signal is increased one _____ at a time until it _____ V_A. At that point, _____ goes LOW to _____ further pulses from _____. This signals the end of conversion, and the digital equivalent of V_A is present at the _____.

B **11-24.** An eight-bit digital-ramp ADC with a 40-mV resolution uses a clock frequency of 2.5 MHz and a comparator with $V_T = 1$ mV. Determine the following values.

(a) The digital output for $V_A = 6.000$ V

(b) The digital output for 6.035 V

(c) The maximum and average conversion times for this ADC

B **11-25.** Why were the digital outputs the same for parts (a) and (b) of Problem 11-24?

D **11-26.** What would happen in the ADC of Problem 11-24 if an analog voltage of $V_A = 10.853$ V were applied to the input? What waveform would appear at the D/A output? Incorporate the necessary logic in this ADC so that an "overscale" indication will be generated whenever V_A is too large.

B   11-27. An ADC has the following characteristics: resolution, 12 bits; full-scale error, 0.03% F.S.; full scale output, +5 V.

(a) What is the quantization error in volts?

(b) What is the total possible error in volts?

C, N   11-28. The quantization error of an ADC such as the one in Figure 11-13 is always positive because the V_{AX} value must exceed V_A in order for the comparator output to switch states. This means that the value of V_{AX} could be as much as 1 LSB greater than V_A. This quantization error can be modified so that V_{AX} would be within $\pm\frac{1}{2}$ LSB of V_A. This can be done by adding a fixed voltage equal to $\frac{1}{2}$ LSB ($\frac{1}{2}$ step) to the value of V_A. Figure 11-35 shows this symbolically for a converter that has a resolution of 10 mV/step. A fixed voltage of +5 mV is added to the D/A output in the summing amplifier, and the result, V_{AY}, is fed to the comparator, which has $V_T = 1$ mV.

For this modified converter, determine the digital output for the following V_A values.

(a) $V_A = 5.022$ V

(b) $V_A = 50.28$ V

Determine the quantization error in each case by comparing V_{AX} and V_A. Note that the error is positive in one case and negative in the other.

FIGURE 11-35   Problems 11-28 and 11-29.

C   11-29. For the ADC of Figure 11-35, determine the range of analog input values that will produce a digital output of 0100011100.

SECTION 11-10

N   **11-30.** Assume that the analog signal in Figure 11-36(a) is to be digitized by performing continuous A/D conversions using an eight-bit digital-ramp converter whose staircase rises at the rate of 1 V every 25 μs. Sketch the reconstructed signal using the data obtained during the digitizing process. Compare it with the original signal, and discuss what could be done to make it a more accurate representation.

FIGURE 11-36   Problems 11-30, 11-31, and 11-41.

C   **11-31.** On the sine wave of Figure 11-36(b), mark the points where samples are taken by a flash A/D converter at intervals of 75 μs (starting at the origin). Then draw the reconstructed output from the D/A converter (connect the sample points with straight lines to show filtering). Calculate the sample frequency, the sine input frequency, and the difference between them. Then compare to the resulting reconstructed waveform frequency.

     11-32. A sampled data acquisition system is being used to digitize an audio signal. Assume the sample frequency F_S is 20 kHz. Determine the output frequency that will be heard for each of the following input frequencies.

       (a) Input signal = 5 kHz

       (b) Input signal = 10.1 kHz

       (c) Input signal = 10.2 kHz

       (d) Input signal = 15 kHz

       (e) Input signal = 19.1 kHz

       (f) Input signal = 19.2 kHz

SECTION 11-11

B   **11-33. DRILL QUESTION**

     Indicate whether each of the following statements refers to the digital-ramp ADC, the successive-approximation ADC, or both.

(a) Produces a staircase signal at its DAC output

(b) Has a constant conversion time independent of V_A

(c) Has a shorter average conversion time

(d) Uses an analog comparator

(e) Uses a DAC

(f) Uses a counter

(g) Has complex control logic

(h) Has an $\overline{EOC}$ output

11-34. Draw the waveform for V_{AX} as the SAC of Figure 11-19 converts $V_A = 6.7$ V.

11-35. Repeat Problem 11-34 for $V_A = 16$ V.

B 11-36. A certain eight-bit successive-approximation converter has 2.55 V full scale. The conversion time for $V_A = 1$ V is 80 μs. What will be the conversion time for $V_A = 1.5$ V?

11-37. Figure 11-37 shows the waveform at V_{AX} for a six-bit SAC with a step size of 40 mV during a complete conversion cycle. Examine this waveform and describe what is occurring at times t_0 to t_5. Then determine the resultant digital output.

FIGURE 11-37
Problem 11-37.

B 11-38. Refer to Figure 11-21. What is the approximate value of the analog input if the microcomputer's data bus is at 10010111 when $\overline{RD}$ is pulsed LOW?

D 11-39. Connect a 2.0-V reference source to $V_{ref}/2$, and repeat Problem 11-38.

C, D 11-40. Design the ADC interface to a digital thermostat using an LM34 temperature sensor and the ADC0804. Your system must measure accurately ($\pm 0.2°$F) from 50 to 101°F. The LM34 puts out 0.01 V per degree F ($0°$F = 0 V).

(a) What should the digital value for 50°F be for the best resolution?

(b) What voltage must be applied to $V_{in}(-)$?

(c) What is the full-scale range of voltage that will come in?

(d) What voltage must be applied to $V_{ref}/2$?

(e) What binary value will represent 72°F?

(f) What is the resolution in °F? In volts?

SECTION 11-12

B   11-41. Discuss how a flash ADC with a conversion time of 1 μs would work for the situation of Problem 11-30.

D   11-42. Draw the circuit diagram for a four-bit flash converter with BCD output and a resolution of 0.1 V. Assume that a +5 V precision supply voltage is available.

DRILL QUESTION

B   11-43. For each of the following statements, indicate which type of ADC—digital-ramp, SAC, or flash—is being described.
(a) Fastest method of conversion
(b) Needs a START pulse
(c) Requires the most circuitry
(d) Does not use a DAC
(e) Generates a staircase signal
(f) Uses an analog comparator
(g) Has a relatively fixed conversion time independent of V_A

SECTION 11-13

B   11-44. **DRILL QUESTION**
For each statement, indicate what type(s) of ADC is (are) being described.
(a) Uses a counter that is never reset to 0
(b) Uses a large number of comparators
(c) Uses a VCO
(d) Is used in noisy industrial environments
(e) Uses a capacitor
(f) Is relatively insensitive to temperature

SECTION 11-14

11-45. A voltage V_A = 3.853 V is applied to the DVM of Figure 11-24. Assume that the counters are initially reset to 0, and sketch the waveforms at V_{AX}, COMP, Q_1, Q_2, and the AND gate output for *two* complete cycles. Use a 100-kHz clock, and assume that both one-shots have t_p = 1 μs.

T   11-46. Refer to the DVM circuit in Figure 11-24. Assume that the circuit works correctly for V_A values up to about 7.99 V. However, when V_A is increased above 8 V, the displays do not change from their previous reading, the staircase waveform at V_{AX} steps up to only about 7.99 V before jumping back down to 0, and COMP remains HIGH. What are some possible causes for this malfunction?

D   11-47. Add the necessary logic to the DVM of Figure 11-24 to turn on an overscale LED whenever the analog input voltage exceeds 9.99 V. The LED should be turned off at the start of each new conversion.

SECTIONS 11-15 AND 11-16

T   11-48. Refer to the sample-and-hold circuit of Figure 11-26. What circuit fault would result in V_{OUT} looking exactly like V_A? What fault would cause V_{OUT} to be stuck at 0?

C, D 11-49. Use the CMOS 4016 IC (Section 8-16) to implement the switching in Figure 11-27, and design the necessary control logic so that each analog input is converted to its digital equivalent in sequence. The ADC is a 10-bit, successive-approximation type using a 50-kHz clock signal, and it requires a 10-μs-duration start pulse to begin each conversion. The digital outputs are to remain stable for 100 μs after the conversion is complete before switching to the next analog input. Choose an appropriate multiplexing clock frequency.

MICROCOMPUTER APPLICATION

C, N, D 11-50. Figure 11-21 shows how the ADC0804 is interfaced to a microcomputer. It shows three control signals, $\overline{CS}$, $\overline{RD}$, and $\overline{WR}$, that come from the microcomputer to the ADC. These signals are used to start each new A/D conversion and to read (transfer) the ADC data output into the microcomputer over the data bus.

Figure 11-38 shows one way the address decoding logic could be implemented. The $\overline{CS}$ signal that activates the ADC0804 is developed from the eight high-order address lines of the MPU address bus. Whenever the MPU wants to communicate with the ADC0804, it places the address of the ADC0804 onto the address bus, and the decoding logic drives the $\overline{CS}$ signal LOW. Notice that in addition to the address lines, a timing and control signal (ALE) is connected to the $\overline{E_2}$ enable input. Whenever ALE is HIGH, it means that the address is potentially in transition, so the decoder should be disabled until ALE goes LOW (at which time the address will be valid and stable). This serves a purpose for timing but has no effect on the address of the ADC.

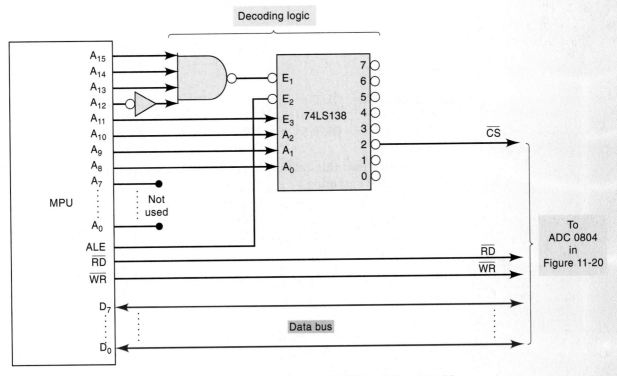

FIGURE 11-38 Problem 11-50: MPU interfaced to the ADC0804 of Figure 11-20.

(a) Determine the address of the ADC0804.

(b) Modify the diagram of Figure 11-38 to place the ADC0804 at address E8XX hex.

(c) Modify the diagram of Figure 11-38 to place the ADC0804 at address FFXX hex.

D   **11-51.** You have available a 10-bit SAC A/D converter (AD 573), but your system requires only eight bits of resolution and you have only eight port bits available on your microprocessor. Can you use this A/D converter, and if so, which of the 10 data lines will you attach to the port?

SECTION 11-18

11-52. The data in Table 11-13 are input samples taken by an A/D converter. Notice that if the input data were plotted, it would represent a simple step function like the rising edge of a digital signal. Calculate the simple average of the four most recent data points, starting with OUT[4] and proceeding through OUT[10]. Plot the values for IN and OUT against the sample number n as shown in Figure 11-39.

TABLE 11-13

Sample n	1	2	3	4	5	6	7	8	9	10
IN[n] (V)	0	0	0	0	10	10	10	10	10	10
OUT[n] (V)	0	0	0							

FIGURE 11-39   Graph format for Problems 11-52 and 11-53.

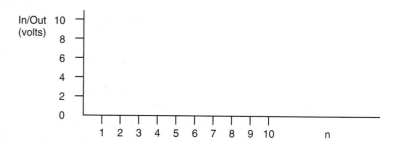

Sample calculations: $\text{OUT}[n] = (\text{IN}[n\text{-}3] + \text{IN}[n\text{-}2] + \text{IN}[n\text{-}1] + \text{IN}[n])/4 = 0$
$\text{OUT}[4] = (\text{IN}[1] \quad + \text{IN}[2] \quad + \text{IN}[3] \quad + \text{IN}[4])/4 = 0$
$\text{OUT}[5] = (\text{IN}[2] \quad + \text{IN}[3] \quad + \text{IN}[4] \quad + \text{IN}[5])/4 = 2.5$

(Notice that this calculation is equivalent to multiplying each sample by $\frac{1}{4}$ and summing.)

11-53. Repeat the previous problem using a weighted average of the last four samples. The weights in this case are placing greater emphasis on recent samples and less emphasis on older samples. Use the weights 0.1, 0.2, 0.3, and 0.4.

$\text{OUT}[n] = 0.1(\text{IN}[n\text{-}3]) + 0.2(\text{IN}[n\text{-}2]) + 0.3(\text{IN}[n\text{-}1]) + 0.4(\text{IN}[n])$
$\text{OUT}[5] = 0.1(\text{IN}[2]) \quad + 0.2(\text{IN}[3]) \quad + 0.3(\text{IN}[4]) \quad + 0.4(\text{IN}[5]) = 4$

11-54. What does the term MAC stand for?

11-55. **DRILL QUESTIONS**

True or false:

(a) A digital signal is a continuously changing voltage.

(b) A digital signal is a sequence of numbers that represent an analog signal.

When processing an analog signal, the output may be distorted due to:

(c) Quantization error when converting analog to digital

(d) Not sampling the original signal frequently enough

(e) Temperature variation in the processor components

(f) The high-frequency components associated with sudden changes in voltage out of the DAC

(g) Electrical noise on the power supply

(h) Alias signals introduced by the digital system

ANSWERS TO SECTION REVIEW QUESTIONS

SECTION 11-1

1. Converts a nonelectrical physical quantity to an electrical quantity
2. Converts an analog voltage or current to a digital representation   3. Stores it; performs calculation or some other operation on it   4. Converts digital data to their analog representation   5. Controls a physical variable according to an electrical input signal

SECTION 11-2

1. 40 μA; 10.2 mA   2. 5.12 mA   3. 0.39 percent   4. 4096   5. 12   6. 8.73 V
7. True   8. It produces a greater number of possible analog outputs between 0 and full scale.

SECTION 11-3

1. It uses only two different sizes of resistors.   2. 640 kΩ 3. 0.5 V
4. Increases by 20 percent

SECTION 11-4

1. Maximum deviation of DAC output from its ideal value, expressed as percentage of full scale   2. Time it takes DAC output to settle to within $\frac{1}{2}$ step size of its full-scale value when the digital input changes from 0 to full scale   3. Offset error adds a small constant positive or negative value to the expected analog output for any digital input.   4. Because of the response time of the op-amp current-to-voltage converter

SECTION 11-8

1. Tells control logic when the DAC output exceeds the analog input   2. At outputs of register   3. Tells us when conversion is complete and digital equivalent of V_A is at register outputs

SECTION 11-9

1. The digital input to a DAC is incremented until the DAC staircase output exceeds the analog input.   2. The built-in error caused by the fact that V_{AX} does not continuously increase but goes up in steps equal to the DAC's resolution. The final V_{AX} can be different from V_A by as much as one step size.   3. If V_A increases, it will take more steps before V_{AX} can reach the step that first exceeds V_A.   4. True

5. Simple circuit; relatively long conversion time that changes with V_A

6. $0010000111_2 = 135_{10}$ for both cases

SECTION 11-10

1. Process of converting different points on an analog signal to digital and storing the digital data for later use   2. Computer generates START signal to begin an A/D conversion of the analog signal. When EOC goes LOW, it signals the computer that the conversion is complete. The computer then loads the ADC output into memory. The process is repeated for the next point on the analog signal.   3. Twice the highest frequency in the input signal   4. An alias frequency will be present in the output.

SECTION 11-11

1. The SAC has a shorter conversion time that doesn't change with V_A.   2. It has more complex control logic.   3. False   4. (a) 8 (b) 0–5 V (c) $\overline{CS}$ controls the effect of the $\overline{RD}$ and $\overline{WR}$ signals; $\overline{WR}$ is used to start a new conversion; $\overline{RD}$ enables the output buffers. (d) When LOW, it signals the end of a conversion. (e) It separates the usually noisy digital ground from the analog ground so as not to contaminate the analog input signal. (f) All analog voltages on $V_{in}(+)$ are measured with reference to this pin. This allows the input range to be offset from ground.

SECTION 11-12

1. True   2. 4095 comparators and 4096 resistors   3. Major advantage is its conversion speed; disadvantage is the number of required circuit components for a practical resolution.

SECTION 11-13

1. It reduces the conversion time by using an up/down counter that allows V_{AX} to track V_A without starting from 0 for each conversion.   2. A VCO
3. Advantages: low cost, temperature immunity; disadvantage: slow conversion time
4. Flash ADC, voltage-to-frequency ADC, and dual-slope ADC   5. One

SECTION 11-14

1. They hold for display the digital values present in the counters at the end of the last conversion.   2. The PGT of OS1 transfers the result of the conversion to the registers; the pulse at OS2 resets the counters to begin a new conversion cycle.
3. Shorter and constant conversion time

SECTION 11-15

1. It takes a sample of an analog voltage signal and stores it on a capacitor.   2. False; they are unity-gain buffers with high input impedance and low output impedance.

SECTION 11-16

1. Uses a single ADC   2. It would become a MOD-8 counter.

SECTION 11-17

1. Displayed waveform can be moved and manipulated; can display portions of the waveform occurring prior to the trigger; indefinite storage time   2. The ADC digitizes the points on the input waveform for storage in memory; the vertical DAC converts the stored data points back to analog voltages to produce the vertical deflection of the electron beam; the horizontal DAC produces a staircase sweep voltage that provides the horizontal deflection of the electron beam.

SECTION 11-18

1. Filtering analog signals   2. An A/D converter   3. To change their dynamic response, you simply change the numbers in the software program, not the hardware components.   4. The Multiply and Accumulate (MAC) unit   5. 3; 7

CHAPTER 12

MEMORY DEVICES

■ OUTLINE

■ OBJECTIVES

Upon completion of this chapter, you will be able to:

- Understand and correctly use the terminology associated with memory systems.
- Describe the difference between read/write memory and read-only memory.
- Discuss the difference between volatile and nonvolatile memory.
- Determine the capacity of a memory device from its inputs and outputs.
- Outline the steps that occur when the CPU reads from or writes to memory.
- Distinguish among the various types of ROMs and cite some common applications.
- Understand and describe the organization and operation of static and dynamic RAMs.
- Compare the relative advantages and disadvantages of EPROM, EEPROM, and flash memory.
- Combine memory ICs to form memory modules with larger word size and/or capacity.
- Use the test results on a RAM or ROM system to determine possible faults in the memory system.

■ INTRODUCTION

A major advantage of digital over analog systems is the ability to store easily large quantities of digital information and data for short or long periods. This memory capability is what makes digital systems so versatile and adaptable to many situations. For example, in a digital computer, the internal main memory stores instructions that tell the computer what to do under *all* possible circumstances so that the computer will do its job with a minimum amount of human intervention.

This chapter is devoted to a study of the most commonly used types of memory devices and systems. We have already become very familiar with the flip-flop, which is an electronic memory device. We have also seen how groups of FFs called *registers* can be used to store information and how this information can be transferred to other locations. FF registers are high-speed memory elements that are used extensively in the internal operations of a digital computer, where digital information is continually being moved from one location to another. Advances in LSI and VLSI technology have made it possible to obtain large numbers of FFs on a single chip arranged in various memory-array formats. These bipolar and MOS semiconductor memories are the fastest memory devices available, and their cost has been continuously decreasing as LSI technology improves.

Digital data can also be stored as charges on capacitors, and a very important type of semiconductor memory uses this principle to obtain high-density storage at low power-requirement levels.

Semiconductor memories are used as the **main memory** of a computer (Figure 12-1), where fast operation is important. A computer's main memory—also called its *working memory*—is in constant communication with the central processing unit (CPU) as a program of instructions is being executed. A program and any data used by the program reside in the main memory while the computer is working on that program. RAM and ROM (to be defined shortly) make up main memory.

Another form of storage in a computer is performed by **auxiliary memory** (Figure 12-1), which is separate from the main working memory. Auxiliary memory—also called *mass storage*—has the capacity to store massive amounts of data without the need for electrical power. Auxiliary memory operates at a much slower speed than main memory, and it stores programs and data that are not currently being used by the CPU. This information is transferred to the main memory when the computer needs it. Common auxiliary memory devices are magnetic disk and compact disk (CD).

We will take a detailed look at the characteristics of the most common memory devices used as the internal memory of a computer. First, we define some of the common terms used in memory systems.

FIGURE 12-1 A computer system normally uses high-speed main memory and slower external auxiliary memory.

12-1 MEMORY TERMINOLOGY

The study of memory devices and systems is filled with terminology that can sometimes be overwhelming to a student. Before we get into any comprehensive discussion of memories, it would be helpful if you had the meaning of some of the more basic terms under your belt. Other new terms will be defined as they appear in the chapter.

- **Memory Cell.** A device or an electrical circuit used to store a single bit (0 or 1). Examples of memory cells include a flip-flop, a charged capacitor, and a single spot on magnetic tape or disk.

- **Memory Word.** A group of bits (cells) in a memory that represents instructions or data of some type. For example, a register consisting of eight FFs can be considered to be a memory that is storing an eight-bit word. Word sizes in modern computers typically range from 8 to 64 bits, depending on the size of the computer.

- **Byte.** A special term used for a group of eight bits. A byte always consists of eight bits. Word sizes can be expressed in bytes as well as in bits. For example, a word size of eight bits is also a word size of one byte, a word size of 16 bits is two bytes, and so on.

- **Capacity.** A way of specifying how many bits can be stored in a particular memory device or complete memory system. To illustrate, suppose that we have a memory that can store 4096 20-bit words. This represents a total capacity of 81,920 bits. We could also express this memory's capacity as 4096 $\times$ 20. When expressed this way, the first number (4096) is the number of words, and the second number (20) is the number of bits per word (word size). The number of words in a memory is often a multiple of 1024. It is common to use the designation "1K" to represent $1024 = 2^{10}$ when referring to memory capacity. Thus, a memory that has a storage capacity of 4K $\times$ 20 is actually a 4096 $\times$ 20 memory. The development of larger memories has brought about the designation "1M" or "1 meg" to represent $2^{20} = 1,048,576$. Thus, a memory that has a capacity of 2M $\times$ 8 is actually one with a capacity of 2,097,152 $\times$ 8. The designation "giga" refers to $2^{30} = 1,073,741,824$.

EXAMPLE 12-1A

A certain semiconductor memory chip is specified as 2K $\times$ 8. How many words can be stored on this chip? What is the word size? How many total bits can this chip store?

Solution

$$2K = 2 \times 1024 = 2048 \text{ words}$$

Each word is eight bits (one byte). The total number of bits is therefore

$$2048 \times 8 = 16,384 \text{ bits}$$

EXAMPLE 12-1B

Which memory stores the most bits: a 5M × 8 memory or a memory that stores 1M words at a word size of 16 bits?

Solution

$$5M \times 8 = 5 \times 1{,}048{,}576 \times 8 = 41{,}943{,}040 \text{ bits}$$
$$1M \times 16 = 1{,}048{,}576 \times 16 = 16{,}777{,}216 \text{ bits}$$

The 5M × 8 memory stores more bits.

■ **Density.** Another term for *capacity*. When we say that one memory device has a greater density than another, we mean that it can store more bits in the same amount of space. It is more dense.

■ **Address.** A number that identifies the location of a word in memory. Each word stored in a memory device or system has a unique address. Addresses always exist in a digital system as a binary number, although octal, hexadecimal, and decimal numbers are often used to represent the address for convenience. Figure 12-2 illustrates a small memory consisting of eight words. Each of these eight words has a specific address represented as a three-bit number ranging from 000 to 111. Whenever we refer to a specific word location in memory, we use its address code to identify it.

■ **Read Operation.** The operation whereby the binary word stored in a specific memory location (address) is sensed and then transferred to another device. For example, if we want to use word 4 of the memory of Figure 12-2 for some purpose, we must perform a read operation on address 100. The read operation is often called a *fetch* operation because a word is being fetched from memory. We will use both terms interchangeably.

■ **Write Operation.** The operation whereby a new word is placed into a particular memory location. It is also referred to as a *store* operation. Whenever a new word is written into a memory location, it replaces the word that was previously stored there.

■ **Access Time.** A measure of a memory device's operating speed. It is the amount of time required to perform a read operation. More specifically, it is the time between the memory receiving a new address input and the data becoming available at the memory output. The symbol t_{ACC} is used for access time.

FIGURE 12-2 Each word location has a specific binary address.

Addresses	
000	Word 0
001	Word 1
010	Word 2
011	Word 3
100	Word 4
101	Word 5
110	Word 6
111	Word 7

- **Volatile Memory.** Any type of memory that requires the application of electrical power in order to store information. If the electrical power is removed, all information stored in the memory will be lost. Many semiconductor memories are volatile, while all magnetic memories are *nonvolatile*, which means that they can store information without electrical power.

- **Random-Access Memory (RAM).** Memory in which the actual physical location of a memory word has no effect on how long it takes to read from or write into that location. In other words, the access time is the same for any address in memory. Most semiconductor memories are RAMs.

- **Sequential-Access Memory (SAM).** A type of memory in which the access time is not constant but varies depending on the address location. A particular stored word is found by sequencing through all address locations until the desired address is reached. This produces access times that are much longer than those of random-access memories. An example of a sequential-access memory device is a magnetic tape backup. To illustrate the difference between SAM and RAM, consider the situation where you have recorded 60 minutes of songs on an audio tape cassette. When you want to get to a particular song, you have to rewind or fast-forward the tape until you find it. The process is relatively slow, and the amount of time required depends on where on the tape the desired song is recorded. This is SAM because you have to sequence through all intervening information until you find what you are looking for. The RAM counterpart to this would be an audio CD, where you can quickly select any song by punching in the appropriate code, and it takes approximately the same time no matter what song you select. Sequential-access memories are used where the data to be accessed will always come in a long sequence of successive words. Video memory, for example, must output its contents in the same order over and over again to keep the image refreshed on the CRT screen.

- **Read/Write Memory (RWM).** Any memory that can be read from or written into with equal ease.

- **Read-Only Memory (ROM).** A broad class of semiconductor memories designed for applications where the ratio of read operations to write operations is very high. Technically, a ROM can be written into (programmed) only once, and this operation is normally performed at the factory. Thereafter, information can only be read from the memory. Other types of ROM are actually read-mostly memories (RMM), which can be written into more than once; but the write operation is more complicated than the read operation, and it is not performed very often. The various types of ROM will be discussed later. *All ROM is nonvolatile* and will store data when electrical power is removed.

- **Static Memory Devices.** Semiconductor memory devices in which the stored data will remain permanently stored as long as power is applied, without the need for periodically rewriting the data into memory.

- **Dynamic Memory Devices.** Semiconductor memory devices in which the stored data will *not* remain permanently stored, even with power applied, unless the data are periodically rewritten into memory. The latter operation is called a *refresh* operation.

- **Main Memory.** Also referred to as the computer's *working memory*. It stores instructions and data the CPU is currently working on. It is the highest-speed memory in the computer and is always a semiconductor memory.

- **Auxiliary Memory.** Also referred to as *mass storage* because it stores massive amounts of information external to the main memory. It is slower in speed than main memory and is always nonvolatile. Magnetic disks and CDs are common auxiliary memory devices.

1. Define the following terms.
   (a) *Memory cell*
   (b) *Memory word*
   (c) *Address*
   (d) *Byte*
   (e) *Access time*
2. A certain memory has a capacity of 8K × 16. How many bits are in each word? How many words are being stored? How many memory cells does this memory contain?
3. Explain the difference between the read (fetch) and write (store) operations.
4. *True or false:* A volatile memory will lose its stored data when electrical power is interrupted.
5. Explain the difference between SAM and RAM.
6. Explain the difference between RWM and ROM.
7. *True or false:* A dynamic memory will hold its data as long as electrical power is applied.

12-2 GENERAL MEMORY OPERATION

Although each type of memory is different in its internal operation, certain basic operating principles are the same for all memory systems. An understanding of these basic ideas will help in our study of individual memory devices.

Every memory system requires several different types of input and output lines to perform the following functions:

1. Select the address in memory that is being accessed for a read or write operation.
2. Select either a read or a write operation to be performed.
3. Supply the input data to be stored in memory during a write operation.
4. Hold the output data coming from memory during a read operation.
5. Enable (or disable) the memory so that it will (or will not) respond to the address inputs and read/write command.

Figure 12-3(a) illustrates these basic functions in a simplified diagram of a 32 × 4 memory that stores 32 four-bit words. Because the word size is four bits, there are four data input lines I_0 to I_3 and four data output lines O_0 to O_3. During a write operation, the data to be stored into memory must be applied

FIGURE 12-3 (a) Diagram of a 32 × 4 memory; (b) virtual arrangement of memory cells into 32 four-bit words.

to the data input lines. During a read operation, the word being read from memory appears at the data output lines.

Address Inputs

Because this memory stores 32 words, it has 32 different storage locations and therefore 32 different binary addresses ranging from 00000 to 11111 (0 to 31 in decimal). Thus, there are five address inputs, A_0 to A_4. To access one of the memory locations for a read or a write operation, the five-bit address code for that particular location is applied to the address inputs. In general, N address inputs are required for a memory that has a capacity of 2^N words.

We can visualize the memory of Figure 12-3(a) as an arrangement of 32 registers, with each register holding a four-bit word, as illustrated in Figure 12-3(b). Each address location is shown containing four memory cells that hold 1s and 0s that make up the data word stored at that location. For example, the data word 0110 is stored at address 00000, the data word 1001 is stored at address 00001, and so on.

The $R/\overline{W}$ Input

This input controls which memory operation is to take place: read (R) or write (W). The input is labeled $R/\overline{W}$; there is no bar over the R, which indicates that the read operation occurs when $R/\overline{W} = 1$. The bar over the W indicates that the write operation takes place when $R/\overline{W} = 0$. Other labels are often used for this input. Two of the more common ones are $\overline{W}$ (write) and $\overline{WE}$ (write enable). Again, the bar indicates that the write operation occurs when the input is LOW. It is understood that the read operation occurs for a HIGH.

A simplified illustration of the read and write operations is shown in Figure 12-4. Figure 12-4(a) shows the data word 0100 being written into the memory register at address location 00011. This data word would have been applied to the memory's data input lines, and it replaces the data previously stored at address 00011. Figure 12-4(b) shows the data word 1101 being read

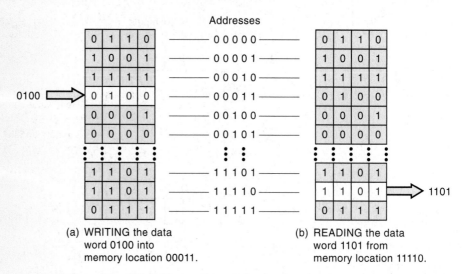

Addresses

(a) WRITING the data
word 0100 into
memory location 00011.

(b) READING the data
word 1101 from
memory location 11110.

FIGURE 12-4   Simplified illustration of the read and write operations on the 32 × 4 memory: (a) writing the data word 0100 into memory location 00011; (b) reading the data word 1101 from memory location 11110.

from address 11110. This data word would appear at the memory's data output lines. After the read operation, the data word 1101 is still stored in address 11110. In other words, the read operation does not change the stored data.

Memory Enable

Many memory systems have some means for completely disabling all or part of the memory so that it will not respond to the other inputs. This is represented in Figure 12-3 as the MEMORY ENABLE input, although it can have different names in the various memory systems, such as chip enable (*CE*) or chip select (*CS*). Here, it is shown as an active-HIGH input that enables the memory to operate normally when it is kept HIGH. A LOW on this input disables the memory so that it will not respond to the address and $R/\overline{W}$ inputs. This type of input is useful when several memory modules are combined to form a larger memory. We will examine this idea later.

EXAMPLE 12-2

Describe the conditions at each input and output when the contents of address location 00100 are to be read.

Solution

Address inputs: 00100

Data inputs: xxxx (not used)

$R/\overline{W}$: HIGH

MEMORY ENABLE: HIGH

Data outputs: 0001

EXAMPLE 12-3

Describe the conditions at each input and output when the data word 1110 is to be written into address location 01101.

Solution

Address inputs: 01101

Data inputs: 1110

$R/\overline{W}$: LOW

MEMORY ENABLE: HIGH

Data outputs: xxxx (not used; usually Hi-Z)

EXAMPLE 12-4

A certain memory has a capacity of 4K × 8.

(a) How many data input and data output lines does it have?

(b) How many address lines does it have?

(c) What is its capacity in bytes?

Solution

(a) Eight of each because the word size is eight.

(b) The memory stores 4K = 4 × 1024 = 4096 words. Thus, there are 4096 memory addresses. Because $4096 = 2^{12}$, it requires a 12-bit address code to specify one of 4096 addresses.

(c) A byte is eight bits. This memory has a capacity of 4096 bytes.

The example memory in Figure 12-3 illustrates the important input and output functions common to most memory systems. Of course, each type of memory may have other input and output lines that are peculiar to that memory. These will be described as we discuss the individual memory types.

REVIEW QUESTIONS

1. How many address inputs, data inputs, and data outputs are required for a 16K × 12 memory?
2. What is the function of the $R/\overline{W}$ input?
3. What is the function of the MEMORY ENABLE input?

12-3 CPU–MEMORY CONNECTIONS

A major part of this chapter is devoted to semiconductor memory, which, as pointed out earlier, makes up the main memory of most modern computers. Remember, this main memory is in constant communication with the central processing unit (CPU). It is not necessary to be familiar with the detailed operation of a CPU at this point, and so the following simplified treatment of the CPU–memory interface will provide the perspective needed to make our study of memory devices more meaningful.

A computer's main memory is made up of RAM and ROM ICs that are interfaced to the CPU over three groups of signal lines or buses. These are shown in Figure 12-5 as the address lines or address bus, the data lines or data bus, and the control lines or control bus. Each of these buses consists of several lines (note that they are represented by a single line with a slash), and the number of lines in each bus will vary from one computer to the next.

FIGURE 12-5 Three groups of lines (buses) connect the main memory ICs to the CPU.

The three buses play a necessary part in allowing the CPU to write data into memory and to read data from memory.

When a computer is executing a program of instructions, the CPU continually fetches (reads) information from those locations in memory that contain (1) the program codes representing the operations to be performed and (2) the data to be operated upon. The CPU will also store (write) data into memory locations as dictated by the program instructions. Whenever the CPU wants to write data to a particular memory location, the following steps must occur:

Write Operation

1. The CPU supplies the binary address of the memory location where the data are to be stored. It places this address on the address bus lines.
2. The CPU places the data to be stored on the data bus lines.
3. The CPU activates the appropriate control signal lines for the memory write operation.
4. The memory ICs decode the binary address to determine which location is being selected for the store operation.
5. The data on the data bus are transferred to the selected memory location.

Whenever the CPU wants to read data from a specific memory location, the following steps must occur:

Read Operation

1. The CPU supplies the binary address of the memory location from which data are to be retrieved. It places this address on the address bus lines.
2. The CPU activates the appropriate control signal lines for the memory read operation.
3. The memory ICs decode the binary address to determine which location is being selected for the read operation.
4. The memory ICs place data from the selected memory location onto the data bus, from which they are transferred to the CPU.

The steps above should make clear the function of each of the system buses:

- **Address Bus.** This *unidirectional* bus carries the binary address outputs from the CPU to the memory ICs to select one memory location.
- **Data Bus.** This *bidirectional* bus carries data between the CPU and the memory ICs.
- **Control Bus.** This bus carries control signals (such as the $R/\overline{W}$ signal) from the CPU to the memory ICs.

As we get into discussions of actual memory ICs, we will examine the signal activity that appears on these buses for the read and write operations.

1. Name the three groups of lines that connect the CPU and the internal memory.
2. Outline the steps that take place when the CPU reads from memory.
3. Outline the steps that occur when the CPU writes to memory.

12-4 READ-ONLY MEMORIES

The read-only memory is a type of semiconductor memory designed to hold data that either are permanent or will not change frequently. During normal operation, no new data can be written into a ROM, but data can be read from ROM. For some ROMs, the data that are stored must be built-in during the manufacturing process; for other ROMs, the data can be entered electrically. The process of entering data is called **programming** or *burning-in* the ROM. Some ROMs cannot have their data changed once they have been programmed; others can be *erased* and reprogrammed as often as desired. We will take a detailed look later at these various types of ROMs. For now, we will assume that the ROMs have been programmed and are holding data.

ROMs are used to store data and information that are not to change during the normal operation of a system. A major use for ROMs is in the storage of programs in microcomputers. Because all ROMs are *nonvolatile,* these programs are not lost when electrical power is turned off. When the microcomputer is turned on, it can immediately begin executing the program stored in ROM. ROMs are also used for program and data storage in microprocessor-controlled equipment such as electronic cash registers, appliances, and security systems.

ROM Block Diagram

A typical block diagram for a ROM is shown in Figure 12-6(a). It has three sets of signals: address inputs, control input(s), and data outputs. From our previous discussions, we can determine that this ROM is storing 16 words because it has $2^4 = 16$ possible addresses, and each word contains eight bits because there are eight data outputs. Thus, this is a 16×8 ROM. Another way to describe this ROM's capacity is to say that it stores 16 bytes of data.

The data outputs of most ROM ICs are tristate outputs, to permit the connection of many ROM chips to the same data bus for memory expansion. The most common numbers of data outputs for ROMs are four, eight, and 16 bits, with eight-bit words being the most common.

∇ = tristate

$\overline{CS}$ (chip select)

Control input

(a)

Word	Address				Data							
	A_3	A_2	A_1	A_0	D_7	D_6	D_5	D_4	D_3	D_2	D_1	D_0
0	0	0	0	0	1	1	0	1	1	1	1	0
1	0	0	0	1	0	0	1	1	1	0	1	0
2	0	0	1	0	1	0	0	0	0	1	0	1
3	0	0	1	1	1	0	1	0	1	1	1	1
4	0	1	0	0	0	0	0	1	1	0	0	1
5	0	1	0	1	0	1	1	1	1	0	1	1
6	0	1	1	0	0	0	0	0	0	0	0	0
7	0	1	1	1	1	1	1	0	1	1	0	1
8	1	0	0	0	0	0	1	1	1	1	0	0
9	1	0	0	1	1	1	1	1	1	1	1	1
10	1	0	1	0	1	0	1	1	1	0	0	0
11	1	0	1	1	1	1	0	0	0	1	1	1
12	1	1	0	0	0	0	1	0	0	1	1	1
13	1	1	0	1	0	1	1	0	1	0	1	0
14	1	1	1	0	1	1	0	1	0	0	1	0
15	1	1	1	1	0	1	0	1	1	0	1	1

(b)

Word	Address $A_3\ A_2\ A_1\ A_0$	Data $D_7 - D_0$
0	0	DE
1	1	3A
2	2	85
3	3	AF
4	4	19
5	5	7B
6	6	00
7	7	ED
8	8	3C
9	9	FF
10	A	B8
11	B	C7
12	C	27
13	D	6A
14	E	D2
15	F	5B

(c)

FIGURE 12-6 (a) Typical ROM block symbol; (b) table showing binary data at each address location; (c) the same table in hex.

The control input $\overline{CS}$ stands for **chip select.** This is essentially an enable input that enables or disables the ROM outputs. Some manufacturers use different labels for the control input, such as *CE* (chip enable) or *OE* (output enable). Many ROMs have two or more control inputs that must be active in order to enable the data outputs so that data can be read from the selected address. In some ROM ICs, one of the control inputs (usually the *CE*) is used to place the ROM in a low-power standby mode when it is not being used. This reduces the current drain from the system power supply.

The $\overline{CS}$ input shown in Figure 12-6(a) is active-LOW; therefore, it must be in the LOW state to enable the ROM data to appear at the data outputs. Notice that there is no $R/\overline{W}$ (read/write) input because the ROM cannot be written into during normal operation.

The Read Operation

Let's assume that the ROM has been programmed with the data shown in the table of Figure 12-6(b). Sixteen different data words are stored at the 16 different address locations. For example, the data word stored at location 0011 is 10101111. Of course, the data are stored in binary inside the ROM, but very often we use hexadecimal notation to show the programmed data efficiently. This is done in Figure 12-6(c).

In order to read a data word from ROM, we need to do two things: (1) apply the appropriate address inputs and then (2) activate the control inputs. For example, if we want to read the data stored at location 0111 of the ROM in Figure 12-6, we must apply $A_3A_2A_1A_0 = 0111$ to the address inputs and then apply a LOW to $\overline{CS}$. The address inputs will be decoded inside the ROM to select the correct data word, 11101101, that will appear at outputs D_7 to D_0. If $\overline{CS}$ is kept HIGH, the ROM outputs will be disabled and will be in the Hi-Z state.

1. *True or false:* All ROMs are nonvolatile.
2. Describe the procedure for reading from ROM.
3. What is *programming* or *burning-in* a ROM?

12-5 ROM ARCHITECTURE

The internal architecture (structure) of a ROM IC is very complex, and we need not be familiar with all of its detail. It is instructive, however, to look at a simplified diagram of the internal architecture, such as that shown in Figure 12-7, for the 16×8 ROM. There are four basic parts: *register array, row decoder, column decoder,* and *output buffers.*

Register Array

The register array stores the data that have been programmed into the ROM. Each register contains several memory cells equal to the word size. In this case, each register stores an eight-bit word. The registers are arranged in a square matrix array that is common to many semiconductor memory chips. We can specify the position of each register as being in a specific row and a specific column. For example, register 0 is in row 0, column 0, and register 9 is in row 1, column 2.

The eight data outputs of each register are connected to an internal data bus that runs through the entire circuit. Each register has two enable inputs (E); both must be HIGH in order for the register's data to be placed on the bus.

Address Decoders

The applied address code $A_3A_2A_1A_0$ determines which register in the array will be enabled to place its eight-bit data word onto the bus. Address bits A_1A_0 are fed to a 1-of-4 decoder that activates one row-select line, and address bits A_3A_2 are fed to a second 1-of-4 decoder that activates one column-select line. Only one register will be in both the row and the column selected by the address inputs, and this one will be enabled.

FIGURE 12-7 Architecture of a 16 × 8 ROM. Each register stores one eight-bit word.

EXAMPLE 12-5

Which register will be enabled by input address 1101?

Solution

$A_3A_2 = 11$ will cause the column decoder to activate the column 3 select line, and $A_1A_0 = 01$ will cause the row decoder to activate the row 1 select line. This will place HIGHs at both enable inputs of register 13, thereby causing its data outputs to be placed on the bus. Note that the other registers in column 3 will have only one enable input activated; the same is true for the other row 1 registers.

EXAMPLE 12-6

What input address will enable register 7?

Solution

The enable inputs of this register are connected to the row 3 and column 1 select lines, respectively. To select row 3, the A_1A_0 inputs must be at 11, and to select column 1, the A_3A_2 inputs must be at 01. Thus, the required address will be $A_3A_2A_1A_0 = 0111$.

Output Buffers

The register that is enabled by the address inputs will place its data on the data bus. These data feed into the output buffers, which will pass the data to the external data outputs, provided that $\overline{CS}$ is LOW. If $\overline{CS}$ is HIGH, the output buffers are in the Hi-Z state, and D_7 through D_0 will be floating.

The architecture shown in Figure 12-7 is similar to that of many IC ROMs. Depending on the number of stored data words, the registers in some ROMs will not be arranged in a square array. For example, the Intel 27C64 is a CMOS ROM that stores 8192 eight-bit words. Its 8192 registers are arranged in an array of 256 rows × 32 registers. ROM capacities range from 256 × 4 to 8M × 8.

EXAMPLE 12-7

Describe the internal architecture of a ROM that stores 4K bytes and uses a square register array.

Solution

4K is actually 4 × 1024 = 4096, and so this ROM holds 4096 eight-bit words. Each word can be thought of as being stored in an eight-bit register, and there are 4096 registers connected to a common data bus internal to the chip. Because $4096 = 64^2$, the registers are arranged in a 64 × 64 array; that is, there are 64 rows and 64 columns. This requires a 1-of-64 decoder to decode six address inputs for the row select, and a second 1-of-64 decoder to decode six other address inputs for the column select. Thus, a total of 12 address inputs is required. This makes sense because $2^{12} = 4096$, and there are 4096 different addresses.

REVIEW QUESTIONS

1. What input address code is required if we want to read the data from register 9 in Figure 12-7?
2. Describe the function of the row-select decoder, the column-select decoder, and the output buffers in the ROM architecture.

12-6   ROM TIMING

There will be a propagation delay between the application of a ROM's inputs and the appearance of the data outputs during a read operation. This time delay, called access time (t_{ACC}) is a measure of the ROM's operating speed. Access time is described graphically by the waveforms in Figure 12-8.

The top waveform represents the address inputs; the middle waveform is an active-LOW chip select, $\overline{CS}$; and the bottom waveform represents the data outputs. At time t_0 the address inputs are all at some specific level, some HIGH and some LOW. $\overline{CS}$ is HIGH, so that the ROM data outputs are in their Hi-Z state (represented by the hatched line).

Just prior to t_1, the address inputs are changing to a new address for a new read operation. At t_1, the new address is valid; that is, each address input is at a valid logic level. At this point, the internal ROM circuitry begins to decode the new address inputs to select the register that is to send its data

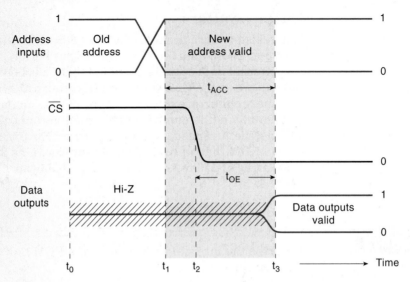

FIGURE 12-8   Typical timing for a ROM read operation.

to the output buffers. At t_2, the $\overline{CS}$ input is activated to enable the output buffers. Finally, at t_3, the outputs change from the Hi-Z state to the valid data that represent the data stored at the specified address.

The time delay between t_1, when the new address becomes valid, and t_3, when the data outputs become valid, is the access time t_{ACC}. Typical bipolar ROMs will have access times in the range from 30 to 90 ns; access times of NMOS devices will range from 35 to 500 ns. Improvements to CMOS technology have brought access times into the 20-to-60-ns range. Consequently, bipolar and NMOS devices are rarely produced in newer (larger) ROMs.

Another important timing parameter is the *output enable time* (t_{OE}), which is the delay between the $\overline{CS}$ input and the valid data output. Typical values for t_{OE} are 10 to 20 ns for bipolar, 25 to 100 ns for NMOS, and 12 to 50 ns for CMOS ROMs. This timing parameter is important in situations where the address inputs are already set to their new values, but the ROM outputs have not yet been enabled. When $\overline{CS}$ goes LOW to enable the outputs, the delay will be t_{OE}.

12-7   TYPES OF ROMs

Now that we have a general understanding of the internal architecture and external operation of ROM devices, we will look at the various types of ROMs to see how they differ in the way they are programmed, erased, and reprogrammed.

Mask-Programmed ROM

The mask-programmed ROM has its storage locations written into (programmed) by the manufacturer according to the customer's specifications. A photographic negative called a *mask* is used to control the electrical interconnections on the chip. A special mask is required for each different set of information to be stored in the ROM. These masks are expensive, so this type of ROM is economical only if you need a very large quantity of the same ROM. Some ROMs of this type are available as off-the-shelf devices preprogrammed with commonly used information or data such as certain mathematical tables and character generator codes for CRT displays. A major disadvantage of this type of ROM is that it cannot be reprogrammed in the event of a design change requiring a modification of the stored data. The ROM would have to be replaced by a new one with the desired data written into it. Several types of user-programmable ROMs have been developed to overcome this disadvantage. Mask-programmed ROMs, however, still represent the most economical approach when a large quantity of identically programmed ROMs are needed.

Mask-programmed ROMs are commonly referred to as just ROMs, but this can be confusing because the term ROM actually represents the broad category of devices that, during normal operation, are only read from. We will use the abbreviation MROM whenever we refer to mask-programmed ROMs.

Figure 12-9 shows the structure of a small MOS MROM. It consists of 16 memory cells arranged in four rows of four cells. Each cell is an N-channel MOSFET transistor connected in the common-drain configuration (input at gate, output at source). The top row of cells (ROW 0) constitutes a four-bit register. Note that some of the transistors in this row (Q_0 and Q_2) have their source connected to the output column line, while others (Q_1 and Q_3) do not.

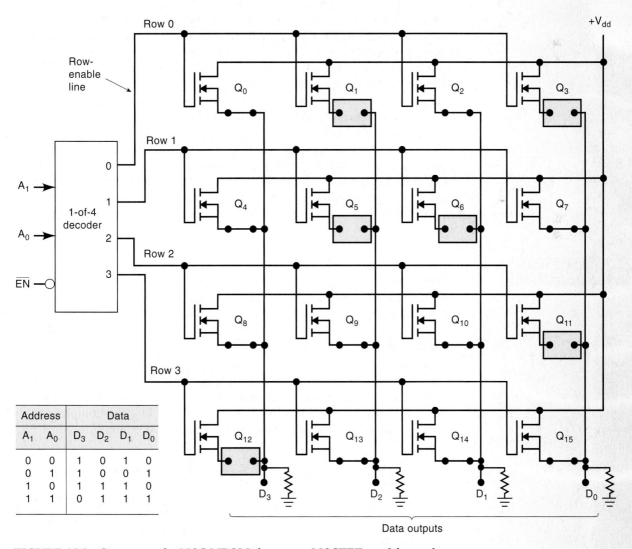

Address		Data			
A_1	A_0	D_3	D_2	D_1	D_0
0	0	1	0	1	0
0	1	1	0	0	1
1	0	1	1	1	0
1	1	0	1	1	1

FIGURE 12-9   Structure of a MOS MROM shows one MOSFET used for each memory cell. An open source connection stores a "0"; a closed source connection stores a "1."

The same is true of the cells in each of the other rows. The presence or absence of these source connections determines whether a cell is storing a 1 or a 0, respectively. The condition of each source connection is controlled during production by the photographic mask based on the customer-supplied data.

Notice that the data outputs are connected to column lines. Referring to output D_3, for instance, any transistor that has a connection from the source (such as Q_0, Q_4, and Q_8) to the output column can switch V_{dd} onto the column, making it a HIGH logic level. If V_{dd} is not connected to the column line, the output will be held at a LOW logic level by the pull-down resistor. At any given time, a maximum of one transistor in a column will ever be turned on due to the row decoder.

The 1-of-4 decoder is used to decode the address inputs A_1A_0 to select which row (register) is to have its data read. The decoder's active-HIGH outputs provide the ROW enable lines that are the gate inputs for the various rows of cells. If the decoder's enable input, $\overline{EN}$, is held HIGH, all of the decoder outputs will be in their inactive LOW state, and all of the transistors in the array will be off because of the absence of any gate voltage. For this situation, the data outputs will all be in the LOW state.

When $\overline{EN}$ is in its active-LOW state, the conditions at the address inputs determine which row (register) will be enabled so that its data can be read at the data outputs. For example, to read ROW 0, the A_1A_0 inputs are set to 00. This places HIGH at the ROW 0 line; all other row lines are at 0 V. This HIGH at ROW 0 turns on transistors Q_0, Q_1, Q_2, and Q_3. With all of the transistors in the row conducting, V_{dd} will be switched on to each transistor's source lead. Outputs D_3 and D_1 will go HIGH because Q_0 and Q_2 are connected to their respective columns. D_2 and D_0 will remain LOW because there is no path from the Q_1 and Q_3 source leads to their columns. In a similar manner, application of the other address codes will produce data outputs from the corresponding register. The table in Figure 12-9 shows the data for each address. You should verify how this correlates with the source connections to the various cells.

EXAMPLE 12-8

MROMs can be used to store tables of mathematical functions. Show how the MROM in Figure 12-9 can be used to store the function $y = x^2 + 3$, where the input address supplies the value for x, and the value of the output data is y.

Solution

The first step is to set up a table showing the desired output for each set of inputs. The input binary number, x, is represented by the address A_1A_0. The output binary number is the desired value of y. For example, when $x = A_1A_0 = 10_2 = 2_{10}$, the output should be $2^2 + 3 = 7_{10} = 0111_2$. The complete table is shown in Table 12-1. This table is supplied to the MROM manufacturer for de-

TABLE 12-1

x		$y = x^2 + 3$			
A_1	A_0	D_3	D_2	D_1	D_0
0	0	0	0	1	1
0	1	0	1	0	0
1	0	0	1	1	1
1	1	1	1	0	0

veloping the mask that will make the appropriate connections within the memory cells during the fabrication process. For instance, the first row in the table indicates that the connections to the source of Q_0 and Q_1 will be left unconnected, while the connections to Q_2 and Q_3 will be made.

MROMs typically have tristate outputs that allow them to be used in a bus system, as we discussed in Chapter 9. Consequently, there must be a control input to enable and disable the tristate outputs. This control input is usually labeled *OE* (for output enable). In order to distinguish this tristate enable input from the address decoder enable input, the latter is usually referred to as a chip enable (*CE*). The chip enable performs more than just enabling the address decoder. When *CE* is disabled, all functions of the chip are disabled, including the tristate outputs, and the entire circuit is placed in a **power-down** mode that draws much less current from the power supply. Figure 12-10 shows a $32K \times 8$ MROM. The 15 address lines (A0–A14) can identify 2^{15} memory locations (32, 767, or 32K). Each memory location holds an eight-bit data value that can be placed on the data lines D7–D0 when the chip is enabled and the outputs are enabled.

FIGURE 12-10 Logic symbol for a $32K \times 8$ MROM.

Programmable ROMs (PROMs)

A mask-programmable ROM is very expensive and would not be used except in high-volume applications, where the cost would be spread out over many units. For lower-volume applications, manufacturers have developed **fusible-link** PROMs that are user-programmable; that is, they are not programmed during the manufacturing process but are custom-programmed by the user. Once programmed, however, a PROM is like an MROM because it cannot be erased and reprogrammed. Thus, if the program in the PROM is faulty or must be changed, the PROM must be thrown away. For this reason, these devices are often referred to as "one-time programmable" (OTP) ROMs.

The fusible-link PROM structure is very similar to the MROM structure because certain connections either are left intact or are opened in order to program a memory cell as a 1 or a 0, respectively. In the MROM of Figure 12-9, these connections were from the MOSFET source lead to the output column. In a PROM, each of these connections is made with a thin fuse link that comes intact from the manufacturer (see Figure 12-11).

FIGURE 12-11 PROMs use fusible links that can be selectively blown open by the user to program a logic 0 into a cell.

The user can selectively *blow* any of these fuse links to produce the desired stored memory data. Typically, a data value is programmed or "burned-into" an address location by: applying the address to the address inputs; placing the desired data at the data pins; and then applying V_{pp}, a high-voltage pulse (10–30 V), to a special programming pin on the IC. Figure 12-11 shows the programming operation. All of the transistors in the selected row (row 0) are turned on, and V_{pp} is applied to their drain leads. Those columns (data lines) that have a logic 0 on them (e.g., Q_1) will provide a high-current path through the fusible link, burning it open and permanently storing a logic 0. Those columns that have a logic 1 (e.g., Q_0) have V_{pp} on one side of the fuse and V_{dd} on the other side, drawing much less current and leaving the fuse intact. Once all address locations have been programmed in this manner, the data are permanently stored in the PROM and can be read over and over again by accessing the appropriate address. The data will not change when power is removed from the PROM chip because nothing will cause an open fuse link to become closed again.

The process of programming a PROM and verifying that the stored data are correct is rarely done manually; rather, it is done automatically by a special apparatus called a PROM **programmer**. Typically, the PROM chip is plugged into a socket on the PROM programmer. The programmer circuitry selects each address of the PROM, burns in the correct data at that address, verifies the data, and sequences to the next address to repeat the process. The data to be burned into the PROM are input to the programmer from a keyboard or from a disk drive, or they are transferred from a computer. The latter operation, called *downloading,* allows the user to develop and test the program/data on a computer and then, when it is finished and working, transfer it from the computer's memory to the PROM programmer, which will "burn it" into the PROM.

Very few bipolar PROMs are still available today. Almost all use MOS technology, with CMOS becoming the most popular. The TMS27PC256 is a very popular CMOS PROM with a capacity of 32K × 8 and a standby power dissipation of only 1.4 mW. It is available with maximum access times ranging from 100 to 250 ns.

Erasable Programmable ROM (EPROM)

An EPROM can be programmed by the user, and it can also be *erased* and reprogrammed as often as desired. Once programmed, the EPROM is a *nonvolatile* memory that will hold its stored data indefinitely. The process for programming an EPROM involves the application of special voltage levels (typically in the 10-to-25-V range) to the appropriate chip inputs for a specified amount of time (typically 50 ms per address location). The programming process is usually performed by a special programming circuit that is separate from the circuit in which the EPROM will eventually be working. The complete programming process can take up to several minutes for one EPROM chip.

The storage cells in an EPROM are MOS transistors with a silicon gate that has no electrical connections (i.e., a *floating gate*). In its normal state, each transistor is off and each cell stores a logic 1. A transistor can be turned on by the application of a high-voltage programming pulse that injects high-energy electrons into the floating-gate region. These electrons remain trapped in this region once the pulse is terminated because there is no discharge path. This keeps the transistor on *permanently* even when power is removed from the device, and the cell is now storing a logic 0. During the programming process, the EPROM's address and data pins are used to select which memory cells will be programmed as 0s and which ones will be left as 1s.

Once an EPROM cell has been programmed, it can be erased by exposing it to ultraviolet (UV) light applied through a window on the chip package. The UV light produces a photocurrent from the floating gate back to the silicon substrate, thereby removing the stored charges, turning the transistor off, and restoring the cell to the logic 1 state. This erasing process typically requires 15 to 20 minutes of exposure to UV rays. Unfortunately, there is no way to erase only selected cells; *the UV light erases all cells at the same time* so that an erased EPROM stores all 1s. Once erased, the EPROM can be reprogrammed.

EPROMs are available in a wide range of capacities and access times; devices with a capacity of 512K × 8 and an access time of 20 ns are commonplace. The 27C64 is an example of a small EPROM. The 27C64 is an 8K × 8 CMOS EPROM that operates from a single +5 V power source during normal operation. Figure 12-12(a) is the logic symbol for the 27C64. Note that it shows 13 address inputs (because $2^{13} = 8192$) and eight data outputs. It has

(b)

Mode	Inputs				Outputs
	$\overline{CE}$	$\overline{OE}$	$\overline{PGM}$	VPP	$D_7 - D_0$
Read	0	0	1	0–5V	DATA$_{out}$
Output Disable	0	1	1	0–5V	High Z
Standby	1	X	X	X	High Z
Program	0	1	0	12.75 V	DATA$_{in}$
PGM Verify	0	0	1	12.75 V	DATA$_{out}$

(c)

FIGURE 12-12 (a) Logic symbol for 27C64 EPROM; (b) typical EPROM package showing ultraviolet window; (c) 27C64 operating modes.

four control inputs. $\overline{CE}$ is the chip enable input that is used to place the device in a standby mode where its power consumption is reduced. $\overline{OE}$ is the output enable and is used to control the device's data output tristate buffers so that the device can be connected to a microprocessor data bus without bus contention. V_{PP} is the special programming voltage required during the programming process. $\overline{PGM}$ is the program enable input that is activated to store data at the selected address.

The 27C64 package in Figure 12-12(b) shows the characteristic "window," which allows the internal circuitry to be exposed to UV light when the complete memory contents are to be erased. A sticker is placed over the window after erasure and reprogramming to protect against accidental erasure from ambient light.

The 27C64 has several operating modes that are controlled by the $\overline{CE}$, $\overline{OE}$, V_{PP}, and PGM pins, as presented in Figure 12-12(c). The program mode is used to write new data into the EPROM cells. This is most often done on a "clean" EPROM, one that has previously been erased with UV light so that all cells are 1s. The programming process writes one eight-bit word into one address location at one time as follows: (1) the address is applied to the address pins; (2) the desired data are placed at the data pins, which function as inputs during the programming process; (3) a higher programming voltage of 12.75 V is applied to V_{PP}; (4) $\overline{CE}$ is held LOW; (5) $\overline{PGM}$ is pulsed LOW for 100 μs and the data are read back. If the data were not successfully stored, another pulse is applied to $\overline{PGM}$. This is repeated at the same address until the data are successfully stored. This process is repeated for all memory locations. If done manually, this can take several hours. Usually, however, it is done automatically with a commercial EPROM programmer much like the PROM programmers described above. A clean EPROM can be programmed in less than a minute once the desired data have been entered, transferred, or downloaded into the EPROM programmer. The 27C512 is a common 64K $\times$ 8 EPROM that operates very much like the 27C64 but offers more storage capacity.

EPROMs were originally intended for use in research and development applications where the need to alter the stored program several times is very common. As they became more reliable and less expensive, they became attractive for inclusion in low- and medium-volume products and systems. Today, millions of EPROMs are still in use. However, they have several major drawbacks that have been overcome by the newer EEPROMs and flash memory devices, so EPROMs are not being used in many new applications and designs. These drawbacks include: (1) they must be removed from their circuit to be erased and reprogrammed, (2) the erase operation erases the entire chip—there is no way to select only certain addresses to be erased, (3) the erase and reprogramming process can typically take 20 minutes or more.

Electrically Erasable PROM (EEPROM)

The disadvantages of the EPROM were overcome by the development of the **electrically erasable PROM (EEPROM)** as an improvement over the EPROM. The EEPROM retains the same floating-gate structure as the EPROM, but with the addition of a very thin oxide region above the drain of the MOSFET memory cell. This modification produces the EEPROM's major characteristic—its electrical erasability. By applying a high voltage (21 V) between the MOSFET's gate and drain, a charge can be induced onto the floating gate, where it will remain even when power is removed; reversal of the same voltage causes a removal of the trapped charges from the floating gate and erases the cell. Because this charge-transport mechanism requires very low currents, the erasing and programming of an EEPROM can be done *in circuit* (i.e., without a UV light source and a special PROM programmer unit).

Another advantage of the EEPROM over the EPROM is the ability to erase and rewrite *individual* bytes (eight-bit words) in the memory array electrically. During a write operation, internal circuitry automatically erases all of the cells at an address location prior to writing in the new data. This byte erasability makes it much easier to make changes in the data stored in an EEPROM.

The early EEPROMs, such as Intel's 2816, required appropriate support circuitry external to the memory chips. This support circuitry included the 21-V programming voltage (V_{PP}), usually generated from a +5 V supply through a dc-to-dc converter, and it included circuitry to control the timing and sequencing of the erase and programming operations. The newer devices, such as the Intel 2864, have integrated this support circuitry onto the same chip with the memory array, so that it requires only a single 5-V power pin. This makes the EEPROM as easy to use as the read/write memory we will be discussing shortly.

The byte erasability of the EEPROM and its high level of integration come with two penalties: density and cost. The memory cell complexity and the on-chip support circuitry place EEPROMs far behind an EPROM in bit capacity per square millimeter of silicon; a 1-Mbit EEPROM requires about twice as much silicon as a 1-Mbit EPROM. So despite its operational superiority, the EEPROM's shortcomings in density and cost-effectiveness have kept it from replacing the EPROM in applications where density and cost are paramount factors.

The logic symbol for the Intel 2864 is shown in Figure 12-13(a). It is organized as an 8K $\times$ 8 array with 13 address inputs (2^{13} = 8192) and eight data I/O pins. Three control inputs determine the operating mode according to the table given in Figure 12-13(b). With $\overline{CE}$ = HIGH, the chip is in its low-power standby mode, in which no operations are being performed on any memory location and the data pins are in the Hi-Z state.

To read the contents of a memory location, the desired address is applied to the address pins; $\overline{CE}$ is driven LOW; and the output enable pin, $\overline{OE}$, is driven LOW to enable the chip's output data buffers. The write enable pin, $\overline{WE}$, is held HIGH during a read operation.

To write into (program) a memory location, the output buffers are disabled so that the data to be written can be applied as inputs to the I/O pins. The timing for the write operation is diagrammed in Figure 12-13(c). Prior to t_1, the device is in the standby mode. A new address is applied at that time. At t_2, the $\overline{CE}$ and $\overline{WE}$ inputs are driven LOW to begin the write operation; $\overline{OE}$ is HIGH so that the data pins will remain in the Hi-Z state. Data are applied to the I/O pins at t_3 and are written into the address location on the rising edge of $\overline{WE}$ at t_4. The data are removed at t_5. Actually, the data are first latched (on the rising edge of $\overline{WE}$) into a FF buffer memory that is part of the 2864 circuitry. The data are held there while other circuitry on the chip performs an erase operation on the selected address location in the EEPROM array, after which the data byte is transferred from the buffer to the EEPROM array and stored at that location. This erase and store operation typically takes 5 ms. With $\overline{CE}$ returned HIGH at t_4, the chip is back in the standby mode while the internal erase and store operations are completed.

The 2864 has an enhanced write mode that allows the user to write up to 16 bytes of data into the FF buffer memory, where it is held while the EEPROM circuitry erases the selected address locations. The 16 bytes of data are then transferred to the EEPROM array for storage at these locations. This process also takes about 5 ms.

Because the internal process of storing a data value in an EEPROM is quite slow, the speed of the data transfer operation can also be slower.

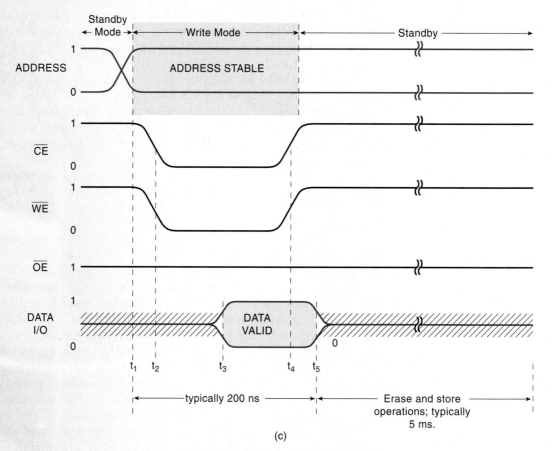

FIGURE 12-13 (a) Symbol for the 2864 EEPROM; (b) operating modes; (c) timing for the write operation.

Consequently, many manufacturers offer EEPROM devices in eight-pin packages that are interfaced to a two- or three-wire *serial* bus. This saves physical space on the system board as opposed to using a 2864 in a 28-pin, wide-DIP package. It also simplifies the hardware interface between the CPU and the EEPROM.

CD ROM

A very prominent type of read-only storage used today in computer systems is the compact disk (CD). The disk technology and the hardware necessary to retrieve the information are the same as those used in audio systems. Only the format of the data is different. The disks are manufactured with a highly reflective surface. To store data on the disks, a very intense laser beam is focused on a *very* small point on the disk. The heat from this beam burns a light-diffracting pit at that point on the disk surface. Digital data (1s and 0s) are stored on the disk one bit at a time by burning or not burning a pit into the reflective coating. The digital information is arranged on the disk as a continuous spiral of data points. The precision of the laser beam allows very large quantities of data (over 550 Mbytes) to be stored on a small, 120-mm disk.

In order to read the data, a much less powerful laser beam is focused onto the surface of the disk. At any point, the reflected light is sensed as either a 1 or a 0. This optical system is mounted on a mechanical carriage that moves back and forth along the radius of the disk, following the spiral of data as the disk rotates. The data retrieved from the optical system come one bit at a time in a serial data stream. The angular rotation of the disk is controlled to maintain a constant rate of incoming data points. If the disk is being used for audio recording, this stream of data is converted into an analog waveform. If the disk is being used as ROM, the data are decoded into parallel bytes that the computer can use. The CD player technology, although very sophisticated, is relatively inexpensive and is becoming a standard way of loading large amounts of data into a personal computer. The major improvements that are occurring now in CD-ROM technology involve quicker access time in retrieving data.

REVIEW QUESTIONS

1. *True or false:* An MROM can be programmed by the user.
2. How does a PROM differ from an MROM? Can it be erased and reprogrammed?
3. *True or false:* A PROM stores a logic "1" when its fusible link is intact.
4. How is an EPROM erased?
5. *True or false:* There is no way to erase only a portion of an EPROM's memory.
6. What function is performed by PROM and EPROM programmers?
7. What EPROM shortcomings are overcome by EEPROMs?
8. What are the major drawbacks of EEPROM?
9. What type of ROM can erase one byte at a time?
10. How many bits are read from a CD-ROM disk at any point in time?

12-8   FLASH MEMORY

EPROMs are nonvolatile, offer fast read access times (typically 120 ns), and have high density and low cost per bit. They do, however, require removal from their circuit/system to be erased and reprogrammed. EEPROMs are nonvolatile, offer fast read access, and allow rapid in-circuit erasure and reprogramming of individual bytes. They suffer from lower density and much higher cost than EPROMs.

The challenge for semiconductor engineers was to fabricate a non-volatile memory with the EEPROM's in-circuit electrical erasability, but with densities and costs much closer to those of EPROMs, while retaining the high-speed read access of both. The response to this challenge was the **flash memory.**

Structurally, a flash memory cell is like the simple single-transistor EPROM cell (and unlike the more complex two-transistor EEPROM cell), being only slightly larger. It has a thinner gate-oxide layer that allows electrical erasability but can be built with much higher densities than EEPROMs. The cost of flash memory is considerably less than for EEPROM, although not yet as close to EPROM as it will be when flash technology improves. Figure 12-14 illustrates the trade-offs for the various semiconductor nonvolatile memories. As erase/programming flexibility increases (from base to apex of the triangle), so do device complexity and cost. MROM and PROM are the simplest and cheapest devices, but they cannot be erased and reprogrammed. EEPROM is the most complex and expensive because it can be erased and reprogrammed in circuit on a byte-by-byte basis.

Flash memories are so called because of their rapid erase and write times. Most flash chips use a *bulk erase* operation in which all cells on the chip are erased simultaneously; this bulk erase process typically requires hundreds of milliseconds compared to 20 minutes for UV EPROMs. Some newer flash memories offer a *sector erase* mode, where specific sectors of the memory array (e.g., 512 bytes) can be erased at one time. This prevents having to erase and reprogram all cells when only a portion of the memory needs to be updated. A typical flash memory has a write time of 10 μs per byte compared to 100 μs for the most advanced EPROM and 5 ms for EEPROM (which includes automatic byte erase time).

Memory manufacturers everywhere are working on developing and improving flash memory devices, and each has its own approach as to what features and performance characteristics are the most important. We will not attempt to survey the vast array of manufacturers and devices here. Instead, we present a representative flash memory IC from one of the leading memory manufacturers and use it as a vehicle for learning the main aspects of flash memory operation. By looking at this representative device, we will also be exposed to the use of *command codes* as a means by which we can control the internal operations of the more complex logic chips. This will serve as a good introduction to the many types of command-controlled ICs that are part of most microprocessor-based systems.

FIGURE 12-14 Trade-offs for semiconductor non-volatile memories show that complexity and cost increase as erase and programming flexibility increases.

The 28F256A CMOS Flash Memory IC

Figure 12-15(a) shows the logic symbol for Intel Corporation's 28F256A CMOS flash memory chip, which has a capacity of 32K $\times$ 8. The diagram shows 15 address inputs (A_0–A_{14}) needed to select the different memory addresses; that is, $2^{15} = 32K = 32,768$. The eight data input/output pins (DQ_0–DQ_7) are used as inputs during memory write operations and as outputs during memory read operations. These data pins float to the Hi-Z state when the chip is deselected ($\overline{CE} = $ HIGH) or when the outputs are disabled ($\overline{OE} = HIGH$) The write enable input ($\overline{WE}$) is used to control memory write operations. Note that the chip requires two power-supply voltages: V_{CC} is the standard +5 V used for the logic circuitry; V_{PP} is the erase/programming power-supply voltage, nominally +12 V, which is needed for the erase and programming (write) operations. Newer flash chips generate V_{PP} internally and require only a single supply. The latest low-voltage devices operate on only 1.8 V.

The control inputs ($\overline{CE}$, $\overline{OE}$, and $\overline{WE}$) control what happens at the data pins in much the same way as for the 2864 EEPROM, as the table in Figure 12-15(b) shows. These data pins are normally connected to a data bus. During a write operation, data are transferred over the bus—usually from the microprocessor—and into the chip. During a read operation, data from inside the chip are transferred over the data bus—usually to the microprocessor. It is important to note that if the V_{PP} pin is not held at a high voltage (over 6.5 V), write operations cannot be performed, and the chip can only be read from; so it acts as a read-only memory (ROM), and the memory contents cannot be altered.

	Inputs			
Mode	$\overline{CE}$	$\overline{OE}$	$\overline{WE}$	Data pins
READ	LOW	LOW	HIGH	DATA$_{OUT}$
STANDBY	HIGH	X	X	High Z
WRITE*	LOW	HIGH	LOW	DATA$_{IN}$

*Note: If $V_{PP} \leq 6.5$ V, a write operation cannot be performed

(a)   (b)

FIGURE 12-15 (a) Logic symbol for the 28F256A flash memory chip; (b) control inputs $\overline{CE}$, $\overline{WE}$, and $\overline{OE}$.

The operation of this flash memory chip can be better understood by looking at its internal structure. Figure 12-16 is a diagram of the 28F256A showing its major functional blocks. You should refer to this diagram as needed during the following discussion. The unique feature of this structure is the *command register*, which is used to manage all of the chip functions. Command codes are written into this register to control which operations take place inside the chip (e.g., erase, erase-verify, program, program-verify). These command codes usually come over the data bus from the microprocessor. State

FIGURE 12-16   Functional diagram of the 28F256A flash memory chip. (Courtesy of Intel Corporation.)

control logic examines the contents of the command register and generates logic and control signals to the rest of the chip's circuits to carry out the steps in the operation. Let's look at how this works for some of the commands.

Read Command

To set up the chip for read operations, it is necessary first to *write* all 0s ($00000000_2 = 00_{16}$) into the command register. This is done by applying 00_{16} to the data pins and pulsing $\overline{WE}$ LOW while $\overline{CE}$ = LOW and $\overline{OE}$ = HIGH [as shown in the table in Figure 12-15(b)]. The address inputs can be in any state for this step because the input data will automatically be routed to the command register. Once this is done, the device is set up to have data *read* from the 262,144-bit (32,768-byte) memory matrix (array). Memory data are read in the usual way: (1) apply the address to be read from; and (2) set $\overline{WE}$ = HIGH, $\overline{CE}$ = LOW, and pulse $\overline{OE}$ = LOW to enable the output buffers to pass data from the cell matrix to the data output pins. The read access time for the 28F256A is 120 ns, maximum. The device remains enabled for memory reads as long as the command register contents remain at 00_{16}.

Set-Up Erase/Erase Commands

To erase the entire contents of the memory array requires two steps: (1) writing the code 20_{16} to the command register to set up the chip for the erase operation and (2) again writing the code 20_{16} to the command register to begin the erase operation. This two-step sequence makes it less likely that the memory contents will be erased accidentally. After the second command code has been written, the 12 V from V_{PP} is used to erase all cells in the array. This process takes about 10 ms.

Erase-Verify Command

After the erase operation, it is necessary to verify that all memory cells have been erased; that is, all bytes = 11111111_2 = FF_{16}. The erase-verify operation must be done for each byte address. It is initiated by writing $A0_{16}$ to the command register. Then a read operation is done on the address to be verified, and the output data value is checked to see that it equals FF_{16}. These two steps are done for each address, and the total time to verify all addresses is about 1 s. If the data at an address do not show up as FF, the chip must be erased again using the erase command sequence above.

Set-Up Program/Program Commands

The device can be set up for byte programming by writing the code 40_{16} to the command register. This is followed by writing the data that are to be programmed to the desired address.

Program-Verify Command

After a byte has been programmed into an address, the content of that address must be verified to assure that it has been programmed successfully. This is done by first writing the code $C0_{16}$ to the command register to set up the verify operation. This is followed by a read operation, which will cause the chip to output the data from the address that was just programmed so that it can be compared to the desired data.

The complete process of programming and verifying each address location takes about 500 ms for the 28F256A.

REVIEW QUESTIONS

1. What is the main advantage of flash memory over EPROMs?
2. What is the main advantage of flash memory over EEPROMs?
3. Where does the word *flash* come from?
4. What is V_{PP} needed for?
5. What is the function of the 28F256A's command register?
6. What is the purpose of an erase-verify command?
7. What is the purpose of the program-verify command?

12-9 ROM APPLICATIONS

With the exception of MROM and PROM, most ROM devices can be reprogrammed, so technically they are not *read-only* memories. However, the term *ROM* can still be used to include EPROMs, EEPROMs, and flash memory because, during normal operation, the stored contents of these devices is not

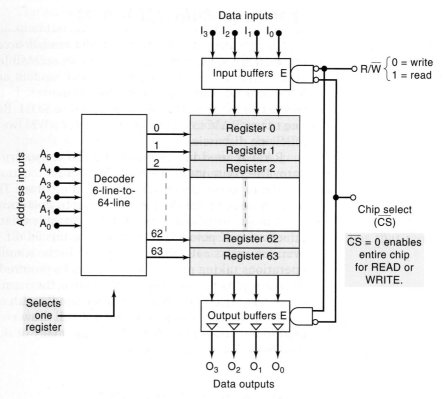

FIGURE 12-19   Internal organization of a 64×4 RAM.

ister will appear at the four data outputs. $R/\overline{W} = 1$ also *disables* the input buffers so that the data inputs do not affect the memory during a read operation.

Write Operation

To write a new four-bit word into the selected register requires $R/\overline{W} = 0$ and $\overline{CS} = 0$. This combination *enables* the input buffers so that the four-bit word applied to the data inputs will be loaded into the selected register. The $R/\overline{W} = 0$ also *disables* the output buffers, which are tristate, so that the data outputs are in their Hi-Z state during a write operation. The write operation, of course, destroys the word that was previously stored at that address.

Chip Select

Most memory chips have one or more *CS* inputs that are used to enable the entire chip or disable it completely. In the disabled mode, all data inputs and data outputs are disabled (Hi-Z) so that neither a read nor a write operation can take place. In this mode, the contents of the memory are unaffected. The reason for having *CS* inputs will become clear when we combine memory chips to obtain larger memories. Note that many manufacturers call these inputs CHIP ENABLE (*CE*). When the *CS* or *CE* inputs are in their active state, the memory chip is said to be *selected*; otherwise, it is said to be *deselected*. Many memory ICs are designed to consume much less power when they are deselected. In large memory systems, for a given memory operation, one or more memory chips will be selected while all others are deselected. More will be said on this topic later.

Common Input/Output Pins

In order to conserve pins on an IC package, manufacturers often combine the data input and data output functions using common input/output pins. The $R/\overline{W}$ input controls the function of these I/O pins. During a read operation, the I/O pins act as data outputs that reproduce the contents of the selected address location. During a write operation, the I/O pins act as data inputs to which the data to be written are applied.

We can see why this is done by considering the chip in Figure 12-19. With separate input and output pins, a total of 18 pins is required (including ground and power supply). With four common I/O pins, only 14 pins are required. The pin saving becomes even more significant for chips with larger word size.

EXAMPLE 12-9

The 2147H is an NMOS RAM that is organized as a 4K × 1 with separate data input and output and a single active-LOW chip select input. Draw the logic symbol for this chip, showing all pin functions.

Solution

The logic symbol is shown in Figure 12-20(a).

FIGURE 12-20 Logic symbols for (a) the 2147H RAM chip; (b) the MCM6206C RAM.

(a)   (b)

EXAMPLE 12-10

The MCM6206C is a CMOS RAM with 32K × 8 capacity, common I/O pins, an active-LOW chip enable, and an active-LOW output enable. Draw the logic symbol.

Solution

The logic symbol is shown in Figure 12-20(b).

In most applications, memory devices are used with a bidirectional data bus like we studied in Chapter 9. For this type of system, even if the memory chip had separate input and output pins, they would be connected together on the same data bus. A RAM having separate input and output pins is referred to as dual-port RAM. These are used in applications where speed is very important and the data in comes from a different device than the data out is going to. A good example is the video RAM on your PC. The RAM must be read repeatedly by the video card to refresh the screen and constantly filled with new updated information from the system bus.

1. Describe the input conditions needed to read a word from a specific RAM address location.
2. Why do some RAM chips have common input/output pins?
3. How many pins are required for the MCM6208C 64K × 4 RAM with one $\overline{CS}$ input and common I/O?

12-12 STATIC RAM (SRAM)

The RAM operation that we have been discussing up to this point applies to a **static RAM**—one that can store data as long as power is applied to the chip. Static-RAM memory cells are essentially flip-flops that will stay in a given state (store a bit) indefinitely, provided that power to the circuit is not interrupted. In Section 12-13, we will describe **dynamic RAMs**, which store data as charges on capacitors. With dynamic RAMs, the stored data will gradually disappear because of capacitor discharge, so it is necessary to **refresh** the data periodically (i.e., recharge the capacitors).

Static RAMs (SRAMs) are available in bipolar, MOS, and BiCMOS technologies; the majority of applications use NMOS or CMOS RAMs. As stated earlier, the bipolars have the advantage in speed (although CMOS is gradually closing the gap), and MOS devices have much greater capacities and lower power consumption. Figure 12-21 shows for comparison a typical bipolar static memory cell and a typical NMOS static memory cell. The bipolar cell contains two bipolar transistors and two resistors, while the NMOS cell contains four N-channel MOSFETs. The bipolar cell requires more chip area than the MOS cell because a bipolar transistor is more complex than a MOSFET,

FIGURE 12-21 Typical bipolar and NMOS static-RAM cells.

Bipolar cell   NMOS cell

and because the bipolar cell requires separate resistors while the MOS cell uses MOSFETs as resistors (Q_3 and Q_4). A CMOS memory cell would be similar to the NMOS cell except that it would use P-channel MOSFETs in place of Q_3 and Q_4. This results in the lowest power consumption but increases the chip complexity.

Static-RAM Timing

RAM ICs are most often used as the internal memory of a computer. The CPU (central processing unit) continually performs read and write operations on this memory at a very fast rate that is determined by the limitations of the CPU. The memory chips that are interfaced to the CPU must be fast enough to respond to the CPU read and write commands, and a computer designer must be concerned with the RAM's various timing characteristics.

Not all RAMs have the same timing characteristics, but most of them are similar, and so we will use a typical set of characteristics for illustrative purposes. The nomenclature for the different timing parameters will vary from one manufacturer to another, but the meaning of each parameter is usually easy to determine from the memory timing diagrams on the RAM data sheets. Figure 12-22 shows the timing diagrams for a complete read cycle and a complete write cycle for a typical RAM chip.

Read Cycle

The waveforms in Figure 12-22(a) show how the address, $R/\overline{W}$, and chip select inputs behave during a memory read cycle. As noted, the CPU supplies these input signals to the RAM when it wants to read data from a specific RAM address location. Although a RAM may have many address inputs coming from the CPU's address bus, for clarity the diagram shows only two. The RAM's data output is also shown; we will assume that this particular RAM has one data output. Recall that the RAM's data output is connected to the CPU data bus (Figure 12-5).

The read cycle begins at time t_0. Prior to that time, the address inputs will be whatever address is on the address bus from the preceding operation. Because the RAM's chip select is not active, it will not respond to its "old" address. Note that the $R/\overline{W}$ line is HIGH prior to t_0 and stays HIGH throughout the read cycle. In most memory systems, $R/\overline{W}$ is normally kept in the HIGH state except when it is driven LOW during a write cycle. The RAM's data output is in its Hi-Z state because $\overline{CS} = 1$.

At t_0, the CPU applies a new address to the RAM inputs; this is the address of the location to be read. After allowing time for the address signals to stabilize, the $\overline{CS}$ line is activated. The RAM responds by placing the data from the addressed location onto the data output line at t_1. The time between t_0 and t_1 is the RAM's access time, t_{ACC}, and is the time between the application of the new address and the appearance of valid output data. The timing parameter, t_{CO}, is the time it takes for the RAM output to go from Hi-Z to a valid data level once $\overline{CS}$ is activated.

At time t_2, the $\overline{CS}$ is returned HIGH, and the RAM output returns to its Hi-Z state after a time interval, t_{OD}. Thus, the RAM data will be on the data bus between t_1 and t_3. The CPU can take the data from the data bus at any point during this interval. In most computers, the CPU will use the PGT of the $\overline{CS}$ signal at t_2 to latch these data into one of its internal registers.

The complete read cycle time, t_{RC}, extends from t_0 to t_4, when the CPU changes the address inputs to a different address for the next read or write cycle.

FIGURE 12-22   Typical timing for static RAM: (a) read cycle; (b) write cycle.

Write Cycle

Figure 12-22(b) shows the signal activity for a write cycle that begins when the CPU supplies a new address to the RAM at a time t_0. The CPU drives the $R/\overline{W}$ and $\overline{CS}$ lines LOW after waiting for a time interval t_{AS}, called the *address setup time*. This gives the RAM's address decoders time to respond to the new address. $R/\overline{W}$ and $\overline{CS}$ are held LOW for a time interval t_W, called the *write time interval*.

During this write time interval, at time t_1, the CPU applies valid data to the data bus to be written into the RAM. These data must be held at the RAM input for at least a time interval t_{DS} prior to, and for at least a time interval t_{DH} after, the deactivation of $R/\overline{W}$ and $\overline{CS}$ at t_2. The t_{DS} interval is called the *data setup time,* and t_{DH} is called the *data hold time.* Similarly, the address inputs must remain stable for the address hold time interval, t_{AH}, after t_2. If any of these setup time or hold time requirements are not met, the write operation will not take place reliably.

The complete write-cycle time, t_{WC}, extends from t_0 to t_4, when the CPU changes the address lines to a new address for the next read or write cycle.

The read-cycle time, t_{RC}, and write-cycle time, t_{WC}, are what essentially determine how fast a memory chip can operate. For example, in an actual application, a CPU will often be reading successive data words from memory one right after the other. If the memory has a t_{RC} of 50 ns, the CPU can read one word every 50 ns, or 20 million words per second; with $t_{RC} = 10$ ns, the CPU can read 100 million words per second. Table 12-2 shows the minimum read-cycle and write-cycle times for some representative static-RAM chips.

TABLE 12-2

Device	t_{RC}(min) (ns)	t_{WC}(min) (ns)
CMOS MCM6206C, 32K × 8	15	15
NMOS 2147H, 4K × 1	35	35
BiCMOS MCM6708A, 64K × 4	8	8

Actual SRAM Chip

An example of an actual SRAM IC is the MCM6264C CMOS 8K × 8 RAM with read-cycle and write-cycle times of 12 ns and a standby power consumption of only 100 mW. The logic symbol for this IC is shown in Figure 12-23. Notice that it has 13 address inputs, because $2^{13} = 8192 = 8$K, and eight data I/O lines. The four control inputs determine the device's operating mode according to the accompanying mode table.

The $\overline{WE}$ input is the same as the $R/\overline{W}$ input that we have been using. A LOW at $\overline{WE}$ will write data into the RAM, provided that the device is selected—both chip select inputs are active. Note that the "&" symbol is used to denote that *both* must be active. A HIGH at $\overline{WE}$ will produce the read operation, provided that the device is selected and the output buffers are enabled by $\overline{OE} = $ LOW. When deselected, the device is in its low-power mode, and none of the other inputs have any effect.

Mode	Inputs				I/O pins
	$\overline{WE}$	$\overline{CS_1}$	CS_2	$\overline{OE}$	
READ	1	0	1	0	DATA$_{OUT}$
WRITE	0	0	1	X	DATA$_{IN}$
Output disable	1	X	X	1	High Z
Not selected (power down)	X	1	X	X	High Z
	X	X	0	X	

X = don't care

FIGURE 12-23 Symbol and mode table for the CMOS MCM6264C.

Because of their simple cell structure, DRAMs typically have four times the density of SRAMs. This increased density allows four times as much memory capacity to be placed on a single board; alternatively, it requires one-fourth as much board space for the same amount of memory. The cost per bit of dynamic RAM storage is typically one-fifth to one-fourth that of static RAMs. An additional cost saving is realized because the lower power requirements of a dynamic RAM, typically one-sixth to one-half those of a static RAM, allow the use of smaller, less expensive power supplies.

The main applications of SRAMs are in areas where only small amounts of memory are needed or where high speed is required. Many microprocessor-controlled instruments and appliances have very small memory capacity requirements. Some instruments, such as digital storage oscilloscopes and logic analyzers, require very high-speed memory. For applications such as these, SRAM is normally used.

The main internal memory of most personal microcomputers (e.g., Windows-based PCs or Macs) uses DRAM because of its high capacity and low power consumption. These computers, however, sometimes use some small amounts of SRAM for functions requiring maximum speed, such as video graphics, look-up tables, and cache memory.

1. What are the main drawbacks of dynamic RAM compared with static?
2. List the advantages of dynamic RAM compared with static RAM.
3. Which type of RAM would you expect to see in battery-operated equipment?

12-14   DYNAMIC RAM STRUCTURE AND OPERATION

The dynamic RAM's internal architecture can be visualized as an array of single-bit cells, as illustrated in Figure 12-25. Here, 16,384 cells are arranged in a 128 $\times$ 128 array. Each cell occupies a unique row and column position within the array. Fourteen address inputs are needed to select one of the cells (2^{14} = 16,384); the lower address bits, A_0 to A_6, select the row, and the higher-order bits, A_7 to A_{13}, select the column. Each 14-bit address selects a unique cell to be written into or read from. The structure in Figure 12-25 is a 16K $\times$ 1 DRAM chip. DRAM chips are currently available in various configurations. DRAMs with a four-bit (or greater) word size have a cell arrangement similar to that of Figure 12-25 except that each position in the array contains four cells, and each applied address selects a group of four cells for a read or a write operation. As we will see later, larger word sizes can also be attained by combining several chips in the appropriate arrangement.

Figure 12-26 is a symbolic representation of a dynamic memory cell and its associated circuitry. Many of the circuit details are not shown, but this simplified diagram can be used to describe the essential ideas involved in writing to and reading from a DRAM. The switches SW1 through SW4 are actually MOSFETs that are controlled by various address decoder outputs and the $R/\overline{W}$ signal. The capacitor, of course, is the actual storage cell.

To write data to the cell, signals from the address decoding and read/write logic will close switches SW1 and SW2, while keeping SW3 and SW4 open. This connects the input data to C. A logic 1 at the data input

FIGURE 12-25   Cell arrangement in a 16K × 1 dynamic RAM.

charges C, and a logic 0 discharges it. Then the switches are open so that C is disconnected from the rest of the circuit. Ideally, C would retain its charge indefinitely, but there is always some leakage path through the off switches, so that C will gradually lose its charge.

To read data from the cell, switches SW2, SW3, and SW4 are closed, and SW1 is kept open. This connects the stored capacitor voltage to the *sense amplifier*. The sense amplifier compares the voltage with some reference value to determine if it is a logic 0 or 1, and it produces a solid 0 V or 5 V for the data output. This data output is also connected to C (SW2 and SW4 are closed) and refreshes the capacitor voltage by recharging or discharging. In other words, the data bit in a memory cell is refreshed each time it is read.

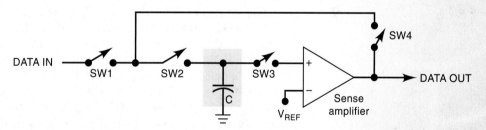

FIGURE 12-26   Symbolic representation of a dynamic memory cell. During a WRITE operation, semiconductor switches SW1 and SW2 are closed. During a read operation, all switches are closed except SW1.

Address Multiplexing

The 16K × 1 DRAM array depicted in Figure 12-25 is obsolete and nearly unavailable. It has 14 address inputs; a 64K × 1 DRAM array would have 16 address inputs. A 1M × 4 DRAM needs 20 address inputs; a 4M × 1 needs 22 address inputs. High-capacity memory chips such as these would require many pins if each address input required a separate pin. In order to reduce the number of pins on their high-capacity DRAM chips, manufacturers utilize **address multiplexing** whereby each address input pin can accommodate two different address bits. The saving in pin count translates to a significant decrease in the size of the IC packages. This is very important in large-capacity memory boards, where you want to maximize the amount of memory that can fit on one board.

In the discussions that follow, we will be describing the order in which the address is multiplexed into the DRAM chips. It should be noted that in older, small-capacity DRAMs, the convention was to present the low-order address first specifying the row, followed by the high-order address specifying the column. The newer DRAMs and the controllers that perform the multiplexing use the opposite convention of applying the high-order bits as the row address and then the low-order bits as the column address. We will describe the more recent convention, but you should be aware of this change as you investigate older systems.

We will use the TMS44100 4M $\times$ 1 DRAM from Texas Instruments to illustrate the operation of DRAM chips today. The functional block diagram of this chip's internal architecture (shown in Figure 12-27) is typical of diagrams you will find in data books. The layout of the memory array in this diagram may appear complicated at first glance, but it can be thought of as just a bigger version of the 16K $\times$ 1 DRAM in Figure 12-25. Functionally, it is an array of cells arranged as 2048 rows by 2048 columns. A single row is selected by address decoder circuitry that can be thought of as a 1-of-2048 decoder. Likewise, a single column is selected by what is effectively a 1-of-2048 decoder. Because the address lines are multiplexed, the entire 22-bit address cannot be presented simultaneously. Notice that there are only 11 address lines and that they go to both the row and the column address registers. Each of the two address registers stores half of the 22-bit address. The row register stores the upper half, and the column register stores the lower half. Two very important **strobe** inputs control when the address information is latched. The **row address strobe** ($\overline{RAS}$) clocks the 11-bit row address register. The **column address strobe** ($\overline{CAS}$) clocks the 11-bit column address register.

A 22-bit address is applied to this DRAM in two steps using $\overline{RAS}$ and $\overline{CAS}$. The timing is shown in Figure 12-27(b). Initially, $\overline{RAS}$ and $\overline{CAS}$ are both HIGH. At time t_0, the 11-bit row address (A_{11} to A_{21}) is applied to the address inputs. After allowing time for the setup time requirement (t_{RS}) of the row address register, the $\overline{RAS}$ input is driven LOW at t_1. This NGT loads the row address into the row address register so that A_{11} to A_{21} now appear at the row decoder inputs. The LOW at $\overline{RAS}$ also enables this decoder so that it can decode the row address and select one row of the array.

At time t_2, the 11-bit column address (A_0 to A_{10}) is applied to the address inputs. At t_3, the $\overline{CAS}$ input is driven LOW to load the column address into the column address register. $\overline{CAS}$ also enables the column decoder so that it can decode the column address and select one column of the array.

At this point the two parts of the address are in their respective registers, the decoders have decoded them to select the one cell corresponding to the row and column address, and a read or a write operation can be performed on that cell just as in a static RAM.

You may have noticed that this DRAM does not have a chip select (CS) input. The $\overline{RAS}$ and $\overline{CAS}$ signals perform the chip select function because they must both be LOW for the decoders to select a cell for reading or writing.

As you can see, there are several operations that must be performed before the data that is stored in the DRAM can actually appear on the outputs. The term **latency** is often used to describe the time required to perform these operations. Each operation takes a certain amount of time, and this amount of time determines the maximum rate at which we can access data in the memory.

FUNCTIONAL BLOCK DIAGRAM

(a)

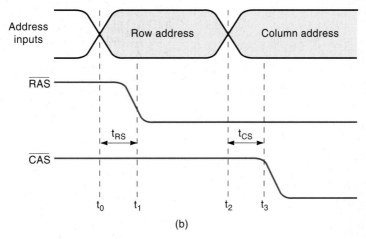

(b)

FIGURE 12-27   (a) Simplified architecture of the TMS44100 4M × 1 DRAM;
(b) $\overline{RAS}/\overline{CAS}$ timing. (Reprinted by Permission of Texas Instruments.)

EXAMPLE 12-12   How many pins are saved by using address multiplexing for a 16M × 1 DRAM?

Solution

Twelve address inputs are used instead of 24; $\overline{RAS}$ and $\overline{CAS}$ are added; no $\overline{CS}$ is required. Thus, there is a net saving of *eleven* pins.

In a simple computer system, the address inputs to the memory system come from the central processing unit (CPU). When the CPU wants to access a particular memory location, it generates the complete address and places it on address lines that make up an **address bus**. Figure 12-28(a) shows this for a small computer memory that has a capacity of 64K words and therefore requires a 16-line address bus going directly from the CPU to the memory.

This arrangement works for ROM or for static RAM, but it must be modified for DRAM that uses multiplexed addressing. If all 64K of the memory is DRAM, it will have only eight address inputs. This means that the 16 address lines from the CPU address bus must be fed into a multiplexer circuit that will transmit eight address bits at a time to the memory address inputs. This is shown symbolically in Figure 12-28(b). The multiplexer select input, labeled *MUX*, controls whether CPU address lines A_0 to A_7 or address lines A_8 to A_{15} will be present at the DRAM address inputs.

The timing of the *MUX* signal must be synchronized with the $\overline{RAS}$ and $\overline{CAS}$ signals that clock the addresses into the DRAM. This is shown in Figure 12-29. *MUX* must be LOW when $\overline{RAS}$ is pulsed LOW so that address lines A_8 to A_{15} from the CPU will reach the DRAM address inputs to be loaded on the NGT of $\overline{RAS}$.

FIGURE 12-28 (a) CPU address bus driving ROM or static-RAM memory; (b) CPU addresses driving a multiplexer that is used to multiplex the CPU address lines into the DRAM.

*MUX = 0 transmits CPU address A_8–A_{15} to DRAM. MUX = 1 transmits A_0–A_7 to DRAM.

(b)

FIGURE 12-29 Timing required for address multiplexing.

A_8-A_{15} latched into DRAM row address register

A_0-A_7 latched into DRAM column address register

Likewise, *MUX* must be HIGH when $\overline{CAS}$ is pulsed LOW so that A_0 to A_7 from the CPU will be present at the DRAM inputs to be loaded on the NGT of $\overline{CAS}$.

The actual multiplexing and timing circuitry will not be shown here but will be left to the end-of-chapter problems (Problems 12-26 and 12-27).

REVIEW QUESTIONS

1. Describe the array structure of a 64K × 1 DRAM.
2. What is the benefit of address multiplexing?
3. How many address inputs would there be on a 1M × 1 DRAM chip?
4. What are the functions of the $\overline{RAS}$ and $\overline{CAS}$ signals?
5. What is the function of the *MUX* signal?

12-15 DRAM READ/WRITE CYCLES

The timing of the read and write operations of a DRAM is much more complex than for a static RAM, and there are many critical timing requirements that the DRAM memory designer must consider. At this point, a detailed discussion of these requirements would probably cause more confusion than enlightenment. We will concentrate on the basic timing sequence for the read and write operations for a small DRAM system like that of Figure 12-28(b).

Dram Read Cycle

Figure 12-30 shows typical signal activity during the read operation. It is assumed that $R/\overline{W}$ is in its HIGH state throughout the operation. The following is a step-by-step description of the events that occur at the times indicated on the diagram.

- t_0: *MUX* is driven LOW to apply the row address bits (A_8 to A_{15}) to the DRAM address inputs.
- t_1: $\overline{RAS}$ is driven LOW to load the row address into the DRAM.
- t_2: *MUX* goes HIGH to place the column address (A_0 to A_7) at the DRAM address inputs.

FIGURE 12-30 Signal activity for a read operation on a dynamic RAM. The $R/\overline{W}$ input (not shown) is assumed to be HIGH.

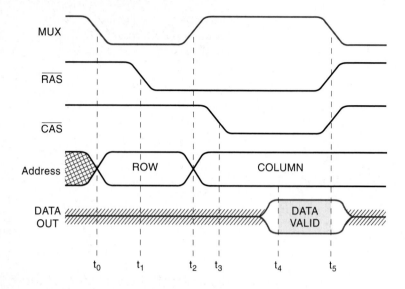

- t_3: $\overline{CAS}$ goes LOW to load the column address into the DRAM.
- t_4: The DRAM responds by placing valid data from the selected memory cell onto the DATA OUT line.
- t_5: MUX, $\overline{RAS}$, $\overline{CAS}$, and DATA OUT return to their initial states.

Dram Write Cycle

Figure 12-31 shows typical signal activity during a DRAM write operation. Here is a description of the sequence of events.

- t_0: The LOW at *MUX* places the row address at the DRAM inputs.
- t_1: The NGT at $\overline{RAS}$ loads the row address into the DRAM.
- t_2: *MUX* goes HIGH to place the column address at the DRAM inputs.

FIGURE 12-31 Signal activity for a write operation on a dynamic RAM.

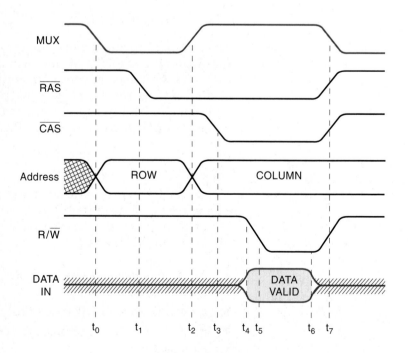

- t_3: The NGT at $\overline{CAS}$ loads the column address into the DRAM.
- t_4: Data to be written are placed on the DATA IN line.
- t_5: $R/\overline{W}$ is pulsed LOW to write the data into the selected cell.
- t_6: Input data are removed from DATA IN.
- t_7: MUX, $\overline{RAS}$, $\overline{CAS}$, and $R/\overline{W}$ are returned to their initial states.

REVIEW QUESTIONS

1. *True or false:*
   (a) During a read cycle, the $\overline{RAS}$ signal is activated before the $\overline{CAS}$ signal.
   (b) During a write operation, $\overline{CAS}$ is activated before $\overline{RAS}$.
   (c) $R/\overline{W}$ is held LOW for the entire write operation.
   (d) The address inputs to a DRAM will change twice during a read or a write operation.
2. Which signal makes sure that the correct portion of the complete address appears at the DRAM inputs?

12-16   DRAM REFRESHING

A DRAM cell is refreshed each time a read operation is performed on that cell. Each memory cell must be refreshed periodically (typically, every 4 to 16 ms, depending on the device) or its data will be lost. This requirement would appear to be extremely difficult, if not impossible, to meet for large-capacity DRAMs. For example, a 1M × 1 DRAM has 10^{20} = 1,048,576 cells. To ensure that each cell is refreshed within 4 ms, it would require that read operations be performed on successive addresses at the rate of one every 4 ns (4 ms/1,048,576 ≈ 4 ns). This is much too fast for any DRAM chip. Fortunately, manufacturers have designed DRAM chips so that

> whenever a read operation is performed on a cell, all of the cells in that row will be refreshed.

Thus, it is necessary to do a read operation only on each *row* of a DRAM array once every 4 ms to guarantee that each *cell* of the array is refreshed. Referring to the 4M × 1 DRAM of Figure 12-27(a), if any address is strobed into the row address register, all 2048 cells in that row are automatically refreshed.

Clearly, this row-refreshing feature makes it easier to keep all DRAM cells refreshed. However, during the normal operation of the system in which a DRAM is functioning, it is unlikely that a read operation will be performed on each row of the DRAM within the required refresh time limit. Therefore, some kind of refresh control logic is needed either external to the DRAM chip or as part of its internal circuitry. In either case, there are two refresh modes: a *burst* refresh and a *distributed* refresh.

In a burst refresh mode, the normal memory operation is suspended, and each row of the DRAM is refreshed in succession until all rows have been refreshed. In a distributed refresh mode, the row refreshing is interspersed with the normal operations of the memory.

The most universal method for refreshing a DRAM is the $\overline{RAS}$-only refresh. It is performed by strobing in a row address with $\overline{RAS}$ while $\overline{CAS}$ and $R/\overline{W}$ remain HIGH. Figure 12-32 illustrates how $\overline{RAS}$-only refresh is used for a burst refresh of the TMS44100. Some of the complexity of the memory array in this chip is there to make refresh operations simpler. Because two banks are lined up in the same row, both banks can be refreshed at the same time, effectively making it the same as if there were only 1024 rows. A refresh counter is used to supply 10-bit row addresses to the DRAM address inputs starting at 0000000000 (row 0). $\overline{RAS}$ is pulsed LOW to load this address into the DRAM, and this refreshes row 0 in both banks. The counter is incremented and the process is repeated up to address 1111111111 (row 1023). For the TMS44100, a burst refresh can be completed in just over 113 μs and must be repeated every 16 ms or less.

While the refresh counter idea seems easy enough, we must realize that the row addresses from the refresh counter cannot interfere with the addresses coming from the CPU during normal read/write operations. For this reason, the refresh counter addresses must be multiplexed with the CPU addresses, so that the proper source of DRAM addresses is activated at the proper times.

In order to relieve the computer's CPU from some of these burdens, a special chip called a **dynamic RAM (DRAM) controller** is often used. At a minimum, this chip will perform address multiplexing and refresh count sequence generation, leaving the generation of the timing for $\overline{RAS}$, $\overline{CAS}$, and *MUX* signals up to some other logic circuitry and the person who programs the computer. Other DRAM controllers are fully automatic. Their inputs look very much like a static RAM or ROM. They automatically generate the refresh sequence often enough to maintain the memory, multiplex the address bus, generate the $\overline{RAS}$ and $\overline{CAS}$ signals, and arbitrate control of the DRAM between the CPU read/write cycles and local refresh operations. In current personal computers, the DRAM controller and other high-level controller circuits are integrated into a set of VLSI circuits that are referred to as a "chip set." As newer DRAM technologies are developed, new chip sets are designed to take advantage of the latest advances. In many cases, the number of existing (or anticipated) chip sets supporting a certain technology in the market determines the DRAM technology in which manufacturers invest.

Most of the DRAM chips in production today have on-chip refreshing capability that eliminates the need to supply external refresh addresses. One

* R/$\overline{W}$ and $\overline{CAS}$ lines held HIGH

FIGURE 12-32   The $\overline{RAS}$-only refresh method uses only the $\overline{RAS}$ signal to load the row address into the DRAM to refresh all cells in that row. The $\overline{RAS}$-only refresh can be used to perform a burst refresh as shown. A refresh counter supplies the sequential row addresses from row 0 to row 1023 (for a 4M $\times$ 1 DRAM).

of these methods, shown in Figure 12-33(a), is called $\overline{CAS}$-*before*-$\overline{RAS}$ *refresh*. In this method, the $\overline{CAS}$ signal is driven LOW first and is held LOW until after $\overline{RAS}$ goes LOW. This sequence will refresh one row of the memory array and increment an internal counter that generates the row addresses. To perform a burst refresh using this feature, $\overline{CAS}$ can be held LOW while $\overline{RAS}$ is pulsed once for each row until all are refreshed. During this refresh cycle, all external addresses are ignored. The TMS44100 also offers "hidden refresh," which allows a row to be refreshed while holding data on the output. This is done by holding $\overline{CAS}$ LOW after a read cycle and then pulsing $\overline{RAS}$ as in Figure 12-33(b).

The self-refresh mode of Figure 12-33(c) fully automates the process. By forcing $\overline{CAS}$ LOW before $\overline{RAS}$ and then holding them both LOW for at least 100 μs, an internal oscillator clocks the row address counter until all cells are refreshed. The mode that a system designer chooses depends on how busy the computer's CPU is. If it can spare 100 μs without accessing its memory, and if it can do this every 16 ms, then the self-refresh is the way to go. However, if this will slow the program execution down too much, it may require some distributed refreshing using $\overline{CAS}$-before-$\overline{RAS}$ or hidden refresh cycles. In any case, all cells must be refreshed within the allotted time or data will be lost.

FIGURE 12-33 TMS44100 refresh modes.

REVIEW QUESTIONS

1. *True or false:*
    (a) In most DRAMs, it is necessary to read only from one cell in each row in order to refresh all cells in that row.
    (b) In the burst refresh mode, the entire array is refreshed by one $\overline{RAS}$ pulse.
2. What is the function of a refresh counter?
3. What functions does a DRAM controller perform?
4. *True or false:*
    (a) In the $\overline{RAS}$-only refresh method, the $\overline{CAS}$ signal is held LOW.
    (b) $\overline{CAS}$-before-$\overline{RAS}$ refresh can be used only by DRAMs with on-chip refresh control circuitry.

12-17　DRAM TECHNOLOGY*

In selecting a particular type of RAM device for a system, a designer has some difficult decisions. The capacity (as large as possible), the speed (as fast as possible), the power needed (as little as possible), the cost (as low as possible), and the convenience (as easy to change as possible) must all be kept in a reasonable balance because no single type of RAM can maximize all of these desired features. The semiconductor RAM market is constantly trying to produce the ideal mix of these characteristics in its products for various applications. This section explains some of the current terms used regarding RAM technology. This is a very dynamic topic, and perhaps some of these terms will be history before this book is printed, but here is the state of the art today.

Memory Modules

With many companies manufacturing motherboards for personal computer systems, standard memory interface connectors have been adopted. These connectors receive a small printed circuit card with contact points on both sides of the edge of the card. These modular cards allow for easy installation or replacement of memory components in a computer. The single-in-line memory module (SIMM) is a circuit card with 72 functionally equivalent contacts on both sides of the card. A redundant contact point on each side of the board offers some assurance that a good, reliable contact is made. These modules use 5-V-only DRAM chips that vary in capacity from 1 to 16 Mbits in surface-mount gull-wing or J-lead packages. The memory modules vary in capacity from 1 to 32 Mbytes.

The newer, 168-pin, dual-in-line memory module (DIMM) has 84 functionally unique contacts on each side of the card. The extra pins are necessary because DIMMs are connected to 64-bit data buses such as those found in modern PCs. Both 3.3-V and 5-V versions are available. They also come in buffered and unbuffered versions. The capacity of the module depends on the DRAM chips that are mounted on it; and as DRAM capacity increases, the capacity of the DIMMs will increase. The chip set and motherboard design that is used in any given system determines which type of DIMM can be used. For compact applications, such as laptop computers, a small-outline, dual-in-line memory module is available (SODIMM).

---

*This topic may be omitted without affecting the continuity of the remainder of the book.

The primary problem in the personal computer industry is providing a memory system that is fast enough to keep up with the ever-increasing clock speeds of the microprocessors while keeping the cost at an affordable level. Special features are being added to the basic DRAM devices to enhance their total bandwidth. A new type of package called the RIMM has entered the market. RIMM stands for Rambus In-line Memory Module. Rambus is a company that has invented some revolutionary new approaches to memory technology. The RIMM is their proprietary package that holds their proprietary memory chips called Direct Rambus DRAM (DRDRAM) chips. Although these methods of improving performance are constantly changing, the technologies described in the following sections are currently being referred to extensively in memory-related literature.

FPM DRAM

Fast page mode (FPM) allows quicker access to random memory locations within the current "page." A page is simply a range of memory addresses that have identical upper address bit values. In order to access data on the current page, only the lower address lines must be changed.

EDO DRAM

Extended data output (EDO) DRAMs offer a minor improvement to FPM DRAMs. For accesses on a given page, the data value at the current memory location is sensed and latched onto the output pins. In the FPM DRAMs, the sense amplifier drives the output without a latch, requiring $\overline{CAS}$ to remain low until data values become valid. With EDO, while these data are present on the outputs, $\overline{CAS}$ can complete its cycle, a new address on the current page can be decoded, and the data path circuitry can be reset for the next access. This allows the memory controller to be outputting the next address at the same time that the current word is being read.

SDRAM

Synchronous DRAM is designed to transfer data in rapid-fire *bursts* of several sequential memory locations. The first location to be accessed is the slowest due to the overhead (latency) of latching the row and column address. Thereafter, the data values are clocked out by the bus system clock (instead of the $\overline{CAS}$ control line) in bursts of memory locations within the same page. Internally, SDRAMs are organized in two banks. This allows data to be read out at a very fast rate by alternately accessing each of the two banks. In order to provide all of the features and the flexibility needed for this type of DRAM to work with a wide variety of system requirements, the circuitry within the SDRAM has become more complex. A command sequence is necessary to tell the SDRAM which options are needed, such as burst length, sequential or interleaved data, and $\overline{CAS}$-before-$\overline{RAS}$ or self-refresh modes. Self-refresh mode allows the memory device to perform all of the necessary functions to keep its cells refreshed.

DDRSDRAM

Double Data Rate SDRAM offers an improvement of SDRAM. In order to speed up the operation of SDRAM, while operating from a synchronous system clock, this technology transfers data on the rising and falling edges of the system clock, effectively doubling the potential rate of data transfer.

FIGURE 12-41 In FIFO, data values are read out of memory (b) in the same order that they were written into memory (a).

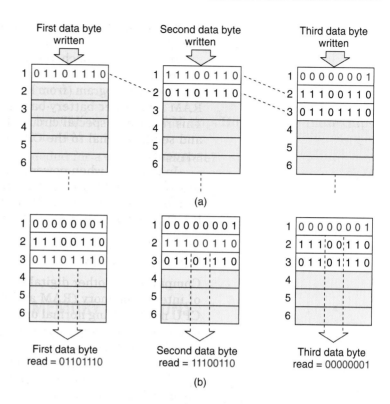

(a)

First data byte
read = 01101110

Second data byte
read = 11100110

Third data byte
read = 00000001

(b)

Figure 12-41(a) shows the sequence of writing three data bytes into the memory block. Note that as each new byte is written into location 1, the other bytes move to the next location. Figure 12-41(b) shows the sequence of reading the data out of the FIFO block. The first byte read is the same as the first byte that was written, and so on. The FIFO operation is controlled by special *address pointer registers* that keep track of where data are to be written and the location from which they are to be read.

A FIFO is useful as a **data-rate buffer** between systems that transfer data at widely different rates. One example is the transfer of data from a computer to a printer. The computer sends character data to the printer at a very high rate, say, one byte every 10 μs. These data fill up a FIFO memory in the printer. The printer then reads out the data from the FIFO at a much slower rate, say, one byte every 5 ms, and prints out the corresponding characters in the same order as sent by the computer.

A FIFO can also be used as a data-rate buffer between a slow device, such as a keyboard, and a high-speed computer. Here, the FIFO accepts keyboard data at the slow and asynchronous rate of human fingers and stores them. The computer can then read all of the recently stored keystrokes very quickly at a convenient point in its program. In this way, the computer can perform other tasks while the FIFO is slowly being filled with data.

Circular Buffers

Data rate buffers (FIFOs) are often referred to as **linear buffers**. As soon as all the locations in the buffer are full, no more entries are made until the buffer is emptied. This way, none of the "old" information is lost. A similar memory system is called a **circular buffer**. These memory systems are used to store the last n values entered, where n is the number of memory locations in the buffer. Each time a new value is written to a circular buffer, it overwrites (replaces) the oldest value. Circular buffers are addressed by a MOD-n address counter. Consequently, when the highest address is reached, the address counter will "wrap around" and the next location will be the lowest address. As you recall from Chapter 11, digital filtering and other DSP operations perform calculations using a group of recent samples. Special hardware included in a DSP allows easy implementation of circular buffers in memory.

REVIEW QUESTIONS

1. What are the various ways to handle the possible loss of critical data when power is interrupted?
2. What is the principal reason for using a cache memory?
3. What does *FIFO* mean?
4. What is a data-rate buffer?
5. How does a circular buffer differ from a linear buffer?

12-20 TROUBLESHOOTING RAM SYSTEMS

All computers use RAM. Many general-purpose computers and most special-purpose computers (such as microprocessor-based controllers and process-control computers) also use some form of ROM. Each RAM and ROM IC that is part of a computer's internal memory typically contains thousands of memory cells. A single faulty memory cell can cause a complete system failure (commonly referred to as a "system crash") or, at the least, unreliable system operation. The testing and troubleshooting of memory systems involves the use of techniques that are not often used on other parts of the digital system. Because memory consists of thousands of identical circuits acting as storage locations, any tests of its operation must involve checking to see exactly which locations are working and which are not. Then, by looking at the pattern of good and bad locations along with the organization of the memory circuitry, one can determine the possible causes of the memory malfunction. The problem typically can be traced to a bad memory IC; a bad decoder IC, logic gate, or signal buffer; or a problem in the circuit connections (i.e., shorts or open connections).

Because RAM must be written to and read from, testing RAM is generally more complex than testing ROM. In this section, we will look at some common procedures for testing the RAM portion of memory and interpreting the test results. We will examine ROM testing in the next section.

Know the Operation

The RAM memory system shown in Figure 12-42 will be used in our examples. As we emphasized in earlier discussions, successful troubleshooting of a relatively complex circuit or system begins with a thorough knowledge of its operation. Before we can discuss the testing of this RAM system, we should first analyze it carefully so that we fully understand its operation.

The total capacity is 4K $\times$ 8 and is made up of four 1K $\times$ 8 RAM modules. A module may be just a single IC, or it may consist of several ICs (e.g., two 1K $\times$ 4 chips). Each module is connected to the CPU through the address and data buses and through the $R/\overline{W}$ control line. The modules have common I/O

EXAMPLE 12-20

The following messages are printed out after the SELF-TEST is run:

```
module-0 test OK
module-1 test OK
address 09A7 faulty at bit 6
module-3 test OK
```

Examine these messages and decide what to do next.

Solution

The faulty address location is in module-2. There is no circuit fault external to module-2 that could cause only that single address to be bad and no others. Thus, it appears that there is an internal fault in module-2. Before we reach this conclusion, however, the SELF-TEST program should be run several more times to see if the same messages appear. The original fault message may have been the result of noise rather than a bad memory cell; if so, additional tests may produce no fault messages or different fault messages. This would point up the need to check for sources of noise in the system such as nonexistent or nonfunctioning power-supply decoupling capacitors, or crosstalk between closely spaced signal lines.

EXAMPLE 12-21

Consider the following messages from SELF-TEST:

```
module-0 test OK
address 0400 faulty at bits 0-7
address 0401 faulty at bits 0-7
address 0402 faulty at bits 0-7

    .    .    .    .    .    .
    .    .    .    .    .    .
    .    .    .    .    .    .

address 07FE faulty at bits 0-7
address 07FF faulty at bits 0-7
module-2 test OK
module-3 test OK
```

Examine these messages, list the possible faults, and describe what to do next.

Solution

Clearly, every address in module-1 is listed as being faulty, and so module-1 is not working at all. Here are several possible faults that could cause this:

- An open connection at the $R/\overline{W}$ input to module-1
- An open path between $\overline{K1}$ and the module-1 chip select
- A faulty decoder output $\overline{K1}$ (stays HIGH)
- The $\overline{K1}/\overline{CS}$ node externally shorted to V_{CC}
- A faulty RAM module-1 (does not respond to $\overline{CS}$)
- An unconnected V_{CC} or ground on module-1

Standard troubleshooting techniques can now be used to isolate the fault.

As these examples have shown, the memory test program is useful in narrowing the problem to a specific area of the circuit. The troubleshooter uses the information provided from the memory test to determine possible faults and then proceeds to locate the actual fault.

It should be pointed out that there are some RAM circuit faults that will prevent the CPU from executing the SELF-TEST program on power-up. For example, any RAM circuit fault that causes a data line or an address line to be stuck LOW or HIGH will cause erroneous operation when the CPU tries to execute the SELF-TEST program in ROM (remember, the ROM shares the address and data buses with the RAM). When this happens, you should check for stuck bus lines and, if one is found, use the standard techniques for determining the exact cause of the stuck line.

REVIEW QUESTIONS

1. What is E's function in the RAM circuit of Figure 12-42?
2. What is the checkerboard test? Why is it used?
3. What is a power-up self-test?

12-21   TESTING ROM

The ROM circuitry in a computer is very similar to the RAM circuitry (compare Figures 12-37 and 12-42). The ROM decoding logic can be tested in the same manner described in the preceding section for the RAM system. The ROM chips, however, must be tested differently from RAM chips because we cannot write patterns into ROM and read them back as we can for RAM. Several methods are used to check the contents of a ROM IC.

In one approach, the ROM is placed in a socket in a special test instrument that is typically microprocessor-controlled. The special test instrument can be programmed to read every location in the test ROM and print out a listing of the contents of each location. The listing can then be compared with what the ROM is supposed to contain. Except for low-capacity ROM chips, this process can be very time-consuming.

In a more efficient approach, the test instrument has the correct data stored in its own *reference ROM* chip. The test instrument is then programmed to read the contents of each location of the test ROM and compare it with the contents of the reference ROM. This approach, of course, requires the availability of a preprogrammed reference ROM.

A third approach uses a **checksum**, a special code placed in the last one or two locations of the ROM chip when it is programmed. This code is derived by adding up the data words to be stored in all of the ROM locations (excluding those containing the checksum). As the test instrument reads the data from each test ROM location, it will add them up and develop its own checksum. It then compares its calculated checksum with that stored in the last ROM locations, and they should agree. If so, there is a high probability that the ROM is good (there is a very small chance that a combination of errors in the test ROM data could still produce the same checksum value). If they do not agree, then there is a definite problem in the test ROM.

The checksum idea is illustrated in Figure 12-45(a) for a very small ROM. The data word stored in the last address is the eight-bit sum of the other

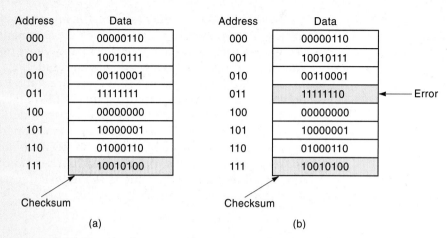

FIGURE 12-45   Checksum method for an 8 × 8 ROM: (a) ROM with correct data; (b) ROM with error in its data.

seven data words (ignoring carries from the MSB). When this ROM is programmed, the checksum is placed in the last location. Figure 12-45(b) shows the data that might actually be read from a faulty ROM that was originally programmed with the data in Figure 12-45(a). Note the error in the word at address 011. As the test instrument reads the data from each location of the faulty ROM, it calculates its own checksum from those data. Because of the error, the calculated checksum will be 10010011. When the test instrument compares this with the checksum value stored at ROM location 111, it sees that they disagree, and a ROM error is indicated. Of course, the exact location of the error cannot be determined.

The checksum method can also be used by a computer or microprocessor-based equipment during an automatic power-up self-test to check out the contents of the system ROMs. Again, as in the self-test used for RAM, the CPU would execute a program on power-up that would do a checksum test on each ROM chip and would print out some type of status message. The self-test program itself will be located in a ROM, and so any error in that ROM could prevent the successful execution of the checksum tests.

REVIEW QUESTION

1. What is a checksum? What is its purpose?

SUMMARY

1. All memory devices store binary logic levels (1s and 0s) in an array structure. The size of each binary word (number of bits) that is stored varies depending on the memory device. These binary values are referred to as *data*.

2. The place (location) in the memory device where any data value is stored is identified by another binary number referred to as an *address*. Each memory location has a unique address.

3. All memory devices operate in the same general way. To write data in memory, the address to be accessed is placed on the address input, the data value to be stored is applied to the data inputs, and the control signals are manipulated to store the data. To read data from memory, the ad-

dress is applied, the control signals are manipulated, and the data value appears on the output pins.

4. Memory devices are often used along with a microprocessor CPU that generates the addresses and control signals and either provides the data to be stored or uses the data from the memory. Reading and writing are *always* done from the CPU's perspective. Writing puts data into the memory, and reading gets data out of the memory.

5. Most read-only memories (ROMs) have data entered one time, and from then on their contents do not change. This storage process is called *programming*. They do not lose their data when power is removed from the device. MROMs are programmed during the manufacturing process. PROMs are programmed one time by the user. EPROMs are just like PROMs but can be erased using UV light. EEPROMs and flash memory devices are electrically erasable and can have their contents altered after programming. CD ROMs are used for mass storage of information that does not need to change.

6. Random access memory (RAM) is a generic term given to devices that can have data easily stored and retrieved. Data are retained in a RAM device only as long as power is applied.

7. Static RAM (SRAM) uses storage elements that are basically latch circuits. Once the data are stored, they will remain unchanged as long as power is applied to the chip. Static RAM is easier to use but more expensive per bit and consumes more power than dynamic RAM.

8. Dynamic RAM (DRAM) uses capacitors to store data by charging or discharging them. The simplicity of the storage cell allows DRAMs to store a great deal of data. Because the charge on the capacitors must be refreshed regularly, DRAMs are more complicated to use than SRAMs. Extra circuitry is often added to DRAM systems to control the reading, writing, and refreshing cycles. On many new devices, these features are being integrated into the DRAM chip itself. The goal of DRAM technology is to put more bits on a smaller piece of silicon so that it consumes less power and responds faster.

9. Memory systems require a wide variety of different configurations. Memory chips can be combined to implement any desired configuration, whether your system needs more bits per location or more total word capacity. All of the various types of ROM and RAM can be combined within the same memory system.

IMPORTANT TERMS

memory cell
memory word
byte
capacity
density
address
read operation
write operation
access time
volatile memory
random-access
  memory (RAM)

sequential-access
  memory (SAM)
read/write memory
  (RWM)
read-only memory
  (ROM)
static RAM (SRAM)
dynamic RAM
main memory
auxiliary memory
address bus
data bus

control bus
programming
chip select
power-down
fusible-link
programmer
electrically erasable
  PROM (EEPROM)
flash memory
firmware
bootstrap program
refresh

JEDEC	latency	FIFO
address multiplexing	$\overline{RAS}$-only refresh	data-rate buffer
strobing	refresh counter	linear buffer
row address strobe	DRAM controller	circular buffer
(RAS)	memory foldback	power-up self-test
column address	memory map	checksum
strobe (CAS)	cache	

PROBLEMS

SECTIONS 12-1 TO 12-3

B
12-1. A certain memory has a capacity of 16K × 32. How many words does it store? What is the number of bits per word? How many memory cells does it contain?

B
12-2. How many different addresses are required by the memory of Problem 12-1?

B
12-3. What is the capacity of a memory that has 16 address inputs, four data inputs, and four data outputs?

B
12-4. A certain memory stores 8K 16-bit words. How many data input and data output lines does it have? How many address lines does it have? What is its capacity in bytes?

DRILL QUESTIONS

B
12-5. Define each of the following terms.
    (a) RAM
    (b) RWM
    (c) ROM
    (d) Internal memory
    (e) Auxiliary memory
    (f) Capacity
    (g) Volatile
    (h) Density
    (i) Read
    (j) Write

B
12-6. (a) What are the three buses in a computer memory system?
    (b) Which bus is used by the CPU to select the memory location?
    (c) Which bus is used to carry data from memory to the CPU during a read operation?
    (d) What is the source of data on the data bus during a write operation?

SECTIONS 12-4 AND 12-5

B
12-7. Refer to Figure 12-6. Determine the data outputs for each of the following input conditions.
    (a) [A] = 1011; CS = 1
    (b) [A] = 0111; CS = 0

B   12-8. Refer to Figure 12-7.

(a) What register is enabled by input address 1011?

(b) What input address code selects register 4?

B   12-9. A certain ROM has a capacity of $16K \times 4$ and an internal structure like that shown in Figure 12-7.

(a) How many registers are in the array?

(b) How many bits are there per register?

(c) What size decoders does it require?

DRILL QUESTION

B   12-10. (a) *True or false:* ROMs cannot be erased.

(b) What is meant by *programming* or *burning* a ROM?

(c) Define a ROM's access time.

(d) How many data inputs, data outputs, and address inputs are needed for a 1024×4 ROM?

(e) What is the function of the decoders on a ROM chip?

SECTION 12-6

C, D 12-11. Figure 12-46 shows how data from a ROM can be transferred to an external register. The ROM has the following timing parameters: $t_{ACC} = 250$ ns and $t_{OE} = 120$ ns. Assume that the new address inputs have been applied to the ROM 500 ns before the occurrence of the TRANSFER pulse. Determine the minimum duration of the TRANSFER pulse for reliable transfer of data.

FIGURE 12-46 Problem 12-11.

C, D 12-12. Repeat Problem 12-11 if the address inputs are changed 70 ns prior to the TRANSFER pulse.

SECTIONS 12-7 AND 12-8

B   12-13. **DRILL QUESTION**

For each item below, indicate the type of memory being described: MROM, PROM, EPROM, EEPROM, flash. Some items will correspond to more than one memory type.

(a) Can be programmed by the user but cannot be erased.

(b) Is programmed by the manufacturer.

(c) Is volatile.

(d) Can be erased and reprogrammed over and over.

(e) Individual words can be erased and rewritten.

(f) Is erased with UV light.

(g) Is erased electrically.

(h) Uses fusible links.

(i) Can be erased in bulk or in sectors of 512 bytes.

(j) Does not have to be removed from the system to be erased and reprogrammed.

(k) Requires a special supply voltage for reprogramming.

(l) Erase time is about 15 to 20 min.

B   12-14. Which transistors in Figure 12-9 will be conducting when $A_1 = A_0 = 1$ and $\overline{EN} = 0$?

12-15. Change the MROM connections in Figure 12-9 so that the MROM stores the function $y = 3x + 5$.

D   12-16. Figure 12-47 shows a simple circuit for manually programming a 2732 EPROM. Each EPROM data pin is connected to a switch that can be set at a 1 or a 0 level. The address inputs are driven by a 12-bit counter. The 50-ms programming pulse comes from a one-shot each time the PROGRAM push button is actuated.

(a) Explain how this circuit can be used to program the EPROM memory locations sequentially with the desired data.

(b) Show how 74293s and a 74121 can be used to implement this circuit.

(c) Should switch bounce have any effect on the circuit operation?

FIGURE 12-47
Problem 12-16.

N   12-17. Figure 12-48 shows a 28F256A flash memory chip connected to a CPU over a data bus and an address bus. The CPU can write to or read from the flash memory array by sending the desired memory address and generating the appropriate control signals to the chip [Figure 12-15(b)]. The CPU can also write to the chip's command register (Figure 12-16) by generating the appropriate control signals and sending the desired command code over the data bus. For this latter operation, the CPU does not have to send a specific memory address to the chip; in other words, the address lines are don't-cares.

(a) Consider the following sequence of CPU operations. Determine what will have happened to the flash memory at the completion of the sequence. Assume that the command register is holding 00_{16}.

   1. The CPU places 20_{16} on the data bus and pulses $\overline{CE}$ and $\overline{WE}$ LOW while holding $\overline{OE}$ HIGH. The address bus is at 0000_{16}.

   2. The CPU repeats step 1.

(b) After the sequence above has been executed, the CPU executes the following sequence. Determine what this does to the flash memory chip.

   1. The CPU places 40_{16} on the data bus and pulses $\overline{CE}$ and $\overline{WE}$ LOW while holding $\overline{OE}$ HIGH. The address bus is at 0000_{16}.

   2. The CPU places $3C_{16}$ on the data bus and 2300_{16} onto the address bus, and it pulses $\overline{CE}$ and $\overline{WE}$ LOW while holding $\overline{OE}$ HIGH.

FIGURE 12-48   Problem 12-17.

SECTION 12-9

N   **12-18.** Another ROM application is the generation of timing and control signals. Figure 12-49 shows a 16×8 ROM with its address inputs driven by a MOD-16 counter so that the ROM addresses are incremented with each input pulse. Assume that the ROM is programmed as in Figure 12-6, and sketch the waveforms at each ROM output as the pulses are applied. Ignore ROM delay times. Assume that the counter starts at 0000.

FIGURE 12-49
Problem 12-18.

D   12-19. Change the program stored in the ROM of Problem 12-18 to generate the D_7 waveform of Figure 12-50.

FIGURE 12-50
Problem 12-19.

D   12-20. Refer to the function generator of Figure 12-17.

(a) What clock frequency will result in a 100-Hz sine wave at the output?

(b) What method could be used to vary the peak-to-peak amplitude of the sine wave?

C   12-21. For the ML2035 of Figure 12-18, assume that a value of 038E (hex) in the latch will produce the desired frequency. Draw the timing diagram for the *LATI, SID*, and *SCK* inputs, and assume that the LSB is shifted in first.

N, C   12-22. The system shown in Figure 12-51 is a waveform (function) generator. It uses four 256-point look-up tables in a 1-Kbyte ROM to store one cycle each of a sine wave (address 000–0FF), a positive slope ramp (address 100–1FF), a negative slope ramp (200–2FF), and a triangle wave (300–3FF). The phase relationship among the three output channels is controlled by the values initially loaded into the three counters. The critical timing parameters are $t_{pd(ck-Q \text{ and } OE - Q \text{ max})}$, counters = 10 ns, latches = 5 ns, and t_{ACC} ROM = 20 ns. Study the diagram until you understand how it operates and then answer the following:

(a) If counter A is initially loaded with 0, what values must be loaded into counters B and C so that A lags B by 90° and A lags C by 180°?

(b) If counter A is initially loaded with 0, what values must be loaded into counters B and C to generate a three-phase sine wave with 120° shift between each output?

(c) What must the frequency of pulses on DAC_OUT be in order to generate a 60-Hz sine wave output?

(d) What is the maximum frequency of the CLK input?

(e) What is the maximum frequency of the output waveforms?

(f) What is the purpose of the function select counter?

FIGURE 12-51

SECTION 12-11

12-23. (a) Draw the logic symbol for an MCM101514, a CMOS static RAM organized as a 256K × 4 with separate data in and data out, and an active-LOW chip enable.

(b) Draw the logic symbol for an MCM6249, a CMOS static RAM organized as a 1M × 4 with common I/O, an active-LOW chip enable, and an active-LOW output enable.

SECTION 12-12

12-24. A certain static RAM has the following timing parameters (in nanoseconds):

$t_{RC} = 100$	$t_{AS} = 20$
$t_{ACC} = 100$	$t_{AH} = $ not given
$t_{CO} = 70$	$t_W = 40$
$t_{OD} = 30$	$t_{DS} = 10$
$t_{WC} = 100$	$t_{DH} = 20$

(a) How long after the address lines stabilize will valid data appear at the outputs during a read cycle?

(b) How long will output data remain valid after $\overline{CS}$ returns HIGH?

(c) How many read operations can be performed per second?

(d) How long should $R/\overline{W}$ and $\overline{CS}$ be kept HIGH after the new address stabilizes during a write cycle?

(e) What is the minimum time that input data must remain valid for a reliable write operation to occur?

(f) How long must the address inputs remain stable after $R/\overline{W}$ and $\overline{CS}$ return HIGH?

(g) How many write operations can be performed per second?

SECTIONS 12-13 TO 12-17

12-25. Draw the logic symbol for the TMS4256, which is a 256K $\times$ 1 DRAM. How many pins are saved by using address multiplexing for this DRAM?

D   12-26. Figure 12-52(a) shows a circuit that generates the $\overline{RAS}$, $\overline{CAS}$, and MUX signals needed for proper operation of the circuit of Figure 12-28(b). The 10-MHz master clock signal provides the basic timing for the computer. The memory request signal ($MEMR$) is generated by the CPU in synchronism with the master clock, as shown in part (b) of the figure. $MEMR$ is normally LOW and is driven HIGH whenever the CPU wants to access memory for a read or a write operation. Determine the waveforms at Q_0, $\overline{Q}_1$, and Q_2, and compare them with the desired waveforms of Figure 12-29.

(a)

(b)

FIGURE 12-52 Problem 12-26.

D   **12-27.** Show how to connect two 74157 multiplexers (Section 9-7) to provide the multiplexing function required in Figure 12-28(b).

12-28. Refer to the signals in Figure 12-30. Describe what occurs at each of the labeled time points.

12-29. Repeat Problem 12-28 for Figure 12-31.

C   **12-30.** The 21256 is a 256K $\times$ 1 DRAM that consists of a 512 $\times$ 512 array of cells. The cells must be refreshed within 4 ms for data to be retained. Figure 12-33(a) shows the signals used to execute a $\overline{CAS}$-before-$\overline{RAS}$ refresh cycle. Each time a cycle such as this occurs, the on-chip refresh circuitry will refresh a row of the array at the row address specified by a refresh counter. The counter is incremented after each refresh. How often should $\overline{CAS}$-before-$\overline{RAS}$ cycles be applied in order for all of the data to be retained?

12-31. Study the functional block diagram of the TMS44100 DRAM in Figure 12-27.

    (a) What are the actual dimensions of the DRAM cell array?

    (b) If the cell array were actually square, how many rows would there be?

    (c) How would this affect the refresh time?

SECTION 12-18

D   **12-32.** Show how to combine two 6206 RAM chips (Figure 12-20) to produce a 32K $\times$ 16 module.

D   **12-33.** Show how to connect two of the 6264 RAM chips symbolized in Figure 12-23 to produce a 16K $\times$ 8 RAM module. The circuit should not require any additional logic. Draw a memory map showing the address range of each RAM chip.

D   **12-34.** Describe how to modify the circuit of Figure 12-37 so that it has a total capacity of 16K $\times$ 8. Use the same type of PROM chips.

D   **12-35.** Modify the decoding circuit of Figure 12-37 to operate from a 16-line address bus (i.e., add A_{13}, A_{14}, and A_{15}). The four PROMs are to maintain the same hex address ranges.

C   **12-36.** For the memory system of Figure 12-38, assume that the CPU is storing one byte of data at system address 4000 (hex).

    (a) Which chip is the byte stored in?

    (b) Is there any other address in this system that can access this data byte?

    (c) Answer parts (a) and (b) by assuming that the CPU has stored a byte at address 6007. (*Hint:* Remember that the EEPROM is not completely decoded.)

    (d) Assume that the program is storing a sequence of data bytes in the EEPROM and that it has just completed the 2048th byte at address 67FF. If the programmer allows it to store one more byte at address 6800, what will be the effect on the first 2048 bytes?

D   **12-37.** Draw the complete diagram for a 256K $\times$ 8 memory that uses RAM chips with the following specifications: 64K $\times$ 4 capacity, common input/output line, and two active-LOW chip select inputs. [*Hint:* The circuit can be designed using only two inverters (plus memory chips).]

SECTION 12-20

12-38. Modify the RAM circuit of Figure 12-42 as follows: change the OR gate to an AND gate and disconnect its output from C; connect the AND output to E_3; connect C to ground. Determine the address range for each RAM module.

C, D 12-39. Show how to expand the system of Figure 12-42 to an 8K × 8 with addresses ranging from 0000 to 1FFF. (*Hint:* This can be done by adding the necessary memory modules and modifying the existing decoding logic.)

T 12-40. A dynamic test is performed on the decoding logic of Figure 12-42 by keeping $E = 1$ and connecting the outputs of a six-bit counter to address inputs A_{10} to A_{15}. The decoder outputs are monitored with an oscilloscope (or a logic analyzer) as the counter is continuously pulsed by a 1-MHz clock. Figure 12-53(a) shows the displayed signals. What are the most probable faults?

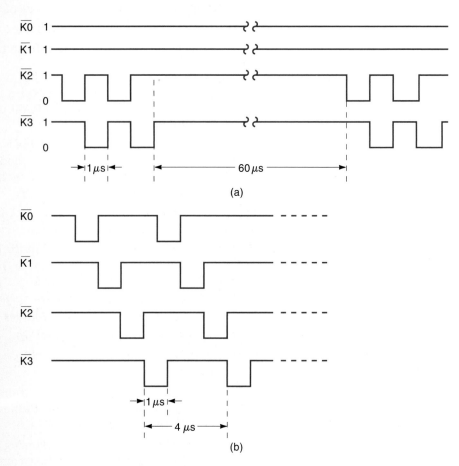

FIGURE 12-53   Problems 12-40 and 12-41.

C, T 12-41. Repeat Problem 12-40 for the decoder outputs shown in Figure 12-53(b).

C, D 12-42. Consider the RAM system of Figure 12-42. The checkerboard pattern test will not be able to detect certain types of faults. For instance, assume that there is a break in the connection to the A input to the decoder. If a checkerboard pattern SELF-TEST is performed on this circuit, the displayed messages will state that the memory is OK.

(a) Explain why the circuit fault was not detected.

(b) How would you modify the SELF-TEST so that faults such as this will be detected?

T   12-43. Assume that the 1K × 8 modules used in Figure 12-42 are formed from two 1K × 4 RAM chips. The following messages are printed out when the power-up self-test is performed on this RAM system:

```
module-0 test OK
module-1 test OK
address 0800 faulty at bits 4-7
address 0801 faulty at bits 4-7
address 0802 faulty at bits 4-7
      .   .   .     .     .   .
      .   .   .     .     .   .
address 0BFE faulty at bits 4-7
address 0BFF faulty at bits 4-7
module-3 test OK
```

Examine these messages and list the possible faults.

T   12-44. The following messages are printed out when the power-up self-test is performed on the RAM system of Figure 12-44.

```
module-0 test OK
module-1 test OK
module-2 test OK
address 0C00 faulty at bit 7
address 0C01 faulty at bit 7
address 0C02 faulty at bit 7
      .   .   .     .   .
      .   .   .     .   .
      .   .   .     .   .
address 0FFE faulty at bit 7
address 0FFF faulty at bit 7
```

Examine these messages and list the possible faults.

T   12-45. What messages would be printed out when a power-up self-test is performed on the RAM system of Figure 12-44 if there is a short between the decoder's $\overline{K2}$ and $\overline{K3}$ outputs?

SECTION 12-21

T   12-46. Consider the 16 × 8 ROM in Figure 12-6. Replace the data word stored at address location 1111 with a checksum calculated from the other 15 data words.

ANSWERS TO SECTION REVIEW QUESTIONS

SECTION 12-1

1. See text.   2. 16 bits per word; 8192 words; 131,072 bits or cells   3. In a read operation, a word is taken from a memory location and is transferred to another device. In a write operation, a new word is placed in a memory location and replaces the one previously stored there.   4. True   5. SAM: Access time is not constant but depends on the physical location of the word being accessed. RAM: Access time

is the same for any address location.   6. RWM is memory that can be read from or written to with equal ease. ROM is memory that is mainly read from and is written into very infrequently.   7. False; its data must be periodically refreshed.

SECTION 12-2

1. 14, 12, 12   2. Commands the memory to perform either a read operation or a write operation   3. When in its active state, this input enables the memory to perform the read or the write operation selected by the $R/\overline{W}$ input. When in its inactive state, this input disables the memory so that it cannot perform the read or the write function.

SECTION 12-3

1. Address lines, data lines, control lines   2. , 3. See text.

SECTION 12-4

1. True   2. Apply desired address inputs; activate control input(s); data appear at data outputs.   3. Process of entering data into ROM

SECTION 12-5

1. $A_3A_2A_1A_0 = 1001$ 2. The row-select decoder activates one of the enable inputs of all registers in the selected row. The column-select decoder activates one of the enable inputs of all registers in the selected column. The output buffers pass the data from the internal data bus to the ROM output pins when the CS input is activated.

SECTION 12-7

1. False; by the manufacturer   2. A PROM can be programmed once by the user. It cannot be erased and reprogrammed.   3. True   4. By exposure to UV light 5. True   6. Automatically programs data into memory cells one address at a time 7. An EEPROM can be electrically erased and reprogrammed without removal from its circuit, and it is byte erasable.   8. Low density; high cost   9. EEPROM 10. One

SECTION 12-8

1. Electrically erasable and programmable in circuit   2. Higher density; lower cost 3. Short erase and programming times   4. For the erase and programming operations   5. The contents of this register control all internal chip functions. 6. To confirm that a memory address has been successfully erased (i.e., data = all 1s) 7. To confirm that a memory address has been programmed with the correct data

SECTION 12-9

1. Microcomputer programs stored in ROM   2. On power-up, the computer executes a small bootstrap program from ROM to initialize the system hardware and to load the operating system from mass storage (disk).   3. Circuit that takes data represented in one type of code and converts it to another type of code 4. Counter, ROM, DAC, low-pass filter   5. They are nonvolatile, fast, reliable, small, and low-power.

SECTION 12-11

1. Desired address applied to address inputs; $R/\overline{W} = 1$; CS or CE activated 2. To reduce pin count   3. 24, including V_{CC} and ground

SECTION 12-12

1. SRAM cells are flip-flops; DRAM cells use capacitors.   2. CMOS   3. RAM 4. CPU   5. Read- and write-cycle times   6. False; when $\overline{WE}$ is LOW, the I/O pins act as data inputs regardless of the state of $\overline{OE}$ (second entry in mode table).

7. A_{13} can remain connected to pin 26. A_{14} must be removed, and pin 27 must be connected to +5 V.

SECTION 12-13

1. Generally slower speed; need to be refreshed 2. Low power; high capacity; lower cost per bit 3. DRAM

SECTION 12-14

1. 256 rows × 256 columns 2. It saves pins on the chip. 3. 1M = 1024K = 1024 × 1024. Thus, there are 1024 rows by 1024 columns. Because $1024 = 2^{10}$, the chip needs 10 address inputs. 4. $\overline{RAS}$ is used to latch the row address into the DRAM's row address register. $\overline{CAS}$ is used to latch the column address into the column address register. 5. *MUX* multiplexes the full address into the row and column addresses for input to the DRAM.

SECTION 12-15

1. (a) True (b) False (c) False (d) True 2. *MUX*

SECTION 12-16

1. (a) True (b) False 2. It provides row addresses to the DRAM during refresh cycles. 3. Address multiplexing and the refresh operation 4. (a) False (b) True

SECTION 12-17

1. No 2. Memory locations with same upper address (same row) 3. Only the column address must be latched. 4. Extended data output 5. *Burst* 6. The system clock

SECTION 12-18

1. Sixteen 2. Four 3. False; when expanding memory capacity, each chip is selected by a different decoder output (see Figure 12-43). 4. True

SECTION 12-19

1. Battery backup for CMOS RAM; flash memory 2. Economics 3. Data are read out of memory in the same order they were written in. 4. A FIFO used to transfer data between devices with widely different operating speeds 5. Circular buffers "wrap around" from the highest address to the lowest, and the newest datum always overwrites the oldest.

SECTION 12-20

1. Prevents decoding glitches by disabling the decoder while the address lines are changing 2. A way to test RAM by writing a checkerboard pattern (first 01010101, then 10101010) into each memory location and then reading it. It is used because it will detect any shorts or interactions between adjacent cells. 3. An automatic test of RAM performed by a computer on power-up

SECTION 12-21

1. A code placed in the last one or two ROM locations that represents the sum of the expected ROM data from all other locations. It is used as a means to test for errors in one or more ROM locations.

CHAPTER 13

PROGRAMMABLE LOGIC DEVICE ARCHITECTURES*†

■ OUTLINE

*Diagrams of the GAL 16V8 device presented in this chapter have been reproduced through the courtesy of Lattice Semiconductor Corporation, Hillsboro, Oregon.

†Diagrams of the MAX7000S and FLEX10K family devices presented in this chapter have been reproduced through the courtesy of Altera Corporation, San Jose, California.

■ OBJECTIVES

Upon completion of this chapter, you will be able to:

- ■ Describe the different categories of digital system devices.
- ■ Describe the different types of PLDs.
- ■ Interpret PLD data book information.
- ■ Define PLD terminology.
- ■ Compare the different programming technologies used in PLDs.
- ■ Compare the architectures of different types of PLDs.
- ■ Compare the features of the Altera MAX7000S and FLEX10K families of PLDs.

■ INTRODUCTION

Throughout the chapters of this book you have been introduced to a wide variety of digital circuits. You now know how the building blocks of digital systems work and can combine them to solve a wide variety of digital problems. More complicated digital systems, such as microcomputers and digital signal processors, have also been briefly described. The defining difference between microcomputer/DSP systems and other digital systems is that the former follow a programmed sequence of instructions that the designer specifies. Many applications require faster response than a microcomputer/DSP architecture can accommodate and in these cases, a conventional digital circuit must be used. In today's rapidly advancing technology market, most conventional digital systems are not being implemented with standard logic device chips containing only simple gates or MSI-type functions. Instead, programmable logic devices, which contain the circuitry necessary to create logic functions, are being used to implement digital systems. These devices are not programmed with a list of instructions, like a computer or DSP. Instead, their internal hardware is configured by electronically connecting and disconnecting points in the circuit.

Why have PLDs taken over so much of the market? With programmable devices, the same functionality can be obtained with one IC rather than using several individual logic chips. This characteristic means less board space, less power required, greater reliability, less inventory, and overall lower cost in manufacturing.

In the previous chapters you have become familiar with the process of programming a PLD using either AHDL or VHDL with MAX+PLUS II. At the same time, you have learned about all the building blocks of digital systems. The PLD implementations of digital circuits up to this point have been presented as a "black box." We have not been concerned with what was going on inside the PLD to make it work. Now that you understand all the circuitry inside the black box, it is time to turn the lights on in there and look at how it works. This will allow you to make the best decisions when selecting and applying a PLD to solve a problem. This chapter will take a look at the various types of hardware available to design digital systems. We will then introduce you to the architectures of various families of PLDs.

13-1　DIGITAL SYSTEMS FAMILY TREE

While the major goal of this chapter is to investigate PLD architectures, it is also useful to look at the various hardware choices available to digital system designers because it should give us a little better perception of today's digital hardware alternatives. The desired circuit functionality can generally be achieved by using several different types of digital hardware. Throughout this book, we have described both standard logic devices as well as how programmable logic devices can be used to create the same functional blocks. Microcomputers and DSP systems can also often be applied with the necessary sequence of instructions (i.e., the application's program) to produce the desired circuit function. The design engineering decisions must take into account many factors, including the necessary speed of operation for the circuit, cost of manufacturing, system power consumption, system size, amount of time available to design the product, etc. In fact, most complex digital designs include a mix of different hardware categories. Many trade-offs between the various types of hardware have to be weighed to design a digital system.

A digital system family tree (see Figure 13-1) showing most of the hardware choices that are currently available can be useful in sorting out the many categories of digital devices. The graphical representation in the figure

FIGURE 13-1　Digital system family tree.

does not show all the details—some of the more complex device types have many additional subcategories, and older, obsolete device types have been omitted for clarity. The major digital system categories include standard logic, application-specific integrated circuits (ASICS) and microprocessor/digital signal processing (DSP) devices.

The first category of **standard logic** devices refers to the basic functional digital components (gates, flip-flops, decoders, multiplexers, registers, counters, etc.) that are available as SSI and MSI chips. These devices have been used for many years (some more than 30 years) to design complex digital systems. An obvious drawback is that the system may literally consist of hundreds of such chips. These inexpensive devices can still be useful if our design is not very complex. As discussed in Chapter 8, there are three major families of standard logic devices: TTL, CMOS, and ECL. TTL is a mature technology consisting of numerous subfamilies that have been developed over many years of use. Very few new designs apply TTL logic, but many, many digital systems still contain TTL devices. CMOS is the most popular standard logic device family today, primarily due to its low power consumption. ECL technology, of course, is applied for higher-speed designs. Standard logic devices are still available to the digital designer, but if the application is very complex, a lot of SSI/MSI chips will be needed. That solution is not very attractive for our design needs today.

The **microprocessor/DSP** category is a much different approach to digital system design. These devices actually contain the various types of functional blocks that have been discussed throughout this text. With microcomputer/DSP systems, devices can be controlled electronically, and data can be manipulated by executing a program of instructions that has been written for the application. A great deal of flexibility can be achieved with microcomputer/DSP systems because all you have to do is change the program. The major downfall with this digital system category is speed. *Using a hardware solution for your digital system design is always faster than a software solution.*

The third major digital system category is called **application-specific integrated circuits (ASICs)**. This broad category represents the modern hardware design solution for digital systems. As the acronym implies, an integrated circuit is designed to implement a specific desired application. Four subcategories of ASIC devices are available to create digital systems: programmable logic devices (PLDs), gate arrays, standard-cell, and full-custom.

PLDs, sometimes referred to as field-programmable logic devices (FPLDs), can be custom-configured to create any desired digital circuit, from simple logic gates to complex digital systems. Many examples of PLD designs have been given in earlier chapters. This ASIC choice for the designer is very different from the other three subcategories. With a relatively small capital investment, any company can purchase the necessary development software and hardware to program PLDs for their digital designs. On the other hand, to obtain a gate array, standard-cell or full-custom ASIC requires that most companies contract with an IC foundry to fabricate the desired IC chip. This option can be extremely expensive and usually requires that your company purchase a large volume of parts to be cost effective.

Gate arrays are ULSI circuits that offer hundreds of thousands of gates. The desired logic functions are created by the interconnections of these prefabricated gates. A custom-designed mask for the specific application determines the gate interconnections, much like the stored data in a mask-programmed ROM. For this reason, they are often referred to as mask-programmed gate arrays (MPGAs). Individually, these devices are less expensive than PLDs of comparable gate count, but the custom programming process by the chip manufacturer is very expensive and requires a great deal of lead time. A newer technology—laser programmed gate array (LPGA)—uses a laser beam to program the IC by cutting unneeded connections before the processed silicon wafer is packaged into individual chips. This eliminates the very expensive mask production needed for MPGAs and offers turnaround times of days instead of weeks.

Standard-cell ASICs use predefined logic function building blocks called cells to create the desired digital system. The IC layout of each cell has been designed previously, and a library of available cells is stored in a computer database. The needed cells are laid out for the desired application, and the interconnections between the cells are determined. Design costs for standard-cell ASICs are even higher than for MPGAs because all IC fabrication masks that define the components and interconnections must be custom designed. Greater lead time is also needed for the creation of the additional masks. Standard cells do have a significant advantage over gate arrays. The cell-based functions have been designed to be much smaller than equivalent functions in gate arrays, which allows for generally higher-speed operation and cheaper manufacturing costs.

Full-custom is considered to be the ultimate ASIC choice. As the name implies, all components (transistors, resistors, and capacitors) and the interconnections between them are custom-designed by the IC designer. This design effort requires a significant amount of time and expense, but it can result in ICs that can operate at the highest possible speed and require the smallest die (individual IC chip) area. Smaller IC die sizes allow for many more die to fit on a silicon wafer, which significantly lowers the manufacturing cost for each IC.

More on PLDs

This chapter is mainly about PLDs, so we will look a little further down that branch of the family tree. The development of PLD technology has advanced continuously since the first PLDs appeared more than 30 years ago. The early devices contained the equivalent of a few hundred gates, and now we have parts available that contain a few million gates. The old devices could handle a few inputs and a few outputs with limited logic capabilities. Now there are PLDs that can handle hundreds of inputs and outputs. Original devices could be programmed only once and, if the design changed, the old PLD would have to be removed from the circuit and a new one, programmed with the updated design, would have to be inserted in its place. With newer devices, the internal logic design can be changed on the fly, while the chip is still connected to a PC board in an electronic system.

Generally, PLDs can be described as being one of three different types: **simple programmable logic devices (SPLDs)**, **complex programmable logic devices (CPLDs)**, or **field programmable gate arrays (FPGAs)**. There are several manufacturers with many different families of PLD devices, so there are many variations in architecture. We will attempt to discuss the general characteristics for each of the types, but be forewarned: the differences are not always clear-cut. The distinction between CPLDs and FPGAs is often a little fuzzy, with the manufacturers constantly designing new, improved architectures and frequently muddying the waters for marketing purposes. Together, CPLDs and FPGAs are often referred to as **high-capacity programmable logic devices (HCPLDs)**. The programming technologies for PLD devices are actually based on the various types of semiconductor memory. As new types of memory have been developed, the same technology has been applied to the creation of new types of PLD devices.

The amount of logic resources available is the major distinguishing feature between SPLDs and HCPLDs. Today, SPLDs are devices that typically contain the equivalent of 600 or fewer gates, while HCPLDs have thousands and hundreds of thousands of gates available. Internal programmable signal interconnect resources are much more limited with SPLDs. SPLDs are generally much less complicated and much cheaper than HCPLDs. Many small digital applications need only the resources of an SPLD. On the other hand, HCPLDs are capable of providing the circuit resources for complete complex digital systems, and larger, more sophisticated HCPLD devices are designed every year.

The SPLD classification includes the earliest PLD devices. The amount of logic resources contained in the early PLDs may be relatively small by today's standards, but they represented a significant technological step in their ability to create easily a custom IC that can replace several standard logic devices. Over the years, numerous semiconductor advances have created different SPLD types. The first PLD type to gain the interest of circuit designers was programmed by literally burning open selected fuses in the programming matrix. The fuses that were left intact in these **one-time programmable (OTP)** devices provided the electrical connections for the AND/OR circuits to produce the desired functions. This logic device was based on the fuse links in PROM memory technology (see Section 12-7) and was most commonly referred to as a programmable logic array (PLA). PLDs didn't really gain widespread acceptance with digital designers until the late 1970s, when a device called a **programmable array logic (PAL)** was introduced. The programmable fuse links in a PAL are used to determine the input connections to a set of AND gates that are wired to fixed OR gates. With the development of the ultraviolet erasable PROM came the EPROM-based PLDs in the mid 1980s, followed soon by PLDs using electrically erasable (EEPROM) technology.

CPLDs are devices that typically combine an array of PAL-type devices on the same chip. The logic blocks themselves are programmable AND/fixed-OR logic circuits with fewer product terms available than most PAL devices. Each logic block (often called a **macrocell**) can typically handle many input variables, and the internal programmable logic signal routing resources tend to be very uniform throughout the chip, producing consistent signal delays. When more product terms are needed, gates may be shared between logic blocks, or several logic blocks can be combined to implement the expression. The flip-flop used to implement the register in the macrocell can often be configured for D, JK, T (toggle), or SC (or SR) operation. Input and output pins for some CPLD architectures are associated with a specific macrocell, and typically additional macrocells are buried (that is, not connected to a pin). Other CPLD architectures may have independent I/O blocks with built-in registers that can be used to latch incoming or outgoing data. The programming technologies used in CPLD devices are all nonvolatile and include EPROM, EEPROM, and flash, with EEPROM being the most common. All three technologies are erasable and reprogrammable.

FPGAs also have a few fundamental characteristics that are shared. They typically consist of many relatively small and independent programmable logic modules that can be interconnected to create larger functions. Each module can usually handle only up to four or five input variables. Most FPGA logic modules utilize a **look-up table (LUT)** approach to create the desired logic functions. A look-up table functions just like a truth table in which the output can be programmed to create the desired combinational function by storing the appropriate 0 or 1 for each input combination. The programmable signal routing resources within the chip tend to be quite varied, with many different path lengths available. The signal delays produced for a design depend on the actual signal routing selected by the programming software. The logic modules also contain programmable registers. The logic modules are not associated with any I/O pin. Instead, each I/O pin is connected to a programmable input/output block that, in turn, is connected to the logic modules with selected routing lines. The I/O blocks can be configured to provide input, output, or bidirectional capability, and built-in registers can be used to latch incoming or outgoing data. A general architecture of FPGAs is shown in Figure 13-2. All of the logic blocks and input/output blocks can be programmed to implement almost any logic circuit. The programmable interconnections are accomplished via lines that run through the rows and columns in the channels between the logic blocks. Some FPGAs include large blocks of RAM memory; others do not.

The programming technologies used in FPGA devices include SRAM, flash, and antifuse, with SRAM being the most common. SRAM-based de-

○   Programmable interconnect
──   Connecting segment
……   Interconnect path

NOTE: Clock inputs may
have special low-skew
interconnect paths.

FIGURE 13-2   FPGA architecture.

vices are volatile and therefore require the FPGA to be reconfigured (programmed) when it is powered-up. The programming information that defines how each logic block functions, which I/O blocks are inputs and outputs, and how the blocks are interconnected is stored in some type of external memory that is downloaded to the SRAM-based FPGA when power is applied. Antifuse devices are one-time programmable and are therefore nonvolatile. Antifuse memory technology is not currently used for memory devices but, as its name implies, it is the opposite of fuse technology. Instead of opening a fuse link to prevent a signal connection, an insulator layer between interconnects has an electrical short created to produce a signal connection. Antifuse devices are programmed in a device programmer either by the end-user or by the factory or distributor.

Differences in architecture between CPLDs and FPGAs, among different HCPLD manufacturers, and among different families of devices from a single manufacturer can affect the efficiency of design implementation for a particular application. You may ask, "Does the architecture of this PLD family provide the best fit for my application?" It is very difficult, however, to predict which architecture may be the best choice to use for a complex digital system. Only a portion of the available gates can be utilized. Who knows how many equivalent gates will be needed for a large design? The basic design of the signal routing resources can affect how much of the PLD's logic resources can be utilized. The segmented interconnects often found in FPGAs can produce shorter delays between adjacent logic blocks, but they may also produce longer delays between the blocks that are further apart than would be produced by the continuous type of interconnect found in most CPLDs. There is no easy answer to your question, but every HCPLD manufacturer will give you an answer anyway: their product is best!

As you can see, the field of PLDs is quite diverse and it is constantly changing. You should now have the basic knowledge of the various types and technologies necessary to interpret PLD data sheets and learn more about them.

REVIEW QUESTIONS

1. What are the three major categories of digital systems?
2. What is the major disadvantage of a microprocessor/DSP design?
3. What does ASIC stand for?
4. What are the four types of ASICs?
5. What are HCPLDs?
6. What are two major differences between CPLDs and FPGAs?
7. What does volatility refer to?

13-2 FUNDAMENTALS OF PLD CIRCUITRY

A simple PLD device is shown in Figure 13-3. Each of the four OR gates can produce an output that is a function of the two input variables, A and B. Each output function is programmed with the fuses located between the AND gates and each of the OR gates.

Each of the inputs A and B feed both a noninverting buffer and an inverting buffer to produce the true and inverted forms of each variable. These are the *input lines* to the AND gate array. Each AND gate is connected to two

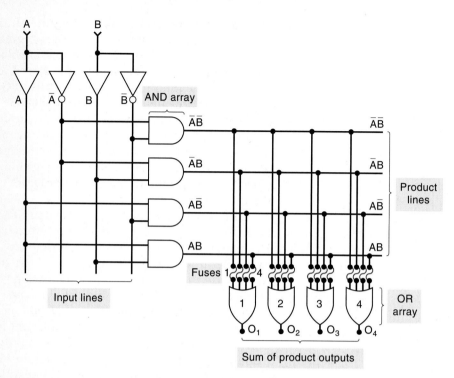

FIGURE 13-3   Example of a programmable logic device.

different input lines to generate a unique product of the input variables. The AND outputs are called the *product lines.*

Each of the product lines is connected to one of the four inputs of each OR gate through a fusible link. With all of the links initially intact, each OR output will be a constant 1. Here's the proof:

$$O_1 = \overline{A}\,\overline{B} + \overline{A}B + A\overline{B} + AB$$

$$= \overline{A}(\overline{B} + B) + A(\overline{B} + B)$$

$$= \overline{A} + A = 1$$

Each of the four outputs O_1, O_2, O_3, and O_4 can be *programmed* to be any function of A and B by selectively blowing the appropriate fuses. PLDs are designed so that a blown OR input acts as a logic 0. For example, if we blow fuses 1 and 4 at OR gate 1, the O_1 output becomes

$$O_1 = 0 + \overline{A}B + A\overline{B} + 0 = \overline{A}B + A\overline{B}$$

We can program each of the OR outputs to any desired function in a similar manner. Once all of the outputs have been programmed, the device will permanently generate the selected output functions.

PLD Symbology

The example in Figure 13-3 has only two input variables and the circuit diagram is already quite cluttered. You can imagine how messy the diagram would be for PLDs with many more inputs. For this reason, PLD manufacturers have adopted a simplified symbolic representation of the internal circuitry of these devices.

FIGURE 13-4 Simplified PLD symbology.

Figure 13-4 shows the same PLD circuit as Figure 13-3 using the simplified symbols. First, notice that the input buffers are represented as a single buffer with two outputs, one inverted and one noninverted. Next, note that a *single line* is shown going into the AND gate to represent all four inputs. Each time the row line crosses a column represents a separate input to the AND gate. The connections from the input variable lines to the AND gate inputs are indicated as dots. A dot means that this connection to the AND gate input is hard-wired (i.e., one that cannot be changed). At first glance, it looks like the input variables are connected to each other. It is important to realize that this is *not* the case because the single row line represents *multiple* inputs to the AND gate.

The inputs to each of the OR gates are also designated by a single line representing all four inputs. An X represents an intact fuse connecting a product line to one input of the OR gate. The absence of an X (or a dot) at any intersection represents a blown fuse. For OR gate inputs, blown fuses (unconnected inputs) are assumed to be LOW, and for AND gate inputs, blown fuses are HIGH. In this example, the outputs are programmed as:

$$O_1 = \overline{A}B + A\overline{B}$$
$$O_2 = AB$$
$$O_3 = 0$$
$$O_4 = 1$$

REVIEW QUESTIONS

1. What is a PLD?
2. What would output O_1 be in Figure 13-3 if fuses 1 and 2 were blown?
3. What does an X represent on a PLD diagram?
4. What does a dot represent on a PLD diagram?

13-3 PLD ARCHITECTURES

The concept of PLDs has led to many different architectural designs of the inner circuitry of these devices. In this section, we will explore some of the basic differences in architecture.

PROMs

The architecture of the programmable circuits in the previous section involves programming the connections to the OR gate. The AND gates are used to decode all the possible combinations of the input variables, as shown in Figure 13-5(a). For any given input combination, the corresponding row is ac-

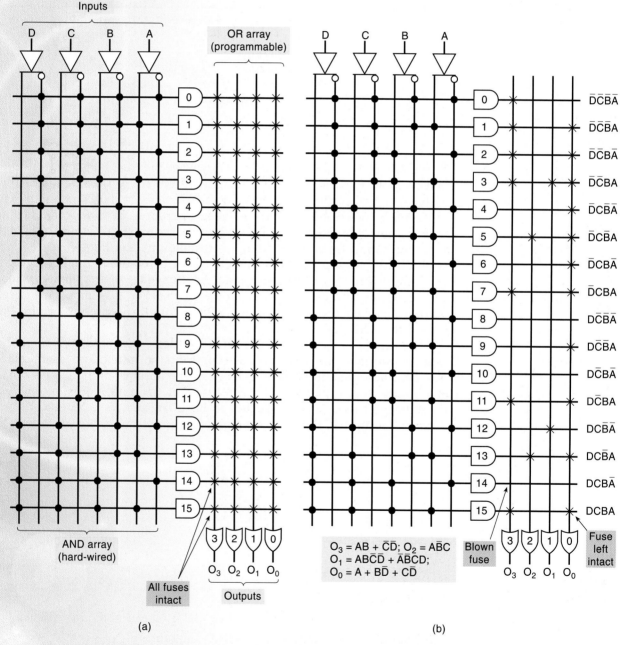

$O_3 = AB + \overline{C}\overline{D}; \quad O_2 = A\overline{B}C$
$O_1 = AB\overline{C}\overline{D} + \overline{A}\overline{B}CD;$
$O_0 = A + B\overline{D} + C\overline{D}$

FIGURE 13-5 (a) PROM architecture makes it suitable for PLDs; (b) fuses are blown to program outputs for given functions.

tivated (goes HIGH). If the OR input is connected to that row, a HIGH appears at the OR output. If the input is not connected, a LOW appears at the OR output. Does this sound familiar? Refer back to Figure 12-9. If you think of the input variables as address inputs and the intact/blown fuses as stored 1s and 0s, you should recognize the architecture of a PROM.

Figure 13-5(b) shows how the PROM would be programmed to generate four specified logic functions. Let's follow the procedure for output $O_3 = AB + \overline{C}\overline{D}$. The first step is to construct a truth table showing the desired O_3 output level for all possible input combinations (Table 13-1).

Next, write down the AND products for those cases where the output is to be a 1. The O_3 output is to be the OR sum of these products. Thus, only the fuses that connect these product terms to the inputs of OR gate 3 are to be left intact. All others are to be blown, as indicated in Figure 13-5(b). This same procedure is followed to determine the status of the fuses at the other OR gate inputs.

The PROM can generate any possible logic function of the input variables because it generates every possible AND product term. In general, any application that requires every input combination to be available is a good candidate for a PROM. However, PROMs become impractical when a large number of input variables must be accommodated because the number of fuses doubles for each added input variable.

Calling a PROM a PLD is really just a semantics issue. You already knew that a PROM is programmable and it is a logic device. This is just a way of using a PROM and thinking of its purpose as implementing SOP logic expressions rather than storing data values in memory locations. The real problem is translating the logic equations into the fuse map for a given PROM. A general-purpose logic compiler (such as CUPL or ABEL) designed to program SPLDs has a list of PROM devices that it can support. If you choose to use any old scavenged EPROM as a PLD, you may need to generate your own bit map (like they used to do it), which is very tedious.

TABLE 13-1

D	C	B	A	O_3		
0	0	0	0	1	→	$\overline{D}\,\overline{C}\,\overline{B}\,\overline{A}$
0	0	0	1	1	→	$\overline{D}\,\overline{C}\,\overline{B}\,A$
0	0	1	0	1	→	$\overline{D}\,\overline{C}\,B\,\overline{A}$
0	0	1	1	1	→	$\overline{D}\,\overline{C}\,B\,A$
0	1	0	0	0		
0	1	0	1	0		
0	1	1	0	0		
0	1	1	1	1	→	$\overline{D}\,C\,B\,A$
1	0	0	0	0		
1	0	0	1	0		
1	0	1	0	0		
1	0	1	1	1	→	$D\,\overline{C}\,B\,A$
1	1	0	0	0		
1	1	0	1	0		
1	1	1	0	0		
1	1	1	1	1	→	$D\,C\,B\,A$

Programmable Array Logic (PAL)

The PROM architecture is well suited for those applications where every possible input combination is required to generate the output functions. Examples are code converters and data storage (look-up) tables that we examined in Chapter 12. When implementing SOP expressions, however, they do not make very efficient use of circuitry. Each combination of address inputs must be fully decoded, and each expanded product term has an associated fuse that is used to OR them together. For example, notice how many fuses were required in Figure 13-5 to program the simple SOP expressions and how many product terms are often not used. This has led to the development of a class of PLDs called programmable array logic (PAL). The architecture of a PAL differs slightly from that of a PROM, as shown in Figure 13-6(a).

The PAL has an AND and OR structure similar to a PROM but in the PAL, inputs to the AND gates are programmable, whereas the inputs to the OR gates are hard-wired. This means that every AND gate can be programmed to generate any desired product of the four input variables and their complements. Each OR gate is hard-wired to only four AND outputs. This limits each output function to four product terms. If a function requires more than four product terms, it cannot be implemented with this PAL; one having more OR inputs would have to be used. If fewer than four product terms are required, the unneeded ones can be made 0.

Figure 13-6(b) shows how this PAL is programmed to generate four specified logic functions. Let's follow the procedure for output $O_3 = AB + \overline{C}\overline{D}$. First, we must express this output as the OR sum of four terms because the OR gates have four inputs. We do this by putting in 0s. Thus, we have

$$O_3 = AB + \overline{C}\overline{D} + 0 + 0$$

Next, we must determine how to program the inputs to AND gates 1, 2, 3, and 4 so that they provide the correct product terms to OR gate 3. We do this term by term. The first term, AB, is obtained by leaving intact the fuses that connect inputs A and B to AND gate 1 and by blowing all other fuses on that line. Likewise, the second term, $\overline{C}\overline{D}$, is obtained by leaving intact only the fuses that connect inputs $\overline{C}$ and $\overline{D}$ to AND gate 2. The third term is a 0. A constant 0 is produced at the output of AND gate 3 by leaving all of its input fuses intact. This would produce an output of $A\overline{A}B\overline{B}C\overline{C}D\overline{D}$, which, as we know, is 0. The fourth term is also 0, so the input fuses to AND gate 4 are also left intact.

The inputs to the other AND gates are programmed similarly to generate the other output functions. Note especially that many of the AND gates have all of their input fuses intact because they need to generate 0s.

An example of an actual PAL integrated circuit is the PAL16L8, which has 10 logic inputs and eight output functions. Each output OR gate is hard-wired to seven AND gate outputs, and so it can generate functions that include up to seven terms. An added feature of this particular PAL is that six of the eight outputs are fed back into the AND array, where they can be connected as inputs to any AND gate. This makes it very useful in generating all sorts of combinational logic.

The PAL family also contains devices with variations of the basic SOP circuitry we have described. For example, most PAL devices have a tristate buffer driving the output pin. Others channel the SOP logic circuit to a D FF input and use one of the pins as a clock input to clock all of the output flip flops synchronously. These devices are referred to as *registered PLDs* because

$$O_3 = AB + \overline{C}\overline{D}; \quad O_2 = A\overline{B}C$$
$$O_1 = AB\overline{C}\overline{D} + \overline{A}\overline{B}CD;$$
$$O_0 = A + B\overline{D} + C\overline{D}$$

FIGURE 13-6 (a) Typical PAL architecture; (b) the same PAL programmed for the given functions.

the outputs pass through a register. An example is the PAL16R8, which has up to eight registered outputs (which can also serve as inputs) plus eight dedicated inputs.

Field Programmable Logic Array (FPLA)

The field programmable logic array (FPLA) was developed in the mid-1970s as the first nonmemory programmable logic device. It used a programmable AND array as well as a programmable OR array. Although the FPLA is more flexible than the PAL architecture, it has not been as widely accepted by engineers. FPLAs are used mostly in state-machine design where a large number of product terms are needed in each SOP expression.

1. Verify that the correct fuses are blown for the O_2, O_1, and O_0 functions in Figure 13-5(b).

2. A PAL has a hard-wired _____ array and a programmable _____ array.

3. A PROM has a hard-wired _____ array and a programmable _____ array.

4. How would the equation for the output of O_1 in Figure 13-5(b) change if all the fuses from AND gate 14 were left intact?

13-4 THE GAL 16V8 (GENERIC ARRAY LOGIC)

The GAL 16V8, introduced by Lattice Semiconductor, has an architecture that is very similar to the PAL devices described in the previous section. Standard, low-density PALs are one-time programmable. The GAL chip, on the other hand, uses an EEPROM array (located at row and column intersections in Figure 13-7) to control the programmable connections to the AND matrix, allowing them to be erased and reprogrammed at least 100 times. In addition to the AND and OR gates used to produce the sum of product functions, the GAL 16V8 contains optional flip-flops for register and counter applications, tristate buffers for the outputs, and control multiplexers used to select the various modes of operation. Consequently, it can be used as a generic, pin-compatible replacement for most PAL devices. Specific locations in the memory array are designated to control the various programmable connections in the chip. Fortunately, it is not necessary to delve into the addresses of each bit location in the matrix. The programming software takes care of these details in a user-friendly manner.

The complete logic diagram of the GAL 16V8 is shown in Figure 13-7. This device has eight dedicated input pins (pins 2–9), two special function inputs (pins 1 and 11), and eight pins (12–19) that can be used as inputs or outputs. The major components of the GAL devices are the input term matrix; the AND gates, which generate the products of input terms; and the output logic macrocells (OLMCs). Notice that the eight inputs (pins 2–9) are each connected directly to a column of the input term matrix. The complement of each of these inputs is also connected to a column of the matrix. These pins must always be specified as inputs when programming the 16V8. A logic level and its complement are also fed from each OLMC back to a column of the input matrix. This accounts for the 32 input variables (columns in the input matrix) that can be programmed as connections to the 64 multiple-input AND gates.

The flexibility of the GAL 16V8 lies in its programmable output logic macrocell. Eight different products (outputs of AND gates) are applied as inputs to each of the eight output logic macrocells. Within each OLMC, the products are ORed together to generate the sum of products (SOP). Recall from Chapter 4 that any logic function can be expressed in SOP form. Within the OLMC, the SOP output may be routed to the output pin to implement a combinational circuit, or it may be clocked into a D flip-flop to implement a registered output circuit.

To understand the detailed operation of the OLMC, refer to Figure 13-8. The figure shows the structure of OLMC(n), where n is a number from 12 to 19. Notice that seven of the products are unconditionally connected to the OR gate inputs. The eighth product term is connected to a two-input product term multiplexer (PTMUX), which drives the eighth input to the OR gate. The eighth product term also connects to one input of a four-input multiplexer (TSMUX). The output of TSMUX enables the tristate inverter that

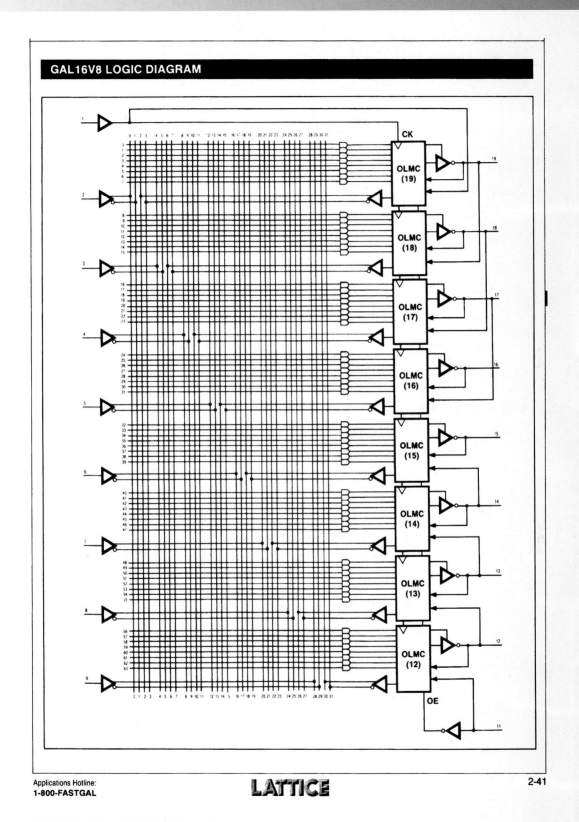

GAL16V8 LOGIC DIAGRAM

FIGURE 13-7 GAL 16V8 logic diagram.

FIGURE 13-8 Output logic macrocell for the GAL 16V8.

drives the output pin [I/O(n)]. The output multiplexer (OMUX) is a two-input MUX that selects between the combinational output (OR gate) and the registered output (the D flip-flop). A fourth MUX selects the logic signal that is fed back to the input matrix. This is called the *feedback* multiplexer (FMUX).

Each of these multiplexers is controlled by programmable bits (AC1 and AC0) in the EEPROM matrix. This is the way that the OLMC configuration can be altered by the programmer. Another programmable bit is the input to the XOR gate. This provides the programmable output polarity feature. Recall that an XOR gate can be used to complement a logic signal selectively, as shown in Figure 13-9. When the control line is a logic 0, the XOR will pass the logic level at input A with no inversion. When the control bit is a logic 1, the XOR will invert the signal so that $X = \overline{A}$. In Figure 13-8, the programmable bit (labeled XOR) is a logic 1 under normal positive logic conditions. This inverts the output of the OR gate, which is inverted again when it passes through the tristate inverting buffer on the output.

We can understand the various configuration options by studying the possible inputs to each multiplexer. The TSMUX controls the tristate buffer's en-

CONTROL	INPUT A	OUTPUT X	
0	0	0	Not inverted (buffered)
0	1	1	
1	0	1	Inverted
1	1	0	

FIGURE 13-9 Using XOR to complement selectively.

able input. If the V_{CC} input is selected, the output is always enabled, like a standard combinational logic gate. If the grounded input is selected, the tristate output of the inverter is always in its high-impedance state (allowing the I/O pin to be used as an input). Another input to the MUX that may be selected comes from the *OE* input, which is pin 11. This allows the output to be enabled or disabled by an external logic signal applied to pin 11. The last possible input selection is a product term from the eighth AND gate. This allows an AND combination of terms from the input matrix to enable or disable the output.

The FMUX selects the signal that is fed back into the input matrix. In this case, there are three possible selections. Selecting the *MUX* input that is connected to an adjacent stage or the *MUX* input connected to its own OLMC I/O pin allows an existing output state to be fed back to the input matrix in some of the modes of operation. This feature gives the GAL 16V8 the ability to implement sequential circuits such as the cross-coupled NAND gate latch circuit described in Chapter 5. This feedback option also allows an I/O pin to be used as a dedicated input as opposed to an output. One of these two feedback paths is chosen, depending on the MODE that the chip is programmed for. The third option, selecting the output from the D flip-flop, allows the present state of the flip-flop (which can be used to determine the next state) to be fed back to the input matrix. This allows synchronous sequential circuits, such as counters and shift registers, to be implemented.

With all of these options, it would seem that there must be a long list of possible configurations. In actual practice, all these configuration decisions are made by the software. Actually, the GAL 16V8 has only three different modes: (1) *simple mode,* which is used to implement simple SOP combinational logic without tristate outputs; (2) *complex mode,* which implements SOP combinational logic with tristate outputs that are enabled by an AND product expression; and (3) *registered mode,* which allows individual OLMCs to operate in a combinational configuration with tristate outputs (similar to the complex mode) or in a synchronous mode with clocked D FFs synchronized to a common clock signal.

The GAL 16V8 is an inexpensive and versatile PLD chip, but what if a design requires more hardware resources than is contained in the 16V8? It may be possible to split the design into smaller blocks that can be implemented in several 16V8 chips. Fortunately, there are other members of the GAL family to choose from. Another popular, general-purpose PLD is the GAL 22V10. This device has 10 output pins and 12 input pins in an architecture that is similar but not identical to the GAL 16V8. Groups of product terms are logically summed with an OR gate, which feeds an OLMC. Unlike the 16V8, however, each OR gate in the 22V10 does not combine the same number of product terms. The number of terms ranges from eight all the way up to 16. To take advantage of the extra terms, you must assign the larger Boolean expressions to the correct output pin. The D flip-flops contained in the OLMCs also have asynchronous reset and synchronous preset capabilities. A newer version of the 22V10—the ispGAL 22V10—is now available. This device is said to be in-system programmable (ISP). Instead of requiring a programmer, as is needed to program PALs and standard GAL chips, a cable from the PC is connected directly to a special set of pins on the ISP device to do the programming.

REVIEW QUESTIONS

1. Name two advantages of GAL devices over PAL devices.
2. Name the three modes of operation for a GAL 16V8.

13-5   THE ALTERA EPM7128S CPLD

We will investigate the architecture of the EPM7128S, an EEPROM-based device in the Altera MAX7000S CPLD family. This device is found on several educational development boards, including the Altera UP2, DeVry eSOC, and RSR PLDT-2. The block diagram for this family is shown in Figure 13-10. The major structures in the MAX7000S are the **logic array blocks (LABs)** and the **programmable interconnect array (PIA)**. A LAB contains a set of 16 macrocells and looks very similar to a single SPLD device. Each macrocell consists of a programmable AND/OR circuit and a programmable register (flip-flop). The macrocells in a single LAB can share logic resources such as common product terms or unused AND gates. The number of macrocells contained in one of the MAX7000S family devices depends on the part number. As shown in Table 13-2, the EPM7128S has 128 macrocells arranged in eight LABs.

Logic signals are routed between LABs via the PIA. The PIA is a global bus that connects any signal source to any destination within the device. All inputs to the MAX7000S device and all macrocell outputs feed the PIA. Up to 36 signals can feed each LAB from the PIA. Only signals needed to produce the required functions for any LAB are fed into that LAB.

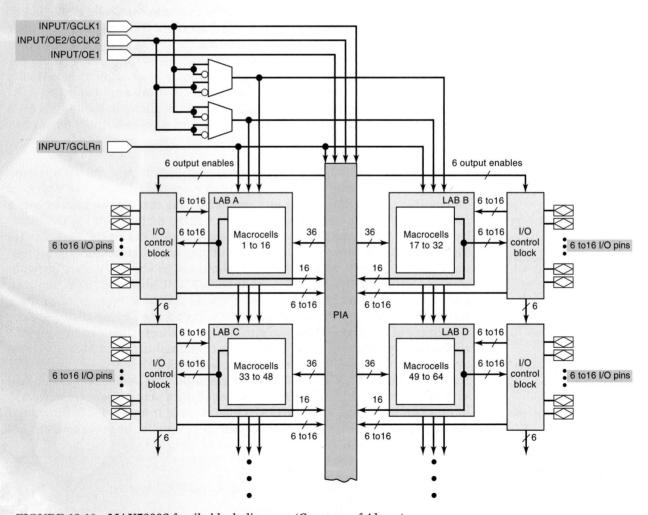

FIGURE 13-10   MAX7000S family block diagram. (Courtesy of Altera)

TABLE 13-2 Altera MAX7000S family device features.

Feature	EPM7032S	EPM7064S	EPM7128S	EPM7160S	EPM7192S	EPM7256S
Usable gates	600	1250	2500	3200	3750	5000
Macrocells	32	64	128	160	192	256
LABs	2	4	8	10	12	16
Maximum number of user I/O pins	36	68	100	104	124	164

I/O pins in the MAX7000S family are connected to specific macrocells. The number of I/O pins available to the user depends on the device package. An EPM7128S in a 160-pin PQFP package has 12 I/Os per LAB plus four additional input-only pins, for a total of 100 pins. On the other hand, in an 84-pin PLCC package, which is included on the above-mentioned development boards, there are eight I/Os per LAB plus the four extras, for a total of 68 I/O pins. The EPM7128S is an in-system programmable (ISP) device. The ISP feature utilizes a joint test action group (JTAG) interface that requires four specific pins to be dedicated to the programming interface and are therefore not available for general user I/O. The target PLD can be programmed in-system via the JTAG pins by connecting them to the parallel port of a PC with driver gates, as shown in Figure 13-11. The JTAG signals are named TDI (test data in), TDO (test data out), TMS (test mode select), and TCK (test clock). This brings the user I/O pin total for an EPM7128SLC84 (an EPM7128S in an 84-pin PLCC package) down to 64 pins. All 68 pins, however, can be used for user I/O if the EPM7128SLC84 is programmed in a PLD programmer instead of in-system. When the design is compiled, you must indicate whether or not the device will use a JTAG interface. In either case,

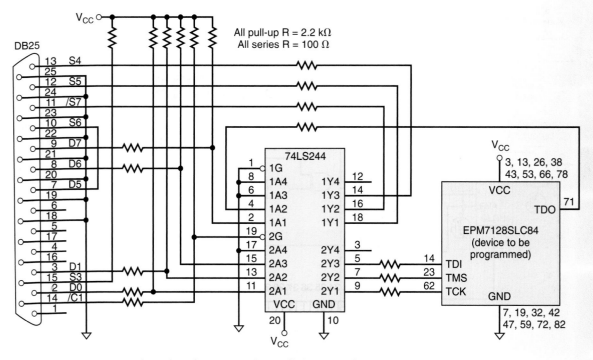

FIGURE 13-11 JTAG interface between PC parallel port and EPM7128SLC84.

register can be clocked in three different modes: (1) with a global clock signal, (2) with a global clock signal when the flip-flop is enabled, or (3) with an array clock signal produced by a buried macrocell or a (nonglobal) input pin. In the EPM7128S, either of the two global clock pins (GCLK1 or GCLK2) can be used to produce the fastest clock-to-Q performance. Either clock edge can be programmed to trigger the flip-flops. Each register can be preset asynchronously or cleared with an active-HIGH or active-LOW product term. Each register may also be cleared with the active-LOW global clear pin (GCLRn). A fast data input path from an I/O pin to the registers, bypassing the PIA, is also available. All registers in the device will be reset automatically at power-up.

MAX7000S devices have a power-saving option that allows the designer to program each individual macrocell for either high-speed (turbo bit turned on) or low-power (turbo bit turned off) operation. Because most logic applications require only a small fraction of all gates to operate at maximum frequency, this feature may produce a significant savings in total system power consumption. Speed-critical paths in the design can run at maximum speed, while the remaining signal paths can operate at reduced power.

REVIEW QUESTIONS

1. What is a macrocell?
2. What is an ISP device?
3. What special control functions are provided with the four input-only pins on a MAX7000S device?
4. What system advantage is achieved by slowing down selected macrocells in a MAX7000S device?

13-6   THE ALTERA FLEX10K FAMILY

The Altera FLEX10K family of programmable logic devices has a very different architecture. Instead of the programmable AND/fixed-OR gate array used in the MAX7000S devices, this family is based on a look-up table (LUT) architecture. The look-up table produces logic functions by storing the function's output results in an SRAM-based memory. It functions essentially like the truth table for the logic function. SRAM technology for PLDs programs much faster than EEPROM-based devices, and it also results in a very high density of storage cells that are used to program the larger PLDs. SRAM-based PLDs that use the LUT architecture are generally classified in the industry as field programmable gate arrays. Unlike most FPGAs, however, Altera has chosen to utilize a programmable signal routing design for the FLEX10K family that looks more like an enhanced version of the PIA found in the CPLD MAX7000S family. As a result, the FLEX10K family has architectural characteristics that are a combination of the two HCPLD classifications. Based on the high-density architecture of the logic cells, the FLEX10K devices are generally classified as FPGAs.

Let us examine the concept of a look-up table. The LUT is the portion of the programmable logic block that produces a combinational function (see Figure 13-14). This function can be used as the output of the logic block or it may be registered (controlled by the internal MUX). The look-up table itself consists of a set of flip-flops that store the desired truth table for our function. LUTs are usually rather small, typically handling four input variables, and so our truth table would have a total of 16 combinations. We will need a flip-flop to store each of the 16 function values (see Figure 13-15). Up to four

FIGURE 13-14 Simplified logic block diagram for FLEX10K device.

FIGURE 13-15 Functional block diagram for an LUT.

input variables in our example LUT will be connected to the data inputs on the decoder block using programmable interconnects. The input combination that is applied will determine which of the 16 flip-flops will be selected to feed the output via the tristate buffers. The look-up table is basically a 16×1 SRAM memory block. All we have to do to create any desired function (of up to four input variables) is to store the appropriate set of 0s and 1s in the LUT's flip-flops. That is essentially what is done to program this type of PLD. Because the flip-flops are volatile (they are SRAM), we need to load the LUT memory for the desired functions whenever the PLD is powered-up. This process is called configuring the PLD. Other portions of the device are also programmed in the same fashion using other SRAM memory bits to store the programming information. This is the basic programming technique for the logic blocks, called **logic elements (LEs)**, found in the FLEX10K devices.

Figure 13-16 shows the block diagram for a FLEX10K logic element. It contains the LUT and programmable register, as well as cascade- and carry-expansion circuitry, programmable control functions, and local and global bus interconnections. The programmable flip-flop can be configured for D, T, JK, or SC or SR operation and will be bypassed for combinational functions. The flip-flop control signals (clock, clear, and preset) can be driven selectively by global inputs, general-purpose I/O pins, or any internally created functions. The LE can produce two outputs to drive local (LAB) and global (FastTrack) interconnects on the chip. This allows the LUT and the register in one LE to be used for unrelated functions. Two types of high-speed data paths—cascade chains and carry chains—connect adjacent LEs without using local interconnects. The cascade-chain expansion allows the FLEX10K architecture to create functions with more than four input variables. Adjacent LUTs can be paralleled together, with each additional LUT providing four more input variables. The carry chain provides a fast carry-forward function between LEs, which allows for efficient implementation of functions that

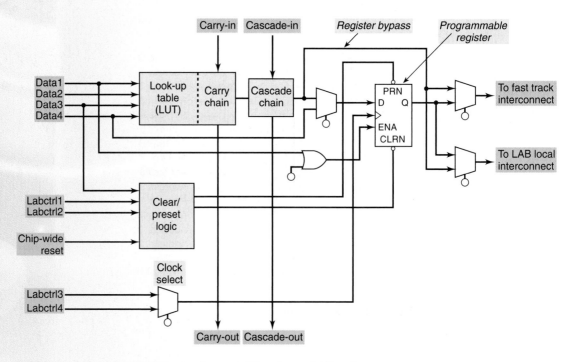

FIGURE 13-16   FLEX10K logic element. (Courtesy of Altera)

build on other functions such as those found in counters, adders, and comparators. In these functions, the upper bits depend on the lower bits. Without an expansion feature like the carry chain, the propagation delays can become quite long for larger circuits. Cascade-chain and carry-chain logic can be created automatically by the compiler software or manually by the designer during design entry. Propagation delays will increase by a small amount when using the expansion options. The MAX+PLUS II Timing Analyzer calculates these added delays for a given design. Intensive use of carry and cascade chains can reduce routing flexibility and should therefore be limited to speed-critical portions of a design.

The logic array block for the FLEX10K family contains eight logic elements and the local interconnect for that LAB (see Figure 13-17). Signals from one LE to another within an LAB are routed with the local interconnect. The row and column interconnects, which Altera has named a FastTrack interconnect, provide the signal pathways between LABs. Each LAB has four control signals available to all eight LEs. Two can be used for register clocks and the other two are for preset or clear.

FIGURE 13-17 FLEX10K logic array block. (Courtesy of Altera)

FIGURE 13-18   FLEX10K device block diagram. (Courtesy of Altera)

The overall block diagram for a FLEX10K device is shown in Figure 13-18. In addition to the logic array blocks and FastTrack interconnects that we have already described, the devices contain I/O elements (IOEs) and embedded array blocks (EABs). The IOEs each contain a bidirectional I/O buffer and a register that can be used for either input or output data storage. Each EAB provides a flexible block of 2048 bits of RAM storage for various internal memory applications. Combining multiple EABs on one chip can create

TABLE 13-3   Altera FLEX10K family device features.

Feature	EPF10K10	EPF10K20	EPF10K30	EPF10K40	EPF10K50	EPF10K70	EPF10K100	EPF10K120	EPF10K250
Typical number of gates	10,000	20,000	30,000	40,000	50,000	70,000	100,000	120,000	250,000
Maximum number of gates	31,000	63,000	69,000	93,000	116,000	118,000	158,000	211,000	310,000
LEs	576	1,152	1,728	2,304	2,880	3,744	4,992	6,656	12,160
LABs	72	144	216	288	360	468	624	832	1,520
EABs	3	6	6	8	10	9	12	16	20
Maximum number of I/O pins	150	189	246	189	310	358	406	470	470

larger blocks of RAM. An EAB can also be used to create large combinational functions by implementing an LUT.

The FLEX10K family contains several different sizes of parts, as shown in Table 13-3. The Altera UP2 educational development board also contains an EPF10K70 device in a 240-pin package. As you can see in the table, this device has a lot of logic resources available!

REVIEW QUESTIONS

1. What is a look-up table?
2. What advantage does SRAM programming technology have over EE-PROM?
3. What disadvantage does SRAM programming technology have compared to EEPROM?
4. What are EABs? What can they be used for?

SUMMARY

1. Programmable logic devices (PLDs) are the key technology in the future of digital systems.

2. PLDs can reduce parts inventory, simplify prototype circuitry, shorten the development cycle, reduce the size and power requirements of the product, and allow the hardware of a circuit to be upgraded easily.

3. The major digital system categories are standard logic, application-specific integrated circuits (ASICs), and microprocessor/digital signal processing (DSP) devices.

4. ASIC devices may be programmable logic devices (PLDs), gate arrays, standard cells, or full-custom devices.

5. PLDs are the least expensive type of ASIC to develop.

6. Simple PLDs (SPLDs) contain the equivalent of 600 or fewer gates and are programmed with fuse, EPROM, or EEPROM technology.

7. High-capacity PLDs (HCPLDs) have two major architectural categories: complex programmable logic devices (CPLDs) and field programmable gate arrays (FPGAs).

8. The most common CPLD programming technologies are EEPROM and flash, both of which are nonvolatile.

9. The most common FPGA programming technology is SRAM, which is volatile.

10. The GAL 16V8 is one of the simplest PLDs available but is still widely used and demonstrates the basic principles behind all PLDs.

11. The Altera EPM7128S CPLD contains 128 macrocells, each of which contains a programmable AND/OR circuit and a programmable register.

12. The EPM7128SLC84 can have up to 68 inputs and outputs.

13. The MAX7000S family of CPLDs is in-system programmable (ISP).

14. The Altera FLEX10K family of devices uses a look-up table (LUT) architecture in a SRAM technology.

15. SRAM programming technology is volatile, meaning that the devices must be reconfigured at power-up.

IMPORTANT TERMS

standard logic
microprocessor
digital signal
    processing
    (DSP)
application-specific
    integrated circuit
    (ASIC)
programmable logic
    device (PLD)
gate array

standard-cell
full-custom
simple PLD (SPLD)
complex PLD (CPLD)
field programmable
    gate array (FPGA)
high-capacity PLD
    (HCPLD)
one-time
    programmable
    (OTP)

programmable array
    logic (PAL)
macrocell
look-up table (LUT)
logic array block
    (LAB)
programmable
    interconnect array
    (PIA)
logic element (LE)

PROBLEMS

SECTION 13-1

13-1. Describe each of the following major digital system categories:

(a) Standard logic

(b) ASICs

(c) Microprocessor/DSP

13-2. Name three factors that are generally considered when making design engineering decisions.

13-3. Why is a microprocessor/DSP system called a software solution for a design?

13-4. What major advantage does a hardware design solution have over a software solution?

13-5. Describe each of the following four ASIC subcategories:

(a) PLDs

(b) Gate arrays

(c) Standard-cell

(d) Full-custom

13-6. What are the major advantages and disadvantages of a full-custom ASIC?

13-7. Name the six PLD programming technologies. Which is one-time programmable? Which is volatile?

13-8. How is the programming of SRAM-based PLDs different from other programming technologies?

SECTION 13-5

13-9. Describe the functions of each of the following architectural structures found in the Altera MAX7000S family:

(a) LAB

(b) PIA

(c) Macrocell

13-10. What two ways can be used to program the MAX7000S family devices?

13-11. What standard device interface is used for in-system programming in the MAX7000S family?

13-12. What are the four input-only pins on the EPM7128SLC84 (by pin number and function)?

13-13. What is the advantage of using one of the global clock inputs for registered operation?

SECTION 13-6

13-14. What is the fundamental architectural difference between the MAX7000S and FLEX10K families? What is the programming technology used by each family? Which family is nonvolatile? Which family contains more logic resources?

ANSWERS TO SECTION REVIEW QUESTIONS

SECTION 13-1

1. Standard logic, ASICs, microprocessor   2. Speed   3. Application-specific integrated circuit   4. Programmable logic devices, gate arrays, standard cells, full custom   5. High-capacity programmable logic device   6. (1) Logic blocks: programmable AND/fixed-OR CPLD versus look-up table FPGA (2) Signal routing resources: uniform CPLD versus varied FPGA   7. Volatility refers to whether a PLD (or memory device) loses stored information when it is powered-down.

SECTION 13-2

1. An IC that contains a large number of gates whose interconnections can be modified by the user to perform a specific function.   2. $O_1 = A$ 3. An intact fuse   4. A hard-wired connection

SECTION 13-3

2. Hard-wired OR; programmable AND   3. Hard-wired AND; programmable OR   4. $O_1 = AB\overline{C}\overline{D} + \overline{A}\,\overline{B}\,CD + \overline{A}\,BCD = AB\overline{C}\overline{D} + \overline{A}CD$

SECTION 13-4

1. Erasable and reprogrammable; has an OLMC   2. Simple, complex, registered

SECTION 13-5

1. A macrocell is the programmable logic block in MAX7000S CPLDs consisting of a programmable AND/OR circuit and a programmable register (flip-flop).   2. An ISP PLD device is in-system programmable, which means that it can be programmed while connected in the circuit.   3. Global clocks, tristate output enables, asynchronous clear   4. Power consumption may be decreased by slowing down macrocells.

SECTION 13-6

1. A look-up table is typically a 16-word by 1-bit SRAM array used to store the desired output logic levels for a simple logic function.   2. SRAM programs faster and has a higher logic cell density than EEPROM.   3. SRAM is volatile and must be reconfigured upon power-up of the device.   4. Embedded array blocks provide RAM storage on the PLD.

GLOSSARY

Access Time Time between the memory's receiving a new input address and the output data's becoming available in a read operation.

Accumulator Principal register of an arithmetic/logic unit (ALU).

Acquisition Time Time required for a sample-and-hold circuit to capture the analog value that is present on its input.

Active-HIGH (LOW) Decoder Decoder that produces a logic HIGH (LOW) at the output when detection occurs.

Active Logic Level Logic level at which a circuit is considered active. If the symbol for the circuit includes a bubble, the circuit is active-LOW. On the other hand, if it doesn't have a bubble, then the circuit is active-HIGH.

Actuator Electrically controlled device that controls a physical variable.

Addend Number to be added to another.

Adder/Subtractor An adder circuit that can subtract by complementing (negating) one of the operands. *See also* Parallel/Adder.

Address Number that uniquely identifies the location of a word in memory.

Address Bus Unidirectional lines that carry the address code from the CPU to memory and I/O devices.

Address Multiplexing Multiplexing used in dynamic RAMs to save IC pins. It involves latching the two halves of a complete address into the IC in separate steps.

Alias A digital signal that results from sampling an incoming signal at a rate less than twice the highest frequency contained in the incoming signal.

Alphanumeric Codes Codes that represent numbers, letters, punctuation marks, and special characters.

Altera Hardware Description Language (AHDL) A proprietary HDL developed by Altera Corporation for programming their programmable logic devices.

Alternate Logic Symbol A logically equivalent symbol that indicates the active level of the inputs and outputs.

Analog Representation Representation of a quantity that varies over a continuous range of values.

Analog System Combination of devices designed to manipulate physical quantities that are represented in analog form.

Analog-to-Digital Converter (ADC) Circuit that converts an analog input to a corresponding digital output.

Analog Voltage Comparator Circuit that compares two analog input voltages and produces an output that indicates which input is greater.

& When used inside an IEEE/ANSI symbol, an indication of an AND gate or AND function.

AND Gate Digital circuit that implements the AND operation. The output of this circuit is HIGH (logic level 1) only if all of its inputs are HIGH.

AND Operation Boolean algebra operation in which the symbol · is used to indicate the ANDing of two or more logic variables. The result of the AND operation will be HIGH (logic level 1) only if all variables are HIGH.

Application-Specific Integrated Circuit (ASIC) An IC that has been specifically designed to meet the requirements of an application. Subcategories include PLDs, gate arrays, standard cells, and full-custom ICs.

ARCHITECTURE Keyword in VHDL used to begin a section of code that defines the operation of a circuit block (ENTITY).

Arithmetic/Logic Unit (ALU) Digital circuit used in computers to perform various arithmetic and logic operations.

ASCII Code (American Standard Code for Information Interchange) Seven-bit alphanumeric code used by most computer manufacturers.

Asserted Term used to describe the state of a logic signal; synonymous with "active."

Astable Multivibrator Digital circuit that oscillates between two unstable output states.

Asynchronous Counter Type of counter in which each flip-flop output serves as the clock input signal for the next flip-flop in the chain.

Asynchronous Inputs Flip-flop inputs that can affect the operation of the flip-flop independent of the synchronous and clock inputs.

Asynchronous Transfer Data transfer performed without the aid of the clock.

Augend Number to which an addend is added.

Auxiliary Memory The part of a computer's memory that is separate from the computer's main working memory. Generally has high density and high capacity, such as magnetic disk.

Backplane Electrical connection common to all segments of an LCD.

Barrel Shifter A shift register that can very efficiently shift a binary number left or right by any number of bit positions.

BCD Adder Special adder containing two four-bit parallel adders and a correction detector circuit. Whenever the addition of two BCD code groups is greater than 1001_2 (9_{10}), the correction detector circuit senses it, adds to the result the correction factor 0110_2 (6_{10}), and generates a carry to the next decimal position.

BCD Counter Binary counter that counts from 0000_2 to 1001_2 before it recycles.

BCD-to-Decimal Decoder Decoder that converts a BCD input into a single decimal output equivalence.

BCD-to-7-Segment Decoder/Driver Digital circuit that takes a four-bit BCD input and activates the required outputs to display the equivalent decimal digit on a 7-segment display.

Behavioral Level of Abstraction A technique of describing a digital circuit that focuses on how the circuit reacts to its inputs.

Bidirectional Data Line Term used when a data line functions as either an input or an output line depending on the states of the enable inputs.

Bilateral Switch CMOS circuit that acts like a single-pole, single-throw (SPST) switch controlled by an input logic level.

Binary-Coded-Decimal Code (BCD Code) Four-bit code used to represent each digit of a decimal number by its four-bit binary equivalent.

Binary Counter Group of flip-flops connected in a special arrangement in which the states of the flip-flops represent the binary number equivalent to the number of pulses that have occurred at the input of the counter.

Binary Digit Bit.

Binary Point Mark that separates the integer from the fractional portion of a binary quantity.

Binary System Number system in which there are only two possible digit values, 0 and 1.

Bipolar DAC Digital-to-analog converter that accepts signed binary numbers as input and produces the corresponding positive or negative analog output value.

Bipolar ICs Integrated digital circuits in which NPN and PNP transistors are the main circuit elements.

BIT In VHDL, the data object type representing a single binary digit (bit).

Bit Digit in the binary system.

Bit Array A way to represent a group of bits by giving it a name and assigning an element number to each bit's position. This same structure is sometimes called a bit vector.

BIT_VECTOR In VHDL, the data object type representing a bit array. *See also* Bit Array.

Boolean Algebra Algebraic process used as a tool in the design and analysis of digital systems. In Boolean algebra only two values are possible, 0 and 1.

Boolean Theorems Rules that can be applied to Boolean algebra to simplify logic expressions.

Bootstrap Program Program, stored in ROM, that a computer executes on power-up.

Bubbles Small circles on the input or output lines of logic-circuit symbols that represent inversion of a particular signal. If a bubble is present, the input or output is said to be active-LOW.

Buffer/Driver Circuit designed to have a greater output current and/or voltage capability than an ordinary logic circuit.

Buffer Register Register that holds digital data temporarily.

Buried Node A defined point in a circuit that is not accessible from outside that circuit.

Bus Group of wires that carry related bits of information.

Bus Contention Situation in which the outputs of two or more active devices are placed on the same bus line at the same time.

Bus Drivers Circuits that buffer the outputs of devices connected to a common bus; used when a large number of devices share a common bus.

Byte Group of eight bits.

C When used as an input label inside an IEEE/ANSI symbol, an indication that the input controls the entry of data into a storage element.

Cache A high-speed memory system that can be loaded from the slower system DRAM and accessed quickly by the high-speed CPU.

Capacity Amount of storage space in a memory expressed as the number of bits or number of words.

Carry Digit or bit that is generated when two numbers are added and the result is greater than the base for the number system being used.

Carry Propagation Intrinsic circuit delay of some parallel adders that prevents the carry bit (C_{OUT}) and the result of the addition from appearing at the output simultaneously.

Carry Ripple *See* Carry Propagation.

CAS (Column Address Strobe) Signal used to latch the column address into a DRAM.

CAS-before-RAS Method for refreshing DRAMs that have built-in refresh counters. When the *CAS* input is driven LOW and held there as *RAS* is pulsed LOW, an internal refresh operation is performed at the row address given by the on-chip refresh counter.

Cascading Connecting logic circuits in a serial fashion with the output of one circuit driving the input of the next, and so on.

CASE A control structure that selects one of several options when describing a circuit's operation based on the value of a data object.

Central Processing Unit (CPU) Part of a computer that is composed of the arithmetic/logic unit (ALU) and the control unit.

Checksum Special data word stored in the last ROM location. It is derived from the addition of all other data words in the ROM, and it is used for error-checking purposes.

Chip Select Input to a digital device that controls whether or not the device will perform its function. Also called *chip enable.*

Circuit Excitation Table Table showing a circuit's possible PRESENT-to-NEXT state transitions and the required *J* and *K* levels at each flip-flop.

Circular Buffer A memory system that always contains the last *n* data values that have been written. Whenever a new data value is stored, it overwrites the oldest value in the buffer.

Circulating Shift Register Shift register in which one of the outputs of the last flip-flop is connected to the input of the first flip-flop.

CLEAR An input to a latch or FF used to make $Q = 0$.

CLEAR State The $Q = 0$ state of a flip-flop.

Clock Digital signal in the form of a rectangular pulse train or a square wave.

Clock Skew Arrival of a clock signal at the clock inputs of different flip-flops at different times as a result of propagation delays.

Clock Transition Times Minimum rise and fall times for the clock signal transitions used by a particular IC, specified by the IC manufacturer.

Clocked D Flip-Flop Type of flip-flop in which the *D* (data) input is the synchronous input.

Clocked Flip-Flops Flip-flops that have a clock input.

Clocked J-K FLip-Flop Type of flip-flop in which inputs *J* and *K* are the synchronous inputs.

Clocked S-C Flip-Flop Type of flip-flop in which the inputs SET and CLEAR are the synchronous inputs.

CMOS (Complementary Metal-Oxide-Semiconductor) Integrated-circuit technology that uses MOSFETs as the principal circuit element. This logic family belongs to the category of unipolar digital ICs.

Combinational Logic Circuits Circuits made up of combinations of logic gates, with no feedback from outputs to inputs.

Comments Text added to any HDL design file or computer program to describe the purpose and operation of the code in general or of individual statements in the code. Documentation regarding author, date, revision, etc., may also be contained in the comments.

Common Anode LED display that has the anodes of all of the segment LEDs tied together.

Common Cathode LED display that has the cathodes of all of the segment LEDs tied together.

Common-Control Block Symbol used by the IEEE/ANSI standard to describe when one or more inputs are common to more than one of the circuits in an IC.

Compiler A program that translates a text file written in a high-level language into a binary file that can be loaded into a programmable device such as a PLD or a computer's memory.

Complement *See* Invert.

Complex PLD (CPLD) Class of PLDs that contain an array of PAL-type blocks that can be interconnected.

COMPONENT A VHDL keyword used at the top of a design file to provide information about a library component.

Computer Word Group of binary bits that form the primary unit of information in a computer.

Concatenate A term used to describe the arrangement or linking of two or more data objects into ordered sets.

Concurrent Events that occur simultaneously (at the same time). In HDL, the circuits generated by concurrent statements are not affected by the order or sequence of the statements in the code.

Concurrent Assignment Statement A statement in AHDL or VHDL that describes a circuit that works concurrently with all other circuits that are described by concurrent statements.

Conditional Signal Assignment A VHDL concurrent construct that evaluates a series of conditions sequentially to determine the appropriate value to assign to a signal. The first true condition evaluated determines the assigned value.

Constants Symbolic names that can be used to represent fixed numeric (scalar) values.

Contact Bounce The tendency of all mechanical switches to vibrate when forced to a new position. The vibrations cause the circuit to make contact and break contact repeatedly until the vibrations settle out.

Contention Two (or more) output signals connected together trying to drive a common point to different voltage levels. *See also* Bus Contention.

Control Bus Set of signal lines that are used to synchronize the activities of the CPU and the separate μC elements.

Control Inputs Input signals synchronized with the active clock transition that determine the output state of a flip-flop.

Control Unit Part of a computer that provides decoding of program instructions and the necessary timing and control signals for the execution of such instructions.

Crystal-Controlled Clock Generator Circuit that uses a quartz crystal to generate a clock signal at a precise frequency.

CUPL Stands for Universal Compiler for Programmable Logic. It is a computer program that takes an input file that describes how a PLD is to operate, and converts it to output files that are used to program the PLD.

Current-Sinking Logic Logic family in which the output of a logic circuit sinks current from the input of the logic circuit that it is driving.

Current-Sinking Transistor Name given to the output transistor (Q_4) of a TTL circuit. This transistor is turned on when the output logic level is LOW.

Current-Sourcing Logic Logic family in which the output of a logic circuit sources, or supplies, current to the input of the logic circuit that it is driving.

Current-Sourcing Transistor Name given to the output transistor (Q_3) of most TTL circuits. This transistor is conducting when the output logic level is HIGH.

Current Transients Current spikes generated by the totem-pole output structure of a TTL circuit and caused when both transistors are simultaneously turned on.

D Flip-Flop *See* Clocked D Flip-Flop.
D **Latch** Circuit that contains a NAND gate latch and two steering NAND gates.

Data Binary representations of numerical values or nonnumerical information in a digital system. Data are used and often modified by a computer program.

Data Acquisition Process by which a computer acquires digitized analog data.

Data Bus Bidirectional lines that carry data between the CPU and the memory, or between the CPU and the I/O devices.

Data Distributors *See* Demultiplexer.

Data-Rate Buffer Application of FIFOs in which sequential data are written into the FIFO at one rate and read out at a different rate.

Data Selectors *See* Multiplexer.

Data Transfer *See* Parallel Data Transfer *or* Serial Data Transfer

Decade Counter Any counter capable of going through 10 different logic states.

Decimal System Number system that uses 10 different digits or symbols to represent a quantity.

Decision Control Structures The statements and syntax that describe how to choose between two or more options in the code.

Decoder Digital circuit that converts an input binary code into a corresponding single numeric output.

Decoding Act of identifying a particular binary combination (code) in order to display its value or recognize its presence.

DEFAULTS An AHDL keyword used to establish a default value for a combinational signal for instances when the code does not explicitly specify a value.

DeMorgan's Theorems (1) Theorem stating that the complement of a sum (OR operation) equals the product (AND operation) of the complements, and (2) theorem stating that the complement of a product (AND operation) equals the sum (OR operation) of the complements.

Demultiplexer (DEMUX) Logic circuit that, depending on the status of its select inputs, will channel its data input to one of several data outputs.

Density A relative measure of capacity to store bits in a given amount of space.

Dependency Notation Method used to represent symbolically the relationship between inputs and outputs of logic circuits. This method employs the use of qualifying symbols embedded near the top center or geometric center of a symbol element.

Differential Inputs Method of connecting an analog signal to an analog circuit's + and − inputs, neither of which is ground, such that the analog circuit acts upon the voltage difference between the two inputs.

Digital Computer System of hardware that performs arithmetic and logic operations, manipulates data, and makes decisions.

Digital Integrated Circuits Self-contained digital circuits made by using one of several integrated-circuit fabrication technologies.

Digital One-Shot A one-shot that uses a counter and clock rather than an RC circuit as a time base.

Digital-Ramp ADC Type of analog-to-digital converter in which an internal staircase waveform is generated and utilized for the purpose of accomplishing the conversion. The conversion time for this type of analog-to-digital converter varies depending on the value of the input analog signal.

Digital Representation Representation of a quantity that varies in discrete steps over a range of values.

Digital Signal Processing (DSP) Method of performing repetitive calculations on an incoming stream of digital data words to accomplish some form of signal conditioning. The data are typically digitized samples of an analog signal.

Digital Storage Oscilloscope Instrument that samples, digitizes, stores, and displays analog voltage waveforms.

Digital System Combination of devices designed to manipulate physical quantities that are represented in digital form.

Digital-to-Analog Converter (DAC) Circuit that converts a digital input to a corresponding analog output.

Digitization Process by which an analog signal is converted to digital data.

Disable Action in which a circuit is prevented from performing its normal function, such as passing an input signal through to its output.

Divide-and-Conquer Troubleshooting technique whereby tests are performed that will eliminate half of all possible remaining causes of the malfunction.

Don't-Care Situation when a circuit's output level for a given set of input conditions can be assigned as either a 1 or a 0.

Down Counter Counter that counts from a maximum count downward to 0.

Downloading Process of transferring output files to a programming fixture.

DRAM Controller IC used to handle refresh and address multiplexing operations needed by DRAM systems.

Driver Technical term sometimes added to an IC's description to indicate that the IC's outputs can operate at higher current and/or voltage limits than a normal standard IC.

Dual-in-Line Package (DIP) A very common IC package with two parallel rows of pins intended to be inserted into a socket or through holes drilled in a printed circuit board.

Dual-Slope Analog-to-Digital Converter Type of analog-to-digital converter that linearly charges a capacitor from a current proportional to V_A for a fixed time interval and then increments a counter as the capacitor is linearly discharged to 0.

Dynamic RAM (DRAM) Type of semiconductor memory that stores data as capacitor charges that need to be refreshed periodically.

ECL Emitter-coupled logic; also referred to as *current-mode logic*.

Edge-Detector Circuit Circuit that produces a narrow positive spike that occurs coincident with the active transition of a clock input pulse.

Edge-Triggered Manner in which a flip-flop is activated by a signal transition. A flip-flop may be either a positive- or a negative-edge-triggered flip-flop.

Electrically Compatible When two ICs from different logic series can be connected directly without any special measures taken to ensure proper operation.

Electrically Erasable Programmable ROM (EEPROM) ROM that can be electrically programmed, erased, and reprogrammed.

Electrostatic Discharge (ESD) The often detrimental act of the transfer of static electricity (i.e., an electrostatic charge) from one surface to another. This impulse of current can destroy electronic devices.

ELSE A control structure used in conjunction with IF/THEN to perform an alternate action in the case that the condition is false. An IF/THEN/ELSE always performs one of two actions.

ELSIF A control structure that can be used multiple times following an IF statement to select one of several options in describing a circuit's operation based on whether the associated expressions are true or false.

Embedded Microcontroller Microcontroller that is embedded in a marketable product such as a VCR or an appliance.

Emitter-Coupled Logic *See* ECL.

Enable Action in which a circuit is allowed to perform its normal function, such as passing an input signal through to its output.

Encoder Digital circuit that produces an output code depending on which of its inputs is activated.

Encoding Use of a group of symbols to represent numbers, letters, or words.

ENTITY Keyword in VHDL used to define the basic block structure of a circuit. This word is followed by a name for the block and the definitions of its input/output ports.

Enumerated Type A VHDL user-defined type for a signal or variable.

Equality Operator A CUPL operator that evaluates bit-by-bit equality between a set of variables and a constant.

Erasable Programmable ROM (EPROM) ROM that can be electrically programmed by the user. It can be erased (usually with ultraviolet light) and reprogrammed as often as desired.

EVENT A VHDL keyword used as an attribute attached to a signal to detect a transition of that signal. Generally, an event means a signal changed state.

Exclusive-NOR (XNOR) Circuit Two-input logic circuit that produces a HIGH output only when the inputs are equal.

Exclusive-OR (XOR) Circuit Two-input logic circuit that produces a HIGH output only when the inputs are different.

Fan-Out Maximum number of standard logic inputs that the output of a digital circuit can reliably drive.

Field Programmable Gate Array (FPGA) Class of PLDs that contain an array of more complex logic cells that can be very flexibly interconnected to implement high-level logic circuits.

Field Programmable Logic Array (FPLA) A PLD that uses both a programmable AND array and a programmable OR array.

Firmware Computer programs stored in ROM.

First-In, First-Out (FIFO) Memory Semiconductor sequential-access memory in which data words are read out in the same order in which they were written in.

555 Timer TTL-compatible IC that can be wired to operate in several different modes, such as a one-shot and an astable multivibrator.

Flash ADC Type of analog-to-digital converter that has the highest operating speed available.

Flash Memory Nonvolatile memory IC that has the high-speed access and in-circuit erasability of EEPROMs but with higher densities and lower cost.

Flip-Flop Memory device capable of storing a logic level.

Floating Bus When all outputs connected to a data bus are in the Hi-Z state.

Floating Input Input signal that is left disconnected in a logic circuit.

Floppy Disk Flexible magnetic disk used for mass storage.

FOR Loop *See* Iterative Loop.

4-to-10 Decoder *See* BCD-to-Decimal Decoder.

Frequency Counter Circuit that can measure and display a signal's frequency.

Frequency Division The use of flip-flop circuits to produce an output waveform whose frequency is equal to the input clock frequency divided by some integer value.

Full Adder Logic circuit with three inputs and two outputs. The inputs are a carry bit (C_{IN}) from a previous stage, a bit from the augend, and a bit from the addend, respectively. The outputs are the sum bit and the carry-out bit (C_{OUT}) produced by the addition of the bit from the addend with the bit from the augend and C_{IN}.

Full-Custom An application-specific integrated circuit (ASIC) that is completely designed and fabricated from fundamental elements of electronic devices such as transistors, diodes, resistors, and capacitors.

Full-Scale Error Term used by some digital-to-analog converter manufacturers to specify the accuracy of a digital-to-analog converter. It is defined as the maximum deviation of a digital-to-analog converter's output from its expected ideal value.

Full-Scale Output Maximum possible output value of a digital-to-analog converter.

Function Generator Circuit that produces different waveforms. It can be constructed using a ROM, a DAC, and a counter.

Function Prototype A text description that contains all the essential defining attributes of a library function or module.

Functionally Equivalent When the logic functions performed by two different ICs are exactly the same.

Fusible Link Conducting material that can be made nonconducting (i.e., open) by passing too much current through it.

Gate Array An application-specific integrated circuit (ASIC) made up of hundreds of thousands of prefabricated basic gates that can be custom interconnected in the last stages of manufacture to form the desired digital circuit.

GENERATE A VHDL keyword used with the FOR construct to iteratively define multiple similar components and to interconnect them.

Glitch Momentary, narrow, spurious, and sharply defined change in voltage.

GSI Giga-scale integration (1,000,000 gates or more).

Half Adder Logic circuit with two inputs and two outputs. The inputs are a bit from the augend and a bit from the addend. The outputs are the sum bit produced by the addition of the bit from the addend with the bit from the augend and the resulting carry (C_{OUT}) bit, which will be added to the next stage.

Hard Disk Rigid metal magnetic disk used for mass storage.

Hardware Description Language (HDL) A text-based method of describing digital hardware that follows a rigid syntax for representing data objects and control structures.

Hexadecimal Number System Number system that has a base of 16. Digits 0 through 9 plus letters A through F are used to express a hexadecimal number.

Hierarchical Design A method of designing a project by breaking it into constituent modules, each of which can be broken further into simpler constituent modules.

Hierarchy A group of tasks arranged in rank order of magnitude, importance, or complexity.

High-Capacity PLD (HCPLD) A PLD with thousands of logic gates and many programmable macrocell resources, along with very flexible interconnection resources.

High-Level Language Computer programming language that utilizes the English language in order to facilitate the writing of a computer program.

Hold Time (t_H) Time interval immediately following the active transition of the clock signal during which the control input must be maintained at the proper level.

Hybrid System System that employs both analog and digital techniques.

IEEE/ANSI Institute of Electrical and Electronics Engineers/American National Standards Institute, both professional organizations that establish standards.

IF/THEN A control structure that evaluates a condition and performs an action if the condition is true or bypasses the action and continues on if the condition is false.

Immediate Addressing Addressing mode in which the part of the instruction following the op code is the actual operand.

Indeterminate Of a logic voltage level, outside the required range of voltages for either logic 0 or logic 1.

Index Another name for the element number of any given bit in a bit array.

Inhibit Circuits Logic circuits that control the passage of an input signal through to the output.

Input Term Matrix Part of a programmable logic device that allows inputs to be selectively connected to or disconnected from internal logic circuitry.

Input Unit Part of a computer that facilitates the feeding of information into the computer's memory unit or ALU.

Instructions Binary codes that tell a computer what operation to perform. A program is made up of an orderly sequence of instructions.

INTEGER In VHDL, the data object type representing a numeric value.

Interfacing Joining of dissimilar devices in such a way that they are able to function in a compatible and coordinated manner; connection of the output of a system to the input of a different system with different electrical characteristics.

Interpolation Filtering Another name for oversampling. Interpolation refers to intermediate values inserted into the digital signal to smooth out the waveform.

Invert Cause a logic level to go to the opposite state.

INVERTER Also referred to as the NOT circuit; logic circuit that implements the NOT operation. An

INVERTER has only one input, and its output logic level is always the opposite of this input's logic level.

Iterative Loop　A control structure that implies a repetitive operation and a stated number of iterations.

Jam Transfer　*See* Asynchronous Transfer.

JEDEC　Joint Electronic Device Engineering Council, which established standards for IC pin assignments and PLD file format.

J-K Excitation Table　Table showing the required *J* and *K* input conditions for each possible state transition for a single J-K flip-flop.

Johnson Counter　Shift register in which the inverted output of the last flip-flop is connected to the input of the first flip-flop.

JTAG　Joint Test Action Group, which created a standard interface that allows access to the inner workings of an IC for testing, controlling, and programming purposes.

Karnaugh Map (K Map)　Two-dimensional form of a truth table used to simplify a sum-of-products expression.

Latch　Type of flip-flop; also, the action by which a logic circuit output captures and holds the value of an input.

Latch-Up　Condition of dangerously high current in a CMOS IC caused by high-voltage spikes or ringing at device input and output pins.

Latency　The inherent delay associated with reading data from a DRAM. It is caused by the timing requirements of supplying the row and column addresses, and the time for the data outputs to settle.

LCD　Liquid-crystal display.

Lead Pitch　The distance between the centers of adjacent pins on an IC.

Least Significant Bit (LSB)　Rightmost bit (smallest weight) of a binary expressed quantity.

Least Significant Digit (LSD)　Digit that carries the least weight in a particular number.

LED　Light-emitting diode.

Libraries　A collection of descriptions of commonly used hardware circuits that can be used as modules in a design file.

Library of Parameterized Modules (LPM)　A set of generic library functions designed to be very flexible in allowing the user to specify the number of bits, mod number, control options, etc.

Linear Buffer　A first-in, first-out memory system that fills at one rate and empties at another rate. After it is full, no data can be stored until data is read from the buffer. *See also* First In, First-Out (FIFO) Memory.

Linearity Error　Term used by some digital-to-analog converter manufacturers to specify the device's accuracy. It is defined as the maximum deviation in step size from the ideal step size.

Literals　In VHDL, a scalar value or bit pattern that is to be assigned to a data object.

Load Operation　Transfer of data into a flip-flop, a register, a counter, or a memory location.

Local Signal　*See* Buried Node.

Logic Array Block (LAB)　A term Altera Corporation uses to describe building blocks of their CPLDs. Each LAB is similar in complexity to an SPLD.

Logic Circuit　Any circuit that behaves according to a set of logic rules.

Logic Elements　A term Altera Corporation uses to describe the building blocks of their FLEX10K family of PLDs. The logic elements are programmed as a ram-based look-up table.

Logic Function Generation　Implementation of a logic function directly from a truth table by means of a digital IC such as a multiplexer.

Logic Level　State of a voltage variable. The states 1 (HIGH) and 0 (LOW) correspond to the two usable voltage ranges of a digital device.

Logic Primitive　A circuit description of a fundamental component that is built into the MAX+PLUS II system of libraries.

Logic Probe　Digital troubleshooting tool that senses and indicates the logic level at a particular point in a circuit.

Logic Pulser　Testing tool that generates a short-duration pulse when actuated manually.

Look-Ahead Carry　Ability of some parallel adders to predict, without having to wait for the carry to propagate through the full adders, whether or not a carry bit (C_{OUT}) will be generated as a result of the addition, thus reducing the overall propagation delays.

Look-Up Table (LUT)　A way to implement a single logic function by storing the correct output logic state in a memory location that corresponds to each particular combination of input variables.

Looping　Combining of adjacent squares in a Karnaugh map containing 1s for the purpose of simplification of a sum-of-products expression.

Low-Power Schottky TTL (LS-TTL)　TTL subfamily that uses the identical Schottky TTL circuit but with larger resistor values.

Low-Voltage Technology　New line of logic devices that operate from a nominal supply voltage of 3.3 V or less.

LSI　Large-scale integration (100 to 9999 gates).

MAC　An abbreviation for Multiply Accumulate Unit, the hardware section of a DSP that multiplies a sample with a coefficient and then accumulates (sums) a running total of these products.

MACHINE　An AHDL keyword used to create a state machine in a design file.

Macrocell　A circuit made up of a group of basic digital components such as AND gates, OR gates, registers, and tristate control circuits that can be interconnected within a PLD via a program.

Macrofunctions　A term used by Altera Corporation to describe the predefined hardware descriptions in their libraries that represent standard IC parts.

Magnetic Disk Memory   Mass storage memory that stores data as magnetized spots on a rotating, flat disk surface.

Magnetic Tape Memory   Mass storage memory that stores data as magnetized spots on a magnetically coated plastic tape.

Magnitude Comparator   Digital circuit that compares two input binary quantities and generates outputs to indicate whether the inputs are equal or, if not, which is greater.

Main Memory   High-speed portion of a computer's memory that holds the program and data the computer is currently working on. Also called *working memory.*

Mask-Programmed ROM (MROM)   ROM that is programmed by the manufacturer according to the customer's specifications. It cannot be erased or reprogrammed.

Mass Storage   Storage of large amounts of data; not part of a computer's internal memory.

Master/Slave Flip-Flops   Obsolete flip-flops that have as their internal structure two flip-flops—a master and a slave.

Maximum Clocking Frequency (f_{MAX})   Highest frequency that may be applied to the clock input of a flip-flop and still have it trigger reliably.

Megafunctions   A complex or high-level building block available in the Altera library.

Memory   Ability of a circuit's output to remain at one state even after the input condition that caused that state is removed.

Memory Cell   Device that stores a single bit.

Memory Foldback   Redundant enabling of a memory device at more than one address range as a result of incomplete address decoding.

Memory Map   Diagram of a memory system that shows the address range of all existing memory devices as well as available memory space for expansion.

Memory Unit   Part of a computer that stores instructions and data received from the input unit, as well as results from the arithmetic/logic unit.

Memory Word   Group of bits in memory that represents instructions or data of some type.

Microcomputer   Newest member of the computer family, consisting of microprocessor chip, memory chips, and I/O interface chips. In some cases, all of the aforementioned are in one single IC.

Microcontroller   Small microcomputer used as a dedicated controller for a machine, a piece of equipment, or a process.

Microprocessor (MPU)   LSI chip that contains the central processing unit (CPU).

Minuend   Number from which the subtrahend is to be subtracted.

MOD Number   Number of different states that a counter can sequence through; the counter's frequency division ratio.

Mode   The attribute of a port in a digital circuit that defines it as input, output, or bidirectional.

Monostable Multivibrator   *See* One-Shot.

Monotonicity   Property whereby the output of a digital-to-analog converter increases as the binary input is increased.

MOSFET   Metal-oxide-semiconductor field-effect transistor.

Most Significant Bit (MSB)   Leftmost binary bit (largest weight) of a binary expressed quantity.

Most Significant Digit (MSD)   Digit that carries the most weight in a particular number.

MSI   Medium-scale integration (12 to 99 gates).

Multiplexer (MUX)   Logic circuit that, depending on the status of its select inputs, will channel one of several data inputs to its output.

Multiplexing   Process of selecting one of several input data sources and transmitting the selected data to a single output channel.

Multistage Counter   Counter in which several counter stages are connected so that the output of one stage serves as the clock input of the next stage to achieve greater counting range or frequency division.

NAND Gate   Logic circuit that operates like an AND gate followed by an INVERTER. The output of a NAND gate is LOW (logic level 0) only if all inputs are HIGH (logic level 1).

NAND Gate Latch   Flip-flop constructed from two cross-coupled NAND gates.

Negation   Operation of converting a positive number to its negative equivalent, or vice versa. A signed binary number is negated by the 2's-complement operation.

Negative-Going Transition   When a clock goes from 1 to 0.

Nested   To have one control structure embedded within another control structure.

Nibble   A group of four bits.

N-MOS (N-Channel Metal-Oxide-Semiconductor)   Integrated-circuit technology that uses N-channel MOSFETs as the principal circuit element.

NODE   A keyword in AHDL used to declare an intermediate variable (data object) that is local to that subdesign.

Noise   Spurious voltage fluctuations that may be present in the environment and cause digital circuits to malfunction.

Noise Immunity   Circuit's ability to tolerate noise voltages on its inputs.

Noise Margin   Quantitative measure of noise immunity.

Nonretriggerable One-Shot   Type of one-shot that will not respond to a trigger input signal while in its quasi-stable state.

Nonvolatile Memory   Memory that will keep storing its information without the need for electrical power.

Nonvolatile RAM   Combination of a RAM array and an EEPROM or flash on the same IC. The EEPROM serves as a nonvolatile backup to the RAM.

NOR Gate   Logic circuit that operates like an OR gate followed by an INVERTER. The output of a NOR gate is LOW (logic level 0) when any or all inputs are HIGH (logic level 1).

NOR Gate Latch Flip-flop constructed from two cross-coupled NOR gates.

NOT Circuit *See* INVERTER.

NOT Operation Boolean algebra operation in which the overbar (‾) or the prime (') symbol is used to indicate the inversion of one or more logic variables.

Objects Various ways of representing data in the code of any HDL.

Observation/Analysis Process used to troubleshoot circuits or systems in order to predict the possible faults before ever picking up a troubleshooting instrument. When this process is used, the troubleshooter must understand the circuit operation, observe the symptoms of the failure, and then reason through the operation.

Octal Number System Number system that has a base of 8; digits from 0 to 7 are used to express an octal number.

Octets Groups of eight 1s that are adjacent to each other within a Karnaugh map.

Offset Error Deviation from the ideal 0 V at the output of a digital-to-analog converter when the input is all 0s. In reality, there is a very small output voltage for this situation.

1-of-10 Decoder *See* BCD-to-Decimal Decoder.

1's-Complement Form Result obtained when each bit of a binary number is complemented.

One-Shot Circuit that belongs to the flip-flop family but that has only one stable state (normally $Q = 0$).

One-Time Programmable (OTP) A broad category of programmable components that are programmed by permanently altering the connections (e.g., melting a fuse element).

Open-Collector Output Type of output structure of some TTL circuits in which only one transistor with a floating collector is used.

Optical Disk Memory Class of mass memory devices that uses a laser beam to write onto and read from a specially coated disk.

OR Gate Digital circuit that implements the OR operation. The output of this circuit is HIGH (logic level 1) if any or all of its inputs are HIGH.

OR Operation Boolean algebra operation in which the symbol + is used to indicate the ORing of two or more logic variables. The result of the OR operation will be HIGH (logic level 1) if one or more variables are HIGH.

Output Logic Macrocell (OLMC) A group of logic elements (gates, multiplexers, flip-flops, buffers) in a PLD that can be configured in various ways.

Output Unit Part of a computer that receives data from the memory unit or ALU and presents it to the outside world.

Overflow When in the process of adding signed binary numbers, a carry of 1 is generated from the MSB position of the number into the sign bit position.

Override Inputs Synonymous with "asynchronous inputs."

Oversampling Inserting data points between sampled data in a digital signal to make it easier to filter out the rough edges of the waveform coming out of the DAC.

PACKAGE A VHDL keyword used to define a set of global elements that are available to other modules.

Parallel Adder Digital circuit made from full adders and used to add all of the bits from the addend and the augend together simultaneously.

Parallel Counter *See* Synchronous Counter.

Parallel Data Transfer Operation by which several bits of data are transferred simultaneously into a counter or a register.

Parallel In/Parallel Out Register Type of register that can be loaded with parallel data and has parallel outputs available.

Parallel In/Serial Out Register Type of register that can be loaded with parallel data and has only one serial output.

Parallel Load *See* Parallel Data Transfer.

Parallel-to-Serial Conversion Process by which all data bits are presented simultaneously to a circuit's input and then transmitted one bit at a time to its output.

Parallel Transmission Simultaneous transfer of all bits of a binary number from one place to another.

Parity Bit Additional bit that is attached to each code group so that the total number of 1s being transmitted is always even (or always odd).

Parity Checker Circuit that takes a set of data bits (including the parity bit) and checks to see if it has the correct parity.

Parity Generator Circuit that takes a set of data bits and produces the correct parity bit for the data.

Parity Method Scheme used for error detection during the transmission of data.

Percentage Resolution Ratio of the step size to the full-scale value of a digital-to-analog converter. Percentage resolution can also be defined as the reciprocal of the maximum number of steps of a digital-to-analog converter.

Pin-Compatible When the corresponding pins on two different ICs have the same functions.

Pixel Small dots of light that make up a graphical image on a display.

P-MOS (P-channel Metal Oxide Semiconductor) Integrated-circuit technology that uses P-channel MOSFETs as the principal circuit element.

PORT MAP A VHDL keyword that precedes the list of connections specified between components.

Positional-Value System System in which the value of a digit depends on its relative position.

Positive-Going Transition (PGT) When a clock signal changes from a logic 0 to a logic 1.

Power-Down Operating mode in which a chip is disabled and draws much less power than when it is fully enabled.

Power-Supply Decoupling Connection of a small RF capacitor between ground and V_{CC} near each TTL integrated circuit on a circuit board.

Power-Up Self-Test Program stored in ROM and executed by the CPU on power-up to test RAM and/or ROM portions of the computer circuitry.

Preprocessor Commands Compiler commands that are processed before the main program code in order to control how the code is interpreted.

Prescaler A counter circuit that takes base reference frequency and scales it by dividing the frequency down to a rate required by the system.

PRESET Asynchronous input used to set $Q = 1$ immediately.

Presettable Counter Counter that can be preset to any starting count either synchronously or asynchronously.

Priority Encoder Special type of encoder that senses when two or more inputs are activated simultaneously and then generates a code corresponding to the highest-numbered input.

PROCESS A VHDL keyword that defines the beginning of a block of code that describes a circuit that must respond whenever certain signals (in the sensitivity list) change state. All sequential statements must occur inside a process.

Product-of-Sums Form Logic expression consisting of two or more OR terms (sums) that are ANDed together.

Program Sequence of binary-coded instructions designed to accomplish a particular task by a computer; also, the act of entering information into a programmable device (e.g., EPROM, PLD).

Program Counter (PC) CPU register that stores the address of the next instruction to be fetched.

Programmable Array Logic (PAL) Class of programmable logic devices. Its AND array is programmable, whereas its OR array is hard-wired.

Programmable Interconnect Array (PIA) A term Altera Corporation uses to describe the resources used to connect the LABs with each other and also with the input/output modules.

Programmable Logic Array (PLA) Class of programmable logic devices. Both its AND and its OR arrays are programmable. Also called a *field programmable logic array (FPLA)*.

Programmable Logic Device (PLD) IC that contains a large number of interconnected logic functions. The user can program the IC for a specific function by selectively breaking the appropriate interconnections.

Programmable Output Polarity Feature of many PLDs whereby an XOR gate with a polarity fuse gives the designer the option of inverting or not inverting a device output.

Programmable ROM (PROM) ROM that can be electrically programmed by the user. It cannot be erased and reprogrammed.

Programmer A fixture used to apply the proper voltages to PLD and PROM chips in order to program them.

Propagation Delays (t_{PLH}/t_{PHL}) Delay from the time a signal is applied to the time when the output makes its change.

Pull-Down Transistor *See* Current-Sinking Transistor.

Pull-Up Transistor *See* Current-Sourcing Transistor.

Pulse-Steering Circuit A logic circuit that can be used to select the destination of an input pulse, depending on the logic levels present at the circuit's inputs.

Quantization Error Error caused by the nonzero resolution of an analog-to-digital converter. It is an inherent error of the device.

Quasi-Stable State State to which a one-shot is temporarily triggered (normally $Q = 1$) before returning to its stable state (normally $Q = 0$).

R/2R Ladder DAC Type of digital-to-analog converter whose internal resistance values span a range of only 2 to 1.

Random-Access Memory (RAM) Memory in which the access time is the same for any location.

RAS (Row Address Strobe) Signal used to latch the row address into a DRAM chip.

RAS-Only Refresh Method for refreshing DRAM in which only row addresses are strobed into the DRAM using the *RAS* input.

Read Term used to describe the condition when the CPU is receiving data from another element.

Read-Only Memory (ROM) Memory device designed for applications where the ratio of read operations to write operations is very high.

Read Operation Operation in which a word in a specific memory location is sensed and possibly transferred to another device.

Read/Write Memory (RWM) Any memory that can be read from and written into with equal ease.

Refresh Counter Counter that keeps track of row addresses during a DRAM refresh operation.

Refreshing Process of recharging the cells of a dynamic memory.

Register Group of flip-flops capable of storing data.

RESET Term synonymous with "CLEAR."

RESET State The $Q = 0$ state of a flip-flop.

Resolution In a digital-to-analog converter, smallest change that can occur in the output for a change in digital input; also called *step size*. In an analog-to-digital converter, smallest amount by which the analog input must change to produce a change in the digital output.

Retriggerable One-Shot Type of one-shot that will respond to a trigger input signal while in its quasi-stable state.

Ring Counter Shift register in which the output of the last flip-flop is connected to the input of the first flip-flop.

Ripple Counter *See* Asynchronous Counter.

Sample-and-Hold Circuit Type of circuit that utilizes a unity-gain buffer amplifier in conjunction with a capacitor to keep the input stable during an analog-to-digital conversion process.

Sampling Acquiring and digitizing a data point from an analog signal at a given instant of time.

Sampling Frequency The rate at which an analog signal is digitized (samples per second).

Sampling Interval   Time window during which a frequency counter samples and thereby determines the unknown frequency of a signal.

SBD   Schottky barrier diode used in all Schottky TTL series.

Schematic Capture   A computer program that can interpret graphic symbols and signal connections and translate them into logical relationships.

Schmitt Trigger   Digital circuit that accepts a slow-changing input signal and produces a rapid, oscillation-free transition at the output.

Schottky TTL   TTL subfamily that uses the basic TTL standard circuit except that it uses a Schottky barrier diode (SBD) connected between the base and the collector of each transistor for faster switching.

Selected Signal Assignment   A VHDL statement that allows a data object to be assigned a value from one of several signal sources depending on the value of an expression.

Sensitivity List   The list of signals used to invoke the sequence of statements in a PROCESS.

Sequential   Occuring one at a time in a certain order. In HDL, the circuits that are generated by sequential statements behave differently, depending on the order of the statements in the code.

Sequential-Access Memory (SAM)   Memory in which the access time will vary depending on the storage location of the data.

Sequential Circuit   A logic circuit whose outputs can change states in synchronism with a periodic clock signal. The new state of an output may depend on its current state as well as the current states of other outputs.

Serial Data Transfer   Transfer of data from one place to another one bit at a time.

Serial In/Parallel Out   Type of register that can be loaded with data serially and has parallel outputs available.

Serial In/Serial Out   Type of register that can be loaded with data serially and has only one serial output.

Serial Transmission   Transfer of binary information from one place to another a bit at a time.

SET   An input to a latch or FF used to make $Q = 1$.

Set   A grouping of concatenated variables or signals.

SET State   The $Q = 1$ state of a flip-flop.

Settling Time   Amount of time that it takes for the output of a digital-to-analog converter to go from 0 to within one-half step size of its full-scale value as the input is changed from all 0s to all 1s.

Setup Time (t_S)   Time interval immediately preceding the active transition of the clock signal during which the control input must be maintained at the proper level.

Shift Register   Digital circuit that accepts binary data from some input source and then shifts these data through a chain of flip-flops one bit at a time.

Sigma (Σ)   Greek letter that represents addition and is often used to label the sum output bits of a parallel adder.

Sigma/Delta Modulation   Method of sampling an analog signal and converting its data points into a bit stream of serial data.

Sign Bit   Binary bit that is added to the leftmost position of a binary number to indicate whether that number represents a positive or a negative quantity.

Sign-Magnitude System   A system for representing signed binary numbers where the most significant bit represents the sign of the number and the remaining bits represent the true binary value (magnitude).

Simple PLD (SPLD)   A PLD with a few hundred logic gates and possibly a few programmable macrocells available.

Simulator   Computer program that calculates the correct output states of a logic circuit based on a description of the logic circuit and on the current inputs.

Speed–Power Product   Numerical value (in joules) often used to compare different logic families. It is obtained by multiplying the propagation delay by the power dissipation of a logic circuit.

Spike   *See* Glitch.

SSI   Small-scale integration (fewer than 12 gates).

Staircase Test   Process by which a digital-to-analog converter's digital input is incremented and its output monitored to determine whether or not it exhibits a staircase format.

Staircase Waveform   Type of waveform generated at the output of a digital-to-analog converter as its digital input signal is incrementally changed.

Standard Cell   An application-specific integrated circuit (ASIC) made of predesigned logic blocks from a library of standard cell designs that are interconnected during the system design stage and then fabricated on a single IC.

Standard Logic   The large assortment of basic digital IC components available in various technologies as MSI, SSI chips.

State Machines   A sequential circuit that advances through several defined states.

State Table   A table whose entries represent the sequence of individual FF states (i.e., 0 or 1) for a sequential binary circuit.

State Transition Diagram   A graphic representation of the operation of a sequential binary circuit, showing the sequence of individual FF states and conditions needed for transitions from one state to the next.

Static Accuracy Test   Test in which a fixed binary value is applied to the input of a digital-to-analog converter and the analog output is accurately measured. The measured result should fall within the expected range specified by the digital-to-analog converter's manufacturer.

Static RAM (SRAM)   Semiconductor RAM that stores information in flip-flop cells that do not have to be periodically refreshed.

STD_LOGIC   In VHDL, a data type defined as an IEEE standard. It is similar to the BIT type, but it offers more possible values than just 1 or 0.

STD_LOGIC_VECTOR In VHDL, a data type defined as an IEEE standard. It is similar to the BIT_VECTOR type, but it offers more possible values than just 1 or 0 for each element.

Step Size *See* Resolution.

Straight Binary Coding Representation of a decimal number by its equivalent binary number.

Strobe Another name for an enable input usually used to latch a value into a register.

Strobing Technique often used to eliminate decoding spikes.

Structural Level of Abstraction A technique for describing a digital circuit that focuses on connecting ports of modules with signals.

SUBDESIGN Keyword in AHDL used to begin a circuit description.

Substrate Piece of semiconductor material that is part of the building block of any digital IC.

Subtrahend Number that is to be subtracted from a minuend.

Successive-Approximation ADC Type of analog-to-digital converter in which an internal parallel register and complex control logic are used to perform the conversion. The conversion time for this type of analog-to-digital converter is always the same regardless of the value of the input analog signal.

Sum-of-Products Form Logic expression consisting of two or more AND terms (products) that are ORed together.

Supercomputers Computers with the greatest speed and computational power.

Surface Mount A method of manufacturing circuit boards whereby ICs are soldered to conductive pads on the surface of the board.

Synchronous Control Inputs *See* Control Inputs.

Synchronous Counter Counter in which all of the flip-flops are clocked simultaneously.

Synchronous Systems Systems in which the circuit outputs can change states only on the transitions of a clock.

Synchronous Transfer Data transfer performed by using the synchronous and clock inputs of a flip-flop.

Syntax The rules defining keywords and their arrangement, usage, punctuation, and format for a given language.

Test Vector Sets of inputs used to test a PLD design before the PLD is programmed.

Timing Diagram Depiction of logic levels as related to time.

Toggle Mode Mode in which a flip-flop changes states for each clock pulse.

Toggling Process of changing from one binary state to the other.

Top-Down A design method that starts at the overall system level and then defines a hierarchy of modules.

Totem-Pole Output Term used to describe the way in which two bipolar transistors are arranged at the output of most TTL circuits.

Transducer Device that converts a physical variable to an electrical variable (for example, a photocell or a thermocouple).

Transmission Gate *See* Bilateral Switch.

Transparent Of a *D* latch, operating so that the *Q* output follows the *D* input.

Trigger Input signal to a flip-flop or one-shot that causes the output to change states depending on the conditions of the control signals.

Tristate Type of output structure that allows three types of output states: HIGH, LOW, and high-impedance (Hi-Z).

Truth Table Logic table that depicts a circuit's output response to the various combinations of the logic levels at its inputs.

TTL (Transistor/Transistor Logic) Integrated-circuit technology that uses the bipolar transistor as the principal circuit element.

2's-Complement Form Result obtained when a 1 is added to the least significant bit position of a binary number in the 1's-complement form.

Type The attribute of a variable in a computer-based language that defines its size and how it can be used.

ULSI Ultra-large-scale integration (100,000 or more gates).

Unasserted Term used to describe the state of a logic signal; synonymous with "inactive."

Undersampling Acquiring samples of a signal at a rate less than twice the highest frequency contained in the signal.

Unipolar ICs Integrated digital circuits in which unipolar field-effect transistors (MOSFETs) are the main circuit elements.

Up Counter Counter that counts upward from 0 to a maximum count.

Up/Down Counter Counter that can count up or down depending on how its inputs are activated.

Up/Down Digital-Ramp ADC Type of analog-to-digital converter that uses an up/down counter to step up or step down the voltage from a digital-to-analog converter until it intersects the analog input.

VARIABLE A keyword in AHDL used to begin a section of the code that defines the names and types of data objects and library primitives. A keyword used in VHDL to declare a local data object within a PROCESS.

Very High Speed Integrated Circuit (VHSIC) Hardware Description Language (VHDL) A hardware description language developed by the Department of Defense to document, simulate, and synthesize complex digital systems.

VLSI Very large-scale integration (10,000 to 99,999 gates).

Volatile Memory Memory requiring electrical power to keep information stored.

Voltage-Controlled Oscillator (VCO) Circuit that produces an output signal with a frequency proportional to the voltage applied to its input.

Voltage-Level Translator Circuit that takes one set of input voltage levels and translates it to a different set of output levels.

Voltage-to-Frequency ADC Type of analog-to-digital converter that converts the analog voltage to a pulse frequency that is then counted to produce a digital output.

Weighted Average An average calculation of a group of samples that assigns a different weight (between 0.0 and 1.0) to each sample.

Wired-AND Term used to describe the logic function created when open-collector outputs are tied together.

Word Group of bits that represent a certain unit of information.

Word Size Number of bits in the binary words that a digital system operates on.

WRITE Term used to describe the condition when the CPU is sending data to another element.

Write Operation Operation in which a new word is placed into a specific memory location.

ZIF Zero-insertion-force IC socket.

ANSWERS TO SELECTED PROBLEMS

CHAPTER 1

1-1. (a) and (e) are digital; (b), (c) and (d) are analog

1-2. (a) 25 (b) 9.5625 (c) 1241.6875

1-4. 1023

1-5. Nine bits

1-6.

1-7. (a) $2^N - 1 = 15$ and $N = 4$; therefore, four lines are required for parallel transmission. (b) Only one line is required for serial transmission.

CHAPTER 2

2-1. (a) 22 (b) 141 (c) 2313 (d) 91 (e) 255
(f) 119 (g) 983 (h) 191

2-2. (a) 100101 (b) 1110 (c) 10111101
(d) 10000000000 (e) 1001101 (f) 110010101
(g) 11001101 (h) 100100001001 (i) 111111111

2-3. 255; 65,535

2-4. (a) 483 (b) 30 (c) 2047 (d) 1024 (e) 117
(f) 5 (g) 175 (h) 644

2-5. (a) 73_8 (b) 564_8 (c) 1627_8 (d) 2000_8
(e) 1403_8 (f) 4411_8 (g) 200000_8 (h) 377_8

2-6. (a) 111100011 (b) 011110 (c) 011111111111
(d) 10000000000 (e) 1110101 (f) 101
(g) 010101111 (h) 001010000100

2-7. (a) 26_8 (b) 215_8 (c) 4411_8 (d) 133_8 (e) 377_8
(f) 167_8 (g) 1727_8 (h) 277_8

2-8. 165, 166, 167, 170, 171, 172, 173, 174, 175, 176, 177, 200

2-9. $100100001001_2 = 4411_8$

2-10. Five

2-11. (a) 146 (b) 422 (c) 14,333 (d) 43,981
(e) 15 (f) 85 (g) 704 (h) 2047

2-12. (a) 4B (b) 13A (c) 800 (d) E (e) 1C4D
(f) 185 (g) 6413 (h) FFF

2-15. (a) 16 (b) 8D (c) 909 (d) 5B (e) FF
(f) 77 (g) 3D7 (h) BF

2-16. (a) 10010010 (b) 000110100110
(c) 0011011111111101 (d) 001011000000
(e) 011111111111

2-17. 280, 281, 282, 283, 284, 285, 286, 287, 288, 289, 28A, 28B, 28C, 28D, 28E, 28F, 290, 291, 292, 293, 294, 295, 296, 297, 298, 299, 29A, 29B, 29C, 29D, 29E, 29F, 2A0

2-18. Five

2-19. (a) 01000111 (b) 100101100010
(c) 000110000111 (d) 0110011100100111
(e) 00010011 (f) 0100001001101000010010110000100111

2-20. 10 bits for binary, 12 bits for BCD

2-21. (a) 9752 (b) 184 (c) 695 (d) 7775 (e) 492
(f) 555

2-22. (a) 64 (b) FFFFFFFF (c) 999,999

2-24. 01011000 (X); 00111101 (=); 00110010 (2); 10110101 (5); 00101111 (/); 11011001 (Y)

2-25. D8, BD, B2, 35, AF, 59

2-26. BEN SMITH

2-27. (a) 101110100 (parity bit on the left)
(b) 000111000 (c) 11000100010000100
(d) 0001001110101 (e) 0000101100101
(f) 11001001000000001

2-28. (a) No single-bit error (b) Single-bit error
(c) Double error (d) No single-bit error

2-30. (a) 10110001001 (b) 11111111 (c) 209
(d) 59,943 (e) 4701 (f) 777 (g) 157 (h) 2254
(i) 1961 (j) 15,900 (k) 640 (l) 952B

(m) 100001100101   (n) 947   (o) 135_{16} (p) 5464_8
(q) 1001010   (r) 01011000 (BCD)
2-31. (a) 100101   (b) 00110111   (c) 25   (d) 0110011
0110111   (e) 45
2-32. (a) Octal   (b) 16   (c) Digit   (d) Gray
(e) Parity; single-bit errors   (f) ASCII   (g) Octal; hex
(h) byte
2-33. (a) 1000   (b) 010001   (c) 1111
2-34. (a) 0110   (b) 001111   (c) 1101
2-35. (a) 10000   (b) 7778   (c) 2001   (d) 2001
(e) A00   (f) 1001
2-36. (a) 7776   (b) 7776   (c) 1777   (d) 1FFF
(e) 9FE   (f) 0FFF
2-37. (a) 1,048,576   (b) Five   (c) 000FF
2-38. (a) 64; 256; 1024   (b) 440,000   (c) 11,363.63
2-39. Eight

CHAPTER 3

3-1.

3-2.

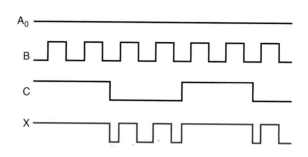

3-3. x will be a constant HIGH.
3-5. 31
3-6. (a) x is HIGH only when A, B, and C are all HIGH.
(b) x is a constant LOW.   (c) x is HIGH only when B
and C are simultaneously HIGH.
3-7. Change the OR gate to an AND gate.
3-8. OUT is always LOW.
3-12. (a) $x = \overline{(\overline{A} + \overline{B})BC}$
(b) $x = \overline{A}\,\overline{B}\,\overline{C} + A\overline{B}\,C + \overline{A}\,\overline{B}\,D$
3-13. X is HIGH for all cases where $E = 1$ except for
$EDCBA = 10101, 10110,$ and 10111.
3-14. $x = D \cdot (\overline{AB + C}) + E$

3-16.

(a)

(b)

3-17.–3-18.

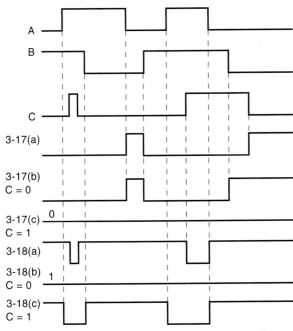

3-19. $x = \overline{(A + B) \cdot (\overline{B}\,C)}$

$x = 0$ only when $A = B = 0, C = 1$.

3-20. X is LOW only when $A = B = 1$ and $C = D = 0$.

3-21. (c)

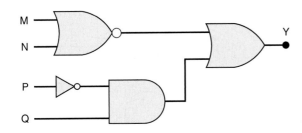

3-22. (a) 1   (b) A (c) 0   (d) C (e) 0   (f) D
(g) D (h) 1   (i) G (j) y

3-24. (a) $MP\overline{N} + \overline{M}\,\overline{P}\,N$ (b) $B\overline{C}$

3-26. (a) $A + \overline{B} + C$ (b) $A \cdot (B + \overline{C})$
(c) $\overline{A} + \overline{B} + CD$ (d) $\overline{A}\,B$ (e) $A + B$ (f) ACD
(g) $\overline{A} + B + \overline{C} + \overline{D}$ (h) $\overline{M}N + M\overline{N}$
(i) $(\overline{A} + \overline{B})C + \overline{D}$

3-27. $A + B + \overline{C}$

3-32. (a) $W = 1$ when $T = 1$ and either $P = 1$ or $R = 0$.

3-34. (a) NOR   (b) AND   (c) NAND

3-35. (a)

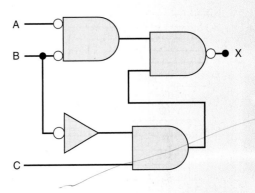

3-36. (a) Z goes HIGH only when $A = B = 0$ and $C = D = 1$.   (b) Z goes LOW when A or B is HIGH or when C or D is LOW.

3-38. X will go HIGH when $E = 1$, or $D = 0$, or $B = C = 0$, or when $B = 1$ and $A = 0$.

3-39. X is LOW for $EDCBA = 01011, 01100, 01101$, and 01111.

3-40. (a) HIGH   (b) LOW

3-41. $\overline{\text{LIGHT}} = 0$ when $A = B = 0$ or $A = B = 1$.

3-42. (a)

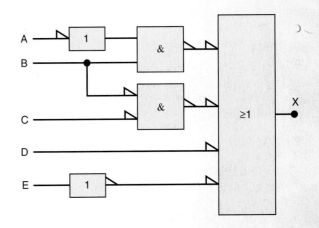

3-43. (a) False   (b) True   (c) False   (d) True
(e) False   (f) False   (g) True   (h) False   (i) True
(j) True

3-44.

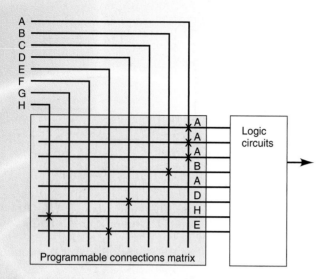

3-45. AHDL and VHDL solutions are on the enclosed CD.

3-46. AHDL and VHDL solutions are on the enclosed CD.

3-47. Put INVERTERs on the A_7, A_5, A_4, A_2 inputs to the 74HC30.

3-49. Requires six 2-input NAND gates.

CHAPTER 4

4-1. (a) $C\overline{A} + CB$ (b) $\overline{Q}R + Q\overline{R}$ (c) $C + \overline{A}$
(d) $\overline{R}\,\overline{S}\,\overline{T}$ (e) $BC + \overline{B}(\overline{C} + A)$
(f) $BC + \overline{B}(\overline{C} + A)$ or $BC + \overline{B}\,\overline{C} + \overline{A}\,\overline{C}$
(g) $\overline{D} + AB\overline{C} + \overline{A}\,\overline{B}C$
(h) $x = ABC + AB\overline{D} + \overline{A}BD + \overline{B}C\overline{D}$

4-2. $Q(M + N)$

4-3. $MN + Q$

4-4. One solution: $\overline{x} = \overline{B}C + AB\overline{C}$. Another:
$x = \overline{A}B + \overline{B}\,\overline{C} + BC$. Another: $BC + \overline{B}\,\overline{C} + \overline{A}\,\overline{C}$

4-5. $x = \overline{A}\,\overline{B} + \overline{A}C + B\overline{C}$

4-6. $x = A\overline{B} + \overline{B}CD$ where A = It's 5 o'clock or later; B = All machines are shut down; C = It's Friday; D = Production for the day is complete.

4-7. $x = \overline{A}_3(A_2 + A_1 A_0)$

4-8. Alarm $= ID + \overline{I}L$

4-9.

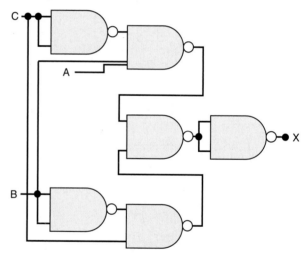

4-11. (a) $x = \overline{A}\,\overline{C} + \overline{B}C + AC\overline{D}$

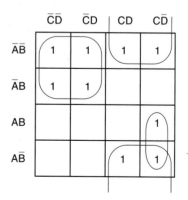

	$\overline{C}\overline{D}$	$\overline{C}D$	CD	$C\overline{D}$
$\overline{A}\overline{B}$	1	1	1	1
$\overline{A}B$	1	1		
AB				1
$A\overline{B}$			1	1

(b) $x = \overline{A}\,\overline{D} + \overline{B}\,C + \overline{B}\,\overline{D}$
(c) $y = \overline{B} + A\overline{C}$

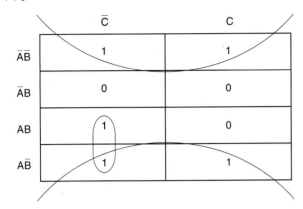

4-12. $x = BC + \overline{B}\,\overline{C} + AC$; or $x = BC + \overline{B}\,\overline{C} + A\overline{B}$
4-13. $y = \overline{D} + \overline{A}\,\overline{B}\,C + A\overline{B}\,\overline{C}$

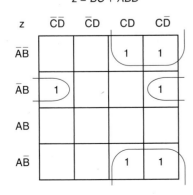

4-14. One possible looping:
$x = \overline{A}\,BD + ABC + AB\overline{D} + \overline{B}\,\overline{C}\,\overline{D}$; another one is:
$x = ABC + BCD + A\overline{C}\,\overline{D} + \overline{B}\,\overline{C}\,\overline{D}$
4-15. $x = \overline{A}_3 A_2 + \overline{A}_3 A_1 A_0$
4-16. Best solution: $x = B\overline{C} + AD$
4-17. $x = \overline{S}_1\overline{S}_2 + \overline{S}_1\overline{S}_3 + \overline{S}_3\overline{S}_4 + \overline{S}_2\overline{S}_3 + \overline{S}_2\overline{S}_4$
4-18. Example 4-3 demonstrated algebraic simplification. Step 3 resulted in the SOP equation $z = \overline{A}\,\overline{B}\,C + \overline{A}\,C\,\overline{D} + \overline{A}\,B\overline{C}\,D + A\overline{B}\,C$. Use a K map to prove that this equation can be simplified further than the answer shown in the example.

$$z = \overline{B}C + \overline{A}B\overline{D}$$

4-20. (a)

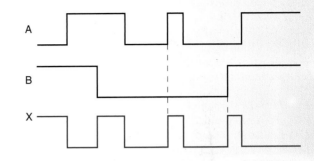

(b) $x = A$ (c) $x = \overline{A}$
4-21. $A = 0, B = C = 1$
4-22. One possibility is shown below.

4-23. Four XNORs feeding an AND gate
4-25. Four outputs where z_3 is the MSB
$z_3 = y_1 y_0 x_1 x_0$
$z_2 = y_1 x_1(\overline{y}_0 + \overline{x}_0)$
$z_1 = y_0 x_1(\overline{y}_1 + \overline{x}_0) + y_1 x_0(\overline{y}_0 + \overline{x}_1)$
$z_0 = y_0 x_0$
4-26. $x = A_3 A_2 + A_3 A_1$
4-27. $x = AB(C \oplus D)$
4-28. $(A + B)(C + D)$
4-29. N–S $= \overline{C}\,\overline{D}(A + B) + AB(\overline{C} + \overline{D})$; E–W $= \overline{\text{N–S}}$
4-30. (a) Add an inverter at output.   (b) Add an inverter at output, or change all XORs to XNORs.
4-33. (a) No   (b) No
4-34. $x = \overline{A}\,\overline{B}\,C$
4-35. $x = A + BCD$
4-36. $x = A + (B \oplus C)$
4-37. $z = A_0\overline{S} + A_1 S$
4-38. $z = x_1 x_0 y_1 y_0 + x_1 \overline{x}_0 y_1 \overline{y}_0 + \overline{x}_1 x_0 \overline{y}_1 y_0 + \overline{x}_1 \overline{x}_0 \overline{y}_1 \overline{y}_0$ No pairs, quads, or octets
4-40. (a) Indeterminate   (b) 1.4–1.8 V   (c) See below.

4-43. Possible faults: faulty V_{CC} or ground on Z2; Z2-1 or Z2-2 open internally or externally; Z2-3 internally open

4-44. Yes: (c), (e), (f).   No: (a), (b), (d), (g).

4-46. Z2-6 and Z2-11 shorted together

4-48. Most likely faults:
faulty ground or V_{CC} on Z1;
Z1 plugged in backwards;
Z1 internally damaged

4-49. Possible faults:
Z2-13 shorted to V_{CC};
Z2-8 shorted to V_{CC};
broken connection to Z2-13;
Z2-3, Z2-6, Z2-9, or Z2-10 shorted to ground

4-50. (a) T, (b) T, (c) F, (d) F, (e) T

4-54. Boolean equation; truth table; schematic diagram

4-56. (a) AHDL: gadgets[7..0]   :OUTPUT;
        VHDL: gadgets   :OUT BIT_VECTOR (7 DOWNTO 0);
      (b) AHDL: buzzer   :OUTPUT;
        VHDL: buzzer   :OUT BIT;
      (c) AHDL: altitude[15..0]   :INPUT;
        VHDL: altitude   :IN INTEGER RANGE 0 TO 65535);
      (d) AHDL: VARIABLE wire2   :NODE;
        VHDL: SIGNAL wire2   :BIT;

4-57. (a) AHDL: H"98"   B"10011000"   152
        VHDL: X"98"   B"10011000"   152
      (b) AHDL: H"254" B" 1001010100" 596
        VHDL: X"254" B" 1001010100" 596
      (c) AHDL: H"3C4" B" 1111000100" 964
        VHDL: X"3C4" B" 1111000100" 964

4-58. AHDL: outbits[3]   =   inbits[1];
              outbits[2]   =   inbits[3];
              outbits[1]   =   inbits[0];
              outbits[0]   =   inbits[2];
      VHDL: outbits(3)   <=   inbits(1);
              outbits(2)   <=   inbits(3);
              outbits(1)   <=   inbits(0);
              outbits(0)   <=   inbits(2);

4-59.
TABLE
    (a,b,c)   =>   y;
    (0,0,0)   =>   0;
    (0,0,1)   =>   0;
    (0,1,0)   =>   1;
    (0,1,1)   =>   1;
    (1,0,0)   =>   1;
    (1,0,1)   =>   0;
    (1,1,0)   =>   1;
    (1,1,1)   =>   1;
END TABLE;

4-60.
WITH in_bits SELECT
    y   <=   '0' WHEN "000",
               '0' WHEN "001",
               '1' WHEN "010",
               '1' WHEN "011",
               '1' WHEN "100",
               '0' WHEN "101",
               '1' WHEN "110",
               '1' WHEN "111";

4-61.
BEGIN
    IF digital_value[] < 10 THEN
        z = VCC; --output a 1
    ELSE z = GND; --output a 0
    END IF;
END;

4-62.
PROCESS (digital_value)
    BEGIN
        IF (digital_value < 10) THEN
            z <= '1';
        ELSE
            z <= '0';
        END IF;
END PROCESS

4-63. The AHDL solution is on the enclosed CD.

4-64. The VHDL solution is on the enclosed CD.

4-65. S = !P#Q&R

4-66. P = D3$D2$D0$D1

4-68. (a) 00 to EF   (b) F0   (c) F1 to FF

CHAPTER 5

5-1.

5-2. Same as 5-1

5-3.

5-6. Z1-4 stuck HIGH

5-7. 20 ns

5-8.

5-9. Assume $Q = 0$ initially.
For PGT FF: Q will go HIGH on first PGT of CLK.
For NGT FF: Q will go HIGH on first NGT of CLK,
LOW on second NGT, and HIGH again on fourth NGT.

5-10.

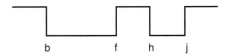

5-11. (a) 5-kHz square wave   (b) 2.5 kHz

5-13.

5-15. 500-Hz square wave

5-16. Q will not change.

5-20.

5-21.

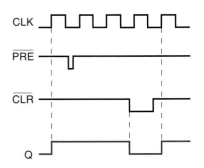

5-23. (a) 200 ns   (b) 7474; 74C74   (c) 30 ns

5-24. 62 ns

5-25. Connect A to J, $\bar{A}$ to K.

5-26. (a) A, B, C

5-27. (a) Connect X to J, $\bar{X}$ to K.   (b) Use
arrangement of Figure 5-41.

5-28. 1011, 1101, 1110, 0111, 1011, 1101, 1110, 0111,
1011

5-29. Connect X_0 to D input of X_2.

5-30. (a) 101;011;000   (b) 001;111;100   (c) 500 kHz

5-33. (a) 10   (b) 1953 Hz   (c) 1024   (d) 12

5-34. (a) 128   (b) 0 to 127

5-36. Put INVERTERs on A_8, A_{11}, and A_{14}.

5-40.

5-41. (a) $\bar{Q}$ will stay LOW.   (b) PGT at $\bar{Q}$ will set $X = 1$.
(c) Make $t_p = 20$ μs.

5-42. (a) A_1 or A_2 must be LOW when a PGT occurs at B.
(b) B and A_2 must be HIGH when a NGT occurs at A_1.

5-44. After 14 pulses, all circuit outputs stop changing
with $A = W = X = Y = Z = 0$ and $B = C = 1$.

5-45. One possibility is $R = 1$ kΩ and $C = 80$ nF.

5-46. One possibility is $R_A = 1$ kΩ, $R_B = 10$ kΩ,
$C = 1800$ pF.

5-48. (a)

5-49. (a) No   (b) Yes   (c) No   (d) No

5-50. Flip-flop X_0 has an open D input.

5-51. (a) Yes   (b) No

5-52. Two cascaded INVERTERs between Q_1 and Q_2
add enough delay to increase the effective t_{PLH} of Q_1
so that by the time Q_2 gets clocked, the Q_1 signal
hasn't reached D_2.

5-54. (a) No   (b) No   (c) Yes   (d) No

5-55. First combination: 101. Second combination:
010.

5-56. (a) No   (b) No   (c) Yes

5-57. (a) NAND and NOR latch   (b) J-K   (c) D latch
(d) D flip-flop   (e) D flip-flop and D latch   (f) All
FFs   (g) Clocked S-C, J-K, D   (h) J-K

5-59. Q = !(C # QBAR); QBAR = !(S # Q);

5-60.

```
SUBDESIGN prob5_60
(    S1BAR, C1BAR, S2, C2, D3, EN3 :INPUT;
     Q1, Q1BAR, Q2, Q2BAR, Q3, Q3BAR
        :OUTPUT; )
BEGIN Q1 = !Q1BAR # !S1BAR; Q1BAR = !Q1 #
      !C1BAR;
      Q2 = ! (C2 # Q2BAR); Q2BAR = ! (S2 #
        Q2);
      Q3 = (D3 & EN3) # !Q3BAR; Q3BAR =
        (!D3 & EN3) # !Q3;
END ;
```

5-61. Remove the inverter operators (!) on lines 13, 14, and 15 in Figure 5-71. Remove the (NOT) operation on lines 17, 18, and 19 of Figure 5-72.

5-63. q = (d & en) # !qbar;
qbar = (!d & en) # !q;

5-64. See Prob5_64.tdf on the enclosed CD.

5-65. See Prob5_65.vhd on the enclosed CD.

5-69. See Prob5_69.tdf on the enclosed CD.

5-70. See Prob5_70.vhd on the enclosed CD.

5-74. See Prob5_74.tdf on the enclosed CD.

5-75. See Prob5_75.vhd on the enclosed CD.

CHAPTER 6

6-1. (a) 10101 (b) 10010 (c) 1111.0101 (d) 1.1010
(e) 100111000

6-2. (a) 00100000 (including sign bit) (b) 11110010
(c) 00111111 (d) 10011000 (e) 01111111
(f) 10000001 (g) 01011001 (h) 11001001
(i) 11111111 (j) 10000000 (k) Can't be done in eight bits (l) 00000000 (m) 01010100
(n) 00000011 (o) 11111101 (p) Can't be done in eight bits

6-3. (a) +13 (b) −3 (c) +123 (d) −103
(e) +127 (f) −64 (g) −1 (h) −127 (i) +99
(j) −39

6-4. (a) −2048 to +2047 (b) 16 bits including sign bit

6-5. -16_{10} to 15_{10}

6-6. (a) 01001001; 10110111 (b) 11110100; 00001100
(c) 00001111; 11110001 (d) 11111111; 00000001
(e) +128 requires 9 bits (f) 01111111, 10000001

6-7. 0 to 1023; −512 to +511

6-9. (a) 00001111 (b) 11111101 (c) 11111011
(d) 10000000 (e) 00000001 (f) 11011110
(g) 00000000 (h) 00010101 (i) and (j) 00000000

6-11. (a) 100011 (b) 1111001 (c) 100011.00101
(d) .10001111

6-12. (a) 11 (b) 111 (c) 101.11 (d) 1111.0011

6-13. (a) 10010111 (BCD) (b) 10010101 (BCD)
(c) 010100100111 (BCD) (d) 010100000011 (BCD)
(e) 0001000000000001 (BCD) (f) 0001001000100010 (BCD)

6-14. (a) 6E24 (b) 100D (c) 18AB (d) 3000
(e) 10FE (f) 17C3C

6-15. (a) 0EFE (b) 229 (c) 02A6 (d) 01FD
(e) 0001 (f) EF00

6-16. 16,849 locations

6-17. (a) 119 (b) +119 (c) 229; −27

6-19. SUM = $A \oplus B$; CARRY = AB

6-21. [A] = 1111, or [A] = 000 (if $C_0 = 1$)

6-22. 200 ns

6-23. Overflow = $\overline{A_3}\overline{B_3}S_3 + A_3 B_3 \overline{S_3}$

6-25. $C_3 = A_2 B_2 + (A_2 + B_2)\{A_1 B_1 + (A_1 + B_1)[A_0 B_0 + A_0 C_0 + B_0 C_0]\}$

6-27. (a) SUM = 0111 (b) SUM = 1010 (c) 1100

6-29.

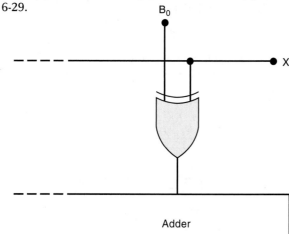

6-30. [S] = 01110; x = 1; [Σ] = 0100

6-31. Yes

6-32. [Σ] = 100001000101

6-33.

	[F]	C_{N+4}	**OVR**
(a)	1001	0	1
(b)	1101	0	0
(c)	0011	1	0

6-34. [S] = 100; [B] = 1111

6-35. (a) 00001100 (b) 11011100

6-37. (a) 0001 (b) 1010 (c) 0010 (d) 1011 (e) 1000

6-39. (a) Yes (b) No (c) Yes (d) No

6-42.
AHDL a [7..0], b[7..0] :INPUT;
         z[7..0]   :OUTPUT;
VHDL
PORT(a, b   :IN   BIT_VECTOR (7 DOWNTO 0);
        z   :OUT   BIT_VECTOR (7 DOWNTO 0));

6-43. (a) 00000100 (b) 10111111 (c) 01000100
(d) 01000000 (e) 00101110

6-44. (a) 0 (b) 1 (c) 0010110

6-45. **AHDL**
        z[6..0] = a[7..1];
        z[7] = a[0];
    VHDL
        z(6..0) <= a(7..1);
        z(7) <= a(0);

6-46.
AHDL ovr <= (a(3) AND b(3) AND (NOTs(3))) OR ((NOT a(3)) AND (NOT b(3)) AND s(3));
VHDL ovr = ((a(3) AND b(3)) AND NOTz(3)) OR (NOTa(3) AND NOTb(3)) AND z(3));
See prob6_46.tdf and prob6_46.vhd on the enclosed CD.

6-47. AHDL: ovr<= c[4] \$ c[3)];
   VHDL: ovr<= c(4) XOR c(3);

6-48. See Prob6_48.tdf and Prob6_48.vhd on the enclosed CD.

6-53. Use D flip-flops. Connect $(\overline{S_3 + S_2 + S_1 + S_0})$ to the D input of the 0 FF; C_4 to the D input of the carry FF; and S_3 to the D input of the sign FF.

6-54. 0000000001001001; 1111111110101110

CHAPTER 7

7-1. (a) 250 kHz; 50% (b) Same as (a)
(c) Frequency at C = 1 MHz (d) 32

7-2. 2 kHz

7-3. 10000_2

7-4. Five FFs are required: Q_0–Q_4 with Q_4 as the MSB. Connect Q_3 and Q_4 outputs to a NAND gate whose output is connected to all CLRs.

7-5.

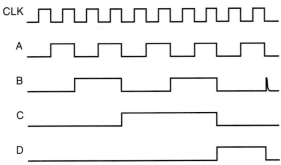

7-6. Glitch at B output on fourteenth NGT of the clock

7-7. Connect Q_5, Q_4, and Q_1 for MOD-50. A MOD-100 cannot be constructed without more logic.

7-8. (b) 000, 001, 010, 100, 101, 110, and repeats

7-10.

7-11. A MOD-15 (from Problem 7-10) driving a MOD-4

7-12. 60 Hz

7-14. (c) 0001

7-15. 100, 011, 010, 001, 000, and repeats

7-16. 1000 and 0000 states never occur.

7-17. (a) 12.5 MHz (b) 8.33 MHz

7-18. (a) Add two FFs (E and F) to Figure 7-17. Connect AND gates below to appropriate FF inputs.
(b) 33 MHz

7-19. 0000, 0001, 0010, 0011, 0100, 0101, 0110, 0111, 1000, 1001, and repeats

7-21. The counter FFs will switch between the 000 and 111 states on each clock pulse.

7-23. (d) Z will not be cleared and the timer cannot be restarted.

7-24. Change the parallel data input to 1010.

7-25. Nine

7-27. (a) Ten (b) Clears counter to 0000 (c) Sets counter to 1001 (d) Up (e) Connect Q_0 to CP_1 and clock signal to CP_0; ground MR_1, MR_2, MS_1, and MS_2. (f) Connect clock signal to CP_1; connect Q_3 to CP_0; ground all MR and MS inputs.

7-28.

7-31.

7-33. (a) When the counter goes from 0111 to 1000

7-34. 12 FFs

7-35. Four; sixteen; thirteen

7-36. (a) $J_A = B\overline{C}, K_A = 1, J_B = CA + \overline{C}\overline{A}, K_B = 1,$ $J_C = \overline{A}B, K_C = B + \overline{A}$
(b) $J_A = B\overline{C}, K_A = 1, J_B = K_B = 1, J_C = K_C = \underline{B}$

7-37. $J_A = K_A = 1, J_B = K_B = \overline{A}, J_C = K_C = \overline{A}B,$ $J_D = K_D = \overline{A}\overline{B}C$

7-38. $J_A = K_A = 1; J_B = K_B = AD + \overline{A}\overline{D};$ $J_C = K_C = ABD + \overline{A}\overline{B}\overline{D}$

7-39.

PRESENT State			NEXT State			Inputs		
Q_2	Q_1	Q_0	Q_2	Q_1	Q_0	D_3	D_2	D_1
0	0	0	0	0	1	0	0	1
0	0	1	0	1	0	0	1	0
0	1	0	0	1	1	0	1	1
0	1	1	1	0	0	1	0	0
1	0	0	1	0	1	1	0	1
1	0	1	1	1	0	1	1	0
1	1	0	1	1	1	1	1	1
1	1	1	0	0	0	0	0	0

7-40.

$Q2.D = \overline{Q2}\,Q1\,Q0 + Q2\,\overline{Q1}\,\overline{Q0} + Q2\,\overline{Q1}\,Q0 + Q2\,Q1\,\overline{Q0}$

$Q1.D = \overline{Q2}\,\overline{Q1}\,Q0 + \overline{Q2}\,Q1\,\overline{Q0} + Q2\,\overline{Q1}\,Q0 + Q2\,Q1\,\overline{Q0}$

$Q0.D = \overline{Q0}$

7-41. See PROB7_41.tdf and PROB7_41.vhd on the enclosed CD.

7-47. Both counters are mod-12, QD is the MSB, MSB signal frequency is 1/12 of the clock frequency, the 74161 counter circuit has an unusual count sequence, duty cycles of QD are different: 33% for 74163 and 50% for 74161 circuit.

7-48. MOD-10 down counter counts from 1010 down to 0001 and recycles. It can produce the same modulus with the 74190 counter but not the same sequence.

7-49. (a)

7-50.

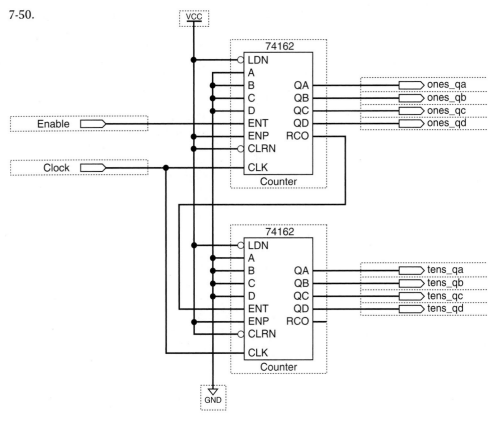

7-51.

Output	X0	X1	X2	X3	X4
Frequency	3 MHz	1.5 MHz	500 kHz	500 kHz	500 kHz
Output	Y0	Y1	Z0	Z1	
Frequency	600 kHz	600 kHz	1.2 MHz	0 Hz	

7-54. Connect Q_0 to D_5 through an inverter.

7-55. (a) 101000　(b) 101101

7-56. t_{310}

7-58. (b) $Q_0 \cdot \overline{Q}_1 + Q_0 \cdot Q_7$　(c) $Q_3 \cdot \overline{Q}_6$

7-59. See figure below.

7-60. (a) *CLR* input is asynchronous.　(b) True
(c)

Q_a	Q_b	Q_c	Q_d	
1	0	1	1	
0	1	1	1	After 1 clock pulse
1	1	1	1	After 2 clock pulses
1	1	1	1	After 3 clock pulses
1	1	1	1	After 4 clock pulses

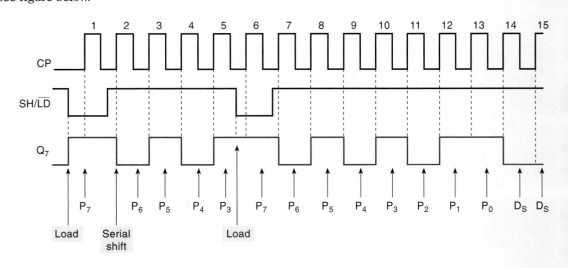

(d)

Q_a	Q_b	Q_c	Q_d	
1	0	1	1	
0	1	0	1	After 1 clock pulse
0	0	1	0	After 2 clock pulses
0	0	0	1	After 3 clock pulses
0	0	0	0	After 4 clock pulses

(e)

Q_a	Q_b	Q_c	Q_d	
0	1	1	0	After 1 clock pulse
0	1	1	0	After 2 clock pulses
0	1	1	0	After 3 clock pulses
0	1	1	0	After 4 clock pulses

(f) Same as (e)

(g)

Q_a	Q_b	Q_c	Q_d	
1	0	1	1	
0	1	1	1	After 1 clock pulse
1	1	1	0	After 2 clock pulses
1	1	0	1	After 3 clock pulses
1	0	1	1	After 4 clock pulses

7-62.

7-63. Counting sequence is: 00000, 10000, 11000, 11100, 11110, 11111, 01111, 00111, 00011, 00001, and repeat.

7-64. Frequency at z is 5 Hz.

7-65. (a) Each light will be ON for 0.5 second of every 4-second interval. (b) Each light will be ON for 4 seconds and OFF for 4 seconds.

7-66. MR_1 to Q_1 is open, or Q_3 or MR_2 is shorted to ground.

7-67. (a) Flip-flop C will toggle on every NGT of the input. (b) Will not count up properly; B and C will not toggle.

7-68. P_1 open or S3 faulty

7-69. Misconnected Q_1 (instead of Q_0) to MR_1.

7-70. See PROB7_70.tdf and PROB7_70.vhd on the enclosed CD.

7-78. (a) Parallel (b) Binary (c) Down counter (d) MOD-10 (e) Asynchronous (f) Ring (g) Johnson (h) All (i) Presettable (j) Up/down (k) Asynchronous (l) BCD (m) Synchronous/parallel

CHAPTER 8

8-1. (a) A; B (b) A (c) A

8-2. (a) 728.3 pJ (b) 459.3 pJ (c) 59.06 pJ (d) 27.41 pJ (e) 67.26 pJ

8-3. (a) 0.9 V (b) 1.4 V

8-4. (a) I_{IH} (b) I_{CCL} (c) t_{PHL} (d) Speed–power product (e) V_{NH} (f) Surface-mount (g) Current sinking (h) Fan-out (i) Totem-pole (j) Sinking transistor (k) 4.75 to 5.25 V (l) 2.5 V; 2.0 V (m) 0.8 V; 0.5 V (n) Sourcing

8-5. (a) 0.7 V; 0.3 V (b) 0.5 V; 0.4 V (c) 0.5 V; 0.3 V

8-6. (b) AND, NAND (c) Unconnected inputs

8-7. (a) 40 (b) 33 (c) 16 (d) 40

8-8. (a) 20 μA/0.4 mA (c) Five

8-9. (a) 30/15 (b) 24 mA

8-11. Fan-out is not exceeded in either case.

8-12. 10 ns

8-13. 60 ns; 38 ns

8-14. (a) 2 kΩ (b) 8 kΩ

8-15. (b) 4.7-kΩ resistor is too large.

8-16. (a) Amplitude too low (b) t_P too small (c) $t_W(L)$ too low

8-18. (a) N-MOSFET (b) P-MOSFET

8-19. a, c, e, f, g, h

8-20. b

8-21. 12.6 mW

8-22. 2.9 V

8-24. Using values at V_{CC} = 6 V, the speed–power product is 1.44 pJ per gate.

8-26. Six AND gates

8-27. $AB + CD + FG$

8-28. Choice (a)

8-29. (a) 5 V (b) R_S = 110 Ω for LED current of 20 mA

8-30. (a) 12 V (b) 40 mA

8-32. (a)

x	y	Data on Bus
0	0	C
0	1	B
1	0	B
1	1	A

(b) Bus contention due to E_A and E_C simultaneously active when $X = Y = 1$.

8-33. Ring counter
8-36. 1.22 V; 0 V
8-37.

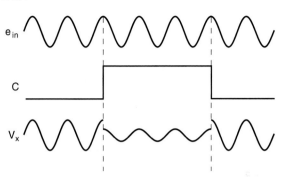

8-38. −1 and −2
8-39. (a) 74HCT   (b) Converts logic voltages
(c) CMOS cannot sink TTL current.   (d) False
8-40. The choice in (c) is a possible reason.
8-41. (a) None   (b) Two
8-44. Fan-out of 74HC00 is exceeded; disconnect pin 3
of 7402 and tie it to ground.
8-46. $R_2 = 1.5$ kΩ, $R_1 = 18$ kΩ
8-49. (b) is a possible fault.
8-50. 0 V to −11.25 V and back up to −6 V

CHAPTER 9

9-1. (a) All HIGH   (b) $\overline{O}_0$ = LOW   (c) $\overline{O}_7$ = LOW
(d) All HIGH
9-2. Six inputs, 64 outputs
9-3. (a) $E_3E_2E_1 = 100$; $[A] = 110$ (b) $E_3E_2E_1 = 100$;
$[A] = 011$ (c) $E_3E_2E_1 = 100$; $[A] = 101$ (d) No input
conditions
9-4.

9-5.

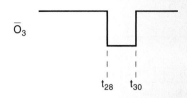

9-6. Connect $\overline{E}_2$ to ground and take output from $\overline{O}_2$.
9-7. Enabled when $D = 0$
9-8.

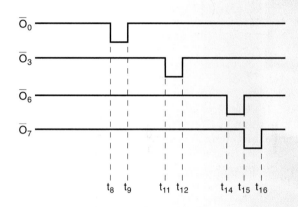

9-10. Resistors are 250 Ω.
9-12.

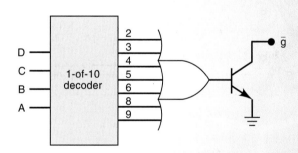

9-13. (a), (b) Encoder   (c), (d), (e) Decoder
9-14. 0111
9-17. The fourth key actuation would be entered into
the MSD register.
9-18. Choice (b)
9-19. Choice (a) or (d)
9-20. (a) Yes   (b) No   (c) No
9-21. A_2 bus line is open between Z2 and Z3.
9-22. E_2 of Z4 is stuck LOW.
9-23. g segment or decoder output transistor would
burn out.
9-24. Decoder D and C inputs are reversed.
9-25. Decoder outputs: a and b are shorted together.
9-26. Connection 'f' from decoder/driver to XOR gate
is open.
9-28. $S_2 = 0$, $S_1 = S_0 = 1$

9-29. A 4-to-1 MUX
9-30. Use nine 74151s.
9-31.

9-32. (b) The total number of connections in the circuit using MUXes is 63, not including V_{CC} and GND, and not including the connections to counter clock inputs. The total number for the circuit using separate decoder/drivers is 66.
9-33.

9-34.

9-35.

A	B	C	
0	0	0	$0 \Rightarrow I_0$
0	0	1	$0 \Rightarrow I_1$
0	1	0	$0 \Rightarrow I_2$
0	1	1	$1 \Rightarrow I_3$
1	0	0	$0 \Rightarrow I_4$
1	0	1	$1 \Rightarrow I_5$
1	1	0	$1 \Rightarrow I_6$
1	1	1	$1 \Rightarrow I_7$

9-36. Connect $I_1, I_5, I_8, I_{11}, I_{14}, I_{15}$ to GND. All other inputs connect to V_{CC}.
9-37. Z = HIGH for $DCBA$ = 0010, 0100, 1001, 1010.
9-38. (b)
$Z = \overline{A}\,B\overline{C} + AB\overline{C} + \overline{A}\,\overline{B}C + A\overline{B}C + \overline{A}\,\overline{B}\,\overline{C}D + ABCD$

9-39. (a) Encoder, MUX   (b) MUX, DEMUX
(c) MUX   (d) Encoder   (e) Decoder, DEMUX
(f) DEMUX   (g) MUX
9-40.

9-41. Each DEMUX output goes LOW, one at a time in sequence.
9-42. (a) All LEDs are off.   (b) Each LED will flash in sequence.   (c) LEDs 2 and 6 will flash.
9-43. Five lines
9-46. (a) Sequencing stops after actuator 3 is activated.   (b) Same as (a)
9-47. Probable fault is short to ground at MSB of tens MUX.
9-48. Q_0 and Q_1 are probably reversed.
9-49. Inputs 6 and 7 of MUX are probably shorted together.
9-50. S_1 stuck LOW
9-51. (a) No   (b) Yes
9-53. Use three 74HC85s
9-54. Final binary output is 1000101.
9-55. A_0 and B_0 are probably reversed.
9-57. $\overline{OE}_C = 0$, $\overline{IE}_C = 1$; $\overline{OE}_B = \overline{OE}_A = 1$; $\overline{IE}_B = \overline{IE}_A = 0$; apply a clock pulse.
9-58. (a) At t_4, each register will have contents of 1011.
9-59.

9-60. (b) ON, ON, ON, OFF (left to right)
9-61. (a) At t_3, each register holds 1001.
9-63. (a) 57FA   (b) 5000 to 57FF   (c) 9000 to 97FF
(d) no
9-65. See Prob9_65.tdf and Prob9_65.vhd on the enclosed CD.
9-68. See Prob9_68.tdf and Prob9_68.vhd on the enclosed CD.
9-70. See Prob9_70.tdf and Prob9_70.vhd on the enclosed CD.

CHAPTER 10

10-1. (d) 20 Hz   (e) Only one LED will be lit at any time.

10-2. 24

10-3. Four states = four steps * 15°/step = 60° of rotation

10-4. Eight states = eight steps * 7.5°/step = 60° of rotation

10-5. Three state transitions * 15°/step = 45° of rotation

10-9.

R3	R2	R1	R0
0	1	1	1
1	0	1	1
1	1	0	1
1	1	1	0

10-10. 1111

10-11. Yes

10-12. (a) 1011   (b) 10_2 (row 2)   (c) 01_2 (row 1)
(d) 1001

10-13. No

10-14. DAV

10-15. The data goes away (hi-Z) before the DAV goes LOW. The hi-Z state is latched.

10-16.

10-17. 60 cycles/sec * 60 sec/min * 60 min/hr * 24 hr/day = 5,184,000 cycles/day. This takes a long time to generate a simulation file.

10-18. When the set input is active, bypass the prescaler and feed the 60-Hz clock directly into the units of seconds counter.

10-22. See Prob10_22.tdf and Prob10_22.vhd on the enclosed CD.

CHAPTER 11

11-1. (f), (g) False

11-2. 3.58 V

11-3. LSB = 20 mV

11-4. 20 mV; approximately 0.4 percent

11-5. Approximately 5 mV

11-6. Eight

11-7. 14.3 percent, 0.286 V

11-9. 250.06 rpm

11-10. 10 mV, 0.1 percent, 6.95 V

11-11.

Binary	BCD
3.88 mV	10 mV
0.392%	1.01%

11-13. 800 Ω; no

11-14. (a) −0.3125 V, −4.6875 V   (b) 4.27 kΩ
(c) −0.0267

11-15. Uses fewer different R values

11-16. 20 μA; yes

11-17. (a) Seven

11-19. 242.5 mV is not within specifications.

11-20. Offset out of spec

11-21. Bit 1 of DAC is open or stuck HIGH.

11-22. Bits 0 and 1 are reversed.

11-24. (a) 10010111   (b) 10010111   (c) 102 μs, 51 μs

11-27. (a) 1.2 mV   (b) 2.7 mV

11-28. (a) 0111110110   (b) 0111110111

11-29. 2.834 to 2.844 V

11-31. Reconstructed waveform frequency is 3.33 kHz.

11-32. (a) 5 kHz   (b) 9.9 kHz   (c) 9.8 kHz
(d) 5 kHz   (e) 900 Hz   (f) 800 Hz

11-33. Digital ramp: *a, d, e, f, h*. SAC: *b, c, d, e, g, h*

11-36. 80 µs

11-37. 100101

11-38. 2.276 V

11-39. 2.869 V

11-40. (a) 00000000   (b) 500 mV   (c) 510 mV
(d) 255 mV   (e) 01101110   (f) 0.2°F; 2 mV

11-45. 386 steps

11-46. The MSB of the MSD is not affecting V_{AX}.

11-48. Switch is stuck closed; switch is stuck open, or capacitor is shorted.

11-50. (a) Address is EA*xx*.

11-51. Yes; Use $D_9..D_0$

11-55. False: a, e, g; True: b, c, d, f, h

CHAPTER 12

12-1. 16,384; 32; 524,288

12-2. 16,384

12-3. 64K × 4

12-4. 16; 16; 13; 16,384

12-6. (a) Data, address, control   (b) Address
(c) Data   (d) CPU

12-7. (a) Hi-Z   (b) 11101101

12-8. (a) Register 11   (b) 0100

12-9. (a) 16,384   (b) Four   (c) Two 1-of-128 decoders

12-11. 120 ns

12-12. 180 ns

12-14. Q_{13}, Q_{14}, Q_{15}

12-15. The following transistors will have open source connections: $Q_0, Q_2, Q_5, Q_6, Q_7, Q_9, Q_{15}$.

12-17. (a) Erases all memory locations to hold FF_{16}
(b) Writes $3C_{16}$ into address 2300_{16}

12-18. The waveform at D_0 is shown below.

12-19. Hex data: 5E, BA, 05, 2F, 99, FB, 00, ED, 3C, FF, B8, C7, 27, EA, 52, 5B

12-20. (a) 25.6 kHz   (b) Adjust V_{ref}.

12-22. (a) [B] = 40 (hex); [C] = 80 (hex)   (b) [B] = 55 (hex); [C] = AA (hex)   (c) 15,360 Hz   (d) 28.6 MHz
(e) 27.9 kHz

12-24. (a) 100 ns   (b) 30 ns   (c) 10 million
(d) 20 ns   (e) 30 ns   (f) 40 ns   (g) 10 million

12-25. Eight

12-26.

12-27.

12-30. Every 7.8 µs

12-31. (a) 4096 columns, 1024 rows   (b) 2048
(c) It would double.

12-34. Add four more PROMs (PROM-4 through PROM-7) to the circuit. Connect their data outputs and address inputs to data and address bus, respectively. Connect AB_{13} to C input of decoder, and connect decoder outputs 4 through 7 to CS inputs of PROMs 4 through 7, respectively.

12-35. Connect AB_{13}, AB_{14}, and AB_{15} to a three-input OR gate, and connect the OR gate output to the C input of the decoder.

12-38. F000–F3FF; F400–F7FF; F800–FBFF; FC00–FFFF

12-39. Add four 1K $\times$ 8 modules and use the four unused outputs of the existing decoder to drive the CS inputs.

12-40. B input of decoder is open or stuck HIGH.

12-41. The OR gate output/C input node is stuck LOW.

12-42. Only RAM modules 1 and 3 are getting tested.

12-43. The RAM chip with data outputs 4 through 7 in module 2 is not functioning properly.

12-44. RAM module 3, output 7 is open or stuck HIGH.

12-46. Checksum = 11101010.

CHAPTER 13

13-2. The necessary speed of operation for the circuit, cost of manufacturing, system power consumption, system size, amount of time available to design the product, etc.

13-4. Speed of operation

13-6. Advantages: highest speed and smallest die area; Disadvantages: design/development time and expense

13-8. SRAM-based PLDs must be configured (programmed) upon power-up.

13-10. In a PLD programmer or in-system (via JTAG interface)

13-12. pin 1—GCLRn (Global Clear)
pin 2—OE2/GCLK2 (Output Enable 2/Global Clock 2)
pin 83—GCLK1 (Global Clock 1)
pin 84—OE1 (Output Enable 1)

13-14. Logic cell in MAX7000S is AND/OR circuit versus look-up table in FLEX10K;
EEPROM (MAX7000S) and SRAM (FLEX10K);
MAX7000S is nonvolatile;
FLEX10K has greater logic resources.

INDEX OF ICS

INDEX

FLIP-FLOPS

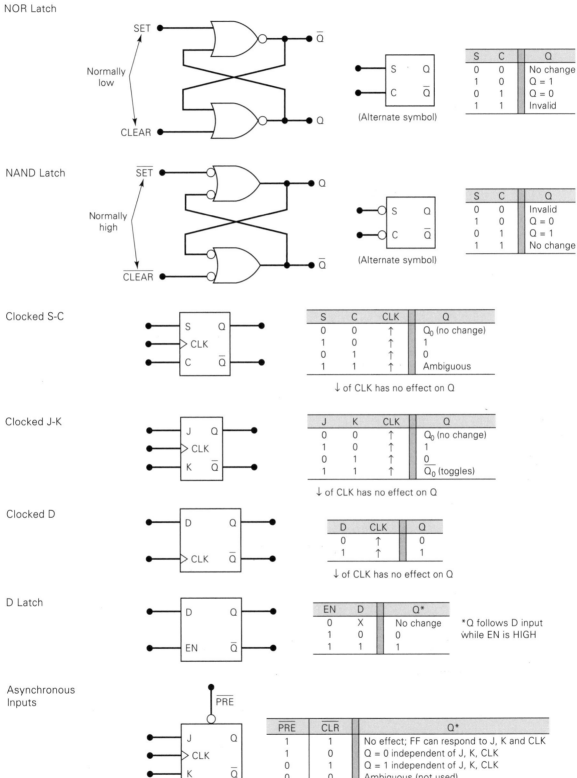

NOR Latch

SET — Normally low — CLEAR — (Alternate symbol)

S	C	Q
0	0	No change
1	0	Q = 1
0	1	Q = 0
1	1	Invalid

NAND Latch

$\overline{SET}$ — Normally high — $\overline{CLEAR}$ — (Alternate symbol)

S	C	Q
0	0	Invalid
1	0	Q = 0
0	1	Q = 1
1	1	No change

Clocked S-C

S	C	CLK	Q
0	0	↑	Q_0 (no change)
1	0	↑	1
0	1	↑	0
1	1	↑	Ambiguous

↓ of CLK has no effect on Q

Clocked J-K

J	K	CLK	Q
0	0	↑	Q_0 (no change)
1	0	↑	1
0	1	↑	0
1	1	↑	$\overline{Q_0}$ (toggles)

↓ of CLK has no effect on Q

Clocked D

D	CLK	Q
0	↑	0
1	↑	1

↓ of CLK has no effect on Q

D Latch

EN	D	Q*
0	X	No change
1	0	0
1	1	1

*Q follows D input while EN is HIGH

Asynchronous Inputs

$\overline{PRE}$

$\overline{CLR}$

$\overline{PRE}$	$\overline{CLR}$	Q*
1	1	No effect; FF can respond to J, K and CLK
1	0	Q = 0 independent of J, K, CLK
0	1	Q = 1 independent of J, K, CLK
0	0	Ambiguous (not used)

*CLK can be in any state